AF847035

CO_2 Sequestration, Biofuels and Depollution

Environmental Chemistry for a Sustainable World

Volume 5

Series Editors

Eric Lichtfouse
INRA, UMR1347 Agroécologie, Dijon, France

Jan Schwarzbauer
Institute of Geology and Geochemistry of Petroleum and Coal, RWTH Aachen University, Aachen, Germany

Didier Robert
Université de Lorraine, ICPEES, Saint-Avold, France

For further volumes:
http://www.springer.com/series/11480

Eric Lichtfouse • Jan Schwarzbauer • Didier Robert
Editors

CO_2 Sequestration, Biofuels and Depollution

Springer

Editors
Eric Lichtfouse
INRA, UMR1347 Agroécologie
Dijon
France

Didier Robert
Université de Lorraine
Saint-Avold
France

Jan Schwarzbauer
RWTH Aachen University
Aachen
Germany

ISSN 2213-7114 ISSN 2213-7122 (electronic)
ISBN 978-3-319-11905-2 ISBN 978-3-319-11906-9 (eBook)
DOI 10.1007/978-3-319-11906-9
Springer Cham Heidelberg New York Dordrecht London

Library of Congress Control Number: 2014957672

© Springer International Publishing Switzerland 2015
This work is subject to copyright. All rights are reserved by the Publisher, whether the whole or part of the material is concerned, specifically the rights of translation, reprinting, reuse of illustrations, recitation, broadcasting, reproduction on microfilms or in any other physical way, and transmission or information storage and retrieval, electronic adaptation, computer software, or by similar or dissimilar methodology now known or hereafter developed. Exempted from this legal reservation are brief excerpts in connection with reviews or scholarly analysis or material supplied specifically for the purpose of being entered and executed on a computer system, for exclusive use by the purchaser of the work. Duplication of this publication or parts thereof is permitted only under the provisions of the Copyright Law of the Publisher's location, in its current version, and permission for use must always be obtained from Springer. Permissions for use may be obtained through RightsLink at the Copyright Clearance Center. Violations are liable to prosecution under the respective Copyright Law.
The use of general descriptive names, registered names, trademarks, service marks, etc. in this publication does not imply, even in the absence of a specific statement, that such names are exempt from the relevant protective laws and regulations and therefore free for general use.
While the advice and information in this book are believed to be true and accurate at the date of publication, neither the authors nor the editors nor the publisher can accept any legal responsibility for any errors or omissions that may be made. The publisher makes no warranty, express or implied, with respect to the material contained herein.

Printed on acid-free paper

Springer is part of Springer Science+Business Media (www.springer.com)

Preface

You will be astonished when I tell you what this curious play of carbon amounts to. A candle will burn some 4, 5, 6, or 7 h. What, then, must be the daily amount of carbon going up into the air in the way of carbonic acid! ... Then what becomes of it? Wonderful is it to find that the change produced by respiration ... is the very life and support of plants and vegetables that grow upon the surface of the earth.

Michael Faraday,
Course of Six Lectures on the Chemical History of a Candle, 1861.

Once a marvel, now a nightmare. What a nice balance that has been observed by Michael Faraday in 1861: burning a candle produces carbon dioxide (CO_2), a greenhouse gas, which in turn feeds plants, algae and later all life. This equilibrium was perfect because at that time candles were made with renewable resources such as beeswax and tallow from beef and mutton fat (http://en.wikipedia.org/wiki/Candle). In other words the amount of carbon entering the atmosphere as CO_2 was balanced by the amount of carbon fixed by plants and algae. Now candles are made with paraffin wax from fossil fuels that are non-renewable on a human time scale. There would be no real issue if only a couple of fossil fuel candles were burning because plants could still absorb the excess atmospheric CO_2. However the burning of fossil fuels by society has increased too fast for plants to capture CO_2, resulting in the fast increase of atmospheric CO_2 concentrations, and, in turn, global warming (http://en.wikipedia.org/wiki/Keeling_Curve). Noteworthy this trend has been foreseen as early as 1889 by the Nobel prize Svante Arrhenius who pointed out a 'greenhouse effect' in which small changes in the concentration of CO_2 in the atmosphere could considerably alter the average temperature of a planet. There is thus an urgent need to use biofuels, to decrease CO_2 emissions and to sequester carbon in plants, waste, soils, sediments, and other materials (Fig. 1).

This book presents advanced reviews on carbon management, pollutant toxicity and remediation. Chapters 2, 3 and 6 review methods to address global warming

Fig. 1 Greenhouse gas emission sampling from a compost pile (*left*, Sánchez et al. This book). Crops for bioenergy: Giant Miscanthus in New Zealand (Renquist and Keckhoffs, DOI 10.1007/978-94-007-5449-2_5, 10.1007_s13593-012-0114-9)

by limiting CO_2, CH_4 and N_2O emissions from organic wastes, using biofuels and photochemical reduction of CO_2. Chapters 1, 5 and 10 discuss water pollution by chlorine by-products and fluoride, then propose remediation tools and alternatives. Agricultural pollution by heavy metals, selenium and transgenes are addressed in chapters 5, 7 and 8. Chapter 9 details air pollution by foundries, and propose control methods.

Thanks for reading

Eric Lichtfouse[1], Jan Schwarzbauer and Didier Robert
Founders of the journal Environmental Chemistry Letters and of the European Association of Environmental Chemistry
E-mail: Eric.Lichtfouse@dijon.inra.fr, Jan.Schwarzbauer@emr.rwth-aachen.de, Didier.Robert@univ-lorraine.fr

[1] Author of the best-selling textbook "Scientific Writing for Impact Factor Journals"

Other Publications by the Editors

Books

Scientific Writing for Impact Factor Journals
http://fr.slideshare.net/lichtfouse/scientific-writing-for-impact-factor-journals

Environmental Chemistry
ISBN: 978-3-540-22860-8 (Print) 978-3-540-26531-3 (Online)

Organic contaminants in riverine and ground water systems. Aspects of the anthropogenic contribution
ISBN: 978-3-540-31169-0 (Print) 978-3-540-31172-0 (Online)

Sustainable Agriculture
Vol. 1: ISBN: 978-90-481-2665-1 (Print) 978-90-481-2666-8 (Online)
Vol. 2: ISBN: 978-94-007-0393-3 (Print) 978-94-007-0394-0 (Online)

Rédiger pour être publié!
ISBN: 978-2-8178-0288-6 (Print) 978-2-8178-0289-3 (Online)

Journals

Environmental Chemistry Letters
ISSN: 1610-3653
http://www.springer.com/10311

Agronomy for Sustainable Development
ISSN: 1774-0746
http://www.springer.com/13593

Book series

Environmental Chemistry for a Sustainable World
ISSN: 2213-7114
http://www.springer.com/series/11480

Sustainable Agriculture Reviews
ISSN: 2210-4410
http://www.springer.com/series/8380

Contents

Contributors

Adriana Artola Department of Chemical Engineering, Universitat Autònoma de Barcelona, Barcelona, Spain

Raquel Barrena Department of Chemical Engineering, Universitat Autònoma de Barcelona, Barcelona, Spain

Prosun Bhattacharya KTH-International Groundwater Arsenic Research Group, Department of Land and Water Resources Engineering, Royal Institute of Technology, Stockholm, Sweden

María Luz Cayuela Centro de Edafología y Biología Aplicada del Segura, Consejo Superior de Investigaciones Científicas, Campus Universitario de Espinardo, Murcia, Spain

Siang-Piao Chai Low Carbon Economy (LCE) Group, Chemical Engineering Discipline, School of Engineering, Monash University, Bandar Sunway, Selangor, Malaysia

Rimeh Daghrir Institut national de la recherche scientifique (Centre Eau, Terre et Environnement), Université du Québec, Québec, QC, Canada

Hossein Delavai Amrei Department of Chemical Engineering, Amirkabir University of Technology, Tehran, Iran

Walelign Demisie Department of Resource Science, College of Environmental and Natural Resource Sciences, Zhejiang University, Hangzhou, China

Department of Dry Land Crop Science, Jigiga University, Jigiga, Ethiopoa

Éva Domokos-Szabolcsy Plant Biotechnology Department, Debrecen University, Debrecen, Hungary

Patrick Drogui Institut national de la recherche scientifique (Centre Eau, Terre et Environnement), Université du Québec, Québec, QC, Canada

Hassan R. El-Ramady Soil and Water Sciences Department, Faculty of Agriculture, Kafrelsheikh University, Kafr El-Sheikh, Egypt

Miklós Fári Plant Biotechnology Department, Debrecen University, Debrecen, Hungary

Xavier Font Department of Chemical Engineering, Universitat Autònoma de Barcelona, Barcelona, Spain

David Gabriel Department of Chemical Engineering, Universitat Autònoma de Barcelona, Barcelona, Spain

Teresa Gea Department of Chemical Engineering, Universitat Autònoma de Barcelona, Barcelona, Spain

Mahdi Gharabaghi School of Mining Engineering, College of Engineering, University of Tehran, Tehran, Iran

Meei Mei Gui Low Carbon Economy (LCE) Group, Chemical Engineering Discipline, School of Engineering, Monash University, Bandar Sunway, Selangor, Malaysia

Mehboob-ur-Rahma National Institute for Biotechnology & Genetic Engineering (NIBGE), Plant Genomics & Molecular Breeding Lab, Faisalabad, Pakistan

Samra Irem National Institute for Biotechnology & Genetic Engineering (NIBGE), Plant Genomics & Molecular Breeding Lab, Faisalabad, Pakistan

R. Krishnaraj Department of Mechanical Engineering, Institute of Technology, Ambo University, Ambo, Ethiopia

Yi Li Department of Resource Science, College of Environmental and Natural Resource Sciences, Zhejiang University, Hangzhou, China

School of Geography and Tourism, Guizhou Normal College, Guiyang, China

Abdul Rahman Mohamed Low Carbon Economy (LCE) Group, School of Chemical Engineering, Universiti Sains Malaysia, Engineering Campus, Nibong Tebal, Pulau Pinang, Malaysia

Claudio Mondini Consiglio per la Ricerca e la Sperimentazione in Agricoltura, Centro di Ricerca per lo Studio delle Relazioni tra Pianta e Suolo, Gruppo di Ricerca di Gorizia, CRA-RPS, Gorizia, GO, Italy

Alireza Moosavi Zenooz Department of Chemical Engineering, Amirkabir University of Technology, Tehran, Iran

Wee-Jun Ong Low Carbon Economy (LCE) Group, Chemical Engineering Discipline, School of Engineering, Monash University, Bandar Sunway, Selangor, Malaysia

József Prokisch Institute of Bio- and Environmental Enegetics, Debrecen University, Debrecen, Hungary

Asunción Roig Centro de Edafología y Biología Aplicada del Segura, Consejo Superior de Investigaciones Científicas, Campus Universitario de Espinardo, Murcia, Spain

Antoni Sánchez Department of Chemical Engineering, Universitat Autònoma de Barcelona, Barcelona, Spain

Miguel Ángel Sánchez-Monedero Centro de Edafología y Biología Aplicada del Segura, Consejo Superior de Investigaciones Científicas, Campus Universitario de Espinardo, Murcia, Spain

Tayyaba Shaheen National Institute for Biotechnology & Genetic Engineering (NIBGE), Plant Genomics & Molecular Breeding Lab, Faisalabad, Pakistan

Department of Bioinformatics and Biotechnology, Government College University, Faisalabad, Pakistan

Jafar Shahrivar Guzullo School of Mining Engineering, College of Engineering, University of Tehran, Tehran, Iran

Tarek A. Shalaby Horticulture Department, Faculty of Agriculture, Kafrelsheikh University, Kafr El-Sheikh, Egypt

Sunil Kumar Singh Department of Chemistry, DDU Gorakhpur University, Gorakhpur, India

Kapil Deo Singh Yadav Department of Chemistry, DDU Gorakhpur University, Gorakhpur, India

Lling-Lling Tan Low Carbon Economy (LCE) Group, Chemical Engineering Discipline, School of Engineering, Monash University, Bandar Sunway, Selangor, Malaysia

Meththika Vithanage Chemical and Environmental Systems Modeling Research Group, Institute of Fundamental Studies, Kandy, Sri Lanka

Meera Yadav Department of Chemistry, DDU Gorakhpur University, Gorakhpur, India

Sudha Yadava Department of Chemistry, DDU Gorakhpur University, Gorakhpur, India

Yusuf Zafar National Institute for Biotechnology & Genetic Engineering (NIBGE), Plant Genomics & Molecular Breeding Lab, Faisalabad, Pakistan

Ming- kui Zhang Department of Resource Science, College of Environmental and Natural Resource Sciences, Zhejiang University, Hangzhou, China

Farzin Zokaee Ashtiani Department of Chemical Engineering, Amirkabir University of Technology, Tehran, Iran

Chapter 1
Chlorine for Water Disinfection: Properties, Applications and Health Effects

Patrick Drogui and Rimeh Daghrir

Contents

Abstract Chlorine compounds are chemicals commonly used as oxidizing agents for drinking-water disinfection and water treatment. Chlorine (Cl) is an halogen element having 7 electrons on its last electronic layer. To reach the stability of 8 electrons chlorine has a tendency to add an electron. This tendency to add an electron gives chlorine special oxidizing properties. Chlorine indeed kills harmful microorganisms, has decolourization properties, and oxidize and modify organic molecules. Chlorine can be used as a bleaching agent and to fight against biological fouling in cooling systems. Chlorine popularity is not only due to lower cost, but also to its higher oxidizing potential, which provides a minimum level of residual chlorine throughout the distribution system and protects against microbial recontamination with a remarkable remnant effect. Chlorine is used in different forms: gaseous, e.g. Cl_2 and ClO_2, liquid, e.g. NaClO, HClO, and NH_2Cl, and solid, e.g. Ca $(OCl)_2$. Chlorine is industrially produced by electrolysis of aqueous sodium chloride (NaCl).

P. Drogui (✉) · R. Daghrir
Institut national de la recherche scientifique (Centre Eau, Terre et Environnement), Université du Québec, 490 rue de la Couronne, Québec, QC G1K 9A9 Canada
e-mail: patrick.drogui@ete.inrs.ca

R. Daghrir
e-mail: rimeh.daghrir@ete.inrs.ca

© Springer International Publishing Switzerland 2015

E. Lichtfouse et al. (eds.), *CO_2 Sequestration, Biofuels and Depollution*, Environmental Chemistry for a Sustainable World 5, DOI 10.1007/978-3-319-11906-9_1

The application of chlorine to water treatment poses risks by producing by-products such as organochlorinated compounds. Disinfection by-products (DBPs) such as trihalomethanes, haloacetic acids, haloacetonitriles and haloketones are formed when chlorine reacts with natural organic matter such as humic acids and phenols. Some organochlorinated compounds are endocrine disruptors involved in brain cancer, immune and reproductive system troubles, and feminization. Modern legislations in all countries impose environmental regulation and health quality standards that become more and more stringent. Emerging regulation limiting the concentration of by-products in drinking water has increased demands for alternative processes that remove organics before disinfection. Processes include coagulation with active carbon, flocculation, sedimentation, sand filtration. Altenrative disinfection is done by advanced oxidation processes including O_3/H_2O_2, UV/O_3, and UV/H_2O_2, membrane filtration and electrochemical disinfection.

Keywords Water treatment · Chlorine · Trihalomethane · Coagulation · Flocculation · Sedimentation · UV · O_3 · H_2O_2

1.1 Chlorine

Due to its low cost, chlorine is a chemical disinfectants widely used in a variety of applications to control the spread of pathogenic organisms (Chowdhury et al. 2011). During oxidation reactions, the different forms of free chlorine include hypochlorous acid (HClO), hypochlorite (OCl^-) and chlorine (Cl_2). According to Nakajima et al. (2004), free chlorine refers to the chlorine that is available for the disinfection and not bound to organic compounds. Chlorine is currently used as post-treatment to maintain a residual concentration of oxidant in the drinking water network and to protect it against microbial recontamination (remnant effect) (Acero et al. 2010). As reported by Luo et al. (2012), chlorine is very effective to oxidize the harmful microorganism present in raw waters (such as river, lakes and groundwater). Above pH = 4, chlorine (Cl_2) is completely hydrolyzed according to the following equation (Eq. 1.1) (Acero et al. 2010).

$$Cl_2 + 2H_2O \Leftrightarrow HClO + Cl^- + H_3O^+ \tag{1.1}$$

The hydrolysis is carried out rapidly in few seconds so that water (at pH around 7) is treated by hypochlorous acid rather than chlorine. The hypochlorous acid resulting from the hydrolysis of chlorine is a weak acid capable of dissociating in water as follows (Eq. 1.2).

$$HClO + H_2O \Leftrightarrow ClO^- + H_3O^+ \tag{1.2}$$

The acidity constant (K_a) of this Eq. (1.2) is given by the following equation:

$$K_a = \frac{[ClO^-].[H_3O^+]}{[HClO]} \tag{1.3}$$

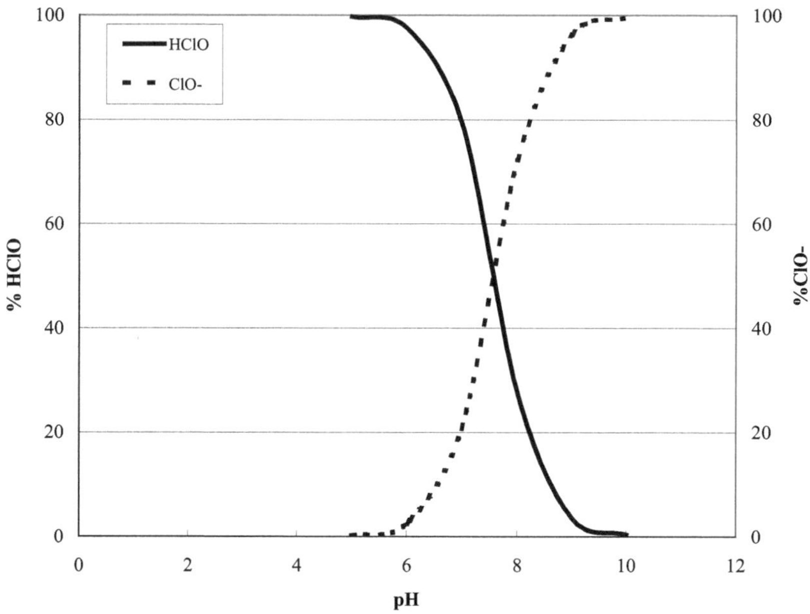

Fig. 1.1 Formation and disappearance of hypochlorite ion and hypochlorous acid as a function of pH

If we consider a pH range where hypochlorous acid and hypochlorite ion quantitatively represent the only entities resulting from the hydrolysis of chlorine in water, the following equation can be written:

$$C_t = [HClO] + [ClO^-] \tag{1.4}$$

In this Eq. 1.4, C_t represents the total concentration of oxidizing entities. The combination of Eqs. (1.3) and (1.4) leads to the Eq. 1.5, which can be exploited to evaluate the proportion of active chlorine in function of pH:

$$\frac{[HClO]}{C_t} = \frac{1}{\left(1 + \frac{K_a}{10^{-pH}}\right)} \tag{1.5}$$

For instance, a proportion of 75% (w/w) of hypochlorite ion and 25% (w/w) of hypochlorous acid are respectively formed when the chlorine is injected in water at pH=8 (Fig. 1.1). However, at neutral pH value (pH around 6.0–7.0), 80% (w/w) of hypochlorous acid and 20% (w/w) hypochlorite ion are generated. Thus, at pH of natural waters, hypochlorous acid is the dominant specie when the chlorine is injected in water. Hypochlorous acid is a powerful oxidant capable of destroying bacteria, oxidizing and modifying the structure of organic molecules and leading to more oxidized and less toxic compounds (Daghrir et al. 2012; Zaviska et al. 2012; Drogui et al. 2007). Owing to its high oxidizing potential and its chemical structure, hypochlorous acid could react with organic compounds through oxidation reactions, electrophilic substitution reactions or addition reactions on unsaturated bonds (Deborde and von Gunten 2008).

1.2 Disinfection By-Products

The application of chlorine in disinfection process reduces the microbial risk but poses chemical risks by producing by-products dependently of bactericidal agent used. Disinfection by-products (DBPs) are formed when chlorine reacts with organic matter (NOM), and/or inorganic substances (i.e. bromide, iodide). Up to now, more than 600–700 DBPs have been identified in drinking water (Krasner et al. 2006; Villanueva et al. 2012). Table 1.1 presents some DBPs formed in function of the disinfectants used. During the different disinfection processes, the concentrations of DBPs may vary (Sadiq and Rodriguez 2004). DBPs are ubiquitous in drinking water (Villanueva et al. 2012). They are classified as contaminants of concern, since human exposure to these by-products has been associated with cancer and reproductive outcomes (Tardiff et al. 2006; Villanueva et al. 2004). Recent epidemiological studies provide that DBPs cause adverse pregnancy outcomes (Bove et al. 2002). Several studies show that DBPs are mutagens, carcinogens, teratogens or developmental toxicants (Villanueva et al. 2004; Muller-pillet et al. 2000; Ahmed et al. 2005).

1.2.1 By-Products from Chlorination Processes

Chlorination is a process commonly used for drinking-water disinfection and water treatment. As indicated above, at pH values close to neutrality, chlorine is essentially in form of hypochlorous acid (HClO). Given the oxidizing properties and the molecular structure characterized by the polarization of the Cl–O bond in the direction $Cl^{\delta+} \rightarrow OH^{\delta-}$, the three main reactions of hypochlorous acid are the following (Doré 1989): (i) the oxidation reactions on the reducing functions; (ii) the addition reactions on unsaturated bonds and; (iii) electrophilic substitution reactions on the nucleophilic sites. Thus, the chlorine reactions towards the organic pollutants in water are limited to a few privileged sites (reducing sites, nucleophilic and unsaturated sites). Therefore, the reactions of hypochlorous acid will be most often reduced to modifying the structure of organic molecules, leading to more oxidized compounds and the formation of organochlorinated molecules (DBPs). The formation of DBPs depends mainly on the initial characteristics of the raw water (type of ion and or organic matter present in waters). As mentioned by Liang and Singer (2003), the physicochemical properties of water such as the nature and the concentrations of organic precursors, pH, and temperature, disinfectant concentration, contact time, influence the formation of DBPs. The levels of these by-products are regulated by USEPA, WHO and other regulatory agencies worldwide (Table 1.2). As reported

Table 1.1 Important groups of diseinfection by-products (DBP) produced during chlorination process

Class of DBPs	Common example	Chlorine	ClO_2	Chloramines
Trihalomethanes (THM)	Chloroform	✓		✓
Other haloalkanes		✓		
Haloalkenes		✓		
Haloacetic acids (HAA)	Chloroacetic acid	✓		✓
Haloaromatic acids		✓		
Other halomonocarboxylic acids		✓		✓
Unsaturated halocarboxylic acids		✓		✓
Halodicarboxylic acids		✓		✓
Halotricarboxylic acids		✓		✓
MX and analogues		✓	✓	✓
Other halofuranones		✓		
Haloketones		✓	✓	
Haloacetonitrile (HAN)	Chloroacetonitrile	✓		
Other halonitrile	Cyanogen chloride	✓		✓
Haloaldehyde	Chloral hydrate	✓		✓
Haloalcohals		✓		✓
Phenols	2-Chlorophenol	✓		
Halonitromethane	Chloropicrin	✓		
Inorganic compounds	Bromate, Hypobromite		✓	
	Chlorite and Chlorate etc.			
Aliphatic aldehyde	Formaldehyde	✓	✓	
Other aldehydes		✓	✓	
Ketones (aliphatic and aromatic)	Acetone	✓	✓	
Carboxylic acids	Acetic acid	✓	✓	
Aromatic acids	Benzoic acid	✓	✓	
Aldo and Ketoacids		✓	✓	
Hydroxy acids		✓		
Others		✓	✓	✓

by previous studies (Yang et al. 2000; Cedergren et al. 2002; Ivancev et al. 2002; Badawy et al. 2012), DBPs have been responsible to the occurrence of cancer, growth retardation, spontaneous abortion, and congenital defects (Table 1.3). During the chlorination process, different types of DBPs can be formed. Those include trihalomethanes (THMs), haloacetic acids (HAAs), haloacetonitriles (HANs), haloketones and others. Among the DBPs, THMs and HAAs are generally well documented, studied and regulated.

Table 1.2 Water quality guidelines and regulations for disinfection by-products

US environmental protection agency (USEPA) regulations	*Maximum contaminants levels (MCL)*[a] (mg/L)
Total THMs	0.08
5HAAs	0.06
Bromate	0.01
Chlorite	1.0
World health organization (WHO) guidelines	*Guideline value*[b] (μg/L)
Chloroform	200
Bromodichloromethane	60
Chlorodibromomethane	100
Bromoform	100
Carbon tetrachloride	4
Chloroacetic acid	20
Dichloroacetic acid	50
Trichloroacetic acid	100
Bromate	10
Chlorite	200
Chloral hydrate (trichloroacetaldehyde)	10
Dichloroacetonitrile	90
Trichloroacetonitrile	1
Dibromoacetonitrile	100
Cyanogen chloride	70
2,4,6-trichlorophenol	200
Formaldehyde	900
N-nitrosodimethylamine (NDMA)	100
European Union (EU) Standards	*Standard values*[c] (μg/L)
Total THMs	100
Bromate	10
Other regulations	*MCL* (ng/L)[d]
N-nitrosodimethylamine (NDMA)	9–10

Total THMs (chloroform, bromoform, bromodichloromethane, chlorodibromomethane), 5HAAs (monochloro-and monobromo-acetic acid (MCAA, MBrAA), dichloro-and trichloroacetic acid (DCAA, TCAA), and dibromoacetic acid (DBAA))

[a] US EPA (2001) Ntional primary drinking water standards, United States Environmental Protection Agency, EPA; 816-F-01-007

[b] WHO guidelines: http://www.who.int/water_sanitation_health/dwq/gdwq3rev/en/

[c] European Union drinking water standards: http://www.nucfilm.com/eu_water_directive.pdf

[d] Government of Ontario 2006

Table 1.3 Toxicological effects of disinfection by-products (DBP)

Chemical compound	DBPs groups	Cancer group	Toxicological effects
Chloroform	Trihalomethanes (THM)	B2	Cancer
			Liver
			Kidney
			Reproductive effects
Dichlorobromomethane (DCBM)		B2	Cancer
			Liver
			Kidney
			Reproductive effects
Dibromochloromethane (DBCM)		C	Nervous system
			Liver
			Kidney
			Reproductive effects
Bromoform (TBM)		B2	Cancer
			Nervous system
			Liver
			Kidney effects
Formaldehyde	Halogenated aldehydes and ketones	B1	Mutagenic
Dichloroacetic acid (DCAA)	Haloacetic acids (HAA)	B2	Cancer
			Reproductive effects
			Developmental effects
Trichloroacetic acid (TCAA)		C	Liver
			Kidney
			Spleen
			Developmental effects
Dichloroacetonitrile (DCA)	Haloacetonitriles	C	–
Trichloroacetonitrile (TCA)		C	Cancer
			Mutagenic effects
			Clastogenic effects
Dibromoacetonitrile (DBA)		C	–
Chloral hydrate	Miscellaneous chlorinated organics	C	–
Chlorite	Inorganic compounds	D	Developmental effects
			Reproductive effects
Bromate		B2	Cancer

Table 1.3 (continued)

Chemical compound	DBPs groups	Cancer group	Toxicological effects
2-Chlorophenol	Halophenol	D	Tumour promoter
			Cancer
2,4-Dichlorophenol		B2	–
2,4,6-Trichlorophenol		D	–
MX	–	B2	–

Group A Human carcinogen
Group B Probable human carcinogen (B1: limited epidemiological evidence; B2: sufficient evidence from animal studies)
Group C Possible human carcinogen
Group D Non classifiable; Control and Elimination of Disinfection By-Products

1.2.1.1 Trihalomethanes

THMs belong to the first family of organic compounds identified as by-products from chlorination (Righi et al. 2012; Cedergren et al. 2002). THMs are formed in water when chlorine reacts with natural organic matter. THMs are a class of compounds that includes chloroform ($CHCl_3$), bromoform ($CHBr_3$), dichlorobromomethane ($CHBrCl_2$) and dibromochloromethane ($CHBr_2Cl$) (Cedergren et al. 2002). The maximum contaminant level of THMs was set at 80 µg/L by EPA (USEPA 2003), while the Swedish regulation limit the total concentration of THMs to 100 µg/L. The high exposure to THMs induces the occurrence of cancer, mainly bladder cancer and reproductive outcomes, especially small for gestational age intrauterine growth retardation (Nieuwenhuijsen et al. 2000, Villanueva et al. 2004; Tardiff et al. 2006; Grellier et al. 2010; Costet et al. 2011). The risks of cancer through ingestion of THMs were highest, followed by inhalation and dermal contact pathways. Both inhalation and dermal contacts contribute to approximately 40% of total cancer risks (Chowdhury et al. 2011). For example, the cancer incidents caused by THMs have been predicted to be 703 per year in Canada (Chowdhury et al. 2011). The Canadian cancer society reported that the bladder cancer incidents caused by THMs in Ontario varied in the range of 1570–1910 with an average of 1660 per year (CCS 2009).

1.2.1.2 Haloacetic Acids

The HAAs are another family of DBPs that are frequently found in drinking water (Krasner et al. 1989). The maximum contaminant level of HAAs was set at 60 µg/L by EPA (USEPA 2003). Nine HAAs compounds are well documented. However, only five species of them are regulated (USEPA 1998). The five HAAs are monochloro-and monobromo-acetic acid (MCAA, MBrAA), dichloro-and trichloroacetic acid (DCAA, TCAA), and dibromoacetic acid (DBAA) (Chowdhury et al. 2011). The maximum contamination level of these five HAAs was set at 60 µg/L in drinking water (USEPA 2006).

The in vivo exposure of rats shows that dichloro-acetic acid and trichloro-acetic acid can induce teratogenic effects (Smith et al. 1992; Epstein et al. 1992). The anomalies observed for both compounds were cardiovascular. Recent studies of Plewa et al. (2002) have focused on the mammalian cell cytotoxicity and genotoxicity of chlorinated and brominated HAAs. They found that the brominated HAAs were more genotoxic and cytotoxic than their chlorinated analogs (Chowdhury et al. 2011).

1.2.1.3 Haloacetonitriles

Among the different types of DBPs, haloacetonitriles (HANs) have received also a special concern due to their cytotoxicity and genotoxicity (Chu et al. 2012). Haloacetonitriles (HANs) compounds are a newer generation of emerging DBPs. According to Richardson (2003), the average concentration of HANs recorded in drinking water was approximately 10 % of THMs. Despite few information reported in the literature for this group of DBPs (compared to those recorded for THMs and HAAs), the toxicity of HANs was much higher than that of the two major by-products (Plewa et al. 2004). HANs have been reported to be more geno-and cytotoxic that THMs and HAAs (Huang et al. 2012; Muellner et al. 2007). The maximum contamination levels were set at 1 and 90 μg/L for trichloroacetonitrile and dichloroacetonitrile, respectively by the World Health Organization (WHO 2000). The hydrolysis of HANs promote the generation of HAAs and halocetamides (Prarat et al. 2013; Olivier 1983; Glezer et al. 1999; Reckhow et al. 2001). Dichloroacetonitrile (DCAN) and dichloroacetamide (DCAcAm) are the two compounds most often detected in waters. These by-products were approximately two orders of magnitude more cytotoxic than dichloroacetic acid (Muellner et al. 2007; Plewa et al. 2008). Recent researches (Reckhow et al. 2001) have reviewed the formation pathway for halonitriles and haloamides during chlorination process (Fig. 1.2). The application of free chlorine (HClO) to amino acid (ex. aspartic acid) promotes the rapid di-chlorination of the α-terminal amine group.

The nitrile and carboxylic acid group present in the intermediate render the methylene carbon acidic and favours its chlorination. The hydrolysis promotes the formation of carbonic acid and dichloroacetonitrile. Subsequently, the hydrolysis of dichloroacetonitrile releases dichloroacetamide and then dichloroacetic acid (Reckhow et al. 2001). To circumvent these drawbacks (formation of DBPs in drinking water), the popular alternative disinfectants used are the chloramines and chlorine dioxide (Sadiq and Rodriguez 2004).

1.2.2 *By-Products from Chloramination Processes*

Chloramination is the treatment of drinking water using monochloramines (NH_2Cl) as disinfectant. The monochloramine is prepared *in situ* from ammonia and chlorine at a relatively high pH (pH around 10) to avoid the formation of dichloramines (precursors of bad taste of water). The monochloramines are not used for their

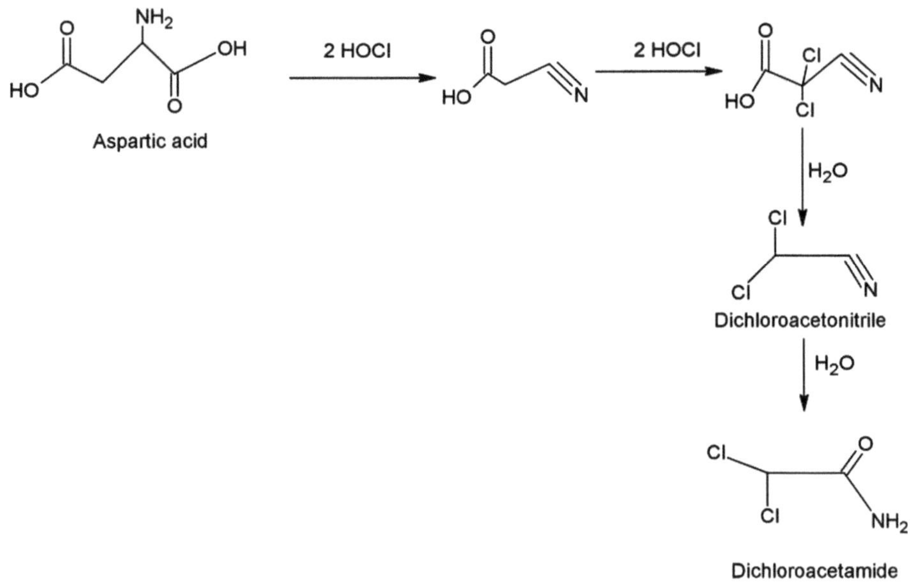

Fig. 1.2 Formation pathways of halonitriles and haloamides

bactericidal effect, but for their remnant effect in the distribution of the relatively warm water (25 °C or more than 25 °C), because they are more stable than chlorine at this temperature. Under specific conditions, monochloramine can be a better biocide for controlling the growth of biofilm through the distribution systems (Norton and Le Chevalier 1997). Nevertheless, in the presence of some microorganisms such as Giardia and Cryptosporidium, chloramines are less effective than the free chlorine (Gagnon et al. 2004). A relatively low concentration of DBPs is generated while using chloramines compared to the use of free chlorine (Singer 1994). Monochloramine (NH_2Cl) is widely used as alternative disinfectant to control the formation of DBPs (Seidel et al. 2005). As reported by Tian et al. (2013), chloramination process forms less amounts of THMs and HAAs than chlorination process. However, the use of chloramination process can lead to the formation of haloacetonitriles and halocetamides, which are more toxic than THMs and HAAs (Huang et al. 2012). NH_2Cl can also react either with Br^- or HOBr to form monobromamine (NH_2Br), bromochloramine (NHClBr) and dibromamine ($NHBr_2$) (Trofe et al. 1980; Lei et al. 2004). These DBPs induce higher health risks than their similar organo-chlorinated compounds (Uyak and Toroz 2007).

Moreover, chloramination process is also reported to increase the formation of nitrogenous DBPs which are higher toxic than the regulated DBPs (Shah et al. 2011; Chu et al. 2011). N-nitrosodimethylamine (NDMA) is the most commonly compound detected in water (Luo et al. 2012). NDMA is a DBPs specially produced by the reaction of monochloramine and DMA (Choi and Valentine 2002). Nitrosamines have been found in many food products and in drinking water (Mills and Alexander 1976). The occurrence of NDMA in drinking water treatment plants

(DWTPs) has been studied in several countries. For instance, Barrett et al. (2003) have observed an increase in the concentrations of NDMA in the drinking water network compared to those measured in DWTPs. The same compound has been also detected in chlorinated and chloraminated drinking water in Canada with a concentration up to 100 ng/L. In Scotland, only 9 ng/L of NDMA was detected in DWTPs using chloramine as disinfectant (Goslan et al. 2009).

N-nitrosamines are an emerging class of nitrogen-containing DBPs classified by the International Agency for Research on Cancer (IARC 1978) in group B2 as a probable human carcinogen in water (USEPA 1997). Nitrosodimethylamine was initially identified in 1989 as a type of DBPs in drinking water in Ontario (Canada) and in 1998 as a contaminant in groundwater at a northern California (USA) (Najm and Trussell 2001). During the 10 last years, the majority of researches focus on DMA because it is the most chemical DBPs detected in drinking water and due to its adverse effects on human health. Its toxicological effects are considered to be stronger than the other mutagenic nitrosamine compounds such as nitrosomethylethylamine (NMEA), nitrosodiethylamine (NDEA), nitrosodi-n-propylamine (NDPA), nitrosodi-n-butylamine (NDBA), nitrosodiphenylamine (NDPhA), nitrosopyrrolidine (NPyr), nitrosopiperidine (NPip) and nitrosomorpholine (NMor) (Luo et al. 2012). In Canada, the Ontario ministry of the Environment established a maximum admissible concentration of 9 ng/L for N-nitrosodimethylamine (Government of Ontario 2002), whereas in USA, the California Department of Public Health established a public health goal of 3 ng/L for N-nitrosodimethylamine (Asami et al. 2009). In Europe, the drinking water inspectorate of England and Wales requires the monitoring of N-nitrosodimethylamine at 1 ng/L and has set an action level for this compound of 10 ng/L. By comparison, the permissible level for NDMA is equal to 10 ng/L in Germany (Boyd et al. 2012).

1.2.3 By-Products from Chlorine Dioxide

Chlorine dioxide (ClO_2) has been widely used as disinfectant in several European countries (Italy, Germany, France and Switzerland) to avoid the formation of THMs during water treatment (Gagnon et al. 2004; Righi et al. 2012). It is worth noting that, at neutral pH values, chlorine dioxide ($E^0_{ClO2/ClO2-} = 0.95$ V) has a redox potential lower than chlorine ($E^0_{Cl2/Cl-} = 1.36$ V). In drinking water treatment, chlorine dioxide is less oxidant than chlorine, which explains its high stability in the distribution network. Likewise, the secondary reactions on the organic matter are lower and allow avoiding the formation of organochlorinated compounds. In the presence of phenolic compounds, there is no formation of chlorophenol. The reaction of chlorine dioxide on phenolic compounds can lead to the formation of quinones and chloroquinones or the formation of aliphatic compounds (Doré 1989). According to Richardson et al. (2010), chlorine dioxide reacts in a lesser extent with natural organic matter and form much fewer halogenated DBPs compared to chlorination. However, chlorine dioxide decomposes rapidly in water and forms inorganic DBPs namely chlorate and chlorite ions. Typically, about 50–70% of the applied chlorine dioxide is ultimately reduced to chlorite (Singer 1994; Righi et al. 2012). When chlorine dioxide reacts with organic matter, chlorite ion is released and induces an unpleasant metallic taste. The USEPA has set a maximum contamination level (MCL) for chlorite (ClO_{2^-}) at 1.0 mg/L (USEPA 1979). Thus, the chlorine dioxide dose should not exceed 1.4 mg/L to ensure that the chlorite limit is met. Generally, a concentration of 0.2 mg/L of chlorine dioxide is required during 15 min to ensure an effective disinfection.

Chlorite ions are suspected to cause haematological damage (haemolytic anaemia and methemoglobinemia) in some animals, when relatively high doses of chlorites are deposited in their drinking waters (Righi et al. 2012). Besides, impairment of neurobehavioral and neurological development, delay in female sexual development, soft tissue anomalies and altered thyroid function were mainly due to chlorite exposure and when it is in association with chlorine dioxide and above all with chlorate exposure (Bull et al. 2009; Righi et al. 2012). Due to its high occurrence and limited toxicity database, the degree of toxicity of chlorate ions is not yet completely understood and is still under investigation. Recently, chlorate has been classified among the most critical unregulated emerging DBPs on which future research should be concentrated (Hebert et al. 2010). The adverse effects of chlorite and chlorate on human health have been scarcely investigated and evidence is still inconclusive (Ouhoummane et al. 2004; Aggazzotti et al. 2004). The application of chlorine dioxide during chlorination process increases the risk for congenital cardiac defect (Cedergren et al. 2002). As reported by Righi et al. (2012), the exposure of chlorite and chlorate in drinking water increases the risk of congenital anomalies. A higher risk of newborns with renal defect, abdominal wall defects and cleft palate were observed when the woman are exposed to a chlorite level concentration (>700 μg/L). By comparison, a higher risk of newborns with obstructive urinary defects, cleft palate and spina bifida were observed by the same research groups when women are exposed to more than 200 μg/L level of chlorate. Owing to its ability to reduce the microbial risk, the application of chlorine for water disinfection remains the best process. Nevertheless, the DBPs generated during the treatment represent the major drawbacks. To this end, practical and economical processes must be developed and applied to avoid the DBPs formation in water.

1.3 Control and elimination of disinfection by-products

The DBPs produced during disinfection process are of great concern due to their adverse effects on human health. In response to the regulations and guidelines of DBPs in drinking water, a great deal of effort has been made by several research groups for developing alternative disinfection processes and technologies (Toor and Mohseni 2007). Thus, an effective strategy for controlling the formation of DBPs is to remove from water the precursors of DBPs formation (Prarat et al. 2013). Natural organic matter (NOM) is one of the main precursors of DBPs (Prarat et al. 2013; USEPA 1998). The colorization, unpleasant odors and unpleasant tastes in water can be attributed to NOM. The NOM is also responsible to bacterial regrowth through the network distribution systems (Yan et al. 2010). Over the years, various technologies and processes have been investigated for reducing the NOM during water treatment. The NOM can be partially removed using a conventional treatment (coagulation, flocculation, sedimentation and filtration) or by using granular activated carbon (GAC) (Sadiq and Rodriguez 2004; Hanigan et al. 2012). Another effective method to control DBPs in drinking water is the use of alternative disinfection processes such as ozonation (Yan et al. 2010), photocatalysis (Murray and Parsons 2006), photo-electro-catalysis (Selcuk and Bekbolet 2008), direct photolysis (UV) (Hansen et al. 2013) and electrochemical disinfection (Matilainen and Sillanpää 2010) among others. Besides, a better control of operational factors (e.g. control of pH or disinfection contact time, concentration of disinfectant) may contribute to reduce the DBPs formation.

1.3.1 Removal of Disinfection By-Product Precursors

The precursors of DBPs are often present in surface waters in form of natural organic matter. The removal of natural dissolved organic matter has been identified as one of the strategies to reduce the risk of DBPs formed during the chlorination water (Tan et al. 2005). Several technologies can be used to remove the precursors of DBPs. Table 1.4 and 1.5 summarizes the recent processes applied to remove NOM.

1.3.1.1 Enhanced Coagulation and Adsorption Process

Coagulation is a safe and effective method of water treatment, commonly used in municipal drinking water treatment. The USEPA has established guidelines for carrying out so-called "enhanced coagulation", in which the metal salt coagulation process is optimized for NOM removal. The most widely used coagulants are aluminum or iron salts (Gao and Yue 2005; Rizzo et al. 2005). However, the release of aluminium in drinking water is suspected to be responsible of Alzheimer disease (Brenner 2013; Rizzo et al. 2008). Therefore, new approaches for improving NOM removal during coagulation will reduce the formation of DBPs and significantly improve the economics of water treatment. An overview of the recent work undertaken using "enhanced coagulation process" has been presented in Table 1.4. Recently, Singer and Bilyk (2002) used alum combined with an ion exchange resin (MIEX technology) to remove organic materials and precursor of DBPs from surface waters. This combination method is more effective than alum used alone. Enhanced coagulation with MIEX was found to be very effective for reducing THM precursors by 60–90 %. Granular activated carbon (GAC) and powdered activated carbon (PAC) can been also used to remove precursors of DBPs from the water (Hanigan et al. 2012). Several researches reported that combination of PAC adsorption combined with a pre-oxidation process increases the effective of coagulation processes (Selcuk et al.

Table 1.4 Removal of disinfection by-product (DBP) precursors using enhanced coagulation processes

Process	Target compound	Matrix	Operating conditions	Results and comments	References
Coagulation-ion exchange resin	THM and HAA precursors	Surface water	Mixing time = 5–60 min; [MIEX] = 0–10 mL/L coagulant agent: alum (10–150 mg/L); flocculation time = 20 min; settling time = 30 min	From 60 to 90 % reduction of THM and HAA precursors	Singer and Bilyk (2002)
Coagulation- pump diffusion flash mixing (PDFM)	Natural organic matter (THM and HAA precursors)	Surface water	Rate of flow = 1000 m^3/day; coagulant = Al_2O_3 (0–30 mg/L); turbidity = 2.3–5.5 NTU; TOC = 3.3–4.8 mg/L; Alkalinity = 43–51 mg/L; pH = 6.9–7.9; Temperature = 2.2–19.2 °C; mixing time = 1–1800 s; settling time = 1 h	*Hydrophobic fraction:* THMFP/DOC = 55.8 μg/mg HAAFP/DOC = 13.8 μg/mg *Hydrophilic fraction:* THMFP/DOC = 35.3 μg/mg HAAFP/DOC = 31.7 μg/mg	Kim and Lee (2006)
Powdered activated carbon and granular activated carbon	N-nitrosodimethylamine precursors	River water and wastewater	PAC doses = 154–210 mg/L; NDMA precursor = 544 ng/L; GAC in rapid small scale column tests	37, 59, and 91 % removal of NDMA precursors with 3, 8 and 75 mg/L of PAC High NDMA precursor reduction ranging from 60 to 90 % was achieved across GAC contactors	Hanigan et al. (2012)
Coagulation-ultrafiltration	NOM	Surface water	pH = 5–10; 0.1 M NaOH; 0.1 M HCl; [Colour] = 68 g/m^3; [TOC] = 9.43 g C/m^3; Abs (254 nm) = 0.425 cm^{-1}; coagulant dose ($Al_2(SO_4)_3 \times n\ H_2O$) = 3.59 g Al/$m^3$; ultrafiltration membrane made of regenerated cellulose and polyethersulphone; transmembrane pressure = 0.1 MPa; effective surface of the membrane = 4.52 10^{-3} m^2	Up to 66.2 % of TOC removal after coagulation/ultrafiltration in the pH range 6–8 and aluminum dose = 3.59 g Al/m^3	Kabsch-Korbutowicz (2005)

Table 1.5 Removal of disinfection by-products (DBP) precursors using different processes

Process	Target compound	Matrix	Operating conditions	Results and comments	References
UV radiation	NOM	Raw surface water	LP UV (254 nm; 185 nm)	42–46 % removal of NOM	Dobrovic et al. (2007)
UV radiation	Low lower mass; hydrophilic NOM	Synthetic solution	H-Lamp 254 nm and 185 nm	DOC removal was 97 % with UV-C dose of 47–48 J cm^{-2}; DOC removal was 58 % with UV-C dose of 21 J cm^{-2}	Bond et al. (2009)
Photo-Fenton	NOM (dihydroxy-benzene)	River water	Dihydroxy-benzene compounds (catechol, resorcinol, hydroquinone); 1.0 mg/L of Fe^{3+}; sunlight irradiation; Volume of reactor = 8 L; flow rate = 9.1 L/min; $[H_2O_2]$ = 35 %	More than 80 % of dihydroxy-benzene mineralization	Moncayo-Lasso et al. (2008)
Photo-Fenton	Humic acid	Synthetic humic acid solution	Fenton reagent; UVA (365 nm)	More than 80 and 90 % of the DOC and UV254 removal, respectively, were achieved over 2.5 h of illumination	Sanly et al. (2007)
Photo-catalysis	NOM (hydrophobic, hydrophilic and transphilic)	Drinking water	N-Pd co-doped TiO_2(0.1 g); solar simulator (375 W. 1.0 W m^{-2}); DOC = 2.38–14.04 mg/L; Turbidity = 0.27–14.6 NTU; pH = 6.39–6.73; Conductivity = 0.12 mS/m	96, 38 and 15 % removal of hydrophobic, hydrophilic and transphilic NOM fractions	Nkambule et al. (2012)
UV/ H_2O_2	NOM	Raw water; synthetic humic acid	HP UV; H2O2 = 0–32.4 mM	90 % of non-purgeable dissolved organic carbon was mineralized	Wang et al. (2006)
MIEX DOC	NOM	Raw water	TOC = 4.44–8.84 mg/L; DOC = 4.21–8.18 mg/L UV_{254} = 4.8–10.2 m^{-1}; UV_{272} = 3.9–8.4 m^{-1}; MIEX resin doses = 5, 10, 15, 20 mL/L	61–91 % removal of DOC	Karpinska et al. (2012)

Table 1.5 (continued)

Process	Target compound	Matrix	Operating conditions	Results and comments	References
Nanofiltration membrane	THM, HAA	Surface water	pH = 7.95 ± 0.09; Turbidity = 3.0 ± 0.5 NTU; Conductivity = 514 ± 3 µS/cm; DOC = 3.4 ± 0.3 mg/L; UVA_{254} = 0.085 ± 0.002; $SUVA_{254}$ = 2.5 ± 0.2 L/mg DOC m; Bromide = 50 ± 10 µg/L; pore sizes ≤ 1000 Da	Reduction above 90 % of THM and HAA formation	Ates et al. (2009)
Ultrafiltration membrane	NOM (hydrophobic: acid, base, neutral, hydrophilic: acid, base, neutral)	River water	Hydrophobic acid = 35 %; hydrophilic neutral = 50 %; pore size of membrane = 0.01–0.02 µm	66 % of DOC; 93 % removal of hydrophobic acid; reduction of 54 and 30 % of THM and HAA formation	Lamsal et al. (2012)

2007; Uyguner et al. 2007). Granular activated carbon has been also suggested due to its greater efficiency in the removal of NOM, THMs, odor, color and other toxic chemicals (Stuart et al. 2000). In particularly, GAC and PAC have been successfully applied by Hanigan et al. (2012) to remove precursors of halogenated DBPs (N-nitrosodimethylamine precursors) from river water and wastewater treatment plants. After 4 h of contact time, NDMA precursors were reduced by 37, 59, and 91 % using 3, 8 and 75 mg/L of PAC. Besides, high NDMA precursor reduction ranging from 60 to 90 % was achieved across GAC contactors. In most cases, NOM removal by GAC cannot be accomplished to any significant degree in a filter/adsorber mode but requires a separate post filtration adsorber bed (Kim and Kang 2008).

Filtration membrane process represents another alternative process that can be combined with coagulation process to reduce the formation of DBPs (Oh and Lee 2005; Kim et al. 2006). It has been found by Kabsch-Korbutowicz that the application of coagulation–ultrafiltration process resulted in higher NOM removal than ultrafiltration used alone (Kabsch-Korbutowicz 2005).

1.3.1.2 Advanced Oxidation Processes

The control and the removal of NOM represent one of the most important approaches that can be applied to reduce the formation of chlorinated DBPs. Ozone is one of the powerful oxidant and disinfectant that can be used to remove DBP precursors (Chiang et al. 2009; Kasprzyk-Horden et al. 2003; Molnar et al. 2012). The decomposition of ozone in water generates hydroxyl radicals ($E_{OH/H2O} = 2.72$ V), which are of the most powerful oxidants that can be used for water treatment. Ozone does not lead to the formation of trihalomethanes or haloforms (Chiang et al. 2009). Ozone is capable of reacting with the aromatic rings that open, which allows the transformation of non-biodegradable compounds in biodegradable products. It is well known that ozone transforms the non-biodegradable organic pollutants into biodegradable compounds. The organic molecules are fragmented into small molecules (Siddiqui et al. 1997). The transformation of the non-biodegradable compounds into biodegradable compounds by ozone leads to the formation of organic acids, aldehydes and ketones. These by-products are a carbon source for bacteria, thereby promoting bacterial regrowth in the network distribution (Siddiqui et al. 1997). That is why today, the ozonation process is not installed at the end of the treatment system, but installed before filtration using granular activated carbon (Chu et al. 2012).

Likewise, the use of ozone as a disinfectant can lead to the formation of toxic by-products. In the presence of bromide ions, ozone oxidizes bromide ions in hypobromite ions (OBr^-) and optionally in the bromate ions (BrO_3^-). Bromate ions are suspected to be carcinogenic (DeAngelo 1998). Their concentrations in drinking water sometimes exceed 10–15 µg/L, whereas the parametric value of the guideline from Canada and United State is 10 µg/L (Legube et al. 2002). To circumvent these drawbacks, advanced oxidation processes (AOPs) have been proposed as alternative methods. Most of the AOPs combine two or three chemical oxidants in order to produce hydroxyl radicals (•OH). The free radicals are species capable of oxidizing numerous complex organics, non-chemically oxidizable or difficulty oxidizable.

They efficiently react with the double bonds of carbon-carbon and attack the aromatic nucleus, which are the major component of DBPs (Chin and Bérubé 2005). The AOP can be divided into four groups: (i) Homogenous chemical oxidation processes (H_2O_2/Fe^{2+} and H_2O_2/O_3); (ii) homogenous/heterogeneous photo catalytic processes (H_2O_2/UV, O_3/UV and Fe^{2+}/H_2O_2/UV; TiO_2/UV); (iii) sonification oxidation processes (ultrasounds oxidation) and; (iv) electrochemical oxidation processes. The system H_2O_2/Fe^{2+} represents the most common and simple AOP, which is often employed for the treatment of industrial effluents. However in drinking water, the system H_2O_2/O_3 is commonly used for pesticide and bio-recalcitrant compounds removal (Gogate and Pandit 2004; Pera-Titus et al. 2004).

The system H_2O_2/UV can be also used to oxidize NOM into smaller and more biodegradable compounds such as aldehydes and carboxylic acids (Toor et and Mohseni 2007). Recently, a medium-pressure (MP) ultraviolet (UV) system was used to investigate the UV photolysis and UV/H_2O_2 oxidation of pharmaceutically active compounds that belong to different therapeutic classes. Overall, MP lamps proved to be more efficient to maximize the bench-scale degradation of the selected group of compounds (ketoprofen, naproxen, carbamazepine, ciprofloxacin, clofibric acid, and iohexol) by both UV photolysis and UV/H_2O_2 oxidation (Pereira et al. 2007).

The removal of natural organic matter (NOM) from waters using hydrogen peroxide and iron-coated pumice particles as heterogeneous catalysts was investigated. Original pumice and peroxide dosed together provided UV absorbance reductions as high as 49 %, mainly due to the presence of metal oxides including Al_2O_3, Fe_2O_3 and TiO_2 in the natural pumice, which are known to catalyze the decomposition of peroxide by forming strong oxidants. Coating the original pumice particles with iron oxides significantly enhanced the removal of NOM with peroxide. A strong linear correlation was found between iron contents of coated pumices and UV absorbance reductions. Peroxide consumption also correlated with UV absorbance reduction (Kitis and Kaplan 2007). As reported in previous studies (Moncayo-Lasso et al. 2008; Goslan et al. 2006), the combination of UV radiation with Fenton reagent (Fe^{2+}/H_2O_2) increases the removal of natural organic matter. For instance, more than 80 % of dissolved organic carbon was achieved over 2.5 h of illumination (365 nm) and in the presence of Fenton reagent (Sanly et al. 2007). The photo-Fenton process using 1.0 mg/L of Fe^{3+} and sunlight radiations promotes the mineralisation of more than 80 % of dihydroxy-benzene compounds (catechol, resorcinol and hydroquinone) from river water (Moncayo-Lasso et al. 2008). Recent research on the adsorption of NOM onto pelletized TiO_2 and the oxidation of the surface of the pellets by UV light have been shown to reduce the dissolved organic carbon concentration of raw water by 70 % (Murray et al. 2007).

1.3.1.3 Ion Exchange Resin Processes

Ion exchange processes have received increased attention in recent years as an efficient alternative technique for removing dissolved organic matter. To remove natural polar organic matter (i.e. aliphatic and carboxylic acid structures), ion exchange is better performing than coagulation processes using inorganic salts (Bolto et al. 2004). Another advantage of employing anion exchange in water treatment is the ability to regenerate resins after the treatment without using a thermal process, by saving energy and by prolonging resin lifetime. Besides, ion exchange processes combined with conventional processes can constitute another strategy to reduce DBP formation (Tan et al. 2005, Humbert et al. 2007). The removal of NOM (specifically humic substances) increased with decrease in resin size. The best-performing ion exchange resins consisted of the smallest resins and/or those with the highest water content.

Anion exchange resins (AERs) have recently received a significant interest in literature since the development of a new AER, the MIEX® resin (Magnetic Ion Exchange resin). Recently, the MIEX process has been applied by Karpinska et al. (2012) to remove natural organic matter from water. The MIEX process revealed to be more efficient in removing DOC (61–91 % of DOC removal) compared to conventional treatment. Performances of AERs for NOM removal are influenced by the inner characteristics of the resins (strong or weak base AER), the water quality (pH, ionic strength, hardness, etc.) and the nature of organic compounds (molecular weight (MW), charge density, polarity) (Humbert et al. 2008).

1.3.1.4 Membrane Filtration

Low-pressure membrane technologies such as microfiltration (MF) and ultrafiltration (UF) are recognized as very attractive processes for producing drinking water. Membrane filtration offers several advantages such as fewer need of chemical agents, good quality of produced water, less production of sludge, compact process and easy automation. (Mijatovid et al. 2004; Shengji et al. 2008). UF membrane process has been successfully applied by Lamsal and his co-workers to remove natural organic compounds from river waters. The removal of 93 % of the hydrophobic acid components through UF treatment resulted in a reduction of trihalomethane (reduction of 54 % of THMs) and haloacetic acid (reduction of 30 % of HAAs) formation (Lamsal et al. 2012). Nanofiltration is a membrane process most often used for water softening and remove DBPs precursors (Hillie and Mbhuti 2007). The NF membrane is very effective to remove NOM from drinking water. NF may compete with other technologies such as GAC adsorption for NOM removal (Lin et al. 2007). NF membrane filtration promotes significant reduction of THM and HAA formation from waters (above 90 % of THM and HAA reduction) (Ates et al. 2009). The flux of NF membranes was related to their hydrophobicity and charges. Fouling by NOM adsorption was an important factor in hydrophobic and positively charged membranes, whereas it was negligible for hydrophilic and negatively charged membranes (Lee et al. 2007). Many factors affect membrane fouling by natural organic matter (NOM), including the nature of the NOM (size, hydrophobicity, and charge), the membrane (hydrophobicity, charge, and surface roughness), the solution (pH, ionic strength, hardness ion concentration) and the hydrodynamic systems (solution flux, surface shear) (Taniguchi et al. 2003; Farahbakhsh et al. 2004). A pretreatment of water using coagulation process is proposed to reduce the level of pore constriction by hydrophobic compounds. The coagulation process is known to preferentially remove the hydrophobic and charged compounds (Gray et al. 2007).

1.3.1.5 Photocatalysis Processes

Over the past 20 years, there has been a growing interest in system employing heterogeneous PC process (Le Clech et al. 2006; Uyguner et al. 2007). PC technology has a great potential due to the low treatment cost, environmental friendly and a sustainable treatment process in the field of wastewater treatment (Chong et al. 2011). This technology using catalyst such as TiO_2, ZnO, etc., and UV light has proved promising results for the degradation of persistent pollutants and producing more degradable and less toxic substances (Vora et al. 2009). Photochemical processes using UV radiation are simple, clean and less expensive. UV radiations absorbed by H_2O molecule allow the generation of powerful oxidizing species such as the hydroxyl radicals (OH°) and the hydrogen peroxide (H_2O_2) (Kim and Tanaka 2009). The direct contact of UV light with contaminated water induced high oxidative degradation. Photochemical processes are able to reduce natural organic matter from raw surface water by 42–46% (Dobrovic et al. 2007). Dissolved organic carbon can be removed up to 97% (DOC) using photochemical processes with UV-C dose of 47–48 J cm^{-2}(Bond et al. 2009). PC process is mainly based on the generation of oxidants reactive species upon absorption of UV light ($\lambda < 368$ nm) by the photo-catalyst. During the photocatalytic process, the pollutant species can be degraded directly on the surface of the semiconductor or indirectly by reacting with oxygen species extremely reactive. The most reactive oxygen species produced during photocatalytic process are hydroxyl radical, superoxide radical, hydrogen peroxide and ozone. According to previous research (Matilainen and Sillanpaa 2010), photocatalytic oxidation process has been proposed to be a good candidate for the degradation of natural organic compounds, and especially for removal of disinfectant by-products and their precursors. The removal of compounds is possible via three ways: oxidation by OH-radicals, reductive dechlorination by superoxide radicals and physical adsorption by TiO_2 (Gerrity et al. 2009; Wiszniowski et al. 2002). Photocatalytic process using Pd-modified N-doped TiO_2 photocatalyst has been successfully applied by Nkambule et al. (2012) to remove natural organic matter fractions from drinking water treatment plant under visible irradiation. Over a period of 120 min of simulated solar irradiation, photocatalytic degradation efficiency of 96, 38 and 15% were recorded for hydrophobic, hydrophilic and transphilic natural organic matter fractions, respectively.

1.4 Future Perspectives and Potential Research Directions

Inadequate protection of water distribution system may lead to increase the exposure to pathogenic microorganisms. As reported by the Water Health Organization (WHO 2000, 2002), approximately 3.4 million of people, mostly children died each year in the developing countries from water-related diseases. Likewise, approximately 1400 deaths and 25.000 hospitalizations were recently recorded in Haiti in 2010 due to Cholera outbreak (HaitiLibre 2010). During water distribution, disinfection process is essential to prevent waterborne diseases and ensure water quality from the water treatment plant outlet to the consumer's tap. Chlorine is widely used for the treatment of drinking water (Gopal et al. 2007). For instance, in Canada, approximately 90% of water treatment systems use chlorine for disinfection (Health Canada 1995; Health Canada 2008). There is no doubt that chlorine is an effective disinfectant against most microorganisms and it provides protection in the distribution network. Until date, chlorine is one of the most inexpensive disinfectants (Clark et al. 1994; Clark et al. 1998; USEPA 2006; Chowdhury et al. 2007). However, chlorination process poses chemical threat to human health due to DBPs (Legay et al. 2011). In the presence of precursors such as NOM, the DBPs are formed (Mesdaghinia et al. 2005). These DBPs (THM, HAAs, etc.) have been a concern since 1974 due to their possible cancer risks to human health and other sub-chronic/chronic health effects such as cardiac anomalies, low birth weight, still birth and pre-term delivery (Richardson et al. 2008; Villanueva et al. 2004; Mills et al. 1998; Wigle 1998; IRIS 2009; King et al. 2000).

The human exposure to DBPs in drinking water occurs through multiple routes including ingestion, inhalation and dermal contact during regular indoor activities such as swimming, bathing and showering (Legay et al. 2011; Chowdhury et al. 2011). To overcome these drawbacks, alternative disinfectants such as ozone, chloramine among others are used to avoid or reduce chlorinated DBPs in waters (Simate et al. 2012). For example, chloramine is a weaker disinfectant which requires greater contact time for disinfection in water treatment plants (Chowdhury et al. 2011). The use of chloramine promotes the formation of regulated and unregulated DBPs including N-nitrosodimethlamine (NDMA) which is more toxic than THMs and HAAs. The maximum level of NDMA set by USEPA was 0.7 ng/L (USEPA 2006). Otherwise, ozone used as disinfectant can form bromate ion (maximum level: 10 µg/L) in the presence of bromide ions, whereas the chlorine dioxide may form chlorite (maximum level: 1.0 mg/L). More detailed concerning the advantages and the drawbacks of ozone and chlorine dioxide are shown in Table 1.6. Despite the several advantage reported by use of these alternatives, inadequate protection of water distribution system could promote the increase of the incidence of water borne diseases as a result of increased exposure to pathogenic microorganisms (Chowdhury et al. 2011).

The removal of DBPs precursors such as natural organic matter (NOM) from water represents one of the effective strategies for controlling the formation of DBPs. Since 1980s, NOM has received an important focus in water treatment throughout the world (Liang and Singer 2003). Recently, AOPs applied to remove NOM during water purification and disinfection have received a great of interest (Matilainen and Sillanpaa 2010; Comninellis et al. 2008; Wang et al. 2006). AOPs have been intensively applied for NOM removal at the laboratory scale for decades. Various combination of oxidants, catalyst and radiation (O_3/H_2O_2, H_2O_2/UV, O_3/UV, UV/TiO_2, Fe^{2+}/H_2O_2, among others) have been developed to remove NOM and organic pollutant from aqueous environment (Le-Clech et al. 2006; Dobrovic et al. 2007;

Table 1.6 Advantages and drawbacks of ozone and chlorine dioxide as alternative disinfectants. (Anderson et al. 1982; Simate et al. 2012; Chowdhury et al. 2011)

Disinfectants	Advantages	Drawbacks
Chlorine dioxide	More effective than chlorine over short contact time	Expensive
	Effective against microbes	Formation of Chlorine, chlorite, chlorate
	High oxidant	Chlorite and chlorate oxidize haemoglobin
	Residual effect	Chlorite is a haemolytic agent
	Good control of taste, odour and colour	
	Removal of iron and manganese	
	No reaction with ammonia or aromatic organics to form THM	
	Forms chlorinated organics less readily than chlorine	
Ozone	Strong oxidizing agent	No remnant effect
	Good control of taste, odour and colour	Expensive
	No THMs were formed	Formation of unknown organic reaction products
	Able to oxidize THMs precursors	
	Can remove pesticides in the presence of UV radiations	

Goslan et al. 2006; Choo et al. 2008). In fact, the hydrophobic and high molar mass organic compounds have been observed to be more suitable for oxidation when the AOPs such as photo-catalysis and UV/H_2O_2 have been applied. Photo-Fenton process has proved to be efficient to remove refractory organic compounds (Murray and Parsons 2004).

Until date, the results of various studies dealing with NOM by AOPs are always varied depending on the water characteristics such as the amount of organic matter. Thus, it is very important to determine the organic characteristic of the treated water before the design and the optimization of the AOPs (Matilainen and Sillanpaa 2010). Besides, the identification of DBPs and the precursors of DBPs are required at the downstream of treatment in order to assess the suitability of AOP in the reduction of NOM (Matilainen and Sillanpaa 2010). The possibility to gain complete oxidation and mineralization of organic contaminants remain the major advantage of AOPs. Hydroxyl radicals are one of the most powerful oxidant produced by AOPs. OH radicals are very non selective oxidants. According to Parsons (2004), the reaction rate constant between OH° radicals and organic species are in the range of 10^8–$10^{10}M^{-1}s^{-1}$. However, the high treatment cost, the operational difficulties, the need to the high degree of pre-treatment and the lack of experience limits the application of AOPs during drinking water treatment at large scale (Matilainen and Sillanpaa 2010).

The combination of AOPs with other treatment technique has been shown to have significant potential to promote higher removal efficiency of NOM (Huang et al. 2008; Toor and Mohseni 2007; Uyguner et al. 2007). For instance, the combination of AOPs with membrane filtration has been shown by several research group's (Yao et al. 2009; Le-Clech et al. 2006) to have significant potential mitigating both NOM fouling of membranes and the formation of DBPs. Photo-catalysis process (UV/TiO_2) combined with membrane filtration has been observed to reduce the fouling of membrane through the mineralization of NOM and even by altering the characteristic of NOM, and therefore to improve membrane performance. Moreover, combining UV/H_2O_2 and filtration technique has been proved to be very robust and reliable barrier against pathogenic microorganism and organic micro-pollutants (Kruithof et al. 2007). The application of AOPs as pre-treatment method changes the structural properties of NOM and thus affects both the coagulation and the adsorption process. According to Uyguner et al. (2007), coagulation prior oxidation methods remove most of the high molar mass and the hydrophilic organic matter. Toor and Mohseni (2007) have also suggested that the combination of AOP and activated carbon offer a suitable reduction of harmful DBPs than the AOPs used alone.

Although more than 500 DBPs have been reported in the literature, researches are continuing to try to uncover those that have not been identified. To this end, the 1998 international workshop "*Identification of New and uncharacterized Disinfection By-products in drinking water*" sponsored by the International life Sciences Institutes (Washington, DC, USA), addressed this topic in order to develop new analytical techniques for the identification, quantification and prioritization of the DBPs in drinking water. The motivation for controlling DBPs in drinking waters is one of the approaches that public health professionals should increasingly develop. The reported concentrations of DBPs formed in aquatic environment are still detected in lower trace levels (from ng/L to µg/L). Given this fact, samples are usually concentrated to allow the detection of DBPs. The concentrations methods commonly applied include solid phase extraction (SPE), solid phase micro-extraction (SPME), liquid-liquid extraction and XAD resin extraction (Richardson 1998).

As far as, analytical method for the DBPs detection should have higher sensitivity, selectivity and specificity. Given this fact, it is so critical to find multi-residual method to detect the different types of DBPs. GC/MS was the primary tool for identifying and measuring DBPs. Large mass spectral libraries which contain more than 200.000 spectra (NSIT and Wiley databases) is able to ensure rapid identification of DBPs. To identify the structural information of new DBPs, GC/MS technique could be also coupled to infrared spectroscopy and chemical ionization (Richardson 2003). For example, iodo-acid (iodoacetic acid, bromoiodoacetic acid, (E)-3-bromo-3 iodopropenoic acid, (Z)-3-bromo-3-iodopropenoic acid, and (E)-2-iodo-3-methylbutenedioic acid) is one of the new DBPs which are discovered in drinking water treated with chloramination (Richardson and Postigo 2012). The information for both the molecular ions and the fragment ions was obtained by high resolution electron ionization (EI)-MS. GC/MS/MS is also popular analytical method for quantifying DBPs. MS-MS proves to be useful analysis technique able to identify and clarify the structure of the compounds in complex matrices.

However, liquid chromatography coupled to mass spectrometry (LC/MS/MS) which provides new opportunities for identify and quantify highly polar DBPs and high molecular weight DBPs, is highly recommended (Zwiener and Richardson 2005). Liquid chromatography coupled to mass spectrometry has been developed and effectively implemented by several research groups for the determination of DBPs. For instance, the first haloquinone DBP (2, 6-dichloro-1, 4-benzoquinone) found in drinking water was discovered using LC/MS/MS technique (Qin et al. 2010). In addition, new method has been developed by Zhao and his co-workers using LC/MS/MS technique for discover two new nitrosamine DBPs (nitrosodiphenylamine and nitrosopiperidine) found in drinking water (Zhao et al. 2006). The detection of nitrosodiphenylamine using LC/MS/MS is essential because it is thermally unstable and cannot be measured by GC/MS. Other DBPs such as chlorate, bromate, iodate and chlorite are present as anions in drinking water. They are not volatile and consequently they cannot be analyzed using GC/MS. The liquid chromatography technique is also unable to ensure a good separation. For that, chlorate, bromate, iodate and chlorite are nicely separate using ion chromatography. Environmental protection agency (EPA) has created several methods for measuring DBPs. The Environmental Protection Agency method 557 uses IC/ESI-MS/MS to measure the commonly occurring chloro-bromo-HAAs and dichloropropanoic acid with detection limits ranging from 0.015 to 0.2 μg/L. In addition, rapid method using IC/ICP-MS (ion chromatography/inductively coupled plasma-mass spectral) has been recently developed by Shi and Adams (2009) for simultaneously analysis of iodoacetic acids, bromoacetic acids, bromate and other related halogenated compounds present in water. Nevertheless, mono, di, and tri-chlorinated compounds are not detected using this technique due to the lower sensitivity of ICP-MS to chlorine. Mainly due to the strict regulations, there have been an increasingly efforts to develop a new analytical methods for the detection of DBPs at lower levels. High-field asymmetric waveform ion mobility spectrometry (FAIMS)/ESI-MS has been successfully applied to measure HAA, bromate, chlorate, iodate and nitrosamine in drinking water with lower ng/L detection limits (Ells et al. 2000; Gabryelski et al. 2003; Yuan et al. 2009; Planas et al. 2008; Barnett et al. 1999).

1.5 Conclusion

Chlorine is one of the oxidizing agents widely used around the world in water treatment. Disinfection by-products are formed when chlorine reacts with natural organic matter. Due to their adverse effect on human health, these disinfection by-products have received a great of concern. In order to meet requirement of new regulatory standards, chloramines, ozone and chlorine dioxide are used as alternative disinfectants. However, the appearance of disinfection by-products from chloramines, ozone and chlorine dioxide and their potential health effect should be further investigated. In addition, it is paramount to determine which exposure routes (inhalation or dermal exposure) are responsible for the adverse human health effects. Recently, AOPs used as alternative to chlorine are among the most studied technologies for water purification and disinfection. The possibility to achieve higher oxidation and mineralization remains the major advantage of AOPs. Thus, current and future research increasingly focuses to combine AOPs with conventional methods to enhance the biodegradability and the mineralization of organic compounds. Taking into account the type and the amount of organic compounds present in water, advanced studies need to be carried out to evaluate their efficiency.

References

Acero JL, Benitez FJ, Real FJ, Roldan G (2010) Kinetics of aqueous chlorination of some pharmaceuticals and their elimination from water matrices. Water Res 44:4158–4170

Aggazzotti G, Righi E, Fantuzzi G, Biasotti B, Ravera G, Kanitz S, Barbone F, Sansebastiano G, Battaglia MA, Leoni V, Fabiani L, Triassi M, Sciacca S (2004) Chlorination by-products (CBPs) in drinking water and adverse pregnancy outcomes in Italy. J Water Health 2:233–247

Ahmed EA, Campbell GA, Jacob S (2005) Neurological impairment in fetal mouse brain by drinking water disinfectant byproducts. Neuro Toxicol 26:633–640

Anderson AC, Reimers RS, Dekernion P (1982) A brief review of the current status of alternatives of chlorine disinfection of water. Am J Public Health 72:1290–1293

Asami M, Oya M, Kosaka K (2009) A nationwide survey of NDMA in raw and drinking water in Japan. Sci Total Environ 407:3540–3545

Ates N, Yilmaz L, Kitis M, Yetis U (2009) Removal of disinfection by-product precursors by UF and NF membranes in low-SUVA waters. J Membr Sci 328:104–112

Badawy MI, Gad-Allah TA, Ali MEM, Yoon Y (2012) Minimization of the formation of disinfection by-products. Chemosphere 89:235–240

Barnett DA, Guevremont R, Purves RW (1999) Determination of parts-per-trillion levels of chlorate, bromate and iodate by electrospray ionization/high-field asymmetric waveform ion mobility spectrometry/mass spectrometry. Appl Spectrosc 53:1367–1374

Barrett S, Hwang C, Guo Y, Andrews SA, Valentine R (2003) Occurrence of NDMA In: Drinking water: a North American survey, 2001–2002, Proceedings of AWWA annual conference, Anaheim, CA

Bolto B, Dixon D, Eldridge R (2004) Ion exchange for the removal of natural organic matter. React Funct Polym 60:171–182

Bond T, Goslan EH, Jefferson B, Roddick F, Fan L, Parsons SA (2009) Chemical and biological oxidation of NOM surrogates and effect on HAA formation. Water Res 43:2615–2622

Bove F, Shim Y, Zeitz P (2002) Drinking water contaminants and adverse pregnancy outcomes: a review. Environ Health Perspect 110:61–74

Boyd JM, Charrois JWA, Hofmann R, Hrudey SE (2012) NDMA and other N-nitrosamines–health risk assessment and management. Disinfection by products and human health. Chapter 7. IWA Publishing, Australia

Brenner S (2013) Aluminum may mediate Alzheimer's disease through liver toxicity, with aberrant hepatic synthesis of ceruloplasmin and ATPase7B, the resultant excess free copper causing brain oxidation, beta-amyloid aggregation and Alzheimer disease. Med Hypotheses 80:326–327. doi: 10.1016/j.mehy.2012.11.036

Bull RJ, Rice G, Teuschler L, Feder P (2009) Chemical measures of similarity among disinfection by-product mixtures. J Toxicol Environ Health 72:482–493

Canadian Cancer Society (2009) Canadian cancer statistics, national cancer institute of Canada 2004–2009. Toronto, pp 835–2976

Cedergren MI, Selbing AJ, Lofman O, Bengt AJ (2002) Chlorination by-products and nitrate in drinking water and risk for congenital cardiac defects. Environ Res 89:124–130

Chiang PC, Chang EE, Chang PC, Huang CP (2009) Effects of pre-ozonation on the removal of THM precursors by coagulation. Sci Total Environ 21:5735–5742
Chin A, Bérubé PR (2005) Removal of disinfection by-product precursors with ozone-uv advanced oxidation process. Water Res 39:2136–2144
Choi J, Valentine RL (2002) Formation of *N*-nitrosodimethylamine (NDMA) from reaction of monochloramine: a new disinfection by-product. Water Res 36:817–824
Chong MN, Jin B, Laera G, Saint CP (2011) Evaluating the photodegradation of carbamazepine in a sequential batch photoreactor system: impacts of effluent organic matter and inorganic ions. Chem Eng Res 174:595–602
Choo KW, Tao R, Kim MJ (2008) Use of photocatalytic membrane reactor for the removal of natural organic matter in water: effect of photoinduced desorption and ferrihydrite adsorption. J Membr Sci 322:368–374
Chowdhury S, Champagne P, Husain T (2007) Fuzzy risk-based decision-making approach for selection of drinking water disinfectants. J Water Supply Res Technol (Aqua) 56:75–93
Chowdhury S, Rodriguez MJ, Sadiq R (2011) Disinfection byproducts in Canadian provinces: associated cancer risks and medical expenses. J Hazard Mater 187:574–584
Chu WH, Gao NY, Deng Y, Templeton MR, Yin DQ (2011) Formation of nitrogenous disinfection by-products from pre-chloramination. Chemosphere 85:1187–1191
Chu W, Gao N, Krasner SW, Templeton MR, Yin D (2012) Formation of halogenated C-, N-DBPs from chlor(am)ination and UV irradiation of tyrosine in drinking water. Environ Pollut 161:8–14
Clark RM, Adams JQ, Lykins BW Jr (1994) DBP control in drinking water: cost and performance. J Environ Eng ASCE 120:759–781
Clark RM, Adams JQ, Sethi V, Sivaganesan M (1998) Control of microbial contaminants and disinfect ion by-products for drinking water in the US: cost and performance. J Water SRT-Aqua 47:255–265
Comninellis C, Kapalka A, Malato S, Parsons SA, Poulios I, Mantzavinos D (2008) Advanced oxidation processes for water treatment: advances and trends for R & D. J Chem Technol Biotechnol 83:769–776
Costet N, Villanueva CM, Jaakkola JJK, Kogevinas M, Cantor KP, King WD, Lynch CF Nieuwenhuijsen MJ, Cordier S (2011) Water disinfection by-products and bladder cancer: is there a European specificity? A pooled and meta-analysis of European case–control studies. Occup Environ Med 68:379–385
Daghrir R, Drogui P, Blais JF, Mercier G (2012) Hybrid process combining electrocoagulation and electrooxidation processes for the treatment of restaurant wastewaters. J Environ Eng 138:1146–1156
DeAngelo AB, George MH, Kilburn SR, Moore TM, Wolf DC (1998) Carcinogenicity of potassium bromate administered in the drinking water to male B6C3F_1 mice and F344/N rats. Toxicol Pathol 26:587–594
Deborde M, Von Gunten U (2008) Reactions of chlorine with inorganic and organic compounds during water treatment—kinetics and mechanisms: a critical review. Water Res 42:13–51
Dobrovic S, Juretic H, Ruzinski N (2007) Photodegradation of natural organic matter in water with UV irradiation at 185 and 254 nm: Importance of hydrodynamic conditions on the decomposition rate. Sep Sci Technol 42:1421–1432
Doré M (1989) Chimie des oxydants et traitement des eaux. TEC. DOC. LAVOISIER, Paris, p 505
Drogui P, Blais JF, Mercier G (2007) Review of electrochemical technologies for environmental applications. Recent Pat Eng 1:257–272
Ells B, Barnett DA, Purves RW, Guevremont R (2000) Detection of nine chlorinated and brominated haloacetic acids at part-per-trillion levels using ESI-FAIMS-MS. Anal Chem 72:4555–4559
Epstein DL, Nolen GA, Randall JL, Christ SA, Read EJ, Stober JA et al (1992) Cardiopathic effects of dichloroacetate in the fetal Long-Evans rat. Teratology 46(3):225–235
Farahbakhsh K, Svrcek C, Guest RK, Smith DW (2004) A review of the impact of chemical pretreatment on low-pressure water treatment membranes. J Environ Eng Sci 3:237–253

Gabryelski W, Wu F, Froese KL (2003) Comaprison of high-field asymmetric waveform ion mobility spectrometry with GC methods in analysis of haloacetic acids in drinking water. Anal Chem 75:2478–2486
Gagnon GA, O'Leary KC, Volk CJ, Chauret C, Stover L, Andrews RC (2004) Comparative analysis of chlorine dioxide, free chlorine and chloramines on bacterial water quality in model distribution systems. J Environ Eng 130:1269–1279
Gao B, Yue Q (2005) Effect of SO_4^{2-}/Al^{3+} ratio and OH^-/Al^{3+} value on the characterization of coagulant poly-aluminum-chloride-sulfate (PACS) and its coagulation performance in water treatment. Chemosphere 61:579–584
Gerrity D, Mayer B, Ryu H, Crittenden J, Abbaszadegan M (2009) A comparison of pilot scale photocatalysis and enhanced coagulation for disinfection by-product mitigation. Water Res 43:1597–1610
Glezer V, Harris B, Tal N, Iosefzon B, Lev O (1999) Hydrolysis of haloacetonitriles: linear free energy relationship, kinetics and products. Wtare Res 33:1938–1948
Gogate PR, Pandit AB (2004) A review of imperative technologies for wastewater treatment II: hybrid methods. Adv Environ Res 8:553–597
Gopal K, Tripathy SS, Bersillon JL, Dubey SP (2007) Chlorination byproducts, their toxicodynamics and removal from drinking water. J Hazard Mater 140:1–6
Goslan EH, Gurses F, Banks J, Parsons SA (2006) An investigation into reservoir NOM reduction by UV photolysis and advanced oxidation processes. Chemosphere 65:1113–1119
Goslan EH, Krasner SW, Bower M, Rocks SA, Holmes P, Levy LS, Parsons SA (2009) A comparison of disinfection by-products found in chlorinated and chloraminated drinking waters in Scotland. Water Res 43:4698–4706
Government of Ontario (2006) Drinking water surveillance program (DWSP) summary report for 2000–2004 Personal communication with dave fellowes, ministry of the environment, Ontario, Canada
Gray SR, Ritchie CB, Tran T, Bolto BA (2007) Effect of NOM characteristics and membrane type on microfiltration performance. Water Res 41:3833–3841
Grellier J, Bennett J, Patelarou E, Smith RB, Toledano MB, Rushton L, Briggs DJ, Nieuwenhuijsen MJ (2010) Exposure to disinfection by-products, fetal growth, and prematurity: a systematic review and meta-analysis. Epidemiol 21:300–313
HaitiLibre: Haiti Cholera Epidemic: Latest Assessment (2010) http://www.haitilibre.com/en/news-1725-haiti-cholera-epidemic-latestassessment-13-people-infected-and-confirmed-each-minute.html. Accessed April 2013
Hanigan D, Zhang J, Herckes P, Krasner SW, Chen C, Westerhoff P (2012) Adsorption of N-Nitrosodimethylamine precursors by powdered and granular activated carbon. Environ Sci Technol 46:12630–12639
Hansen KMS, Zortea R, Piketty A, Vega SR, Andersen HR (2013) Photolytic removal of DBPs by medium pressure UV in swimming pool water. Sci Total Environ 443:850–856
Health Canada (1995) A national survey of disinfection byproducts in Canadian drinking water (95-EHD-197), Environmental Health Directorate, Ottawa, Ontario
Health Canada (2008) Guidelines for Canadian drinking water quality, prepared by the Federal-Provincial-Territorial committee on health and the Environment, 2008
Hebert A, Forestier D, Lene D, Benanou D, Jacob S, Arfi C, Lambolez L, Levi Y (2010) Innovative method for prioritizing emerging disinfection by-products (DBPs) in drinking water on the basis of their potential impact on public health. Water Res 44:3147–3165
Hillie T, Hlophe M (2007) Nanotechnology and the challenge of clean water. Nat Nanotech 2:663–664
Huang X, Leal M, Li Q (2008) Degrdation of natural organic matter by TiO2 photocatalytic oxidation and its effect on fouling of low pressure membranes. Water Res 42:1142–1150
Huang H, Wu QY, Hu HY, Mitch WA (2012) Dichloroacetonitrile and Dichloroacetamide Can Form Independently during Chlorination and Chloramination of Drinking Waters, Model Organic Matters, and Wastewater Effluents. Environ Sci Technol 46:10624–10631

Humbert H, Gallard H, Jacquemet V, Croué JP (2007) Combination of coagulation and ion exchange for the reduction of UF fouling properties of a high DOC content surface water. Water Res 41:3803–3811
Humbert H, Gallarda H, Sutyb H, Croué J (2008) Natural organic matter (NOM) and pesticides removal using a combination of ion exchange resin and powdered activated carbon (PAC). Water Res 42:1635–1643
International Agency for research on Cancer (IARC) (1978) N-nitrosodimethylamine. In some N-nitroso compounds. IARC Monogr Eval Carcinog Risks Hum 17:125–175
IRIS, The Integrated Risk Information System Online Database, US Environmental Protection Agency (2009) http://www.epa.gov/iris/subst/index.html
Ivancev VT, Dalmacijam B, Tamas Z, Karlovic E (2002) The effect of different drinking water treatment processes on the rate of chloroform formation in the reactions of natural organic matter with hypochlorite. Water Res 33:3715–3722
Kabsch-Korbutowicz M (2005) Application of ultrafiltration integrated with coagulation for improved NOM removal. Desalination 174:13–22
Karpinska AM, Boaventura RAR, Vilar VJP, Bilyk A, Molczan M (2012) Applicability of MIEX DOC process for organics removal from NOM laden water. Environ Sci Pollut Res. doi:10.1007/s11356–012-1334-x
Kasprzyk-Hordern B, Ziolek M, Nawrocki J (2003) Catalytic ozonation and methods of enhancing molecular ozone reactions in water treatment. Appl Catal B 46:639–669
Kim J, Kang B (2008) DBPs removal in GAC filter-adsorber. Water Res 82:145–152
Kim HC, Lee S (2006) Pump diffusion flash mixing (PDFM) for improving coagulation process in drinking water treatment. Sep Purif Technol 52:117–125
Kim I, Tanaka H (2009) Photodegradation characteristics of PPCPs in water with UV treatment. Environ Int 35:793–802
Kim HC, Hong JH, Lee S (2006) Fouling of microfiltration membranes by natural organic matter after coagulation treatment: a comparison of different initial mixing conditions. J Membr Sci 283:266–272
King WD, Dodds L, Allen AC (2000) Relation between stillbirths and specific chlorinated byproducts in public water supplies. Environ Health Perspect 108:883–886
Kitis M, Kaplan SS (2007) Advanced oxidation of natural organic matter using hydrogen peroxide and iron-coated pumice particles. Chemosphere 68:1846–1853
Krasner SW, McGuire MJ, Jacangelo JG, Patania NL, Reagan KM, Aieta EM (1989) The occurrence of disinfection by-products in US drinking water. J Am Water Works Assoc 81:41–53
Krasner SW, Weinberg HS, Richardson SD, Pastor SJ, Chinn R, Sclimenti MJ, Onstad GD, Thruston AD Jr (2006) Occurrence of a new generation of disinfection byproducts. Environ Sci Technol 40:7175–7185
Kruithof JC, Kamp PC, Martijin BJ (2007) UV/H_2O_2 treatment: a practical solution for organic contaminant control and primary disinfection. Ozone Sci Eng 29:273–280
Lamsal R, Montreuil KR, Kent FC, Walsh ME, Gagnon GA (2012) Characterization and removal of natural organic matter by an integrated membrane system. Desalination 303:12–16
Le-Clech P, Lee EK, Chen V (2006) Hybrid photocatalysis/membrane treatment for surface waters containing low concentrations of natural organic matters. Water Res 40:323–330
Lee S, Kim S, Cho J, Hoek EMV (2007) Natural organic matter fouling due to foulant–membrane physicochemical interactions. Desalination 202:377–384
Legay C, Rodriguez MJ, Sadiq R, Sérodes JB, Levallois P, Proulx F (2011) Spatial variations of human health risk associated with exposure to chlorination by-products occurring in drinking water. J Environ Manage 92:892–901
Legube B, Parinet B, Berne F, Croue JP (2002) Bromate surveys in French drinking water works. Ozone Sci Eng J Int Ozone Assoc 24:293–304
Lei H, Marinas BJ, Minear RA (2004) Bromamine decomposition kinetics in aqueous solutions. Environ Sci Technol 38:2111–2119
Liang L, Singer PC (2003) Factors Influencing the Formation and Relative Distribution of Haloacetic Acids and Trihalomethanes in Drinking Water. Environ Sci Technol 37:2920–2928

Lin YL, Chiang PC, Chang EE Removal of small trihalomethane precursors from aqueous solution by nano filtration. J Hazard Mater 146:20–29
Luo Q, Wang D, Wang Z (2012) Occurences of nitrosamines in chlorinated and chloraminated drinking water in three representative cities, China. Sci Total Environ 437:219–225
Matilainen A, Sillanpää M (2010) Removal of natural organic matter from drinking water by advanced oxidation processes. Chemosphere 80:351–365
Mesdaghinia A, Rafiee MT, Vaezi F, Mahvi AH (2005) Evaluation of ferric chloride and alum efficiencies in enhanced coagulation for TOC removal and related residual metal concentrations. Iran J Environ Health Sci Eng 2:189–194
Mijatovid I, Matogid M, Cerneha BH, Bratulid D (2004) Removal of natural organic matter by ultrafiltration and nanofiltration for drinking water production. Desalination 169:223–230
Mills AL, Alexander M (1976) N-Nitrosamine formation by cultures of several microorganisms. Appl Environ Microbiol 31:892–895
Mills CJ et al (1998) Health risks of drinking water chlorination by-products. Chronic Dis Can 19:91–102
Molnar J, Agbaba J, Dalmacija B, Klašnja M, Watson M, Kragulj M (2012) The effects of ozonation and catalytic ozonation on the removal of natural organic matter from groundwater. J Environ Eng 138:804–808
Moncayo-Lasso A, Pulgarin C, Benitez N (2008) Degradation of DBPs' precursors in river water before and after slow sand filtration by photo-Fenton process at pH 5 in a solar CPC reactor. Water Res 42:4125–4132
Muellner MG, Wagner ED, McCalla K, Richardson SD, Woo YT, Plewa MJ (2007) Haloacetonitriles vs. regulated haloacetic acids: are nitrogen-containing DBPs more toxic? Environ Sci Technol 4:645–651
Muller-Pillet V, Joyeux M, Ambroise D, Hartemann P (2000) Genotoxic activity of five haloacetonitriles: comparative investigations in the single cell gel electrophor1esis (comet) assay and the Ames-fluctuation test. Environ Mol Mutagen 36:52–58
Murray CA, Parsons SA (2004) Removal of NOM from drinking water: Fenton's and photo-Fenton's processes. Chemosphere 54:1017–1023
Murray CA, Parsons SA (2006) Preliminary laboratory investigation of disinfection by-product precursor removal using an advanced oxidation process. Water Environ J 20:123–129
Murray CA, Goslan EH, Parsons SA (2007) TiO_2/UV: single stage drinking water treatment for nom removal. J Environ Eng Sci 6:311–317
Najm I, Trussell RR (2001) NDMA formation in water and wastewater. J Am Water Works Assoc 93:92–99
Nakajima N, Nakano T, Harada F, Taniguchi H, Yokoyama I, Hirose J, Daikoku E, Sano K (2004) Evaluation of disinfective potential of reactivated free chlorine in pooled tap water by electrolysis. J Microbiol Methods 57:163–173
Nieuwenhuijsen MJ, Taledano MB, Eaton NE, Fawell J, Elliott P (2000) Chlorination disinfection byproducts in water and their association with adverse reproductive outcomes: a review. Occup Environ Med 57:73–85
Nkambule TI, Kuvarega AT, Krause RWM, Haarhoff J, Mamba BB (2012) Synthesis and characterisation of Pd-modified N-doped TiO_2 for photocatalytic degradation of natural organic matter (NOM) fractions. Environ Sci Pollut Res 16:4120–4132
Norton CD, Le chevallier MW (1997) Chloramination: its effect on distribution system water quality. J Am Water Works Assoc 89:66–77
Oh Jl, Lee SH (2005) Influence of streaming potential on flux decline of microfiltration with in-line rapid pre-coagulation process for drinking water production. J Membr Sci 254:39–47
Olivier BG (1983) Dihaloacetonitriles in drinking water: algae and fulvic acid as precursors. Environ Sci Technol 17:80–83
Ouhoummane N, Levallois P, Gingras S (2004) Thyroid function of newborns and exposure to chlorine dioxide by-products. Arch Environ Health 59:582–587
Parsons S (2004) Advanced oxidation processes for water and wastewater treatment. IWA Publishing, London

Pera-Titus M, Garcia-Molina V, Banos MA, Giménez J, Esplugas S (2004) Degradation of chlorophenols by means of advanced oxidation processes: a general review. Appl Catal B 47:219–256
Pereira VJ, Linden KG, Weinberg HS (2007) Evaluation of UV irradiation for photolytic and oxidative degradation of pharmaceutical compounds in water. Water Res 41:4413–4423
Planas C, Palacios O, Ventura F, Rivera J, Caixach J (2008) Analysis of nitrosamines in water by automated spe and isotope dilution gc/hrms-occurrence in the different steps of a drinking water treatment plant, and in chlorinated samples from a reservoir and a sewage treatment plant effluent. Talanta 76:906–913
Plewa MJ, Kargalioglu Y, Vankerk D, Minear RA, Wagner ED (2002) Mammalian cell cytotoxicity and genotoxicity analysis of drinking water disinfection by-products. Environ Mol Mutagen 40:134–142
Plewa MJ, Wagner ED, Jazwierska P, Richardson SD, Chen P H, McKague AB (2004) Halonitromethane drinking water disinfection by products: chemical characterization and mammalian cell cytotoxicity and genotoxicity. Environ Sci Technol 38:62–68
Plewa MJ, Wagner ED, Muellner MG, Kang-Mei Hsu KM, Richardson SD (2008) Comparative mammalian cell toxicity of N-DBPs and C-DBPs. ACS Symp Ser 995:36–50
Prarat P, Ngamcharussrivichai C, Khaodhiar S, Punyapalakul P (2013) Removal of haloacetonitriles in aqueous solution through adsolubilization process by polymerizable surfactant-modified mesoporous silica. J Hazard Mater 244–245:151–159
Qin F, Zhao YY, Zhao Y, Boyd JM, Zhou W, Li XF (2010) A toxic disinfection by-product, 2,6-dichloro-1,4-benzoquinones, identified in drinking water. Angew Chem Int Ed 49:790–792
Reckhow DA, MacNeill AL, Platt TL, McClellan JN (2001) Formation and degradation of dichloroacetonitrile in drinking waters. J Water Suppl Res Technol AQUA 50:1–13
Richardson SD (1998) Drinking water disinfection by-products. In: Meyers RA (ed) The encyclopedia of environmental analysis and remediation, vol 3. Wiley, New York, pp 1398–1421
Richardson SD (2003) Disinfection by-products and other emerging contaminants in drinking water. Trac-Trend Anal Chem 22(10):666–684
Richardson SD, Postigo C (2012) Drinking water disinfection by-products. Emerg Org Contam Hum Health Hand Environ Chem 93–137
Richardson SD, Fasano F, Ellington JJ, Crumley FG, Buettner KM, Evans JJ, Blount BC, Silva LK, Waite TJ, Luther GW, Mckague AB, Miltner RJ, Wagner ED, Plewa PJ (2008) Occurrences and mammalian cell toxicity of iodinated disinfection byproducts in drinking water. Environ Sci Technol 42:8330–8338
Richardson SD, DeMarini DM, Kogevinas M, Fernandez P, Marco E, Lourencetti C, Ballesté C, Heederick D, Meliefste K, Bruce McKague A, Marcos R, Fon-Ribera L, Grimalt JO, Villanueva CM (2010) What's in the pool? A comprehensive identification of disinfection by-products and assessment of mutagenicity of chlorinated and brominated swimming pool water. Environ Health Perspect 118:1523–1530
Righi E, Bechtold P, Tortorici D, Lauriola P, Calzolari E, Astolfi G, Nieuwenhuijsen MJ, Fantuzzi G, Aggazzotti G (2012) Trihalomethanes, chlorite, chlorate in drinking water and risk of congenital anomalies: a population-based case-control study in Northern Italy. Environ Res 116:66–73
Rizzo L, Belgiorno V, Gallo M, Meriç S (2005) Removal of THM precursors from a high-alkaline surface water by enhanced coagulation and behaviour of THMFP toxicity on *D. magna*. Desalination 176:177–188
Rizzo L, Di Gennaro A, Gallo M, Belgiorno V (2008) Coagulation/chlorination of surface water: a comparison between chitosan and metal salts. Sep Purif Technol 62:79–85
Sadiq R, Rodriguez MJ (2004) Disinfection by-products (DBPs) in drinking water and predictive models for their occurrence: a review. Sci Total Environ 321:21–46
Sanly, Lim M, Chiang K, Amal R, Fabris R, Chow C, Drikas M (2007) A study on the removal of humic acid using advanced oxidation process. Sep Sci Technol 42:1391–1404
Seidel CJ, McGuire MJ, Summers RS, Via S (2005) Have utilities switched to chloramines? J Am Water Works Assoc 97:87–97

Selcuk H, Bekbolet M (2008) Photocatalytic and photoelectrocatalytic humic acid removal and selectivity of TiO_2 coated photo-anode. Chemosphere 73:854–858

Selcuk H, Rizzo L, Nikolaou AN, Meric S, Belgiorno V, Bekbolet M (2007) DBPs formation and toxicity monitoring in different origin water treated by ozone and alum/PAC coagulation. Desalination 210:31–43

Shah AD, Dotson AD, Linden KG, Mitch WA (2011) Impact of UV disinfection combined with chlorination/chloramination on the formation of halonitromethanes and haloacetonitriles in drinking water. Environ Sci Technol 45:3657–3664

Shengji X, Xing L, Ji Y, Bingzhi D, Juanjuan Y (2008) Application of membrane techniques to produce drinking water in China. Desalination 222:497–501

Siddiqui MS, Amy GL, Murphy BD (1997) Ozone enhanced removal of natural organic matter from drinking water sources. Water Res 31:3098–3106

Simate GS, Iyuke SE, Ndlovu S, Heydenrych M, Walubita LF (2012) Human health effects of residual carbon nanotubes and traditional water treatment chemicals in drinking water. Environ Int 39:38–49

Singer P (1994) Control of disinfection by products in drinking water. J Environ Eng 120:727–744

Singer PC, Bilyk K (2002) Enhanced coagulation using a magnetic ion exchange resin. Water Res 36:4009–4022

Smith MK, Randall JL, Read EJ, Stober JA. (1992) Developmental toxicity of dichloroacetate in the rat. Teratology 46:217–223

Stuart B, Zhanj L, Wang S (2000) Quenching of chlorination disinfection byproducts formulation in drinking water by hydrogen peroxide. Water Res 34:1652–1658

Tan Y, Kilduff JE, Kitis M, Karanfil T (2005) Dissolved organic matter removal and disinfection byproduct formation control using ion exchange. Desalination 176:189–200

Taniguchi M, Kilduff JE, Belfort G (2003) Modes of Natural Organic Matter Fouling during Ultrafiltration. Environ Sci Technol 37:1676–1683

Tardiff RG, Carson ML, Ginevan ME (2006) Updated weight of evidence for an association between adverse reproductive and developmental effects and exposure to disinfection by-products. Regul Toxicol Pharmacol 45:185–205

Tian C, Liu R, Guo T, Liu H, Luo Q, Qu J (2013) Chlorination and chloramination of high bromide natural water: DBPs species transformation. Sep Purif Technol 102:86–93

Toor R, Mohseni M (2007) UV-H_2O_2 based AOP and its integration with biological activated carbon treatment for DBP reduction in drinking water. Chemosphere 66:2087–2095

Trofe TW, Inman GW Jr, Johnson JD (1980) Kinetics of monochloramine decomposition in the presence of bromide. Environ Sci Technol 14:544–549

USEPA (1979) Methods for chemical analysis of waters and wastes. US Environmental Protection Agency, Washington

USEPA (1997) Mercury emissions from the disposal of fluorescent lamps. Final report. Washington:1v. http://www.epa.gov/epaoswer/hazwaste/id/merc-emi/merc-pgs/merc-rpt.pdf. Accessed April 2013

USEPA (1998) Environmental protection agency. 40 CFR Parts 9, 141, and 142. National primary drinking water regulations: interim enhanced surface water treatment; final rule. Federal Regist 63:69398–69476

USEPA (2003) National primary drinking water regulations: stage 2 disinfectants and disinfection byproducts (D/DBP). Final rule 2003, 68, 159

USEPA (2006) National primary drinking water regulations: stage 2 disinfectants and disinfection byproducts rule: final rule. Federal Regist 71(2):387–493

Uyak V, Toroz I (2007) Investigation of bromide ion effects on disinfection by-products formation and speciation in an Istanbul water supply. J Hazard Mater 149:445–451

Uyguner CS, Suphandag SA Kerc A, Bekbolet M (2007) Evaluation of adsorption and coagulation characteristics of humic acids preceded by alternative advanced oxidation techniques. Desalination 2010:183–193

Villanueva CM, Cantor KP, Cordier S, Jaakkola JJ, King WD, Lynch CF, Porru S, Kogevinas M (2004) Disinfection by products and bladder cancer: a pooled analysis. Epidemiology 15:357–367

Villanueva CM, Castano-Vinyals G, Moreno V, Carrasco-Turigas G, Aragonés N, Boldo E, Ardanaz E, Toledo E, Altzibar JM, Zaldua I, Azpiroz L, Goni F, Tardon A, Molina AJ, Martin V, Lopez-Rojo C, Jiménez-Moleon JJ, Capelo R, Gomez-Acebo I, Peiro R, Ripoll M, Garcia-Lvedan E, Nieuwenhujsen MJ, Rantakokko P, Goslan EH, Pollàn M, Kogevinas M (2012) Concentrations and correlations of disinfection by-products in municipal drinking water from an exposure assessment perspective. Environ Res 114:1–11

Vora JJ, Chauhan SK, Parmar KC, Vasava SB, Sharma S, Bhutadiya LS (2009) Kinetic Study of application of ZnO as a photocatalyst in heterogeneous medium. J Chem 6:531–536

Wang GS, Liao CH, Chen HW, Yang HC (2006) Characteristics of natural organic matter degradation in water by UV/H_2O_2 treatment. Environ Technol 27:277–287

WHO (World Health Organization) (2000) Disinfectants and disinfectant by-products. WHO (World Health Organization), Geneva

WHO (World Health Organization) (2002) Water and sanitation, facts and figures. http://www.who.int/watersanitationhealth/General/factsandfigures.htm. Accessed April 2013

Wigle DT (1998) Safe drinking water: a public health challenge. Chronic Dis Can 19:103–107

Wiszniowski J, Robert D, Surmacz-Gorska D, Miksch K, Weber JV (2002) photocatalytic decomposition of humic acids on TiO_2 part I: Discussion of adsorption and mechanism. J Photochem Photobiol A 152:267–273

Yan M, Wang D, Ma X, Ni J, Zhang H (2010) THMs precursor removal by an integrated process of ozonation and biological granular activated carbon for typical Northern China water. Sep Purif Technol 72:263–268

Yang CY, Cheng BH, Tsai SS, Wu TN, Lin MC, Lin KC (2000) Association between chlorination of drinking water and adverse pregnancy outcome in Taiwan. Environ Health Perspect 108:765–768

Yao P, Choo K, Kim M (2009) A hydridized photocatalysis-microfiltration system with iron oxide-coated membranes for the removal of natural organic matter in water treatment: effects of iron oxide layers and colloids. Water Res 43:4238–4248

Yuan YZ, Liu X, Boyd JM, Qin F, Li J, Li XF (2009) Identification of N-nitroamines in treated drinking water using nanoelectrospray ionization high field asymmetric waveform ion mobility spectrometry with quadrupole time of flight mass spectrometry. J Chromatogr Sci 47:92–96

Zaviska F, Drogui P, Pablo G (2012) Statistical optimization of active chlorine production from a synthetic saline effluent by electrolysis. Desalination 296:16–23

Zhao YY, Boyd J, Hrudey SE, Li XF (2006) Characterization of new nitrosamines in drinking water using liquid chromatography tandem mass spectrometry. Environ Sci Technol 40:7636–7641

Zwiener C, Richardson SD (2005) Drinking water disinfection by-product analysis by LC/MS and LC/MS/MS. Trends Anal Chem 24:613–621

Chapter 2
Greenhouse Gas from Organic Waste Composting: Emissions and Measurement

Antoni Sánchez, Adriana Artola, Xavier Font, Teresa Gea, Raquel Barrena, David Gabriel, Miguel Ángel Sánchez-Monedero, Asunción Roig, María Luz Cayuela and Claudio Mondini

Contents

Abstract There is actually common consensus to use biological technologies for the treatment of organic wastes. For instance composting involving the aerobic biological stabilization of organic wastes is gaining popularity. The amount of materials and the variety of wastes composted is increasing fast. However

A. Sánchez (✉) · A. Artola · X. Font · T. Gea · R. Barrena · D. Gabriel
Department of Chemical Engineering, Universitat Autònoma de Barcelona, 08193-Cerdanyola del Vallès, Barcelona, Spain
e-mail: Antoni.Sanchez@uab.cat

M. Á. Sánchez-Monedero · A. Roig · M. L. Cayuela
Centro de Edafología y Biología Aplicada del Segura, Consejo Superior de Investigaciones Científicas, Campus Universitario de Espinardo, 30100-Espinardo, Murcia, Spain

C. Mondini
Consiglio per la Ricerca e la Sperimentazione in Agricoltura, Centro di Ricerca per lo Studio delle Relazioni tra Pianta e Suolo, Gruppo di Ricerca di Gorizia, CRA-RPS, 34170 Gorizia, GO, Italy

© Springer International Publishing Switzerland 2015

E. Lichtfouse et al. (eds.), *CO_2 Sequestration, Biofuels and Depollution*, Environmental Chemistry for a Sustainable World 5, DOI 10.1007/978-3-319-11906-9_2

composting is a process emitting gases some of which being greenhouse gases (GHG) that favour global warming. In particular carbon dioxide (CO_2), methane (CH_4) and nitrous oxide (N_2O^-) are responsible for the global warming potential of composting. A part of these gases can be abated by low-cost biological technologies such as biofiltration. This review compiles all the points related to the emission of GHG from composting processes, from detection and measurement to minimization and abatement. We focus on measurements of GHG to obtain reliable emission factors for designing composting technologies. This will help to compare waste treatment options based on integrated tools such as Life Cycle Assessment (LCA). A chapter discusses C and N dynamics in the compost, and implications on emitted C and N gases. Finally we review the best available practices to minimize the GHG emissions from composting. We also present the final treatment of composting gases.

Keywords Composting · Anaerobic digestion · Greenhouse gas (GHG) · Environmental impact · Life cycle assessment (LCA) · Carbon dioxide · Methane · Nitrous oxide · Volatile organic compounds (VOCs) · Biofiltration

2.1 Introduction

The sustainable use of resources and wastes, including waste minimisation and valorisation, is a common objective of the plans, directives and rules published in the last few decades. One example is the Sixth Programme of Community Action in the field of Environment ("Environment 2010: the future is in our hands") published by the European Union for the period 2001–2012 (European Union 2008a). The Sixth Program of Action includes the implementation of seven thematic strategies and, among them, specifically waste prevention and recycling, with the objective to reduce the negative environmental impacts during the whole life cycle of wastes, from their production to their elimination, including their recycling. One of the results of all these legislation efforts was the publication of the Waste Framework Directive in 2008. This Directive considers waste not only as a potential source of pollution, but also as a resource that can be used. Specifically, in the case of biodegradable wastes, the Directive 1999/31/CE on landfilling of wastes encourages the diversion of these wastes to other treatment technologies involving the recycling and energy recovery from wastes, where composting will have a great importance (European Union 2008b; Commission of the European Communities 2008). Nevertheless, the environmental impact assessment during the whole life cycle of wastes lacks of data obtained directly at full-scale waste treatment facilities operating in different locations, thus limiting the quality and reliability of these analyses necessary for the decision-making process.

A direct consequence of the above mentioned plans and directives has been the proliferation of a large number of new waste treatment plants installed in Europe and all over the world in the last years, as well as the modification and adaptation of the existing ones. In particular, composting and anaerobic digestion are the more widely accepted processes for organic waste treatment. Composting plants are typically operated either in piles or tunnels, whereas anaerobic digestion can take place either in wet or dry digesters, typically followed by composting of the digested sludge with the aim of ensuring its sanitation and stabilization (Ponsá et al. 2008). The anaerobic decomposition process that is carried out in anaerobic digestion facilities allows energy recovery from wastes in the form of biogas. All these treatments also allow the valorisation of wastes by their use in agriculture or as soil organic amendments.

Waste treatment facilities can be the origin of public complaints, most of them associated to annoyances caused by odour emissions generated during the process. Biological treatment plants are a clear example of this problem. Odours generated from this type of treatment plants are mainly associated to the emissions of volatile organic compounds (terpenes, alcohols, ketones, sulphur compounds, amines, etc.) and ammonia (Goldstein 2002; Komilis et al. 2004). Some of the annoyances caused by these emissions are often magnified because of the lack of real data from operating plants that would contribute to have an objective and scientific base to analyse these problems. Such lack of data also represents a problem for the design of mitigation measures such as the use of biofilters. In addition to this, greenhouse gases (GHG) emission inventories evidence the increase in the amount of these compounds that are emitted from waste treatment facilities. Emission of CO_2, CH_4 and N_2O are the main responsible of this increase (Colón et al. 2012).

Emissions generated in waste treatment plants, in particular those based on biological treatments, are related to the type of technology, the type of wastes treated and the operational conditions of the plant. For this reason, it is very important to relate the emissions to the performance of the biological treatment plants and also to the wastes being treated, since each treatment technology and waste will give rise to different end products quality and organic matter stabilisation degrees. The use of respirometric indices to monitor the stability of the organic matter has been one of the main research topics in the last years (Barrena et al. 2005; Barrena et al. 2006; Barrena et al. 2009a; Barrena et al. 2009b; Ponsá et al. 2008).

Although ammonia is not considered a GHG, its emissions during composting are usually studied because it causes acid rain and from the point of view of the conservation of nitrogen in the end-product because of the potential use of compost in agriculture as organic fertiliser, as well as for the determination of the efficiency of the systems for gas emission treatment, such as scrubbers and biofilters. Ammonia emissions are affected by the C/N ratio of the initial composting mixture, by the temperature reached during the process and by the aeration (Pagans et al. 2006b; Raviv et al. 2002; Sánchez-Monedero et al. 2001). Biofilters have shown to be an efficient equipment for the reduction of ammonia emissions in enclosed waste treatment plants (Hong and Park 2004; Pagans et al. 2006b), although for long periods ammonia tends to reduce the efficiency of this technology (Baquerizo et al. 2005).

An important part of the published literature in the field of gaseous emissions is related to odours, mainly by means of dynamic olfactometry, in both composting plants and mechanical-biological treatment plants (MBT). As already mentioned, a number of laboratory-scale experiments have been performed with the aim of determining the compounds that more significantly contribute to odour pollution. Thus, Goldstein (2002) identified terpenes, alcohols, aldehydes, fatty acids, ammonia and a range of sulphur compounds as the main responsible of odour emissions at composting plants. Other authors have studied the effect of some operational conditions, such as ventilation and turning, in these emissions (Szanto et al. 2007). Gage (2003) proposed a number of managing practices aimed at reducing the annoyances generated by odour emissions; for instance, preparation of an optimal initial mixture and the maintenance of high levels of porosity to assure aerobic conditions in the pile (Ruggieri et al. 2009). Enclosing the composting operations and the use of biofilters are among the main mitigation strategies for both odours and GHG.

The importance of GHG emissions generated during the biological treatment of wastes has been also stated by several authors. CO_2 emissions coming from biological process are not considered to contribute to global warming since this carbon has a biogenic origin, i.e., this carbon has been previously fixed biologically. Regarding other gases, He et al. (2001) measured the emissions of N_2O and CH_4 during the composting of food wastes under laboratory conditions in a closed system with forced aeration. Although generated in small amounts, N_2O and CH_4 have a great contribution to global warming since they have a warming potential 25 (CH_4) and 296 (N_2O) times higher than that of CO_2.

There are some scientific publications that provide gaseous emissions data generated during the biological treatment of organic wastes, mainly for manures and sewage sludge. However, the number of published papers dealing with municipal solid wastes is scarce (Colón et al. 2012). The works carried out by Eitzer (1995) and Staley et al. (2006) are very important for the characterization of the emissions generated during the biological treatment of wastes and the identification of specific compounds. In 1995, Eitzer performed a comprehensive characterisation of the volatile organic compounds (VOC) generated in composting plants treating municipal solid wastes and its possible relation to the process performance. On the other hand, Staley et al. (2006) studied the VOC emissions originated during the aerobic treatment of wastes and also during the anaerobic biodegradation process. These works highlighted the importance that forced aeration, used in the biological processes, had on the total emissions (Delgado-Rodriguez et al. 2012). Terpenes and ketones are shown to be the most abundant compounds. These experimental works were performed under laboratory conditions, which would limit their extrapolation to full-scale plants. Pagans et al. (2006a) also evaluated the effect of the type of waste (industrial and municipal solid waste) on VOC emissions under laboratory conditions. Komilis et al. (2004) identified the main VOC emitted during composting of pruning residues (mainly terpenes, alkyl benzenes, ketones and alkanes) and also during composting of food wastes (sulphides, organic acids and alcohols), as well as the stages of the process that generated the highest emissions (thermophilic phase).

The determination of emission factors for different wastes and different treatment technologies will be a useful tool for the calculation of global emissions at facilities operating with a technology already studied in other treatment plants. Emission factors for VOC, NH_3 or GHG are usually expressed per ton of treated waste or per amount of obtained compost (Amlinger et al. 2008).

Sampling and measurement protocols for the determination of emissions have been also studied (Sironi and Botta 2001). Even though there are several published papers about this topic (Sommer et al. 2004), there is a lack of information on the measurement of emissions from surface sources, in both non-aerated (composting piles with natural aeration by convection) and those with a common source that will be later spread in an outlet surface (biofilters).

The main factors controlling a composting process are those characteristics of an aerobic biological process such as oxygen concentration, temperature, moisture, pH and C/N ratio. The optimum values for the C/N ratio range from 15 to 30, even though it is possible that composting takes place in a wider range of values (Haug 1993). For this reason, adjusting the optimum C/N ratio of the starting mixture is recommended. The use of different organic wastes or some selected additives could also be satisfactory (Charest and Beauchamp 2002). Nevertheless, the amounts of carbon and nitrogen used for the calculations should be referred to the amounts that are ready available for the microorganisms when considering the C/N ratio as a parameter to be optimised (Puyuelo et al. 2011). This specific point is very important for the potential practical implications in the preparation of starting composting mixtures. In relation to pH, recent studies have demonstrated its effect on the emissions of odours (Sundberg et al. 2013).

In this context, respirometric methodologies have been shown to be suitable and reliable for the determination of the amount of biodegradable organic matter in wastes of different origin and characteristics. There are two types of respirometric analysis for this purpose: dynamic and static determinations, being the dynamic methods the most widely accepted and recommended (Adani et al. 2004; Barrena et al. 2006; Gea et al. 2004). The measurement of the CO_2 produced during the respirometric test is also used as a measurement of the biodegradability of the organic matter (Cooper 2004) and, consequently, of the biodegradable organic C.

Other researchers have worked on the emissions generated during the composting process of agricultural wastes (Komilis et al. 2004; Cayuela et al. 2006; Mondini et al. 2006; Mondini et al. 2007; Sánchez-Monedero et al. 2008; Szanto et al. 2007). In the USA, other studies are focused on VOC and NH_3 emissions during the composting of biowaste (Büyüksönmez and Evans 2007).

This review is a compilation of the different works dealing with the measurement, detection, minimization and treatment of the GHG emitted during the composting process of a wide variety of organic wastes.

2.2 Composting

2.2.1 The Specific Role of Composting in Greenhouse Gas (GHG) Emissions

Composting is an environmentally friendly waste treatment process where organic matter is biologically degraded. Although the benefits of composting are evident, GHG can be generated and emitted to the atmosphere during this process contributing to global warming.

In this context, composting of organic waste contributes (composting process) and avoids (compost application) at the same time to GHG emissions. GHG are released from composting facilities due to degradation of organic matter and the use of electricity and fuels in management waste operations. The use of compost in agriculture has a positive effect in GHG emissions since its application as an organic amendment provokes that carbon stays bound to soil, although the content of other nutrients (N, P, etc.) is typically low. GHG emissions from composting processes depend on the waste type and composition, the technology systems used (static and dynamic process, open and closed systems, presence or not of gas treatment units) and the final use of compost.

Benefits of compost application have to be assessed together with a real knowledge about the amount of GHG such as N_2O and CH_4 generated during the composting process. The relation of GHG with some operational conditions and the technology used must be also considered. Data on GHG emissions from full-scale composting facilities are necessary to improve the knowledge about the contribution to the composting in GHG emissions. In the last years, there has been an increase in the number of scientific publications studying GHG emissions during composting (Amlinger et al. 2008; Boldrin et al. 2009; Sánchez-Monedero et al. 2010; Cayuela et al. 2012; Colón et al. 2012; Deportes 2012).

GHG emissions from composting processes are highly dependent on the waste type and composition. The composition and characteristics of the feedstock are key parameters for the design and operation of the composting facilities and for the final quality of the compost (Haug 1993).

Wastes with a low C/N ratio and high water content have a great potential for generating GHG emissions both during the storage and the composting process. In fact, wastes lacking of nutrients, porosity and structure, or presenting low biodegradability can hamper the correct evolution of the process, increasing the GHG emission. In order to minimize these emissions, optimal conditions for the initial mixture are required.

For some wastes, pretreatment operations before composting are required. This is the case of municipal solid wastes, especially when a source-separation system is not implemented. The production of high-quality compost from MSW may require a lot of energy because of the use of heavy machinery that makes GHG emissions unavoidable (Lou and Nair 2009). Other materials, such biosolids or manure, have a poor structure and an excess of water content and require the use of a bulking agent.

Grinding and mixing this bulking agent are operations that require energy that again contribute to GHG emissions.

Composting technologies can be open and closed systems. In open systems, composting is performed in facilities where, in general, gaseous emissions are neither collected nor treated. However, when the composting process takes place in an enclosed system usually the exhaust gases are treated. As expected, concentrations of GHG reported in facilities when the gas treatment systems are well-implemented were lower (Colón et al. 2012) than those of open systems. Effects of forced aeration and turning in GHG emissions have been also studied. Szanto et al. (2007) observed lower N_2O and CH_4 emissions in turned piles than in static systems. They related these emissions to the prevalence of anaerobic regions in the static systems, as other similar studies (Parkinson et al. 2004). Amlinger et al. (2008) proposed that high aeration and effective stripping of NH_3 during the early stages of composting can reduce N_2O formation. Ermolaev et al. (2012) studied the effects of different aeration and temperature settings on the emission of CH_4, N_2O and CO_2 during windrow composting with forced aeration following three different control strategies. However, they found that the emissions of CH_4 and NO_2 were low regardless the amount of ventilation. The oxygen concentration, temperature profile and moisture content are factors controlling GHG emissions. Nowadays, in the composting field, the technology that allows the control of these parameters is available.

Regarding CO_2, its emissions in composting derived from the organic matter biodegradation are not taken into account in their contribution to global warming since this carbon has a biogenic origin. The CO_2 that contributes to GHG emissions is generated by composting facilities as a result of operational activities. In composting, the main GHG that can contribute to global warming are CH_4 and N_2O. Both are related to a lack of oxygen during the composing process and consequently they depend on the management of the composting process (Cayuela et al. 2012; Colón et al. 2012). These gases, although they are generated in small amounts, have a great contribution to global warming since they have a warming potential of 25 (CH_4) and 296 (N_2O) times higher than that of CO_2.

Several authors reported that even in well-aerated process CH_4 was emitted (He et al. 2000; Clemens and Cuhls 2003) while Beck-Friis et al. (2000) observed a rapid decrease when the oxygen supply was increased. The production of N_2O can be due to an incomplete ammonium oxidation or incomplete denitrification (Beck-Friis et al. 2000). Emissions of N_2O have been reported at different stages of the process. Some authors reported high emissions at the beginning of composting (He et al. 2000; Parkinson et al. 2004). Other studies reported the production of N_2O during the mesophilic and maturation phases (Beck-Friis et al. 2000; Hao et al. 2004) when the readily available carbon sources has been depleted (He et al. 2000). According to Cayuela et al. (2012), N_2O formation will be hampered if there are conditions to inhibit nitrification (such as low available NH_4^+ in the pile or high pH). Beck-Friis et al. (2000) and Fukumoto et al. (2003) related N_2O emissions to the temperature of the process and CH_4 emissions to the size of the pile (both works were performed at full-scale, using windrows and forced aeration systems, respectively), the structure of the material and the time of the process. Higher emissions were measured in larger piles, with a poor structure and longer composting times. Monitoring of CH_4 emissions showed a large experimental fluctuation in all works.

Several authors have reported the GHG emissions generated during the biological treatment of several typologies of wastes. Most of them were calculated from laboratory and pilot scale processes, although interesting data at industrial scale have been also reported (Boldrin et al. 2009; Colón et al. 2012; Ermolaev et al. 2012). There are an important number of studies that quantify CH_4 and N_2O emissions from animal manures (Fukumoto et al. 2003; Hao et al. 2004; Szanto et al. 2007). However, less published works dealing with municipal solid wastes can be found, and even less works studying the GHG emissions of different composting systems have been published.

Colón et al. (2012) evaluated four different full-scale facilities treating the source-separated organic fraction of municipal solid wastes (OFMSW). They reported a range of CH_4 and N_2O emissions between 0.34 and 4.37 kg CH_4 Mg $OFMSW^{-1}$ and 0.035 and 0.251 kg CH_4 Mg $OFMSW^{-1}$, respectively. Regarding CH_4, the highest values were found in facilities without gas treatment units. Also, Boldrin et al. (2009) presented a study where several technologies for municipal solid waste treatment were evaluated. They reported CH_4 and N_2O emissions ranging from 0.02 to 1.8 kg CH_4 Mg $OFMSW^{-1}$ and 0.0075 and 0.252 kg CH_4 Mg $OFMSW^{-1}$.

As previously commented, although ammonia is not considered a GHG, its emission during composting plays an important role. Ammonia emissions are affected by the C/N ratio of the initial composting mixture, by the temperature reached during the process and the aeration (Pagans et al. 2006b). High loads of ammonia can reduce the optimal use of the biofilter system in enclosed facilities (Amlinger et al. 2008). Moreover the conservation of nitrogen in the end-product improves compost use in agriculture as organic fertiliser. Consequently, from a global warming point of view, less use of chemical fertilizers will be required (Favoino and Hogg 2008).

In the role played by composting in GHG emissions it is important to bear in mind the role of compost as an end-product. The use of compost as an organic amendment can contribute to mitigate GHG in several forms.

Compost utilization can reduce the need of chemical fertilisers and pesticides, which implies the reduction of GHG emissions associated with their production and application. Also, a positive effect in soil structure is produced with compost application by improving tillage and workability. Improved structure of soils associated with the application of organic matter can help to reduce requirements for water irrigation in periods of drought and to increase the potential of soils to retain moisture (Favoino and Hogg 2008).

One of the aspects associated with compost utilization that more attention has received in the last years is the potential for sequestration of carbon in agricultural soils (Mondini et al. 2007; Favoino and Hogg 2008). By applying compost, biogenic carbon is held in soils for a period of time before carbon is released, increasing carbon uptake and storage within the plant and removing CO_2 from the atmosphere.

2.2.2 GHG Emitted During Composting and Their Relationship to C and N Dynamics

Microbial transformations involved in the formation of CH_4 and N_2O in composting piles are similar to those taking place in other environments such as soil, water bodies, wastewater treatment plants, etc. However, the microbial gas production and the final emission to the atmosphere will be affected by the particular environmental conditions of composting piles (such as temperatures up to 70 °C, high organic matter content, easily available organic compounds, rich and active microbial population and limited amount of oxygen, etc.) and composting management operations (turning, watering, pile size and geometry, etc.). All these variables represent a characteristic environment affecting not only the microbial gas production in the pile, but also its transport within the pile and the final emission to the atmosphere. In the following sections the impact of the C and N dynamics on GHG emissions during composting will be also discussed.

2.2.2.1 Carbon Dioxide (CO_2)

As previously mentioned, there are two main sources of CO_2 emissions from composting facilities, biogenic and non-biogenic CO_2. Biogenic CO_2 emissions derive from the biological degradation of the organic matter, mostly as a consequence of aerobic decomposition and, to a lesser extent, from anaerobic processes or the oxidation of CH_4 by aerobic methanotrophic bacteria. This emission accounts for the highest amount of gas generated during the process, since between 40 and 70 % of the original organic matter can be degraded during composting (Haug 1993). However, the global warming potential of these emissions are not taken into account in the environmental impact of composting operations since this biological CO_2 is considered to be carbon neutral (IPCC 2006). The exclusion from the inventories has reduced the number of papers studying CO_2 emissions and this gas is only studied from the point of view of establishing mass balances of composting operations (Boldrin et al. 2009) or as an index of the overall microbial activity of the pile, reflecting the progress of the process (Hobson et al. 2005; Sánchez-Monedero et al. 2010) and the evaluation of the stability of the end-product (Barrena et al. 2006).

Non-biogenic CO_2 from composting includes the emissions associated to energy and fuel consumption in the composting facility. These emissions are dependent on the technology of the plant and the machinery used such as shredders, front-loaders, turning equipment, screenings, and other processing activities. These emissions are beyond the scope of this review, but updated information can be found elsewhere (Boldrin et al. 2009; Scheutz et al. 2009; Lou and Nair 2009; Brown et al. 2008).

2.2.2.2 Methane (CH_4)

Methane emissions derived from organic waste composting have attracted the attention of researchers as a considerable contributor to global warming since this greenhouse gas has a global warming potential 25 times greater than carbon dioxide over a time horizon of 100 years (IPCC 2006).

The optimum growing conditions for methanogenic bacteria are a lack of oxygen (strict anaerobic microorganisms), a redox potential below −200 mV, neutral pH and the presence of nutrients and substrates rich in organic matter (Kebreab et al. 2006). These conditions can be temporally found at the early stages of the composting process, where large amounts of nutrients and available sources of organic compounds stimulate microbial growth, depleting the oxygen levels in the pile. Accordingly, most of CH_4 emissions have been recorded during the initial weeks of the process, at the beginning of the thermophilic phase (Beck-Friis et al. 2000; Sánchez-Monedero et al. 2010). The high temperatures reached at the beginning of the process reduce oxygen solubility (Pel et al. 1997), facilitating the creation of anaerobic spots within the pile. However, there are other variables such as high concentration of ammonia, which may inhibit the activity of methanogens at $pH > 9$ (Kebreab et al. 2006), or the presence of electron acceptors such as sulphates, which reduce their activity by competition with sulphate reducing bacteria (Hao et al. 2005). Sánchez-Monedero et al. (2010) reported that the high ammonia levels generated by the hydrolysis of urea, used as nitrogen source, inhibited the production of CH_4 in olive mill waste composting piles.

The emission of CH_4 from composting piles is governed by the biological activity of the pile (Hao et al. 2001) and also by other factors affecting gas transport from the anaerobic spots to the pile surface, such as gas diffusion within the pile and the presence of methanotrophic bacteria. Methanotrophs are aerobic microorganisms colonising the surroundings of anaerobic zones and pile surface, which are able to oxidise between 46 and 98 % of the CH_4 generated in the pile (Jäckel et al. 2005). Methanotrophic bacteria also play an important role in the production and consumption of other relevant GHG emitted during composting, such as N_2O and CO (Topp and Hanson 1991). Sánchez-Monedero et al. (2011) performed a 4-year interannual evaluation of the GHG emissions from a composting plant treating olive mill wastes and found a reduction of CH_4 emissions associated to the improvement of the management of the composting plants (watering and turning frequencies).

Kebreab et al. (2006) and Brown et al. (2008) reviewed the topic of GHG emissions from livestock and composting operations and they highlighted the importance of the composting feedstock, the height and shape of the pile, the control of moisture content and turning frequency as the main factors governing CH_4 emissions during the process, since these variables will affect both the oxygen availability and gas diffusion in the composting pile. The presence of manure can also increase the methane emissions due to the incorporation of anaerobic microorganisms, as observed by He et al. (2000) and Sánchez-Monedero et al. (2010) in composting piles treating food and olive mill wastes, respectively.

2.2.2.3 Nitrous Oxide (N_2O)

There is an increasing awareness about the emission of N_2O from composting operations due to the high global warming potential of this gas (296 times higher than that of CO_2 over a 100 year horizon, IPCC 2006) and its impact on the ozone layer (Smith et al. 2010). Despite the relatively small amounts of N_2O released during composting, its contribution to the global N_2O budget in waste management or livestock agriculture cannot be discarded due to the impact of composting operations treating manures or other N-rich organic wastes (de Klein et al. 2010).

The biological production of N_2O during composting is a complex process since there are different microbial pathways involved in the formation of N_2O (nitrification, nitrifier denitrification and denitrification among others), which may simultaneously occur at different locations within the pile (Czepiel et al. 1996; Kebreab et al. 2006; Maeda et al. 2011). For this reason, the identification of N_2O sources as well as the microorganisms involved in these processes still remains a key research topic (Maeda et al. 2011).

Nitrification is one of the main microbial processes leading to the emission of N_2O during composting. Aerobic nitrification involves the initial transformation of ammonia to nitrite by different genera of ammonia-oxidising bacteria (AOB), such as *Nitrosomonas* and *Nitrososporas*, according to the following equation:

$$2NH_3 + 3O_2 \rightarrow 2NO_2^- + 2H^+ + 2H_2O$$

and the oxidation of nitrite to nitrate by nitrite-oxidising bacteria (NOB), such as *Nitrobacter* (Kowalchuk et al. 1999; Maeda et al. 2010):

$$2NO_2^- + O_2 \rightarrow 2NO_3^-$$

NH_4^+ is the main precursor of nitrification. NH_4^+ is generated by ammonification of OM at early stages of the process (Sánchez-Monedero et al. 2001). Typical alkaline pHs found in composting matrices favour the transformation of this soluble NH_4^+ into NH_3, which is then initially oxidised by AOB into NO_2^- and then transformed to NO_3^- by nitrite-oxidising bacteria. N_2O is produced during the initial step of the oxidation of NH_4^+, as an intermediate between NH_2OH and NO_2^- (Czepiel et al. 1996). Ammonia-oxidising *archaea* (AOA) have been recently suggested to be actively involved in nitrification in composting piles, but the contribution of AOA to the total amount of N_2O still remains unclear (Yamamoto et al. 2010; Zeng et al. 2012).

Denitrification has traditionally represented the main source of N_2O, especially in the case of manures (Kebreab et al. 2006). Denitrification is an anoxic process carried out by denitrifiers, which are heterotrophic microorganisms that can use NO_3^- as the electron acceptor, causing the reduction of NO_3^- to N_2 according to the following steps:

$$NO_3^- \rightarrow NO_2^- \rightarrow NO \rightarrow N_2O \rightarrow N_2$$

In absence of O_2, NO_3^- is reduced to N_2 without appreciable N_2O production, but N_2O production can increase as the concentration of O_2 increases in the pile (Czepiel et al. 1996). In this case, nitrifier denitrification (denitrification coupled to an incomplete nitrification at low O_2 concentrations) can be the responsible of the generation of N_2O during the initial step of ammonia oxidation and also as a consequence of NO_2^- reduction. This mechanism has been studied in agricultural soils (Wrage et al. 2001), but there is only limited information during composting (He et al. 2001; Hobson et al. 2005). Fukumoto and Inubushi (2009) observed that the addition of NOB reduced the emission of N_2O during composting of pig manure, suggesting that the accumulation of NO_2^- in the pile could be a significant source of N_2O, due to the reduction of NO_2^- to N_2O (under limited O_2 conditions) rather than the final oxidation to NO_3^- (with no O_2 limitation). Under these conditions, when available C was depleted, nitrifier denitrification would be the main mechanisms leading to N_2O emissions, as observed by He et al. (2000), who found an increase in the N_2O emission when the ratio between water-soluble C and dissolved N was lower than 5.

Nitrifiers and denitrifiers show their optimal growth under different environmental conditions. Nitrifiers require aerobic conditions, mesophilic temperatures (below 40 °C), pH values above 5 and the presence of NH_4^+, whereas denitrifiers need anaerobic conditions, or at least low O_2 concentration, the presence of sources of available C and the presence of NO_3^-, NO_2^- or NO as electron acceptors (Kebreab et al. 2006). Due to the heterogeneity of the composting materials, both environmental conditions (aerobic and anaerobic zones) can coexist simultaneously in the composting mass, since different oxygen concentration gradients are created along the pile (Beck-Friis et al. 2000; Hao et al. 2001). Denitrifiers may colonise the inner part of the pile whereas nitrifiers, which require oxygen concentrations in the range within 1 and 10 % (Béline et al. 1999), may colonise the aerobic pile surface. The relative contribution of nitrifiers and denitrifiers to the N_2O emission is governed by the oxygen concentration and moisture of the pile (Hwang and Hanaki 2000). These authors reported that denitrification was the main source of N_2O at moisture levels between 40–60 % and oxygen concentrations around 10 %, whereas nitrification became more dominant at higher oxygen concentrations.

Similarly to those of CH_4, N_2O emissions can be affected not only by the biological activity of the composting mixture, but also by the N availability and gas diffusion within the pile (Hao et al. 2001). Several authors reported peak N_2O emissions either at early stages of the process or after the thermophilic phase of composting, when the environmental conditions of the pile (temperatures below 40 °C) favour the growth of nitrifying bacteria (He et al. 2001; Kebreab et al. 2006; Sánchez-Monedero et al. 2010). Once NO_3^- has been generated, the mixing of the composting matrix facilitates the transport of nitrates from the surface to the interior of the pile where they can be reduced to N_2 and N_2O by denitrifiers. The use of urea as N source can enhance N_2O emissions up to levels similar to those found in N-rich manure heaps due to the increase of available N from the hydrolysis of urea (Sánchez-Monedero et al. 2010). Vermicomposting also increases the N_2O emissions by stimulating denitrification and nitrification processes, due to the increase of N availability and the transport of N facilitated by the activity of earthworms (Frederickson and Howell 2003; Hobson et al. 2005).

Gas exchange within the pile also plays an important role since the generation of N_2O by both nitrifiers and denitrifiers is enhanced at low oxygen concentrations (Czepiel et al. 1996). N_2 is obtained as the final product of denitrification in absence of O_2, but significant amounts of N_2O are generated as the concentration of O_2 increases in the pile. In addition, pure cultures of *Nitrosomona* bacteria responsible of the initial step of ammonia oxidation have been shown to significantly increase the production of N_2O under limited oxygen conditions (Goreau et al. 1980). Since these factors are highly dependent on the composting material and the process performance, the specific characteristics of the starting materials will determine the environmental conditions for N transformation during composting.

2.2.2.4 Other Relevant Greenhouse Gases

There are other gases generated in small amounts during organic waste composting that have been studied due to their impact on global warming. Carbon monoxide (CO) and nitrogen oxides different than N_2O (NO_x) have small direct global warming potential, but they both lead to indirect radiative effects by increasing CH_4 lifetime and elevating concentrations of tropospheric O_3 (IPCC 2006). The calculation of their contribution to global warming is subject to large uncertainties due to the short lifetime and reactivity of these gases in the atmosphere. According to IPCC (2006) the global warming potential, over a 100-year horizon, is likely to be 1–3 for CO, and in the order of 5 for surface NO_X emissions.

The emission of CO occurs during the aerobic decomposition of the organic wastes during composting by a mixture of physical processes and biological activity (Hellebrand and Halk 2001; Hellebrand and Shade 2008). These authors found the maximum CO-flux rates at the beginning of the composting process, probably due to physicochemical generation, and then the levels decreased during periods of high biological activity, reflecting the temperature dependence of CO emissions and also the impact of oxygen availability and the oxidation to CO_2. CO emissions only represent a minor GHG source in green waste and livestock waste (Hellebrand and Shade 2008) and in urban wastes, where CO-C emissions varied from 0.07 to 0.13 kg Mg^{-1} of wet feedstock, which represents approximately about 0.04–0.08 % of the total C emitted (Andersen et al. 2010a, b). CO emissions have been also investigated as a potential health risk to workers in enclosed facilities treating municipal solid wastes (Phillip et al. 2011).

From the two gases composing NO_x ($NO+NO_2$), only NO is generated during composting, either as by-product or intermediate of microbial nitrification and denitrification (Del Prado et al. 2006; Hao et al. 2001). Fukumoto et al. (2011) studied the NO emissions from swine manure composting and observed a similar trend to that of N_2O, characterised by a peak after the thermophilic phase of composting (coinciding with the activity of nitrifiers) and a decreasing trend towards the end of the process. Total NO emissions only represented one-tenth of the magnitude of N_2O emission, approximately 3 % of total N losses.

2.2.3 *Greenhouse Gas Production for Different Typologies of Organic Wastes*

There is a wide range of organic wastes that can be used as composting substrates such as manures, municipal solid wastes, garden and yard wastes, agricultural crop residues, sewage sludge and other industrial sludge, etc. The characteristics of these starting materials will affect the physicochemical properties of the pile and, consequently, will govern the microbial processes leading to the formation of GHG and also their diffusion and transport within the pile. As already discussed in the description of the main pathways of CH_4 and N_2O generation, the main variables affecting GHG emissions are the moisture content and porosity, which control the oxygen availability and gas diffusion, and the presence of nutrients and organic compounds to be used as substrates for the microorganisms participating in gas production. The composting technology used for the aeration (forced aeration or windrowing), the size of the piles and pile temperature also represent key variables affecting GHG generation and emission.

2.2.3.1 Manures

Manures represent one of the most important and studied substrates for composting (Kebreab et al. 2006). Manures are N-rich organic materials characterised by high moisture contents that make them to be considered as wet feedstock for composting (Haug 1993). The treatment of manures through composting permits the reduction of volume and moisture, their sanitisation and organic matter stabilisation, giving rise to a valuable end-product that can be safely used in agriculture. However, manure characteristics favour GHG emission during composting. The large amounts of easily available N compounds enhance the microbial activity of the pile and can serve as substrates for the nitrification and denitrification processes leading to the emission of N_2O. Furthermore, high moisture together with enhanced microbial activity at early stages of the process can lead to the creation of anaerobic spots for the formation of CH_4.

A summary of the amounts of CH_4 and N_2O generated during manure composting is shown in Table 2.1. The amounts of CH_4 emitted during composting are

Table 2.1 Summary of CH_4 and N_2O emissions reported in the literature for different typologies of wastes and composting technologies

Feedstock	Bulking agent	System	N_2O	CH_4	Reference
Manures					
Swine	Cardboard	Windrow	0.1 % of initial N	–	Kuroda et al. 1996
Swine	Bedding (straw)	Static	9.8 % of initial N	12.6 % of initial OM	Szanto et al. 2007
		Windrow	0.5 % of initial N	0.5 % of initial OM	
Swine	Sawdust	Static	3.7–4.7 % of initial N	1.0–1.9 kg CH_4 ton^{-1}	Fukumoto et al. 2003
Swine	Sawdust	Static	3.0–9.3 % of initial N	–	Fukumoto and Inubushi 2009
Swine	Barley straw	–	0.058 kg N_2O-N ton^{-1}	0.19 kg CH_4-C ton^{-1}	Sommer and Moller 2000
Swine	Sawdust	Static	1.6 μg m^{-2} s^{-1}	5.2 μg m^{-2} s^{-1}	Park et al. 2011
		Windrow	7.9 μg m^{-2} s^{-1}	7.5 μg m^{-2} s^{-1}	
Cattle	Bedding (straw)	Static	0.11 kg N_2O-N ton^{-1}	6.3 kg CH_4-C ton^{-1}	Hao et al. 2001
		Windrow	0.19 kg N_2O-N ton^{-1}	8.1 kg CH_4-C ton^{-1}	
Cattle	Bedding (straw)	Windrow	0.077 kg N_2O-N ton^{-1}	8.92 kg CH_4-C ton^{-1}	Hao et al. 2004
	Bedding (wood chips)		0.084 kg N_2O-N ton^{-1}	8.93 kg CH_4-C ton^{-1}	
Cattle (dairy)	House wrap	Static	0.370 kg N_2O-N ton^{-1}	1.14 kg CH_4-C ton^{-1}	Pattey et al. 2005
Cattle (beef)			0.103 kg N_2O-N ton^{-1}	0.11 kg CH_4-C ton^{-1}	
Cattle (dairy)	Bedding (straw)	Static	0.046 kg N_2O-N ton^{-1}	–	El Kader et al. 2007
	Bedding (straw)	Windrow	0.070 kg N_2O-N ton^{-1}	–	
Turkey	Wood shaving & straw	Static	0.091 kg N_2O-N ton^{-1}	–	
Cattle and horse (50:50)	Bedding (hay)	Windrow	0.32 kg N_2O-N ton^{-1}	–	Czepiel et al. 1996
Cattle (dairy)	Bedding	Windrow	0.90 g N_2O m^{-2} d^{-1}	13.5 g CH_4 m^{-2} d^{-1}	Leytem et al. 2011
MSW					
Food waste	Biochip	Static	1.36 kg N_2O-N ton^{-1}	–	He et al. 2001
MSW	–	Static	0–0.24 kg N_2O-N ton^{-1}	4.5–9 kg CH_4-C ton^{-1}	Clemens and Cuhls 2003
Separated organic household waste		Static	0.02–0.11 kg N_2O-N ton^{-1}	0.04–0.8 kg CH_4-C ton^{-1}	Amlinger et al. 2008

Table 2.1 (continued)

Feedstock	Bulking agent	System	N_2O	CH_4	Reference
		Windrow			
Organic fraction source-separated MSW	Wood chips or pruning wastes	In vessel	0.048 kg N_2O-N ton^{-1}	0.26 kg CH_4-C ton^{-1}	Colón et al. 2012
		Confined windrow	0.048 kg N_2O-N ton^{-1}	1.26 kg CH_4-C ton^{-1}	
		Turned windrow	0.160 kg N_2O-N ton^{-1}	3.28 kg CH_4-C ton^{-1}	
		Home composting	0.430 kg N_2O-N ton^{-1}	0.12 kg CH_4-C ton^{-1}	
Source-separated MSW	Yard waste	Windrow	261 mg N_2O-N m^{-2} d^{-1}	35 g CH_4 m^{-2} d^{-1}	Beck-Friis et al. 2000
Other wastes					
Cattle mortalities	Barley straw	windrow	0.97 % of initial N	0.55 % of initial C	Hao et al. 2009
	Cattle manure & barley straw	windrow	0.85 % of initial N	0.13 % of initial C	
Animal meals	Straw & cotton	windrow	0.07–0.11 % of initial N	–	Cayuela et al. 2012
Biosolids	Wood ash	static	0.32 kg N_2O-N ton^{-1}	–	Czepiel et al. 1996
Grass	Soil	static	0.054 kg N_2O-N ton^{-1}	5 kg CH_4-C ton^{-1}	Hellebrand 1998
Hens mortalities	–	windrow	0.003–0.004 kg N_2O-N ton^{-1}	52–120 kg CH_4 ton^{-1}	Dong et al. 2011
Garden waste	–	windrow	0.05 kg N_2O-N ton^{-1}	1.9 kg CH_4-C ton^{-1}	Andersen et al. 2010a
Olive mill waste	Diverse manures & olive pruning	windrow	0.01–3.36 g N-N_2O m^{-2} d^{-1}	1–147 g CH_4-C m^{-2} d^{-1}	Sánchez-Monedero et al. 2010
Default emissions factors for CH_4 and N_2O emissions from waste composting from IPCC 2006					
On a dry basis			0.6 g (0.2–1.6) N_2O per kg waste treated	10 g (0.08–20) CH_4 per kg waste treated	IPCC 2006
On a wet basis			0.3 g (0.06–0.6) N_2O per kg waste treated	4 g (0.03–8) CH_4 per kg waste treated	

within 0.1 and 8.93 kg of CH_4 per ton of treated manure. This wide range may be affected by the pre-treatment of manure prior to composting (manure storage can represent an important source of CH_4) and also by the aeration system, windrow vs. forced aeration (agitation favours CH_4 emissions) (Kebreab et al. 2006). The levels of N_2O emitted from manure composting piles varied from 0.046 up to 0.370 kg N_2O-N per ton of treated manure depending on the composting system. Aerated static piles usually increase the emissions of N_2O by preventing ammonia losses, which can be later oxidised to nitrates generating N_2O. The emission of N_2O-N from manure composting can represent up to 9.8% of the initial N. These experimental results have been used by IPCC (2006) to propose default emission factors of 4 kg CH_4 ton^{-1} and 0.3 kg N_2O ton^{-1} (Table 2.1) from the biological treatment of organic wastes (for different types of feedstock and composting operations).

2.2.3.2 Municipal Solid Wastes (MSW)

Municipal solid wastes also represent a major source of organic wastes suitable for composting. This group includes not only mixed MSW, but also other materials such as the organic fraction of the source separated MSW, garden and yard wastes, food wastes, etc. This type of composting substrates is characterised by lower organic matter, nitrogen and moisture content than manures. For this reason the impact on GHG emissions is expected to be different, since lower amounts of organic C and N in the feedstock would lead to reduced GHG emissions (Brown et al. 2008; Büyüksönmez 2012).

Amounts of CH_4 emitted during MSW composting varied from 0.12 up to 9 kg CH_4 per ton of treated waste (Table 2.1). This large variability in gaseous emissions reflects the impact of the feedstock, the composting system and the efficiency of the composting facility on GHG emissions (Colón et al. 2012). The levels of N_2O emitted from MSW composting ranged from 0 to 0.430 kg N_2O-N per ton of treated waste, which represents values generally lower than those registered from the biological degradation of manure. In the case of MSW, where most of the composting piles are operated with little amounts of water, the small amount of CH_4 generated in the pile is most likely oxidised when it reaches the aerobic surface, considering CH_4 emissions to be essentially zero from a practical point of view, as far as life cycle assessments are concerned (US EPA 2006).

2.2.3.3 Other Organic Wastes

Table 2.1 shows the CH_4 and N_2O emissions for a range of organic wastes used as feedstock for composting. The impact of the different wastes will depend on their physical-chemical composition. Organic wastes such as biosolids, characterised by high N and moisture contents are expected to have a similar behaviour than manures, whereas other wastes such as cattle and hens mortalities or olive mill wastes can have different behaviour depending on their physical-chemical characteristics.

Sánchez-Monedero et al. (2010) studied GHG emissions from composting piles prepared with olive mill wastes and different N sources and bulking agent observing that the peculiar characteristics of these wastes, characterised by a low degradation rate and low N levels, reduced the emission of GHG.

2.3 Methodologies to Determine GHG Emissions in Composting Processes

2.3.1 Closed and Open Chambers

The most widely used method to sample GHG emitted from composting piles is based on the use of chambers. This technique is based on enclosures, generally inverted boxes or cylinders of known dimensions, that are placed over a part of the pile surface or encompasses all the composting pile and the measurement of the concentration of gases emitted from the composting material by several instrumental techniques.

This section describes the two main versions of this technique, namely closed and open chambers, and discusses the advantages and disadvantages related to each of them.

2.3.1.1 Closed Chambers

Closed chambers involve the sealing of the compost surface (or the entire pile) with an enclosure of known dimensions for determined periods. This method has been successfully used for measuring surface GHG emissions in piles prepared with manure (Hao et al. 2001), source-separated organic household wastes (Beck-Friis et al. 2000), lignocellulosic wastes (Andersen et al. 2010a; Cayuela et al. 2012) and olive mill wastes (Sánchez-Monedero et al. 2010). There are several kinds of chambers varying on shape, materials and dimensions. It is recommended that flux chambers should be fabricated of non-reactive materials (stainless steel, aluminium, PVC, polypropylene, polyethylene, or plexiglas) and the material should be white or coated with reflective material (mylar or painted) (Parkin and Venterea 2010). The most widely used shapes are cylinders or boxes (parallelepipeds) and the volume of the chamber may vary from 10 to 400 l.

When operating in this mode, gases are sampled at intervals during the closing period, which duration varies depending also from the instrumental technique used to measure the gas concentration. In the specific literature, this time ranges from few minutes to 30 min, with the latest as the most frequently used enclosure time with a sampling frequency of every 10 or 15 min. Gas analysis can be performed on site or off-site. In the first case air samples from the chamber are pumped through the measurement cell of the instrument (IRGA, Photoacustic analyzer, FTIR) and

a b

Fig. 2.1 Example of greenhouse gas (GHG) sampling using vacutaineers: **a** pile sampling; **b** gas sampling for analysis

then back into the chamber to avoid pressure changes. This chamber design is called dynamic closed chamber and represents a variation of the closed chamber method.

In the second case, few millilitres of air are withdrawn from the chamber headspace through a septum or a sampling port fitted into the chamber top. Samples are stored in syringes, vials, vacutaineers, Tedlar bags or metal tubes and then analyzed in the laboratory, usually by gas chromatography (Fig. 2.1). Operations can be facilitated by automation of chamber operation and sampling.

Gas fluxes are calculated from concentrations by assuming a steady state gradient in the underlying windrows. Emission fluxes are determined by fitting the experimental data to a second-order polynomial equation (gas concentration vs. time). The flux at time 0 is calculated by taking derivates of the second-order polynomial (Hao et al. 2001).

The advantage of the closed chamber method is that it is an easy and cheap technique that does not require sophisticated instruments. Further strengths of closed chambers are the versatility, as their design and deployment protocol can be adapted to a wide range of situations, and the capability to measure very low gas fluxes. On the other hand this method suffers for several disadvantages and limitations. The first limitation is due to the fact that the chamber may induce pressure gradients between compost pore space and chamber headspace resulting in convective transfer and biased flux estimates (Rochette 2011). This problem can be overcome by inserting a properly designed vent tube on the chamber to equalize barometric pressure inside and outside the chamber. For instance, Sommer et al. (2004) inserted a vent tube with an internal diameter of 1.6 mm and a length of 17.4 cm. However, care need to be taken as wind passing over the vent tube can cause a continue depressurization of the chamber due to the Venturi effect, thus resulting in much larger measurement errors than could be found with a chamber without vent.

Another problem of emission measurements from compost with closed chambers is the variation in the rate of diffusion of gas from compost to the headspace of the chamber. In undisturbed composts, diffusion is driven by a very large difference in concentration between low values in the atmosphere and high values in the compost (one to two orders of magnitude higher). The increase in concentration in the headspace of the closed chamber, especially for that concerning CO_2 in the bio-oxidative phase of the composting process, may decrease the rate of diffusion of gases resulting in an underestimation of compost emission. However, there are studies demonstrating that the critical CO_2 concentrations in the chamber affecting the rate of CO_2 diffusion ranged from 1000 to 1500 ppm_v (Bekku et al. 1997). Therefore care should be taken in order to utilize enclosure times that does not allow for the build-up of CO_2 concentrations leading to a decrease of the rate of diffusion.

A further problem is related to the fact that the chamber may lead to a rise in the temperature causing a perturbation in the rate of diffusion of the gases. This limitation can be overcome, at least partially, by constructing the chamber with insulated or reflective material and limiting the enclosure time.

Finally, gas fluxes can be affected by the height of the chamber. High chambers decrease the error in headspace volume determination and problems associated with high headspace gas concentration (e.g., leakage; feedback on gas flux), but on the other side high chambers may not allow adequate mixing of headspace air (Rochette 2011). To overcome this problem, some authors used a fan positioned within the chamber (Czepiel et al. 1996), but other authors do not agree with this solution, as it has been observed that fans can induce pressure perturbations within the chamber (Parkin and Venterea 2010).

2.3.1.2 Open Chambers

The open chamber method still includes the presence of an enclosure, but the main difference with the closed chamber is that it involves a continuous flow, through the chamber, of outside air (Ahn et al. 2011). Calculation of the flux is related to the difference in gas concentration between the incoming and out coming air.

Conversely to closed chambers, which are generally used to measure flux from limited surfaces of the composting materials, open chambers are usually designed as large-scale chambers that entirely encompass the compost piles and therefore allow to capture the whole flux of gases generated by the compost. Large-scale chambers have been utilized by Ahn et al. (2011), Amlinger et al. (2008), Fukumoto et al. (2003) and Park et al. (2011).

Open chambers systems are generally coupled with portable automatic gas analyzers and therefore they present the capacity to measure emissions on site; this allows for timely variations in the sampling strategy in order to capture increased gas fluxes due to changes in management operations and/or in environmental conditions.

On the other side, open chambers are more expensive and require higher technical skills to be operated. Moreover, these systems present the limitation that the accuracy of the measure strongly depends on the rate of air flux and consequently are greatly affected by variations in the environmental conditions. Therefore, they need a strict flow control and a continuous correction for changes in temperature and atmospheric pressure.

2.3.1.3 Chamber Methods and GHG Emissions from Composting Processes at Plant Scale

Chamber used in composting processes at laboratory and pilot-scale may be constructed in such a way that they cover all the composting material and consequently they can account for the total flux of gases generated by the compost. This is not possible in composting process at full-scale. A potential problem with the application of flux chambers for this purpose is that the large spatial and temporal variations in gas emissions increase the difficulty in estimating the whole-plant emissions. Investigations performed on composting processes at full-scale showed an extremely high spatial and temporal variability in GHG flux dynamics and this make particularly difficult to accurately estimate emissions using chamber methods (Andersen et al. 2010a).

Nevertheless, Amlinger et al. (2008) stated that a manual discontinuous analysis of N_2O and CH_4 with closed chambers from single air samples is acceptable for measurements over short-term durations. However, it is extremely relevant to consider the fact that gaseous emissions in windrow composting tend to be concentrated in the windrow top, the so-called chimney effect. The chimney effect was thoroughly investigated by Andersen et al. (2010a) by placing small chambers at nine different locations across the section of a windrow. The highest fluxes were observed at monitoring points located near the windrow top and nearly all other points showed insignificant gas fluxes. The investigation indicates that most (>85%) of the gases vented through a narrow (1 m wide), chimney-like area in the top of the windrow. Andersen et al. (2010a) also found the same flux pattern for all gases measured at different points across the windrow section. Their results are contradicted by those of Sommer et al. (2004), who found that a significant amount of CH_4 and CO_2 were emitted from the top, while N_2O was emitted preferably from the sides of the pile, indicating that the spatial emission patterns of the three gases were not related. Such evidences highlight the fact that the sampling strategies need to be carefully planned in terms of chamber size, position and sampling frequencies in order to capture emissions from all areas and relevant phases of the process exhibiting different emission behaviour.

In conclusion, results from several studies indicated the need to further validate the chamber flux technique for estimation of GHG emissions from composting plants (Amlinger et al. 2008; Andersen et al. 2010a).

2.3.2 *Other Specific Methodologies*

There are other methodologies reported in literature that have been used to determine GHG emission rates.

The methodology proposed by Cadena et al. (2009) is based on direct sampling and simultaneous measurement of airflow rate. This methodology has been applied for plants where the composting process takes place in enclosed facilities with air collection and treatment, where it is supposed that all emissions are finally released

Fig. 2.2 Image of fictitious partition of a biofilter surface to undertake the measurement of air velocity and gaseous emissions

through a biofilter, and for plants using open air composting processes (i.e. turned or aerated windrows). This methodology is based on measuring the dimensions of the emitting surface (windrow or biofilter) and gas sampling and airflow rate measurement in different points of this surface (Fig. 2.2 and 2.3). When outlet air velocity is under the anemometer detection limit, a Venturi tube can be used to accelerate air velocity (Veeken et al. 2002). The product of compound concentration (mg m^{-3}) and air velocity (m s^{-1}) results in the mass flow of a given compound released per surface area unit (mg s^{-1} m^{-2}). Thus, multiplying the pollutant mass flow per area unit by the corresponding total surface results in the compound mass flow. Then, the sum of the different quantities obtained corresponds to the total mass flow of a pollutant (g s^{-1}) (Fig. 2.4). Measures of gaseous emissions must be repeated at different days during the composting process to determine the evolution of the emission of each compound. The periodicity of sampling must be established as a function of plant operation and the development of the composting process. Pollutant mass flows obtained for each sampling day are represented versus process time. The area below the curve obtained corresponds to the total mass of a given pollutant emitted throughout the composting process analysed.

The Funnel method proposed by Andersen et al. (2010a) is based on the use of a funnel shape instrument made of aluminium. In fact, it can be considered a modification of the chamber methods. The instrument is placed upside-down on the windrow to let the gases escape through the pipe of the funnel. Measurement of air velocity and gas sampling are performed in the pipe of the funnel. The funnel covers 1 m^2 of the windrow top and the flow mass of the contaminant is calculated multiplying the contaminant concentration by the air velocity. The obtained flow is corrected with the ratio between the pipe area and the funnel area.

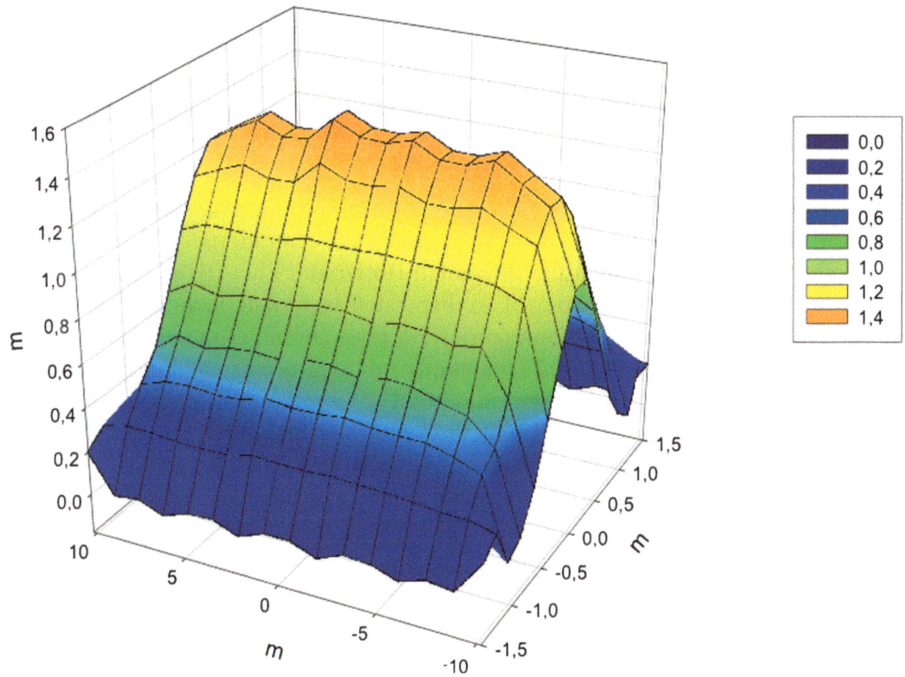

Fig. 2.3 Example of a threedimentional representation of a composting pile

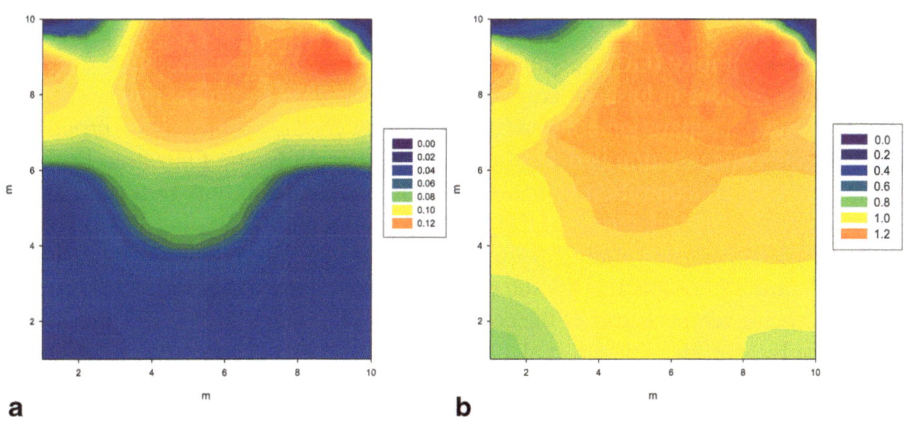

Fig. 2.4 Image of gaseous emissions from a superior view of a biofilter. Axes correspond to the width and length of the biofilter (in m) and colours to the intensity of gas (in concentration): **a** ammonia; **b** total volatile organic compounds (VOC)

The Dynamic Plume method was designed for measuring emissions in landfills (Galle et al. 2001) and it has also been applied to manure storage emission in farms (Skiba et al. 2006). However, Andersen et al. (2010a) applied the dynamic plume method to a composting facility. This method was proven to be the most effective when compared to the Funnel and the Dynamic Chamber methods. The method is based on the release of a tracer gas (SF_6, CO or N_2O, depending on the target gas) and its measure in the downwind plume. According to this, the dynamic plume method should be applied to the determination of the whole emissions of the studied facility. The ratio of the emission rates of tracer gas and contaminant are the same as the ratio of the concentrations of the tracer and the contaminant measured downwind, allowing the calculation of the contaminant emission rate. Measures upwind must be also performed to determine background concentrations.

Mathematical models such the Integrated Horizontal Flux (IHF) method or the backward Lagrangian stochastic (bLS) dispersion technique have been also used to measure emissions of GHG in composting piles (Sommer et al. 2004; Leytem et al. 2011). IHF method is based on a number of measurements along the vertical (downwind and upwind of the emission point) of contaminant concentration and windspeed (Wilson et al. 1983). bLS method allows estimating the emissions within the downwind plume from measurements of wind speed and contaminant concentration at specific heights and distances downwind (Flesch et al. 1995).

Finally, some optical methods developed for natural gas and other very specific fields have been applied to the measurement of fugitive emission of VOCs. Although not extended in the waste area of study, they are used in some cases when the precision need is high (e.g. Differential Absorption Lidar: DIAL) and they have been reported in some recent works (Steffens et al. 2009).

2.3.3 Specific Analytical Methods and Sensors

Gas chromatography (GC) has been commonly used to determine N_2O, CH_4 and CO_2 concentrations in gaseous emissions from biological treatment processes. Different detectors, separation columns and operating conditions have been proposed.

Methane concentration is determined by GC using a Flame Ionization Detector (FID) and CO_2 concentration by means of a Thermal Conductivity Detector (TCD). Sommer et al. (2004) determined CH_4 and CO_2 simultaneously using TCD in series with a FID and a 2 m × 3 mm SS Poropack QS 80/100 pre-column followed by a 0.5 m × 3 mm SS Poropack N 80/100 column. Carrier gas was helium at 30 ml min^{-1} flow rate being temperatures of the column oven, TCD and FID, 55, 130 and 230 °C respectively. Martínez-Blanco et al. (2010) used a HP-Plot Q column for CH_4 ensuring a detection limit of 1 ppm_v. The gas chromatography operation conditions were as follows: oven temperature, 60 °C, injector temperature, 240 °C, FID temperature, 250 °C; carrier gas, N_2. The same gas detectors were also used by Börjesson and Svenson (1997) when measuring gaseous emissions from a landfill and by Hobson et al. (2005) in CH_4 determination in the final phase of household waste composting. Methods for CO_2 and CH_4 in biogas samples require different sensibility because of the concentration range in which both compounds are found in biogas (Ward et al. 2011).

N_2O was measured by GC using an Electron capture detector (ECD). Czepiel et al. (1996) dried gas samples across $CaSO_4$ before being injected into the GC injection loop. These authors used a 2 m × 3 mm Porapak Q column with a mixture of Ar (95 %) and CH_4 (5 %) as carrier gas. ECD calibration was performed over the range of 310 ppb_v to 100 ppm_v using N_2O in N_2 standard gases. He et al. (2001) used the same column and carrier gas, but preceded by gaseous samples cleaning-up across two glass made columns packed with magnesium perchlorate and AS-CARITE II (Thomas Scientific) to remove moisture and CO_2, respectively Carrier gas flow rate was of 40 ml min^{-1} and temperatures of the detector and oven were 340 and 80°C respectively. Sommer et al. (2004) also used a pre-column (0.5 m × 3 mm SS Poropak N 80/100) and a 2 m × 3 mm SS HayeSep D 80/100 column with a Ar (90 %)-CH_4 (10 %) mixture as carrier gas at 30 ml min^{-1}. Temperature of column oven and detector were 55 and 330 °C, respectively. Colón et al. (2012) used a HP-Plot Q column (30 m × 0.53 mm × 40 µm) with N_2 as carrier gas being operation temperatures: 60 °C (column oven), 120 °C (injector) and 345 °C (detector). Detection limit was established in 50 ppb_v.

Some authors simultaneously determine the concentrations of CO_2, CH_4 and NO_2. Mondini et al. (2010) when measuring soil respiration used a micro-GC especially designed for continuous gas analysis. The GC was equipped with two capillary columns, PoraPlot Q (fused silica, 10 m length, 0.25 mm ID, 8 µm Df) and Molsieve (fused silica, 10 m length, 0.32 mm ID, 30 µm Df), in which head pressure and temperature could be electronically programmed. The measure of CO_2, CH_4 and N_2O was accomplished by means of a TCD detector. The concentration operating range is from 1 ppm_v to 100 %, with a linear dynamic range of 10^6. Operative conditions of micro-GC were: 30 s sampling time, 30 ms injection time, 120 s running time, 40 and 60 °C column temperature for PoraPlot Q and Molsieve, respectively, 30 °C injector temperature. The chromatograph was calibrated by injecting a mixture of pure standard gases of CO_2, N_2O and CH_4 at a concentration of 5000, 50 and 1000 ppm_v, respectively. The detection limits of the system for CO_2, CH_4 and N_2O were 2, 4 and 1 ppm_v, respectively. Cayuela et al. (2012) used the same procedure in the determination of GHG emissions during composting of lignocellulosic residues.

In addition to GC determination, other techniques have been presented, most of them with the advantage of being able to determine gas concentration on field without gaseous samples capture, transport and storage.

The photoacoustic field multi gas monitor has been widely used to determine GHG concentration. Andersen et al. (2010a) performed CO_2, CH_4, N_2O and CO determination in gaseous emissions from home composting of organic household waste using an INNOVA 1312 model (Lumasense Technologies A/S, 2750 Ballerup, Denmark). These authors state that this instrument measures real-time concentrations, provides high accuracy over a broad concentration range and only one calibration is necessary per year (calibration is performed by the manufacturer). The equipment requires a water filter to ensure that no moisture was transferred to the measuring chamber. Calibration ranges were 1.5–10,000 ppm_v for CO_2, 0.4–20,000 ppm_v for CH_4 and 0.03–50 ppm_v for N_2O (range for CO was 0.2–50 ppm_v).

Also Fukumoto et al. (2003), Tamura and Osada (2006) and Ahn et al. (2011) used the same technique when measuring the presence of these compounds in gaseous emissions from swine manure, dairy manure and farm manure composting respectively.

Photoacoustic spectroscopy is the measurement of the effect of adsorbed electromagnetic energy on matter by means of acoustic detection. Laser-based photoacoustic detectors are able to monitor trace gases concentrations under atmospheric conditions with sensitivity of orders of magnitude better than conventional scientific instrumentation in non-invasively and on-line way under dynamic conditions (Harren et al. 2000).

CO_2 has also been measured in gaseous emissions from full-scale composting plants using an infrared sensor (Abd et al. 2007). FTIR (Fourier Transform Infrared) absorption spectroscopy was also used by Andersen et al. (2010b) to measure GHG emission from windrow composting of garden waste including particular gases such as N_2O, CH_4 or CO. Manios et al. (2007) used a mobile gas analyzer to measure the volumetric composition of CO_2 and CH_4 in gaseous emissions from olive oil mill sludge composting in windrow piles. The equipment (GA2000, Geotechnical Instruments) was an infrared gas analyzer able to detect simultaneously CO_2, CH_4, O_2, CO and H_2S, originally used for landfill gas composition determination. This equipment is actually replaced by the GA5000 portable gas analyzer. Although not a GHG, VOCs have recently studied by on-line monitoring, in a clear advance with respect to current techniques (Shen et al. 2012).

2.4 Reduction of GHG Emitted from Composting

2.4.1 Best Practices for the Minimization of GHG Emissions

GHG emissions from composting can be minimized through diverse actions undertaken from different points of view: the material to be composted and the process performance.

2.4.1.1 Feedstock and Initial Mixture

The effect of the composition of the mixture of wastes to be composted is critical in the amount and type of emissions derived from the process. High moisture content and high bulk density has been related to higher GHG emissions. An excess of water reduces free air space (FAS) and creates anaerobic sites where methane can be formed (Tamura and Osada 2006). A correct level of FAS ensures the proper aeration of the composting material both in forced and natural aerated systems and prevents anaerobiosis (Ruggieri et al. 2009).

The biochemical composition of the material to be composted also plays an important role on gaseous emissions, especially the C/N ratio. However, the bioavailability of these nutrients determines the carbon and nitrogen dynamics along the process and the derived emissions (Cayuela et al. 2012). Consequently, the C/N ratio assessment should be based on the biodegradable content (Puyuelo et al. 2011). Co-composting of complementary wastes to obtain a balanced initial mixture with a balanced porosity and biodegradable C/N ratio should significantly reduce the GHG emissions of the subsequent composting process.

2.4.1.2 Composting Process

The composting process can be undertaken in different industrial systems. A general classification is made as open and closed systems. Contrary to open systems, closed systems present the collection of the exhaust gases to a gas treatment system.

Closed systems include closed reactors such as rotatory drums and composting tunnels, but also confined piles (with textile cover) or composting piles inside closed buildings with a gas management system. Plants with gas treatment systems present much lower environmental impact because process emissions are not released to the atmosphere (Colón et al. 2012). Discussion on how to treat GHG emissions is presented below. In this sense and according to published conclusions (Colón et al. 2012), a first technical recommendation to minimize GHG emissions would be to include gas treatment systems wherever possible.

Another important process parameter to consider is process temperature. Higher temperatures enhance volatile compounds volatilization resulting in higher gaseous emissions (Pagans et al. 2006b; Cayuela et al. 2012).

Open systems as static piles, turned piles and aerated windrows at open air have been studied to better understand gaseous emissions dynamics related to aeration strategies: airflow and pile turning. Different authors have highlighted the importance of airflow in gaseous emissions from composting. It is considered that a high airflow increases oxygen availability, avoiding anaerobic pockets and consequent methane formation, and dilutes gaseous emissions. On the negative side, a high airflow strips gaseous compounds present in the composting mass facilitating their volatilization (Pagans et al. 2006a). Jiang et al. (2011) reported that an increase in the aeration rate reduced methane emissions, but increased NH_3 and N_2O emissions. Pile turning enhances the composting process by providing matrix homogenization (moisture and microorganisms redistribution) and particle size reduction. It also provides punctual oxygenation of the solid material and compaction correction. From a biodegradation point of view, turning is recommended to enhance the process. However, pile turning has been shown to have a negative effect on gaseous emissions, including GHG (Colón et al. 2012). The turning itself releases the entrapped gases within the pile. Ahn et al. (2011) reported that CO_2, CH_4 and N_2O fluxes increased after turning due to greater gas diffusion rates resulting from porosity increased after turning. They recommend avoiding pile turning in the first stage of the process if the oxygen concentration and temperature of the pile are in an appropriate range. In a second stage, when oxygen levels within the pile increase the formed methane is oxidized to CO_2. These authors suggest considering a turning plan to minimize CH_4 emissions and maximize CH_4 oxidation within the pile. Park et al. (2011) also reported higher emissions in turned systems than in aerated systems. When considering methane and nitrous oxide as CO_2 equivalents, the non-aerated system provided the higher process emissions, followed by the turned system, the system aerated by natural convection and finally the forced aerated system, which presented the lowest process emissions. However, as pointed by the authors, when approaching the problem from an overall impact assessment, the energy consumed to aerate the pile contributes to total CO_2 non-biogenic emissions. The operational activities can contribute to GHG of composting process more than the decomposition process itself (Lou and Nair 2011).

LCA tools impute the impact of both process emissions and emissions related to energy consumption (operational activities, aeration, turning and mass displacement within the plant) to assess the comparison of different waste management systems. In this sense, turned pile composting systems resulted in an overall higher impact than aerated systems (confined aerated windrows and tunnel) because of fuel consumption and turning that implies the above mentioned increase in gas emissions (Colón et al. 2012; Kong et al. 2012).

2.4.1.3 Final Recommendations to Minimize GHG Emissions

From the text above it can be stated that a critical point for the success of the composting process with minimal gaseous emissions is the disposal of the material in piles with a suitable size and porosity to favor homogeneous oxygen distribution. In non-aerated systems, this would enhance natural convection. In aerated systems it is recommended to adjust forced aeration to ensure aerobic conditions without providing air in excess. High air flows beyond oxygen needs can be justified to avoid the emissions increase due to high temperatures. To overcome these problems, new advanced controllers have been proposed to base the aeration on the oxygen uptake rate measured on-line (Puyuelo et al. 2010).

Besides the physical structure of the matrix, the mixture to be composted should present appropriate moisture content and a balanced biodegradable C/N ratio.

Despite of whether the composting system is open or closed, the operational activities that imply electricity or fuel consumption must be optimized to reduce the overall environmental impact of the process.

Finally, gas treatment (by biofiltration or other technologies) is recommended when possible as the final solution to minimize gaseous emissions to the atmosphere.

2.4.2 Treatment of GHG Emissions

A variety of technologies are available nowadays for treating emissions from composting processes. Selection of the best available technologies depends essentially on the composition and gas flow rate to be treated. Amongst such technologies, chemical scrubbing combining acidic plus caustic scrubbers coupled to biological processes such as biofilters are the most common technologies installed in full-scale facilities (Artola et al. 2009). However, current reactors design and operation focuses on treatment of VOCs and ammonia as main pollutants contained in composting emissions while low attention has been paid to GHG treatment. In any case, biological systems still are the preferred alternative from an economical and environmental point of view for GHG removal since the low concentrations of GHG in composting emissions make existing physical-chemical technologies non-viable from an economical point of view.

Acidic scrubbers preceding biofilters are installed to reduce the large ammonia loads often generated during composting. Otherwise, ammonia may inhibit AOB and/or NOB that, concurrently, would hinder the performance of the biofilters (Gabriel et al. 2007). Caustic scrubbers are often installed to remove acid gases such as hydrogen sulfide and to absorb highly soluble VOCs emitted such as alcohols. Biofilters have demonstrated to work well as end-of-pipe systems to treat a variety of odorant compounds found in off-gases from composting systems.

Design and operating conditions of chemical scrubbers and biofilters do not provide suitable conditions for the treatment of GHG. Dimensionless gas-liquid Henry coefficients for N_2O, NO, CH_4 and CO of 1.7, 21.5, 29.2 and 43.1 (Sander 1999), respectively, indicate that GHG are sparingly soluble in water. Except for N_2O, which can be considerate as moderately soluble in water, large gas contact times in the treatment system are required to solubilise significant amounts of NO, CH_4 and CO which, consequently, leads to large reactor volumes and investment costs. In addition, the relatively low concentrations of GHG in the gas phase provide low driving force for GHG mass transfer from the gas to the liquid/biofilm phase. Chemical scrubbers generally operate at gas contact times below 2–3 s and retain large amounts of water within the packed bed to facilitate the absorption of soluble compounds (Gabriel and Deshusses 2003). Instead, biofilters are generally operated at gas contact times between 20 and 40 s for the treatment of composting off-gases with low to no external supply of water to improve sorption of poorly soluble compounds (Gabriel et al. 2007; Pagans et al. 2006a). Altogether leads to reduced elimination capacities of GHG in both systems in conventional chemical scrubbers and biofilters.

A short number of references exist about GHG treatment capacities in biofilters from composting emissions, even if several references exist about CH_4 removal by biofiltration. The latter has been addressed by several authors and shown as an effective technology for biofiltration of landfill biogas or gaseous emissions from the piggery industry (Nikiema et al. 2007; Girard et al. 2012). Moderate-to-large CH_4 concentrations of such gases are partly responsible for such effectiveness and treatment capacities. In composting facilities with biofilters, where much lower methane concentrations are found, removal efficiencies between 33 and 100 % have been reported for CH_4 (Boldrin et al. 2009).

In the case of N_2O emissions, Amlinger et al. (2008) reported that additional N_2O may be synthesized during the oxidation of NH_3. Also, Maia et al. (2012) found a clear correlation between the NH_3 load and the N_2O production in a compost biofilter demonstrating that NH_3 removal was a trigger for N_2O production. Clemens and Cuhls (2003) studied the emission of direct and indirect greenhouse gases in a MBT facility. They also found that biofilters had no net effect on CH_4 and approximately 26 % of the NH_3 that was removed in the biofilter was transformed into N_2O when NH_3 was the exclusive nitrogen source due to nitrifier denitrification.

Regarding NO emissions, Barnes et al. (1995) showed that removal efficiencies up to 90 % can be achieved in a compost biofilter for NO concentrations of 500 ppm_v at a gas contact time of 60 s if an external carbon and energy source were added. Similarly, Yang et al. (2007) found that NO concentrations in the range of 200 to 500 ppm_v can be treated in aerobic and anoxic biofilters with a strong influence of the O_2 percentage on NO removal. Even if hardly difficult to implement in composting facilities, anoxic conditions were reported to almost double NO removal compared to aerobic biofilters.

It is interesting to notice that almost no studies exist concerning CO biofiltration. Prado et al. (2008) reported CO removal efficiencies higher than 80 % for low (40 ppm_v) CO concentrations from synthetic-resin producing industries in a biofilter operated at above 30 s gas contact time. Further optimization showed that a maximum elimination capacity of 33 g CO m^{-3} h^{-1} could be obtained with a mixture of lava rock and peat as packing material with more than 85 % removal efficiency at gas contact times of 3 min or more suggesting that biofiltration offers potential for the biological removal of CO from polluted gas streams (Jin et al. 2009).

Reported data on GHG removal in chemical scrubbers is inexistent. However, one can infer from biofilters design, operating conditions and performance that GHG removal efficiencies in chemical scrubbers are probably very close to zero mostly due to the extremely reduced gas contact time of the gas in the scrubber coupled to the reduced solubility of most GHG.

Most of the research efforts on biological processes for GHG removal have been directed towards the use of existing bioreactor configurations (bioscrubbers, biotrickling filters or biofilters) while improving methane solubility using other solvents different to water. As reviewed by Muñoz et al. (2007), two-phase partitioning bioreactors (TPPBs) provide a non-aqueous phase (e.g. hexadecane, silicone oil) to an aqueous phase that contains the microorganisms responsible for degrading CH_4. Larger CH_4 mass transfer coefficients are encountered in TPPBs compared to conventional biofilters. Thus, improved solubilisation of hydrophobic compounds and, concomitantly, CH_4 elimination capacities are found. Rocha-Rios et al. (2009) reported increases of 131 and 41 % in the specific and volumetric CH_4 elimination capacity, respectively, in a biotrickling filter when silicone oil was added compared to the elimination capacities without silicone oil addition. However, silicon oil is expensive and difficult to manipulate which may hinder its use in full-scale systems. Alternatively, non-ionic surfactants do not pose the abovementioned problems and have shown to improve CH_4 elimination capacities in biofilters, even if some growth problems may exist leading to decreased biomass accumulation in the packed bed due to their detergent character (Ramirez et al. 2012). Similarly, ionic liquids have shown to largely improve non-methane-VOCs absorption in biological reactors without much toxicological issues (Quijano et al. 2010; Darracq et al. 2012). Such ionic liquids can be specifically designed based on the characteristics of the gaseous compound to be selectively separated (Carvalho and Coutinho 2011), which provides potential application for improving CH_4 absorption in biofilters and biotrickling filters.

Overall, there are a number of opportunities to improve GHG removal by means of biological reactors. While CH_4, CO and NO can be treated to a certain extent in conventional biofilters already installed in full-scale composting facilities, N_2O has been shown to be generated rather than removed in biofiltration systems. Thus, research efforts should be directed towards reducing N_2O generation during the composting process and improving biofiltration conditions to reduce its production. Also, proper characterization of current biofiltration systems installed in composting facilities in terms of GHG treatment capacities is necessary to gain specific knowledge. Finally, design and operating conditions of end-of-pipe systems should not be only based on odours and ammonia removal, but also GHG loads should be taken into consideration.

2.5 Conclusions

GHG from composting are an important issue for research and for improvement in real-scale composting facilities. From this review, it is evident that now GHG can be accounted, measured and properly characterized. However, it is clear that the disparities of emissions factors for the different GHG that can be found in scientific literature are due to several factors:

1. The diversity of wastes and technologies used for the composting of organic wastes.
2. The absence of a consensus in the way to measure GHG, especially in open systems, where the flow is not controlled.
3. There is wide margin to minimize the GHG emissions from composting, by changing or updating the current facilities and by improving the performance of the treatment technologies.
4. The beneficial uses of compost must be also investigated, since it is not clear if the GHG emitted during the process are compensated by this compost utilization in the long term.
5. From a Life Cycle Assessment perspective, it is necessary to have experimental data both on the GHG emissions and the efficiency of the process, to have a fair evaluation of the environmental impacts of composting.

Further research is necessary to solve these limitations and to provide reliable emissions factors for composting processes and, in general, for any biological technology for waste treatment.

Acknowledgements The authors wish to thank the financial support of the Spanish *Ministerio de Ciencia y Tecnologia* (Project CTM2009-14073-C02-01/02).

References

Abd N, Robin P, Paillat JM, Leterme P (2007) Turning, compacting and the addition of water as factors affecting gaseous emissions in farm manure composting. Bioresour Technol 98:2619–2628

Adani F, Confalonieri R, Tambone F (2004) Dynamic respiration index as a descriptor of the biological stability of organic wastes. J Environ Qual 33:1866–1876

Ahn HK, Mulbry W, White JW, Kondrad SL (2011) Pile mixing increases greenhouse gas emissions during composting of dairy manure. Bioresour Technol 102:2904–2909

Amlinger F, Peyr S, Cuhls C (2008) Green house gas emissions from composting and mechanical biological treatment. Waste Manage Res 26:47–60

Andersen JK, Boldrin A, Samuelsson J, Christensen TH, Scheutz C (2010a) Quantification of greenhouse gas emissions from windrow composting of garden waste. J Environ Qual 39:713–724

Andersen JK, Boldrin A, Christensen TH, Scheutz C (2010b) Greenhouse gas emissions from home composting of organic household waste. Waste Manage 30:2475–2482

Artola A, Barrena R, Font X, Gabriel D, Gea T, Mudhoo A, Sánchez A (2009) Composting from a sustainable point of view: respirometric indices as a key parameter. Dyn Soil Dyn Plant 3:1–16

Baquerizo G, Maestre JP, Sakuma TA, Deshusses MA, Gamisans X, Gabriel D, Lafuente J (2005) A detailed model of a biofilter for ammonia removal: model parameters analysis and model validation. Chem Eng J 113:205–214

Barnes JM, Apel WA, Barrett KB (1995) Removal of nitrogen oxides from gas streams using biofiltration. J Hazard Mater 41:315–326

Barrena R, Vázquez F, Gordillo MA, Gea T, Sánchez A (2005) Respirometric assays at fixed and process temperatures to monitor composting process. Bioresour Technol 96:1153–1159

Barrena R, Vázquez F, Sánchez A (2006) The use of respiration indices in the composting process: a review. Waste Manage Res 24:37–47

Barrena R, d'Imporzano G, Ponsá S, Gea T, Artola A, Vázquez F, Sánchez A, Adani F (2009a) In search of a reliable technique for the determination of the biological stability of the organic matter in the mechanical-biological treated waste. J Hazard Mater 162:1065–1072

Barrena R, Artola A, Vázquez F, Sánchez A (2009b) The use of composting for the treatment of animal by-products: experiments at lab scale. J Hazard Mater 161:380–386

Beck-Friis B, Pell M, Sonesson U, Jönsson H, Kirchmann H (2000) Formation and emission of N_2O and CH_4 from compost heaps of organic household waste. Environ Monit Assess 62:317–331

Bekku Y, Koizumi H, Oikawa T, Iwaki H (1997) Examination of four methods for measuring soil respiration. Appl Soil Ecol 5:247–254

Béline F, Martínez J, Chadwick D, Guiziou F, Coste CM (1999) Factors affecting nitrogen transformations and related nitrous oxide emissions from aerobically treated piggery slurry. J Agric Eng Res 73:235–243

Boldrin A, Andersen KJ, Moller J, Christensen HT, Favoino E (2009) Composting and compost utilization: accounting of greenhouse gases and global warming contributions. Waste Manage Res 27:800–812

Borjesson G, Svensson BH (1997) Seasonal and diurnal methane emissions from a landfill and their regulation by methane oxidation. Waste Manage Res 15:33–54

Brown S, Kruger C, Subler S (2008) Greenhouse gas balance for composting operations. J Environ Qual 37:1396–1410

Büyüksönmez F (2012) Full-scale VOC emissions from green and food waste windrow composting. Compost Sci Util 20:57–62.
Büyüksönmez F, Evans J (2007) Biogenic emissions from green waste and comparison to the emissions resulting from composting part ii: volatile organic compounds. Compost Sci Util 15:191–199
Cadena E, Colón J, Sánchez A, Font X, Artola, A (2009) A methodology to develop an inventory of gaseous emissions in a composting plant. Waste Manage 29:2799–2807
Carvalho PJ, Coutinho JAP (2011) The polarity effect upon the methane solubility in ionic liquids: a contribution for the design of ionic liquids for enhanced CO_2/CH_4 and H_2S/CH_4 selectivities. Energy Environ Sci 4:4614–4619
Cayuela ML, Sánchez-Monedero MA, Roig A (2006) Evaluation of two different aeration systems for composting two-phase olive mill wastes. Process Biochem 41:616–623
Cayuela ML, Sánchez-Monedero MA, Roig A, Sinicco T, Mondini C (2012) Biochemical changes and GHG emissions during composting of lignocellulosic residues with different N-rich by-products. Chemosphere 88:196–203
Charest MH, Beauchamp CJ (2002) Composting of de-inking paper sludge with poultry manure at three nitrogen levels using mechanical turning: behavior of physico-chemical parameters. Bioresour Technol 81:7–17
Clemens J, Cuhls C (2003) Greenhouse gas emissions from mechanical and biological waste treatment of municipal waste. Environ Technol 24:745–754
Colón J, Cadena E, Pognani M, Barrena R, Sánchez A, Font X, Artola A (2012) Determination of the energy and environmental burdens associated to the biological treatment of source-separated municipal solid wastes. Energy Environ Sci 5:5731–5741
Commission of the European Communities (2008) Green paper on the management of bio-waste in the European Union. http://ec.europa.eu/environment/waste/compost/pdf/green_paper_en.pdf. Accessed July 2012
Cooper BJ (2004) Stability (Biodegradability) Horizontal-7 WP4. Energy research center of the Netherlands. http://www.ecn.nl/docs/society/horizontal/hor_desk_7_stability.pdf. Accessed July 2012
Czepiel P, Douglas E, Harriss R, Crill P (1996) Measurements of N_2O from composted organic wastes. Environ Sci Technol 30:2519–2525
Darracq G, Couvert A, Couriol C, Thomas D, Amrane A, Dumont E, Andres Y, Le Cloirec P (2012) Optimization of the volumefraction of the NAPL, silicone oil, and biodegradation kinetics of toluene and DMDS in a TPPB. Int Biodeterior Biodegrad 71:9–14
de Klein CAM, Eckard RJ, van der Weerden TJ (2010) Nitrous oxide emissions from the nitrogen cycle in livestock agriculture: estimation and mitigation. In: Smith K (ed) Nitrous oxide and climate change. Earthscan, Edinburg, pp 107–142
del Prado A, Merino P, Estavillo JM, Pinto M, González-Murua C (2006) N_2O and NO emissions from different N sources and under a range of soil water contents. Nutr Cycl Agroecosyst 74:229–243
Delgado-Rodriguez M, Ruiz-Montoya M, Giraldez I, Lopez R, Madejon E, Diaz MJ (2012) Effect of aeration rate and moisture content on the emissions of selected VOCs during municipal solid waste composting. J Mater Cycles Waste Manage 14:371–378
Deportes I (2012) Programme de recherche de l'Ademe sur les emissions atmospheriques du compostage. Connaissances acquises et synthese bibliographique (in French). ADEME (Agence de l'Environnement et de la Maîtrise de l'Energie)
Dong H, Zhu Z, Xi J, Xi H (2011) Ammonia and greenhouse gas emissions from co-composting of dead hens with manure as affected by forced aeration rate. Am Soc Agric Biol Eng Annual Int Meeting 7:5765–5775
Eitzer BD (1995) Emissions of volatile organic chemicals from municipal solid waste composting facilities. Environ Sci Technol 29:896–902
El Kader NA, Robin P, Paillat JM, Leterme P (2007) Turning, compacting and the addition of water as factors affecting gaseous emissions in farm manure composting. Bioresour Technol 98:2619–2628

Ermolaev E, Pell M, Smars S, Sundberg C, Jonsson H (2012) Greenhouse gas emission from covered windrow composting with controlled ventilation. Waste Manage Res 30:155–160
European Union (2008a) Europa: summaries of EU legislation, agriculture, environment: sixth environment action programme. http://europa.eu/legislation_summaries/agriculture/environment/l28027_en.htm. Accessed July 2012
European Union (2008b) Europa: a gateway to the European Union. Environment: biodegradable waste. http://ec.europa.eu/environment/index_en.htm. Accessed July 2012
Favoino E, Hogg D (2008) The potential role of compost in reducing greenhouse gases. Waste Manage Res 26:61–69
Flesch TK, Wilson JD, Yee E (1995) Backward-time Lagrangian stochastic dispersion models and their application to estimate gaseous emissions. J Appl Meteorol 34:1320–1332
Frederickson J, Howell G (2003) Large-scale vermicomposting: emission of nitrous oxide and effects of temperature on earthworm populations. Pedobiologia 47:724–730
Fukumoto Y, Inubushi K (2009) Effect of nitrite accumulation on nitrous oxide emission and total nitrogen loss during swine manure composting. Soil Sci Plant Nutr 55:428–434
Fukumoto Y, Osada T, Hanajima D, Haga K (2003) Patterns and quantities of NH_3, N_2O and CH_4 emissions during swine manure composting without forced aeration- effect of compost pile scale. Bioresour Technol 89:109–114
Fukumoto Y, Suzuki K, Kuroda K, Waki M, Yasuda T (2011) Effects of struvite formation and nitratation promotion on nitrogenous emissions such as NH_3, N_2O and NO during swine manure composting. Bioresour Technol 102:1468–1474
Gabriel D, Deshusses MA (2003) Retrofitting existing chemical scrubbers to biotrickling filters for H_2S emission control. Proc Natl Acad Sci U S A 100:6308–6312
Gabriel D, Maestre JP, Martín L, Gamisans X, Lafuente J (2007) Characterisation and performance of coconut fiber as packing material in the removal of ammonia in gas-phase biofilters. Biosyst Eng 97:481–490
Gage J (2003) Checklist for odor management at compost facilities. Biocycle 44:42–47
Galle B, Samuelsson J, Svensson BH, Borjesson G (2001) Measurements of methane emissions from landfills using a time correlation tracer method based on FTIR absorption spectroscopy. Environ Sci Technol 35:21–25
Gea T, Barrena R, Artola A, Sánchez A (2004) Monitoring the biological activity of the composting process: oxygen uptake rate (OUR), respirometric index (RI) and respiratory quotient (RQ). Biotechnol Bioeng 88:520–527
Girard M, Viens P, Ramirez AA, Brzezinski R, Buelna G, Heitz M (2012) Simultaneous treatment of methane and swine slurry by biofiltration. J Chem Technol Biotechnol 87:697–704
Goldstein N (2002) Getting to know the odor compounds. Biocycle 43:42–44
Goreau TJ, Kaplan WA, Wofsy SC, McElroy MB, Valois FW, Watson SW (1980) Production of NO_2^- and N_2O by nitrifying bacteria at reduced concentrations of oxygen. Appl Environ Microbiol 40:526–532
Hao X, Chang C, Larney FJ, Travis GR (2001) Greenhouse gas emissions during cattle feedlot manure composting. J Environ Qual 30:376–386
Hao X, Chang C, Larney FJ (2004) Carbon, nitrogen balances, and greenhouse gas emissions during cattle feedlot manure composting. J Environ Qual 33:37–44
Hao X, Larney FJ, Chang C, Travis GR, Nichol CK, Bremer E (2010) The effect of phosphogypsum on greenhouse gas emissions during cattle manure composting. J Environ Qual 34:774–781
Harren FJM, Cotti G, Oomens J, Hekkert SL (2000) Photoacoustic spectroscopy in trace gas monitoring. In: Meyers RA (ed) Encyclopedia of analytical chemistry. Wiley, Chichester, pp 2203–2226
Haug RT (1993) The practical handbook of compost engineering. Lewis Publishers, Boca Raton
He Y, Inamori Y, Mizuochi M, Kong H, Iwami N, Sun T (2000) Measurements of N_2O and CH_4 from the aerated composting of food waste. Sci Total Environ 254:65–74
He Y, Inamori Y, Mizuochi M, Kong H, Iwami N, Sun T (2001) Nitrous oxide emissions from aerated composting of organic waste. Environ Sci Technol 35:2347–2351

Hellebrand HJ (1998) Emission of nitrous oxide and other trace gases during composting of grass and green waste. J Agric Eng Res 69:365–375
Hellebrand HJ, Kalk WD (2001) Emission of carbon monoxide during composting of dung and green waste. Nutr Cycl Agroecosyst 60:79–82
Hellebrand HJ, Schade GW (2008) Carbon monoxide from composting due to thermal oxidation of biomass. J Environ Qual 37:592–598
Hobson AM, Frederickson J, Dise NB (2005) CH_4 and N_2O from mechanically turned windrow and vermicomposting systems following in-vessel pre-treatment. Waste Manage 25:345–352
Hong JH, Park KJ (2004) Wood chip biofilter performance of ammonia gas from composting manure. Compost Sci Util 12:25–30
Hwang S, Hanaki K (2000) Effects of oxygen concentration and moisture content of refuse on nitrification, denitrification and nitrous oxide production. Bioresour Technol 71:159–165
IPCC (Intergovernmental Panel on Climate Change) (2006) IPCC guidelines for National Greenhouse Gas Inventories. http://www.ipcc-nggip.iges.or.jp/public/2006gl/index.html. Accessed July 2012
Jäckel U, Thummes K, Kämpfer P (2005) Thermophilic methane production and oxidation in compost. FEMS Microbiol Ecol 52:175–184
Jiang T, Schuchardt F, Li G, Guo R, Zhao Y (2011) Effect of C/N ratio, aeration rate and moisture content on ammonia and greenhouse gas emission during the composting. J Environ Sci 23:1754–1760
Jin YM, Guo L, Veiga MC, Kennes C (2009) Optimization of the treatment of carbon monoxide-polluted air in biofilters. Chemosphere 74:332–337
Kebreab E, Clark K, Wagner-Riddle C, France J (2006) Methane and nitrous oxide emissions from Canadian animal agriculture: a review. Can J Anim Sci 86: 135–158
Komilis DP, Ham RK, Park JK (2004) Emission of volatile organic compounds during composting of municipal solid wastes. Water Res 38:1707–1714
Kong D, Shan J, Iacoboni M, Maguin SR (2012) Evaluating greenhouse gas impacts of organic waste management options using life cycle assessment. Waste Manage Res 30:800–812
Kowalchuk GA, Naoumenko ZS, Derikx PJL, Felske A, Stephen JR, Arkhipchenko IA (1999) Molecular analysis of ammonia-oxidizing bacteria of the β subdivision of the class proteobacteria in compost and composted materials. Appl Environ Microbiol 65:396–403
Kuroda K, Osada T, Yonaga M, Kanematu A, Nitta T, Mouri S, Kojima T (1996) Emissions of malodorous compounds and greenhouse gases from composting swine feces. Bioresour Technol 56:265–271
Leytem AB, Dungan RS, Bjorneberg DL, Koehn AC (2011) Emissions of ammonia, methane, carbon dioxide, and nitrous oxide from dairy cattle housing and manure management systems. J Environ Qual 40:1383–1394
Lou XF, Nair J (2009) The impact of landfilling and composting on greenhouse gas emissions—a review. Bioresour Technol 100:3792–3798
Maeda K, Toyoda S, Shimojima R, Osada T, Hanajima D, Morioka R, Yoshida N (2010) Source of nitrous oxide emissions during the cow manure composting process as revealed by isotopomer analysis of and amoA abundance in Betaproteobacterial ammonia-oxidizing bacteria. Appl Environ Microbiol 76:1555–1562
Maeda K, Hanajima D, Toyoda S, Yoshida N, Morioka R, Osada T (2011) Microbiology of nitrogen cycle in animal manure compost. Microbial Biotechnol 4:700–709
Maia GDN, Day V GB, Gates RS, Taraba JL (2012) Ammonia biofiltration and nitrous oxide generation during the start-up of gas-phase compost biofilters. Atmos Environ 46:659–664
Manios T, Manidakis K, Boutzakis P, Nazaridis Y, Lasaridi K, Markakis G, Stentiford EI (2007) Methane and carbon dioxide emission in a two-phase olive oil mill sludge windrow pile during composting. Waste Manage 27:1092–1098
Martínez-Blanco J, Colón J, Gabarrell X, Font X, Sánchez A, Artola A, Rieradevall J (2010) The use of life cycle assessment for the comparison of biowaste composting at home and full scale. Waste Manage 30:983–994.

Mondini C, Sánchez-Monedero MA, Sinicco T, Leita L (2006) Evaluation of extracted organic carbon and microbial biomass as stability parameters in ligno-cellulosic waste composts. J Environ Qual 35:2313–2320
Mondini C, Cayuela ML, Sinicco T, Cordaro F, Roig A, Sánchez-Monedero MA (2007) Greenhouse gas emissions and carbon sink capacity of amended soils evaluated under laboratory conditions. Soil Biol Biochem 39:1366–1374
Mondini C, Sinicco T, Cayuela ML, Sanchez-Monedero MA (2010) A simple automated system for measuring soil respiration by gas chromatography. Talanta 81:849–855
Muñoz R, Villaverde S, Guieysse B, Revah S (2007) Two-phase partitioning bioreactors for treatment of volatile organic compounds. Biotechnol Adv 25:410–422
Nikiema J, Brzezinski R, Heitz M (2007) Elimination of methane generated from landfills by biofiltration: a review. Rev Environ Sci Technol 6:261–284
Pagans E, Font X, Sánchez A (2006a) Emission of volatile organic compounds from composting of different solid wastes: abatement by biofiltration. J Hazard Mater B 131:179–186
Pagans E, Barrena R, Font X, Sánchez A (2006b) Ammonia emissions from the composting of different organic wastes. Dependency on process temperature. Chemosphere 62:1534–1542
Park KH, Jeon JH, Jeon KH, Kwag JH, Choi DY (2011) Low greenhouse gas emissions during composting of solid swine manure. Anim Feed Sci Technol 166:550–556
Parkin TB, Venterea RT (2010) USDA-ARS GRACEnet project protocols, chapter 3. Chamber-based trace gas flux measurements. In: Follett RF (ed) Sampling protocols. Beltsville, MD p 1–39. http://www.ars.usda.gov/SP2UserFiles/Program/212/Chapter%203.%20GRACEnet%20Trace%20Gas%20Sampling%20Protocols.pdf. Accessed July 2012
Parkinson R, Gibbs P, Burchett S, Misselbrook T (2004) Effect of turning regime and seasonal weather conditions on nitrogen and phosphorus losses during aerobic composting of cattle manure. Bioresour Technol 91:171–178
Pattey E, Trzcinski MK, Desjardins RL (2005) Quantifying the reduction of greenhouse gas emissions as a result of composting dairy and beef cattle manure. Nutr Cycl Agroecosyst 72:173–187
Pel R, Oldenhuis R, Brand W, Vos A, Gottschal JC, Zwart KB (1997) Stable-isotope analysis of a combined nitrification-denitrification sustained by thermophilic methanotrophs under low-oxygen conditions. Appl Environ Microbiol 63:474–481
Phillip EA, Clark OG, Londry K, Yu S, Leonard J (2011) Emission of carbon monoxide during composting of municipal solid waste. Compost Sci Util 19:170–177
Ponsá S, Gea T, Alerm L, Cerezo J, Sánchez A (2008) Comparison of aerobic and anaerobic stability indices through a MSW biological treatment process. Waste Manage 28:2735–2742
Prado OJ, Veiga MC, Kennes C (2008) Removal of formaldehyde, methanol, dimethylether and carbon monoxide from waste gases of synthetic resin-producing industries. Chemosphere 70:1357–1365
Puyuelo B, Gea T, Sánchez A (2010) A new control strategy for composting process based on the oxygen uptake rate. Chem Eng J 165:161–169
Puyuelo B, Ponsá S, Gea T, Sánchez A (2011) Determining C/N ratios for typical organic wastes using biodegradable fractions. Chemosphere 85:653–659
Quijano G, Couvert A, Amrane A (2010) Ionic liquids: applications and future trends in bioreactor technology. Bioresour Technol 101:8923–8930
Ramirez AA, Garcia-Aguilar BP, Jones JP, Heitz M (2012) Improvement of methane biofiltration by the addition of non-ionic surfactants to biofilters packed with inert materials. Process Biochem 47:76–82
Raviv N, Medina S, Krasnovsky A, Ziadna H (2002) Conserving nitrogen during composting. Biocycle 43:48–51
Rocha-Rios J, Bordel S, Hernández S, Revah S (2009) Methane degradation in two-phase partition bioreactors. Chem Eng J 152:289–292
Rochette P (2011) Towards a standard non-steady-state chamber methodology for measuring soil N_2O emissions. Anim Feed Sci Technol 166:141–146

Ruggieri L, Gea T, Artola A, Sánchez A (2009) Air filled porosity measurements by air pycnometry in the composting process: a review and a correlation analysis. Bioresour Technol 100:2655–2666

Sánchez-Monedero MA, Roig A, Paredes C, Bernal MP (2001) Nitrogen transformation during organic waste composting by the Rutgers system and its effects on pH, EC and maturity of the composting mixtures. Bioresour Technol 78:301–308

Sánchez-Monedero MA, Cayuela ML, Mondini C, Serramiá N, Roig A (2008) Potential of olive mill wastes for soil C sequestration. Waste Manage 28:767–773

Sánchez-Monedero MA, Serramia N, García-Ortiz Civantos C, Fernández-Hernández A, Roig A (2010) Greenhouse gas emissions during composting of two-phase olive mill wastes with different agroindustrial by-products. Chemosphere 81:18–25

Sánchez-Monedero MA, Cayuela ML, Serramiá N, García-Ortiz Civantos C, Fernández-Hernández A, Roig A (2011) Emission of CO_2, CH_4 and N_2O during composting of two-phase olive mill wastes: a four-year study. Dyn Soil Dyn Plant 5:93–97

Sander R (1999) Compilation of Henry's law constants for inorganic and organic species of potential importance in environmental chemistry (3rd version). http://www.mpch-mainz.mpg.de/~sander/res/henry.html. Accessed July 2012

Scheutz C, Kjeldsen P, Gentil E (2009) Greenhouse gases, radiative forcing, global warming potential and waste management—an introduction. Waste Manage Res 27:716–723

Shen YJ, Chen TB, Gao D, Zheng GD, Liu HT, Yang QW (2012) Online monitoring of volatile organic compound production and emission during sewage sludge composting. Bioresour Technol 123:463–470

Sironi S, Botta D (2001) Biofilter efficiency in odor abatement at composting plants. Compost Sci Util 9:149–155

Skiba U, DiMarco C, Hargreaves K, Sneath R, McCartney L (2006) Nitrous oxide emissions from a dung heap measured by chambers and plume methods. Agric Ecosyst Environ 112:122–128

Smith K, Crutzen P, Mosier A, Winiwarter W (2010) The global nitrous oxide budget: a reassessment. In: Smith K (ed) Nitrous oxide and climate change. Earthscan,London, pp 63–84

Sommer SG, Møller HB (2000) Emission of greenhouse gases during composting of deep litter from pig production—effect of straw content. J Agric Sci 134:327–335

Sommer SG, McGinn SM, Hao X, Larney FJ (2004) Techniques for measuring gas emissions from a composting stockpile of cattle manure. Atmos Environ 38:4643–4652

Staley BF, Xu F, Cowie SJ, Barlaz MA, Hater GR (2006) Release of trace organic compounds during the decomposition of municipal solid waste components. Environ Sci Technol 40:5984–5991

Steffens J, Landulfo E, Guardani R, Ollera CN, Moreira A (2009) Measurements of air quality using lidar system. Comput Aided Chem Eng 2009:1443–1448

Sundberg C, Yu D, Franke-Whittle I, Kauppi S, Smars S, Insam H, Romantschuk M, Jonsson H (2013) Effects of pH and microbial composition on odour in food waste composting. Waste Manage 33:204–211

Szanto GL, Hamelers HVM, Rulkens WH, Veeken AHM (2007) NH_3, N_2O and CH_4 emissions during passively aerated composting of straw-rich pig manure. Bioresour Technol 98:2659–2670

Tamura T, Osada T (2006) Effect of moisture in pile-type composting of dairy manure by adding wheat straw on greenhouse gas emission. Int Congr Ser 1293:311–314

Topp E, Hanson RS (1991) Metabolism of radiatively important trace gases by methane-oxidizing bacteria. In: Rogers JE, Whitmann W (eds) Microbial production and consumption of greenhouse gases: methane, nitrogen oxides and halometanes. American Society for Microbiology, Washington, pp 71–90

US EPA (2006) Solid waste management and greenhouse gases. A life-cycle assessment of emissions and sinks, 3rd Edition, Report EPA530-R-02-006. http://epa.gov/climatechange/wycd/waste/downloads/fullreport.pdf. Accessed July 2012

Veeken A, Wilde V de, Hamelers B (2002) Passively aerated composting of straw-rich pig manure: effect of compost bed porosity. Compost Sci Util 10:114–128

Ward AJ, Bruni E, Lykkegaard MK, Feilberg A, Adamsen APS, Jensen AP, Poulsen AK (2011) Real time monitoring of a biogas digester with gas chromatography, near-infrared spectroscopy, and membrane-inlet mass spectrometry. Bioresour Technol 102:4098–4103

Wilson JD, Catchpole VR, Denmead OT, Thurtell GW (1983) Verification of a simple micrometeorological method for estimating the rate of gaseous mass transfer from the ground to the atmosphere. Agric Meteorol 29:183–189

Wrage N, Velthof GL, van Beusichem ML, Oenama O (2001) Role of nitrifier denitrification in the production of nitrous oxide. Soil Biol Biochem 33:1723–1732

Yamamoto N, Otawa K, Nakai Y (2010) Diversity and abundance of ammonia-oxidizing bacteria and ammonia-oxidizing *archaea* during cattle manure composting. Microb Ecol 60:807–815

Yang WF, Hsing HJ, Yang YC, Shyng JY (2007) The effects of selected parameters on the nitric oxide removal by biofilter. J Hazard Mater 148:653–659

Zeng Y, De Guardia A, Ziebal C, De Macedo FJ, Dabert P (2012) Nitrification and microbiological evolution during aerobic treatment of municipal solid wastes. Bioresour Technol 110:144–152

Chapter 3
CO_2 Photocatalytic Reduction: Photocatalyst Choice and Product Selectivity

Meei Mei Gui, Lling-Lling Tan, Wee-Jun Ong, Siang-Piao Chai and Abdul Rahman Mohamed

Contents

Abstract Carbon dioxide (CO_2) emission is one of the most well-known causes of global warming. Conversion of CO_2 into useful chemical products is an attractive approach to sequestrating CO_2 as stable liquids and solids. Among sequestration methods the photocatalytic reduction of CO_2 is promising. CO_2 photocatalytic reduction involves radical-chain reactions that form proton and anion radicals from electron (e^-) and proton (h^+) transfer between metal oxide photocatalysts and the

S.-P. Chai (✉) · M. M. Gui · L.-L. Tan · W.-J. Ong
Low Carbon Economy (LCE) Group, Chemical Engineering Discipline, School of Engineering, Monash University, Jalan Lagoon Selatan, 46150 Bandar Sunway, Selangor, Malaysia
e-mail: chai.siang.piao@monash.edu

A. R. Mohamed
Low Carbon Economy (LCE) Group, School of Chemical Engineering, Universiti Sains Malaysia, Engineering Campus, Seri Ampangan, 14300 Nibong Tebal, Pulau Pinang, Malaysia

© Springer International Publishing Switzerland 2015
E. Lichtfouse et al. (eds.), *CO_2 Sequestration, Biofuels and Depollution,*
Environmental Chemistry for a Sustainable World 5, DOI 10.1007/978-3-319-11906-9_3

reactants. Therefore the product distribution of a particular photocatalytic reduction process is difficult to predict. In general, the CO_2 photocatalytic reduction process is controlled by several conditions such as reactor configuration, photocatalyst type and the nature of reducing agents. Here we review the parameters that control the photocatalytic reduction activity of CO_2. We lists the different photocatalysts for the reduction. All types of photocatalysts exhibit specific behaviours which lead to different product distribution. Metal and non-metal dopants improve the photoactivity of a photocatalyst. The dopants also modify the product distribution by altering the active species. Finally, we identify key factors ruling the photocatalytic activity of CO_2 reduction under UV or visible light irradiation.

Keywords Photocatalysis · Carbon dioxide reduction · Redox chemistry · Semiconductor photocatalyst · Synthesis design · Aqueous-phase reactions · Gas-phase reactions · Photocatalytic efficiency

Abbreviations

CO_2	Carbon dioxide
IPCC	Intergovernmental Panel on Climate Change
syngas	Synthesis gas
CO	Carbon monoxide
CH_3OCH_3	Dimethyl ether
UV	Ultraviolet
TiO_2	Titanium dioxide
N-TNT	Nitrogen-doped TiO_2 nanotubes
MWCNTs	Multi-walled carbon nanotubes
SEG	Solvent-exfoliated graphene
Ch	*Carboxydothermus hydrogenoformans*
Ga_2O_3	Gallium oxide
ZrO_2	Zirconium oxide
ZnO	Zinc oxide
TaO_3	Tantalum oxide
•OH	Hydroxyl radicals
H^+	Hydrogen ions
Ag	Silver

3.1 Background

3.1.1 The Need for CO_2 Capture and Utilization

Carbon dioxide (CO_2) emission has brought upon anthropogenic climate change worldwide, leading to extreme climate change in recent years. The discussion on CO_2 emission control has been a hot topic for the past few decades. Till now, efforts

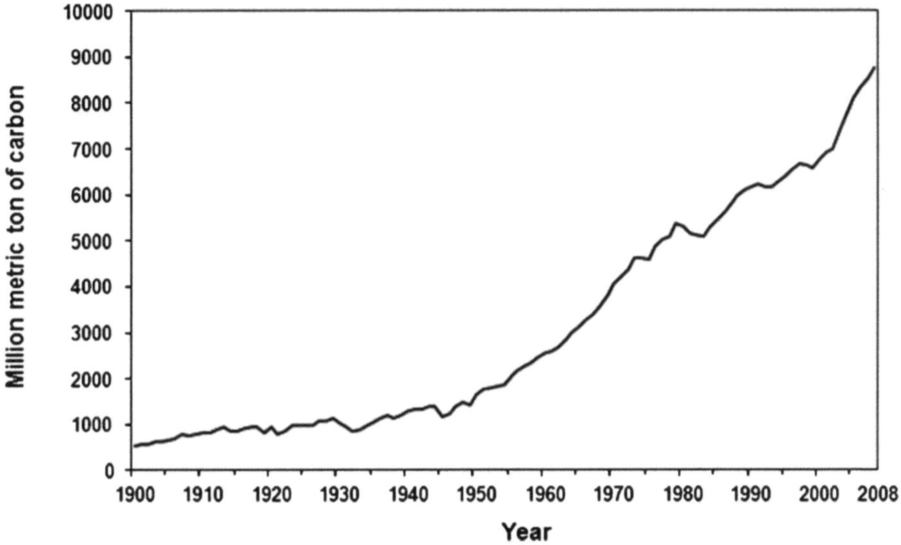

Fig. 3.1 Global CO_2 emission trend from 1900 to 2008. Data extracted from CDIAC (2011). (*Other includes commercial/public services, agricultural/forestry, fishing, energy industries other than electricity and heat generation, and other emissions not specified elsewhere)

are still being carried out to reduce regional and global CO_2 emissions but there is yet any specific solution to control and limit the CO_2 emission to the atmosphere.

Figure 3.1 shows the global emission of CO_2 from 1900 to 2010. As shown in the chart, an escalating increase in global CO_2 emissions was observed since 1950. The increment is mainly due to the continuous increase in population and human technology global wide that eventually lead to alarming increase of energy consumption. It is well known that the consumption of energy is the main contributor to the global CO_2 emissions. The two sectors, electricity and heat generation and transport, contributed nearly two-thirds of global CO_2 emissions in 2009, as shown in Fig. 3.2. Both sectors are mainly involved in the generation of energy to facilitate

Fig. 3.2 Global CO_2 emissions by sector in year 2009, adapted from IEA report 2011 (IEA 2011)

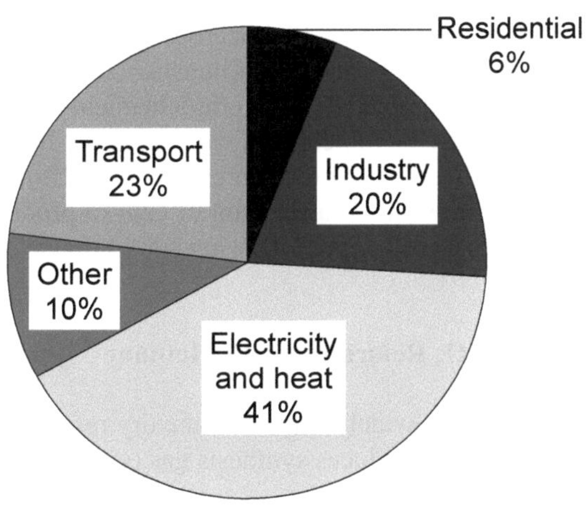

human activities. Generation of electricity and heat is by far the largest sector that contributes to global CO_2 emissions and is responsible for 41 % of world CO_2 emissions. Evidently, the temperature of the Earth's surface has risen by approximately 0.6 K in the past century and a more significant warming trend was observed over the past two decades according to the Intergovernmental Panel on Climate Change (IPCC) (Usubharatana et al. 2006). With the current technology, energy generation still relies heavily on fossil fuels such as coal and petroleum which are high carbon-containing energy resources (IEA 2011).

3.1.2 *Utilization of CO_2 for the Generation of Useful Products*

Much effort has been made to reduce CO_2 emissions global wide. Conventional methods for the removal of CO_2 are mainly through pre- or post-combustion CO_2 capture. This includes physical/chemical solvent scrubbing, separation using membrane or adsorbent; followed by compression and geological sequestration (Davison and Thambimuthu 2005; Li et al. 2010). These processes are energy intensive and thus costly. In addition, there are many uncertainties with regard to long-term storage of CO_2 in geological formations.

In order to execute carbon capture in a more realistic and cost-effective manner, recent research trend mainly emphasizes on alternative ways to convert CO_2 into useful chemicals or renewable energy resources; in other words, chemical sequestration. The abundantly available CO_2 can be recycled and utilized for the production of chemicals and fuel products at minimal cost. Moreover, the availability of CO_2 is promising as it is continuously generated along with energy consumption. Various methods such as electrochemical, photocatalytic and photoelectrochemical processes have been investigated to convert CO_2 into useful products. Among the products generated are methane, methanol, formic acids, aldehydes and various short chain hydrocarbon compounds (Centi and Perathoner 2004; Centi and Perathoner 2010; Olah et al. 2011).

CO_2 is a relatively inert and stable chemical compound. High energy input and reactive catalysts are required to break the strong C-O bonding to further convert it into other hydrocarbon compounds. Therefore, the reactions associated with the conversion of CO_2 into hydrocarbon products must be endothermic and involve a positive change of enthalpy (Chunshan 2006; Jiang et al. 2010). The major routes in the utilization of CO_2 are thermochemical reaction processes. The reaction of the highly stable CO_2 molecule is made possible by treating it in extreme environment such as high temperature and pressure, with the aid of highly efficient catalysts. The different routes in the utilization of CO_2 to produce useful fuels or chemical products are discussed in the following subsections.

3.1.2.1 CO_2 Reforming with Methane

This process is widely-known as the dry reforming process. CO_2 reacts with methane (CH_4) and produces synthesis gas (syngas), as shown in Eq. 3.1. Subsequently, these products are used in chemical energy transmission systems or in the production of liquid hydrocarbon (Centi and Perathoner 2009).

$$CO_2 + CH_4 \rightarrow 2CO + 2H_2 \quad (3.1)$$

CO_2 reforming of methane is a highly endothermic reaction process. To achieve reasonable conversions, extreme reaction temperatures of 700 °C and above are required to supply sufficient energy. CO_2 is typically present along with methane in natural gas deposits. Hence, the dry reforming process is always coupled with Fischer-Tropsch reaction to produce liquid hydrocarbon, which is easier to be transported for distribution (Centi and Perathoner 2009).

The concept of methane reforming with CO_2 could also be applied in the "tri-reforming" process, which couples the processes of CO_2 reforming of methane, steam reforming of methane and the partial/complete oxidation of methane. The "tri-reforming" process is superior compared to the steam reforming and CO_2 reforming processes. It is cost effective as it directly utilizes flue gas instead of purified CO_2 (Chunshan and Pan 2004; Chunshan 2006; Jiang et al. 2010).

3.1.2.2 CO_2 Hydrogenation

A wide range of hydrocarbon compounds, particularly higher molecular weight alkanes, alkenes and alcohols; such as dimethyl ether, formic acid and acetic acid can be synthesized via CO_2 hydrogenation. In general, the reaction process to convert CO_2 into hydrocarbon is a multi-step process. For example, dimethyl ether (CH_3OCH_3) can be produced from the following route: CO_2 is first hydrogenated into carbon monoxide (CO) and water, followed by the hydrogenation of CO into dimethyl ether (Eq. 3.2–3.4) (Schaub et al. 2004; Olah et al. 2008; Centi and Perathoner 2009).

$$CO_2 + H_2 \rightleftharpoons CO + H_2O \quad (3.2)$$

$$CO + 4H_2 \rightleftharpoons CH_3OCH_3 + H_2O \quad (3.3)$$

$$3CO + 3H_2 \rightleftharpoons CH_3OCH_3 + CO_2 \quad (3.4)$$

Direct hydrogenation of CO_2 into hydrocarbons is also possible. The hydrogenation of CO_2 into methanol is not a newly discovered process. On the contrary, in the commercial production of methanol, the syngas used for the methanation reaction is usually associated with small quantities of CO_2. The direct conversion of CO_2 into methanol is described by Eq. 3.5 (Jiang et al. 2010).

$$CO_2 + 3H_2 \rightleftharpoons CH_3OH + H_2O \quad (3.5)$$

In some applications, ethanol is more preferable compared to methanol in terms of safer handling and transport. Although ethanol can be produced from the hydrogenation of syngas, the process does not seem economically viable when compared to ethanol formation from biomass (Centi and Santen 2007; Centi and Perathoner

2009). Nevertheless, the direct synthesis of ethanol from the hydrogenation of CO_2 is of great interest. The reaction process, which is somewhat similar to that of methanol, is described in Eq. 3.6.

$$2CO_2 + 6H_2 \rightleftharpoons C_2H_5OH + 3H_2O \tag{3.6}$$

3.1.2.3 Reaction of CO_2 with Hydrocarbon for Formation of Hydrocarbon Products

Long chain hydrocarbons can be derived from the conversion of CO_2 with shorter chain hydrocarbons, such as methane and methanol. Possible products from the reaction of CO_2 with hydrocarbons are dimethyl carbonate, polypropylene glycol, cyclic carbonate, salicylic acids and urea. For example, CO_2 can further react with methanol and propylene glycol to produce dimethyl carbonate and cyclic carbonates, respectively (Ma et al. 2009).

In view of the limitations of the thermochemical properties of CO_2 conversion, a route chosen for the photoreduction of CO_2 into useful products must be viable in terms of energy consumption. The energy required for such routes should not exceed the energy level which can be subsequently generated by the fuel products. Another main concern for the conversion of CO_2 is the availability of sacrificial agents for the reaction, such as hydrogen. Taking the hydrogenation of CO_2 into methanol and ethanol (Eq. 3.5, 3.6) as an example, the availability of hydrogen must be promising in order to ensure the sustainability of the entire process.

To overcome the high energy consumption of the thermochemical reduction of CO_2, more research has been extended to the photocatalytic reduction of CO_2. This process possesses unique advantages of utilizing energy for light source instead of energy generated from fuel resources. Also, photocatalytic reactions promote thermodynamically unfavourable reactions at low temperature leading to lessenergy consumption and reduced catalyst deactivation.

3.1.3 Photocatalytic Conversion of CO_2 into Useful Chemicals or Fuel Products

Photocatalytic reduction is an ideal method to convert CO_2 into useful products due to its low cost with energy input from cheap and abundant sources (Li et al. 2010). Photocatalysis uses photosensitive semiconductor materials such as TiO_2, ZrO_2, V_2O_5, ZnO, CeO_2, and NbO_5, as photocatalysts to absorb energy in the light source for the subsequent reactions. When a semiconducting material is irradiated by ultraviolet (UV) or visible light with energy levels exceeding its band gap energy, photoexcitation occurs, which consequently generates electron-hole pairs in the photocatalyst. The photoinduced electrons will then reduce CO_2 with the reducing agent absorbed on the photocatalyst surface, forming energy-bearing products such as carbon monoxide, methane, methanol, formaldehyde, formic acid and so on (Lo et al. 2007; Zhang et al. 2009; Li et al. 2010).

UV light is the most commonly used light source for photocatalysis as it exhibits high energy content which can effectively excite most of the photocatalysts. As a result, most of the processes reported on the photocatalytic reduction of CO_2 are still replying on artificial UV light generated from high power lamps (Zhao et al. 2007; Zhao et al. 2009a, b). However, the UV light content in natural sunlight is only 3%, making the entire process not viable for natural sunlight applications. In a recent research trend, more focus has been put on the direct utilization of visible light from both artificial source and natural sunlight. The use of visible light is more promising as compared to UV light because it can be readily available from sunlight. However, the energy content in visible light is not as competitive when compared to UV light. Therefore, visible light might not be able to provide sufficient energy for the photoexcitation of catalysts in a photocatalytic reduction. Evidently, the utilization of visible light or natural sunlight for photocatalysis still remains a challenge (Jia et al. 2009).

Photocatalysis is a chain reaction involving multiple steps of reactions: formation of new compounds or radicals with the consumption of electrons, generation of radicals and reaction between radicals in every stage. Therefore, the product distribution of a photocatalytic reaction is strongly affected by a wide range of reaction conditions, such as reactor geometry, catalyst type, sacrificial reagents used and even illumination type.

Generally, the photocatalytic conversion of CO_2 into fuel products can be categorized into two types, i.e. aqueous system and gas phase system. In an aqueous system, the photocatalytic conversion of CO_2 is mainly limited by batch processes, which involve a suspension of solid photocatalysts and the bubbling of CO_2 gas into the system. Alkaline solutions such as NaOH solution are often employed to enhance the solubility of CO_2. Furthermore, the high content of OH^- ions in the aqueous solution could serve as strong hole scavengers to form •OH radicals. This would in turn suppress the recombination of electron-hole pairs, leading to higher utilization of surface electrons to take part in the photoreduction of CO_2 (Srinivas et al. 2011).

On the other hand, gas phase systems involve the bubbling of CO_2 into water to induce water (the sacrificial reagent) content in the reactant. Subsequently, the reactant mixture is fed into the reactor which is irradiated by UV or visible light. Water is not the sole sacrificial reagent that can be used in the photoreduction of CO_2. There are other kinds of sacrificial reagents such as methane (Shi et al. 2004; Yoshida and Maeda 2010; Woolerton et al. 2010), ethane (Wang et al. 2007), methanol (Ulagappan and Frei 2000; Qin et al. 2011), 2-propanol (Dey et al. 2004) and hydrogen (Teramura et al. 2008, 2010). It should be noted that the sacrificial reagents chosen will manipulate the reaction product distribution.

The type of photocatalyst used also plays an important role in controlling the product distribution and the efficiency of the photocatalytic reduction process. In this work, the different types of photocatalysts reported in literature will be reviewed in detail. Their applications in different reactor configurations, reaction product yield and product distribution, as well as related mechanisms will also be extensively discussed.

3.2 Choice of Photocatalyst

3.2.1 *Pure TiO_2 Photocatalyst*

Among all metal oxides, titanium dioxide (TiO_2) has appeared to be the novel catalyst used in most of the photocatalytic processes, regardless of the reactor geometry and the nature of the reaction process. This is due to its low band gap energy values of approximately 3.0 and 3.2 eV for rutile and anatase, respectively (Tan et al. 2006b). Most of the research work reported on the photocatalytic reduction of CO_2 mainly utilizes TiO_2 anatase as photocatalyst due to its unique properties, such as strong oxidizing power, non-toxicity and long-term photostability. Thus, TiO_2 anatase is highly active for photocatalytic reactions and is frequently used as the catalyst in various applications such as air purification and water purification (Kaneco et al. 1999; Dey et al. 2004; Kočí et al. 2009). Pure TiO_2 photocatalyst is suitable for photocatalysis in both aqueous and gas phase systems under UV light irradiation. CO_2 can be derived into methane using TiO_2 pellets in an aqueous system with CO_2 being saturated in a mixture of 2-propanol solution and TiO_2 (Dey et al. 2004). Methanol, formaldehyde and formic acid are some of the main reaction products formed in the aqueous solution. In addition, pure TiO_2 can also be used for the photocatalytic reduction of CO_2 in gas phase. In this system, methane is produced from the photocatalytic reduction of CO_2 with water vapour contacted via TiO_2 pellets under UV light irradiation (Tan et al. 2006b). In a similar reaction set-up, similar hydrocarbon products such as ethylene and carbon monoxide can also be derived from CO_2 in a gas phase photoreduction system with hydrogen being used as the sacrificial reagent (Lo et al. 2007).

Particle size is one of the important factors that may affect the performance of TiO_2 in photocatalytic reaction processes. In some of the research work reported, TiO_2 pellet was chosen in place of TiO_2 powder with the justification that pellet type TiO_2 exhibits higher adsorption capacity of CO_2 than other forms of TiO_2 (Tan et al. 2006b, 2007). However, some argued that TiO_2 nanocomposites are preferable due to their high surface areas and possibly low electron-hole pair recombination, leading to improved photocatalytic performance of TiO_2 (Pathak et al. 2004; Chen et al. 2009; MacFarlane and Scott 2012). Nevertheless, in an aqueous reaction system, TiO_2 nanocomposites can be dispersed homogeneously in the reaction medium, allowing better exposure to the light source. This leads to a more effective utilization of irradiation light and higher surface area of nanoparticle exposed for reactions (Pathak et al. 2004).

On the other hand, Koči and co-workers (2009) reported the effect of TiO_2 particle size on the photocatalytic reduction of CO_2 in an aqueous system. The optimum particle size reported was 14 nm, which was a result of competing effects of specific surface area, charge-carrier dynamics and light absorption efficiency. They also reported that the specific surface area is not the most decisive parameter that controls the photocatalytic conversion of TiO_2. Although small particle sizes with high surface areas might improve the photocatalytic activity, there are several other constraints that should not be overlooked such as surface electron-hole recombination rate, particle aggregation and size quantization effect (Kočí et al. 2009).

TiO_2 powders can also be used in the photocatalytic reduction of CO_2 in supercritical CO_2 systems. In such systems, formic acid can eventually be produced through the protonation of reaction intermediate on TiO_2 powders in solution. The reaction was considered as a homogeneous system with CO_2 gas being pressurized into liquid form in supercritical conditions. However, the use of sacrificial reagents such as water or acid is still necessary due to the nature of the photocatalytic reaction (Kaneco et al. 1999).

In most of the research studies conducted utilizing TiO_2 photocatalyst; anatase is the most commonly chosen crystal phase as it presents higher photoactivity than the rutile (Romero-Gomez et al. 2011). However, the high photoactivity of the anatase phase is only limited to UV wavelength. This is the primary reason that causes low efficiencies in TiO_2 photocatalytic reactions conducted under visible light irradiation. On the other hand, rutile phase TiO_2 displays visible light excitation, with low photoactivity due to the high recombination rate of its photo generated electrons and holes. Due to these reasons, Chen et al. (2009) reported the use of TiO_2 nanocomposites in a mixed phase of anatase-rutile for the photocatalytic reduction of CO_2 under visible light irradiation. The composites were well-aligned in a column shape using DC-magnetron sputtering deposition. An improved light absorption range of up to 550 nm wavelength was observed in the mixed phase TiO_2 nanocomposites, which far exceeded that of rutile which achieved a wavelength of 410 nm. The mixed phase TiO_2 nanocomposites exhibited highly reactive behaviour which successfully converted CO_2 into methane under visible light irradiation (Tan et al. 2006b; Chen et al. 2009). The enhancement in the activity of mixed-phase TiO_2 photocatalysts is a result of three factors: (i) the smaller band gap of rutile extends the light absorption range into the visible region, (ii) the improved charge separation by electron transfer from rutile to anatase at the interface and (iii) the relatively small size of rutile crystallites assist this transfer, creating catalytic hot spots at the interface of rutile/anatase (Hurum et al. 2003).

3.2.2 Modified TiO_2 Photocatalyst

There is also active research involved in the modification of TiO_2 for the enhancement of its photocatalytic activity, so that it can be applied in reactions utilizing both UV and visible light irradiation. Photoactivity of the TiO_2 can be improved in a way that drives towards favourable characteristics, particularly: (i) higher surface area, (ii) higher absorption of light with wider absorption wavelength, (iii) smaller band gap and (iv) higher electron-hole pair separation with minimal recombination rates. There are various approaches in modifying TiO_2. Among all, metal oxide doping is the most commonly used method as it can be easily done via impregnation, precipitation or sol-gel synthesis.

3.2.2.1 TiO_2 Modified with Metal/Metal Oxides

Transition metal oxides with various redox states are always the best choices as metal oxide photocatalysts. Among all, copper/copper oxide is one of the most popular species for the modification of TiO_2. Copper/copper oxide has been frequently loaded onto TiO_2 to produce highly reactive photocatalysts via the improved impregnation method (Slamet et al. 2005; Qin et al. 2011) or improved sol-gel process using a homogeneous hydrolysis technique (Tseng et al. 2002, 2004). The photoactivity of the copper-modified TiO_2 was reported to be significantly higher than pure TiO_2 in the photocatalytic CO_2 reduction conducted in both aqueous phase (Tseng et al. 2002, 2004; Slamet et al. 2005; Yang et al. 2011) and gas phase systems (Nguyen and Wu 2008a; Wu 2009; Yang et al. 2010) under UV light irradiation. In the photocatalytic CO_2 reduction on a gas-solid interface, the products generated are mainly gaseous compounds, such as carbon monoxide, methane and ethylene (Nguyen and Wu 2008b; Li et al. 2010).

Copper oxide-modified TiO_2 is a highly efficient photocatalyst for CO_2 reduction because copper is an effective electron trapper that is able to reduce the recombination of electron-hole pairs in the photocatalyst (Tseng et al. 2002; Yang et al. 2011). However, excess loading of copper may cause a shading effect on the catalyst which will eventually reduce the TiO_2 photoexcitation capacity. Tseng et al. (2002) reported an optimum Cu loading of 2.0 wt% for the highest dispersion among catalysts. Hence, optimum loading of copper is vital in order to obtain best photocatalytic activities (Slamet et al. 2005; Yang et al. 2011). Other than copper/copper oxides, several other types of metal oxides such as Ag, Al_2O_3, Pd, Pt, SiO_2 and ZnO (Subrahmanyam et al. 1999; Sasirekha et al. 2006; Zhang et al. 2009; Kočí et al. 2010, 2011b; Uner and Oymak 2012) have also been reported in the enhancement of TiO_2 photocatalyst.

The addition of ZnO and Au on the TiO_2 was reported to improve significantly the CO_2 photoreduction activity of the TiO_2 by enhancing the adsorption of CO_2, OH radicals and sacrificial reagents on the TiO_2 surface (Mei et al. 2013). In view that CO_2 photoreduction is a heterogeneous catalysis reaction, adsorption of the reactants on the photocatalysis surface is vital to boost the electrons and holes transfer between the photoactivated TiO_2 and the reactants.

To date, the research reported on the photocatalytic reduction of CO_2 utilizing visible light source is not as common as compared to those using UV light irradiation. Most of the work related to the photocatalytic reduction of CO_2 using visible light are relatively new (Pan and Chen 2007; Sato et al. 2007, 2010; Li et al. 2011; Fu et al. 2012). In general, applications of TiO_2 photocatalyst for the photoreaction with visible light irradiation are still feasible. TiO_2 photocatalyst, which is initially only active under UV light irradiation, can be activated for visible light absorption ($\lambda > 400$ nm) upon addition of various types of dopants. The dopants can either be metal oxide species or non-metal oxide species.

Co-catalysts developed from co-doping TiO_2 with a metal/metal oxide species could be a better choice for the photocatalytic reduction of CO_2 under visible light irradiation. In these co-catalysts, one of the metal oxide species plays the crucial role of reducing the band gap of the TiO_2 photocatalyst while the other improves

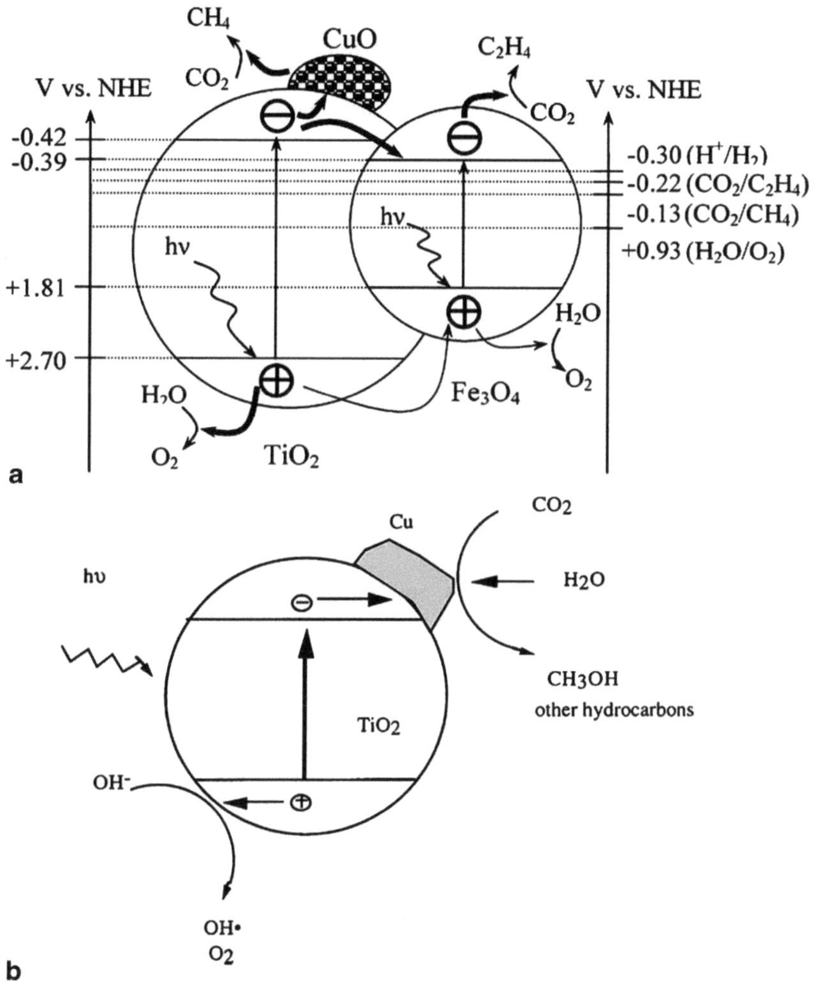

Fig. 3.3 Reaction mechanism of the CO_2 photoreduction over: **a** Fe-Cu/TiO_2 (Nguyen and Wu 2008a) and **b** Cu/TiO_2 (Tseng et al. 2002) photocatalysts

the optical absorption of the photocatalyst. Examples of the dual metal oxide co-catalysts include Fe co-doped $LaCoO_3$ (Jia et al. 2009), Fe-Cu/TiO_2 (Nguyen and Wu 2008a, b) and CdSe-Pt/TiO_2 (Wang et al. 2010a). Figure 3.3 displays the reported reaction mechanism of the CO_2 photoreduction with Cu-TiO_2 and Fe-Cu/TiO_2 photocatalysts. Fe and Cu, upon being doped on TiO_2 photocatalyst, could form a dopant energy level above the valence band of the transition metal oxide. This narrows the overall bandgap energy of the photocatalyst and improves the electron-hole pair separation (Jia et al. 2009). Nguyen and Wu, the pioneer in optical fiber reactor for the photocatalytic reduction of CO_2, has also proven the excellent effect of Fe-doping in improving the optical absorption and reactivity of TiO_2 under the irradiation of concentrated natural sunlight (Nguyen and Wu 2008b). CdSe has also been reported to exhibit similar roles to Fe in the CdSe-Pt/TiO_2 photocatalyst, in which CdSe acts as a photosensitizer and improves the optical absorption of Pt/TiO_2 during the photoreaction process (Wang et al. 2010a).

On the other hand, Feng and co-workers (2011) reported excellent CO_2 photoreduction activity under sunlight irradiation by merely using Pt nanoparticles loaded on TiO_2 nanotube arrays. Pt was reported to be an effective species for reducing water into hydrogen. In addition, the authors also proposed that the yield of hydrocarbons from the photoreduction of Pt/TiO_2 photocatalysts can be further improved by the addition of Cu species, which is more active in reducing CO_2 (Feng et al. 2011). Similar photocatalyst (Pt/TiO_2 nanotube and nanoparticle) was studied by Zhang et al. (2009) for photocatalytic reduction of CO_2 with H_2O vapour under UV light at mild conditions of low temperature and pressure. They reported that the photocatalytic activity of Pt/TiO_2 nanotube was more active in obtaining higher methane yield than Pt/TiO_2 nanoparticle catalyst.

3.2.2.2 Coupling of Two Semiconductor Systems

The use of nanostructured AgBr/TiO_2 and NiO/$InTaO_3$ were also reported for the photocatalytic reduction of CO_2 with water under visible light irradiation (Pan and Chen 2007; Wang et al. 2010b; Abou Asi et al. 2011). Metallophtalocyanine is also a potential metal species for the photoreduction process under visible light irradiation. TiO_2-supported zinc-phthalocyanine and TiO_2-supported cobalt-phthalocyanine were also found to give positive results for the photocatalytic reduction of CO_2 with water in an aqueous system under visible sunlight irradiation, producing formic acid as the main product (Zhao et al. 2007, 2009a, b).

Most recently, Truong et al. (2012) reported the synthesis of $FeTiO_3$/TiO_2 photocatalyst using a simple hydrothermal method. Under both visible and UV-Vis light illumination, $FeTiO_3$/TiO_2 was demonstrated to possess excellent photocatalytic performance in the photoreduction of CO_2. The maximum CH_3OH yield was approximately 0.46 μmol $g^{-1} h^{-1}$, which was 3 times higher than that of pure TiO_2. The significant enhancement was attributed to the unique band structure and the efficient transfer of charges between both semiconductors. The narrow band gap of $FeTiO_3$ was also said to contribute to the high efficiency of the photocatalytic system.

3.2.2.3 TiO_2 Modified with Silica

Silica plays an important role as one of the common photocatalysts for various reactions. It can be used as a support for the immobilization of photocatalyst as well as improve its photoactivity; particularly in aqueous phase systems. The development of pure TiO_2 or modified TiO_2 supported on silica was reported using various approaches such as one pot sol-gel method (Li et al. 2010), impregnation method (Sasirekha et al. 2006), multistep impregnation method (Shi et al. 2004), as well as thin film coating (Ikeue et al. 2002; Shioya et al. 2003).

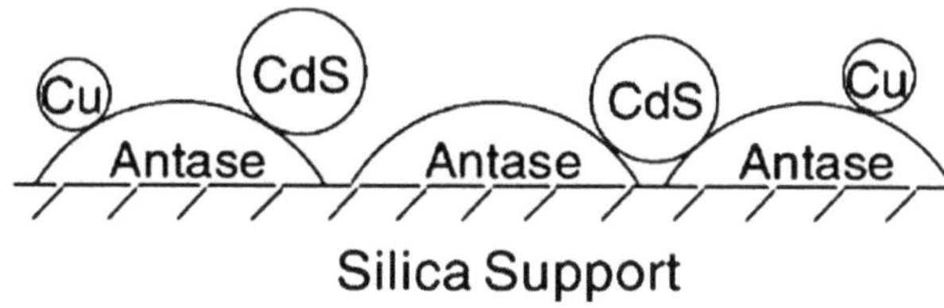

Fig. 3.4 Proposed structure of silica supported photocatalyst. Original diagram adapted from (Shi et al. 2004)

Loading of photocatalysts on mesoporous silica substrate that has high surface area can greatly enhance the CO_2 photoreduction by improving TiO_2 dispersion and increasing the absorption of CO_2 and H_2O on the surface of the photocatalyst (Li et al. 2010). Figure 3.4 illustrates a possible structure of silica supported photocatalyst. The structure was proposed by Shi and co-workers (2004) based on several characterization analysis on photocatalysts developed from Cu-CdS-SiO_2 mixture. High dispersion of active metals, in this case Cu and CdS, can be achieved by the uniform dispersion of these species over the entire silica substrate through the formation of Si-O-Si structure.

The chemical and physical properties of silica may be altered by various kinds of treatment. For example, thermal treatment can be adopted to change the porosity of silica whereas the introduction of additives can induce its chemical properties. The presence of silica changes the nature of the titanium catalyst. The hydrophilic and hydrophobic properties of a titanium catalyst can be altered by supporting the titanium species on different types of zeolites. Titanium with high hydrophilicity can be developed by incorporating titanium on β-zeolite by means of hydrothermal treatment with OH^- anions (Yamashita et al. 2002). On the other hand, titanium with high hydrophobicity can be achieved by incorporating titanium on zeolite through a hydrothermal treatment with F^- anions (Ikeue et al. 2001a). The hydrophobic nature of the titanium catalyst could bring important effects on the product distribution in the CO_2 photoreduction process. These effects are discussed in detail in Sect. 3.0.

Surface geometry of the titanium photocatalyst can be aligned with the aid of a silica support. Instead of the conventional powdered form, solvent evaporation techniques can be employed to produce transparent and self-standing titanium on porous silica frameworks of mesoporous thin films. The titanium-containing porous silica thin film was reported to exhibit superior photocatalytic activity due to its high surface area, well-structured surface and unique hexagonal and cubic pore structure. Mesostructures of the silica film can be obtained by controlling the solvent evaporation process, especially during the condensation stage with water (Ikeue et al. 2002; Shioya et al. 2003).

Sasirekha et al. (2006) reported the modification of TiO_2 by doping it with Ru and mounting them on a silica support for the photocatalytic reduction of CO_2 in an aqueous system. Among the products formed were methanol, formic acid and formaldehyde. Ru was doped on TiO_2 to enhance the photocatalytic activity of bare TiO_2, while the silica support was used to immobilize the photocatalyst. It was reported that 0.5 wt% Ru doping can improve the photoactivity of TiO_2. The silica support was believed to form Ti-O-Si bridging bonds which could increase the efficiency of

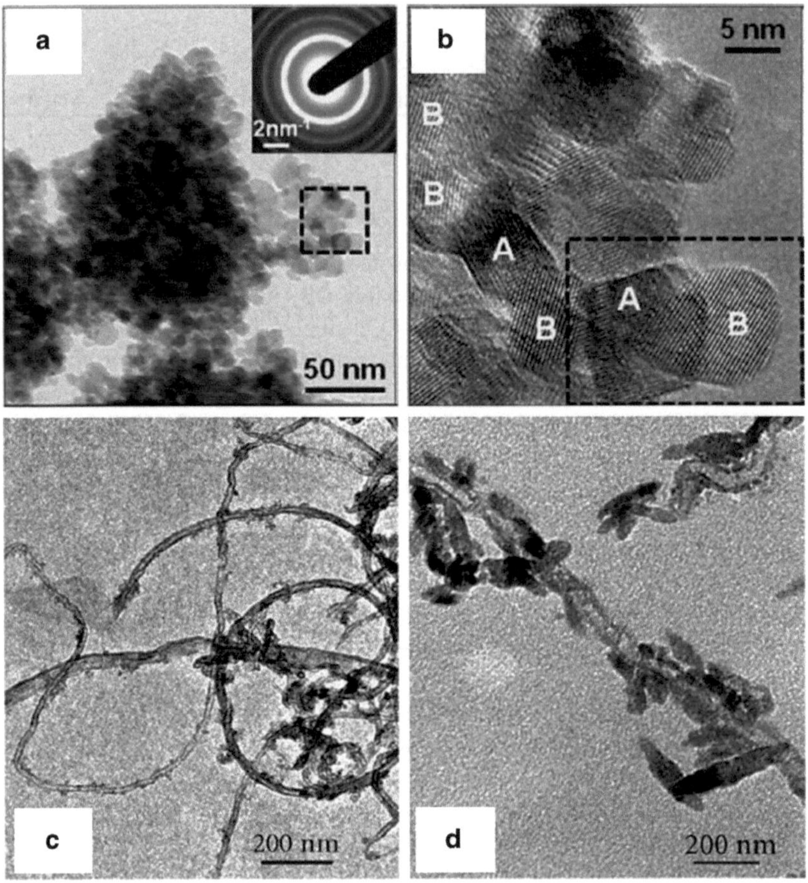

Fig. 3.5 Images of selected modified TiO_2 photocatalysts with: (**a, b**) iodine (Ozcan et al. 2007). (**c, d**) MWCNTs (Xia et al. 2007). (**e**) graphene ((Liang et al. 2011) and (**f**) enzyme CODH1(Woolerton et al. 2010)

electron charge separation. At the same time, the difficulty of separating the products from the liquid-solid mixture, which is a commonly faced problem in aqueous reaction systems, could be overcome (Sasirekha et al. 2006).

3.2.2.4 TiO_2 Modified with Non-metal Oxide Compounds

Studies on the enhancement of TiO_2 with non-metal oxide components are still limited as compared to those modified with metal doping. Although this method is not as popular as metal oxide modification, the outcome is superior, especially under visible light irradiation. Figure 3.5 shows the images of selected modified TiO_2 photocatalysts with non-metal oxides.

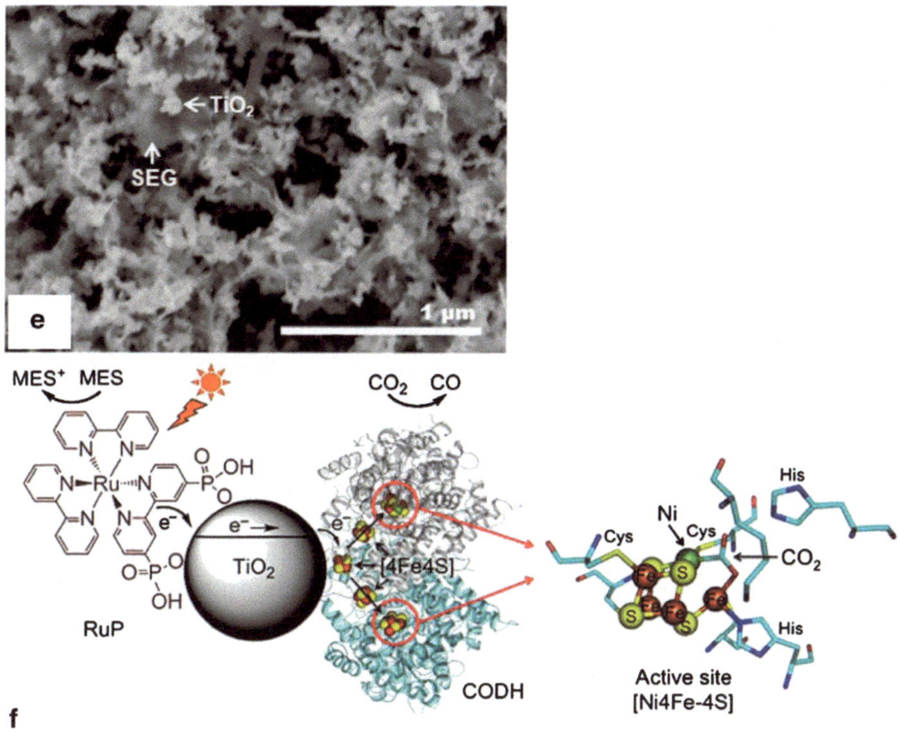

Fig. 3.5 Continued

Dye-sensitized TiO_2 is one of the most commonly used non-metal oxide-modified TiO_2 photocatalyst. Addition of ruthenium dye on TiO_2 can induce full absorption of visible light on the photocatalyst. This results in an efficient charge transfer in the dye-TiO_2 system, leading to superior photocatalytic activity as compared to bare TiO_2 photocatalyst (Ozcan et al. 2007; Nguyen et al. 2008). Iodine and nitrogen doping on TiO_2 nanoparticles were also reported by several researchers (Fig. 3.5a). The modifications greatly improved the optical absorption of TiO_2 by widening the absorption band and shifting the optical absorption edge to the low-energy, visible light region (Varghese et al. 2009; Zhang et al. 2011). Despite enabling the photoexcitation of modified-TiO_2 under visible light irradiation, nitrogen doping was found to create localized states favouring carrier recombination (Varghese et al. 2009).

On the contrary, sulphur doping could generate oxygen defects on the lattice of anatase TiO_2 particles, which is important to induce visible light absorption and in turn, enhance the photoreactivity of the modified TiO_2 (Hussain et al. 2009). In addition to the conventionally employed TiO_2 nanoparticles, nitrogen-doped TiO_2 nanotubes (N-TNTs) were prepared via a hydrothermal method (Zhao et al. 2012). The loaded TiO_2 nanotubes displayed superior photocatalytic activities towards the photoreduction of CO_2. The authors claimed that the doping of nitrogen led to the substitution of N atoms for lattice oxygen atoms in the TiO_2 structure. This narrowed the band gap of the TNT to the visible light region, ultimately enhancing the activity of the photocatalytic system.

Multi-walled carbon nanotubes (MWCNTs) supported TiO_2 composites could also be used in the photocatalytic reduction of CO_2 (Fig. 3.5b). The presence of the MWCNTs in the TiO_2 composites can help to mitigate the agglomeration of the TiO_2 particles. Electron transfer on the photocatalyst surface is unique with the use of MWCNTs. The electrons generated during the photoexcitation process are transported along the tubes, hence reducing the electron-hole pair recombination rates and improving the photocatalytic activity (Xia et al. 2007).

Graphene, with its unique electronic and optical properties, has also been regarded as an attractive component for the tailoring of TiO_2 photocatalysts. Liang and co-workers (2011) recently reported on the synthesis of solvent-exfoliated graphene (SEG)–TiO_2 nanocomposites for the photocatalytic reduction of CO_2 (Fig. 3.5c). The optimized composite was demonstrated to yield a 7-fold improvement for the reduction of CO_2 to CH_4, as compared to pure TiO_2 under visible light illumination. In addition, the less-defective SEG was also shown to be a better photocatalytic promoter as compared to reduced graphene oxide. The improved electrical mobility of SEG enabled a longer electronic mean path, which in turn allowed farther diffusion of electrons from the graphene-TiO_2 interface. This decreases the probability of electron-hole recombination, thus improving photocatalytic performance (Liang et al. 2011).

Modification of TiO_2 with non-metal oxide compounds not only improves the photocatalytic performance of TiO_2, but also alters the nature of the photocatalyst. Yamashita and co-workers (2004) reported the development of hydrophobic Ti-containing mesoporous silica by the fluorination of the photocatalyst. The changes on the hydrophobicity of the catalyst surface will subsequently affect the product distribution of the reaction process, which will be extensively discussed in Sect. 3.0 (Yamashita et al. 2002; Yamashita et al. 2004).

In the photocatalytic reduction of CO_2 incorporated with biochemistry process, TiO_2 nanoparticles modified with photosensitizer and CO_2-reducing enzyme CODH1 from the anaerobic microbe *Carboxydothermus hydrogenoformans* (*Ch*) provides an extraordinary catalyst with high efficiency, even under visible light irradiation (Fig. 3.5d). In the enzyme-catalyzed process, CO_2 can be readily converted into carbon monoxide with TiO_2 playing an essential role as an electron relay between the photosensitizer and CODH1, rather than acting as a common support or photocatalyst (Woolerton et al. 2010).

3.2.3 Other Metal Oxide Photocatalysts

TiO_2 is not the sole metal oxide that is active for the photocatalytic reduction of CO_2. Other metal oxides reported include gallium oxide (Ga_2O_3), zirconium oxide (ZrO_2), zinc oxide (ZnO), tantalum oxide (TaO_3)—based catalysts and mixed oxides of zinc, copper, aluminium etc.

ZnO, nickel oxide, ZrO_2, Ga_2O_3 and magnesium oxide are proven to act as photocatalysts in the photocatalytic reduction of CO_2 (Kohno et al. 2000; Lo et al. 2007). Studies showed that Ga_2O_3 on its own can actively catalyze the photoreduction of CO_2 in a gaseous reaction system with several reductants such as methane and hydrogen (Teramura et al. 2008; Ahmed et al. 2011). However, the product gas obtained is merely carbon monoxide instead of hydrocarbon components. Ga_2O_3 can also be used with co-catalysts as well. For example, Ga_2O_3 co-doped with zinc is a transparent and conductive material, which is useful in various applications such as water splitting and wastewater treatment (Yan et al. 2010). Combination of Ga_2O_3 with zinc, copper and aluminium layered hydroxides were also reported for the photocatalytic reduction of CO_2 into methanol in aqueous-phase systems (Ahmed et al. 2011).

In a recent research by Teramura and co-workers (2010), TaO_3-based catalysts were used in the photocatalytic reduction of CO_2 with H_2 gas. TaO_3 added with lithium, sodium or potassium showed positive results for the photocatalytic reduction of CO_2 to carbon monoxide in the presence of H_2 under UV light irradiation (200–400 nm). On the other hand, Wang et al. (2007) reported on the photocatalytic reduction of CO_2 into hydrocarbon oxygenates in a gas-phase system under UV light illumination using Pd-MoO_3 catalyst supported on silica prepared from incipient wetness impregnation method (Wang et al. 2007). TaO_3 modified with NiO and indium was also reported to be viable for the photoreduction of CO_2 under visible light irradiation (Wang et al. 2010b).

Cobalt-supported photocatalyst is another choice of metal oxide for the photocatalytic reduction of CO_2 (Zhao et al. 2009b). Cobalt corrin was found to act as a photocatalyst for the conversion of CO_2 into carbon monoxide and formic acid in an aqueous system containing acetonitrile and methanol with p-terphenyl as a photosensitizer and triethylamine as a reductive quencher (Grodkowski and Neta 2000). The photocatalytic activity of cobalt can be enhanced by doping with co-catalysts such as carbon and Fe or a combination of both. In such hybrid-doped photocatalysts, the carbon doping which substitutes for lattice oxygen atoms in $LaCoO_3$ plays a role in shifting the optical absorption edge to the visible light region and narrowing the band gap. Meanwhile, Fe doping could form a dopant energy level above the valence band of $LaCoO_3$, which releases electrons in this interband during light irradiation and thus initiates the photocatalytic reaction (Jia et al. 2009).

Based on the research works summarized above, it can be observed that metal oxides are capable of exhibiting positive results on the photocatalytic reaction, perhaps as good as TiO_2 in spite of the significant difference in product distribution and product yield.

3.2.4 Metal Complex Photocatalysts

Metal complex/ligand photocatalysts are one of the most frequently reported substrates in the photoreduction of CO_2. Metal complexes are suitable for processes involving solar irradiation due to their excellent photochemical properties. Metal complexes, which mainly consist of transition metal compounds, have multiple accessible redox states that can be easily photoexcited into unstable and reactive species upon receiving energy during light irradiation.

There are generally two categories of metal complex photocatalysts. The first category is associated with a photosensitized CO_2 reduction process. This type of catalyst usually involves light absorption by a photosensitizer during irradiation, typically ruthenium (II) trisbipyridine, to initiate the radical chain reaction process which eventually reduces CO_2 into hydrocarbons (Morris et al. 2009). Examples of such photocatalysts are cobalt and nickel tetraaza-macrocyclic compounds and supramolecular compounds with covalent attachment of photosensitizer with coordinations capable for CO_2 reduction. The second type of metal complex photocatalysts involves only a single compound as light absorber and the catalyst. For example, metalloporphyrins and related metallomacrocycles compounds such as metallopthalocyanines metallocorroles and metallocorrins (Morris et al. 2009).

Metal complex catalysts are usually only reported for the photoreduction of CO_2 in aqueous reaction systems (Takeda et al. 2008). The most popular metal complexes for the photocatalytic reduction of CO_2 are ruthenium Ru(II) (Hirose et al. 2003; Jansen and Pitter 2004; Ying-min et al. 2006) and rhenium Re(I) (Hori et al. 2002; Takeda et al. 2008; Cheung et al. 2009) derived metal complexes. Most of the reactive metal complex photocatalysts are mainly derived from these important compounds (Gholamkhass et al. 2005; Sato et al. 2007). Other potential metal complexes active for the photoreduction of CO_2 include carbine-supported copper(I) boryl complex (Laitar et al. 2005), cobalt and iron corroles (Grodkowski et al. 2002) and hexatitanate complex, $K_2Ti_6O_{13}$ loaded with platinum, copper and iron (Guan et al. 2003a, b).

3.3 Influence of Photocatalysts on Product Distribution

Photocatalytic reduction of CO_2 involves chain reactions with the generation and combination of intermediate radicals and the formation of a wide range of reaction product distribution. The reaction products derived from a photoreduction process are strongly dependent on the type of photocatalyst used. In general, the reaction products are distributed between C1–C3 hydrocarbon chains, ranging from carbon monoxide to larger molecules such as formic acid, formaldehyde, methanol and ethanol.

Table 3.1 compares the product distribution to the type of photocatalyst used in both aqueous and gas phase reaction systems. The type and nature of photocatalyst used can manipulate the product distribution using the same reaction system. In the photoreduction of CO_2 conducted in an aqueous system using $AgBr/TiO_2$ and $NiO/InTaO_3$, these two photocatalysts favour the formation of less oxygenic hydrocarbons such as methane, methanol and ethanol (Pan and Chen 2007; Wang et al. 2010b; Abou Asi et al. 2011). On the other hand, high oxygenated compounds such as formic acid and formaldehyde were favourably obtained from reactions utilizing

Table 3.1 Summary of photocatalyst type and corresponding product distribution in aqueous and gas phase systems

References	Photocatalyst used	Products produced	Light source	Maximum product yield (total yield)/ efficiency
Aqueous phase system:				
(Abou Asi et al. 2011)	$AgBr/TiO_2$ nanocomposite	Methane, methanol, ethanol, CO	Visible light, λ 420 nm. 150 W Xe lamp	Total yield = 251.85 μmol/g
				Methane = 128.56 μmol/g
				Methanol = 77.87 μmol/g
				Ethanol = 13.28 μmol/g
				CO = 32.14 μmol/g
(Dey et al. 2004)	TiO_2 anatase (325 mesh)	Methane	UV light. λ = 350 nm	Total yield = 200×10^{-8} mol
(Hussain et al. 2009)	S-TiO_2	Methanol, ethanol, propanol	UV light and sunlight	Yield = 5.26 %
(Jia et al. 2009)	C and Fe co-doped $LaCoO_3$	Formic acid and formaldehyde	Visible light. 125 W Xenon lamp. λ > 400 nm	Total yield = 145 μmol/g/h
(Kaneco et al. 1999)	TiO_2 powders (anatase)	Formic acid	UV light. λ = 340 nm	Yield = 8.8 μmol/g
(Kočí et al. 2009)	Pure TiO_2 anatase	Methane, methanol, CO	UV light. 8 W Hg lamp, peak intensity 254 nm	Total yield ≈ 0.18 μmol/g (24 h irradiation)
(Kočí et al. 2011a)	ZnS-MMT (nanoparticles)	Methane, methanol, CO, H_2	UV light, λ = 300–600 nm	Total yield ≈ 220 μmol/g
(Luo et al. 2011)	Cu-Ce/TiO_2	Methanol	125 W UV lamp. λ = 365 nm	Total yield = 180.3 μmol/g (16 h irradiation)
(Pan and Chen 2007)	$NiO/InTaO_4$	Methanol	Visible light	Yield = 1.394 μmol/g/h
(Qin et al. 2011)	CuO-TiO_2 composite	Methyl formate	UV light. 250 W high pressure mercury lamp. λ = 365 nm	Yield = 1600 μmol/g/h
(Srinivas et al. 2011)	Cu/TiO_2 supported on molecular sieve 5A	Oxalic acid, acetic acid, methanol	250 W high pressure Hg lamp.	Total yield = 78.4 μmol/g/h
				Oxalic acid = 65.6 μg/g/h
				Acetic acid = 12 μg/g/h
				Methanol = 0.8 μg/g/h
				(20 h irradiation)

Table 3.1 (continued)

References	Photocatalyst used	Products produced	Light source	Maximum product yield (total yield)/efficiency
(Subrahmanyam et al. 1999)	Various metal oxides: TiO_2/Pd/SiO_2, TiO_2/Pd/Al_2O_3, CuO/ZnO/MgO, Li_2O-TiO_2/MgO	C1–C3	UV light	Total yield ≈ 16.5 μmol/h
(Truong et al. 2012)	$FeTiO_3$/TiO_2	Methanol	500 W high pressure Xe lamp. λ > 300 nm	Total yield = 0.462 μmol/g/h
(Tseng et al. 2002)	TiO_2 and Cu-TiO_2	Methanol	UV light. λ = 254 nm	Yield = 250 μmol/g catalyst (20h irradiation)
(Tseng et al. 2004)	Cu/TiO_2	Methanol	UVA (λ = 254 nm) and UVC (λ = 365 nm)	Yield = 600 μmol/g (30 h irradiation)
(Xia et al. 2007)	MWCNTs supported TiO_2 composite	Ethanol and Formic acid	15 W UV lamp. λ = 365 nm	Total yield ≈ 450.77 μmol/g (5 h irradiation)
(Yang et al. 2009)	Cu/TiO_2/SBA-15	Methanol	400 W medium pressure metal halide lamp, peak intensity 365 nm	Total yield = 627 μmol/g/h
(Yang et al. 2011)	Cu/TiO_2 or TiO_2 powders	Formic acid, formaldehyde, methanol	UV light. Mercury lamp	Total yield = 3.52 mmol/g/h
(Zhao et al. 2007)	Titania-supported zinc-phthalocyanine	Formic acid (main product), CO, methane	Visible light. 500 W tungsten-halogen lamp	Total yield = 978.6 μmol/g catalyst (10 h irradiation)
(Zhao et al. 2009a)	CoPc/TiO_2 nanocomposite	Formic acid, methanol, formaldehyde, methane	Visible light. Tungsten-halogen lamp	Total yield = 406.65 μmol/g
(Zhao et al. 2009b)	CoPc/TiO_2 nanocomposite	Formic acid, methanol, formaldehyde	Visible light. Tungsten-halogen lamp	Total yield = 1714.9 μmol/g (10 h irradiation)
(Zhao et al. 2012)	N-TiO_2 nanotube	Formic acid, methanol, formaldehyde	500 W tungsten-halogen visible lamp	Total yield = 14530.0 μmol/g
				Formic acid = 12475.8 μmol/g
				Methanol = 1132.6 μmol/g
				Formaldehyde = 921.6 μmol/g

Table 3.1 (continued)

References	Photocatalyst used	Products produced	Light source	Maximum product yield (total yield)/ efficiency
Gas phase system:				
(Ahmed et al. 2011)	Zinc-copper-M (M = aluminium, gallium)	CO, methanol	UV-visible light	CO = 620 nmol/g/h
				Methanol = 170 nmol/g/h
(Ikeue et al. 2001b; Yamashita et al. 2002)	Ti-beta zeolites; Ti-beta(OH) and Ti-beta(F)	Methane, methanol	UV light. $\lambda > 250$nm	Total yield ≈ 6.5 μmol/g/h
(Ikeue et al. 2002)	Ti-Silica thin films	Methane, methanol, CO and O_2 (minor)	UV light	Quantum yield ≈ 0.28 %
(Li et al. 2010)	Cu/TiO_2 nanocomposites supported on mesoporous silica	CO (using TiO_2-SiO_2 catalysts), methane (using Cu/TiO_2/SiO_2 catalysts)	Xe Arc lamp $\lambda = 250$–400 nm	Total yield = 70 μmol/g/h
				Methane = 10 μmol/g/h
				CO = 60 μmol/g/h
(Lo et al. 2007)	TiO_2 (Degussa, P25) and ZrO_2	Methane, CO, ethane	15 W UV (maximal spectral wavelength 365 and 254 nm) and near-UV fluorescent lamp	Total yield (TiO_2) = 8.69 μmol/g
				Total yield (ZrO_2) = 1.24 μmol/g
(Nguyen and Wu 2008a)	Cu/TiO_2 and Cu-Fe/TiO_2	Methane, ethylene	UVA	Quantum yield = 0.024 %
			$\lambda = 254$ nm	Methane = 0.91 μmol/g/h
				Ethylene = 0.58 μmol/g/h
(Nguyen and Wu 2008b)	TiO_2-SiO_2 Cu-Fe/ TiO_2-SiO_2	Methane, ethylene	Natural sunlight and UVA artificial light	Overall quantum efficiency = 0.0182 %
(Nguyen et al. 2008)	Ru dye sensitized Cu/ Fe-TiO_2	Methane, ethylene	Natural sunlight	Total yield = 0.617 μmol/g/h
(Shi et al. 2004)	Cu/CdS-TiO_2/ SiO_2	Acetone, ethane, CO	UV light	Conversion: Methane = 1.47 %; CO_2 = 0.74 %
(Shioya et al. 2003)	Ti-silica thin film	Methane, methanol	UV light	Quantum yield ≈ 0.28 %
(Tan et al. 2006b)	TiO_2 pellets	CO, methane	UV light, UVC ($\lambda = 253.7$ nm), NEC black light blue fluorescent lamp	200 ppm

Table 3.1 (continued)

References	Photocatalyst used	Products produced	Light source	Maximum product yield (total yield)/ efficiency
(Tan et al. 2007)	TiO_2 pellet	Methane, H_2	UVC lamp	Methane =0.25 μmol/h;
				Hydrogen =0.16 μmol/h (24 h irradiation)
(Teramura et al. 2008)	Ga_2O_3	CO	UV light	Conversion of CO_2 =3%
(Teramura et al. 2010)	$LiTaO_3$, $NaTaO_3$, $KTaO_3$	CO	λ=200–2500 nm All ranges	Yield=0.42 μmol/g
(Tsuneoka et al. 2010)	MgO, CaO, ZrO_2, Ga_2O_3, Al_2O_3 But main discussion focus on Ga_2O_3	CO	UV light	Conversion=7.3% (48 h irradiation)
(Varghese et al. 2009)	N-doped TiO_2	Methane and other hydrocarbons	Natural sunlight	Total yield =111 ppm/cm²/h or 160 μL/g/h
(Wang et al. 2007)	Pd-MoO_3/SiO_2	Propanol, ethanol, acetaldehyde	UV light	Conversion: Ethane=1.1%
(Wang et al. 2010a)	CdSe/Pt/TiO_2	Methane, methanol, hydrogen, CO	Visible light	Total yield ≈51.3 ppm/g/h
(Wu and Lin 2005)	Cu/TiO_2	Methanol	UV light. λ=365nm	Yield=0.45 μmol/g/h
(Yoshida and Maeda 2010)	Ga_2O_3	Hydrocarbons (C_2H_6, C_2H_4, C_3H_8, C_4H_{10})	Xe Lamp	Total yield =9 μmol/0.2 g catalyst (3 h irradiation)
(Zhang et al. 2009)	Pt/TiO_2 nanotube	Methane	UV light. 300 W high pressure mercury lamp	4.8 μmol/g/h
(Zhang et al. 2011)	I-TiO_2	CO	Visible light. UV light for comparison	2.4 μmol/g/h

metallophthalocyanine or $LaCoO_3$ photocatalyst (Pan and Chen 2007; Zhao et al. 2007; Jia et al. 2009; Abou Asi et al. 2011).

As the photocatalytic reduction of CO_2 derives a variety of products, there is a need to investigate the relation between the type of photocatalyst used and the product distribution. The product distribution can be highly dependent on the acidic/basic characteristic of the metal oxide photocatalyst (Subrahmanyam et al. 1999;

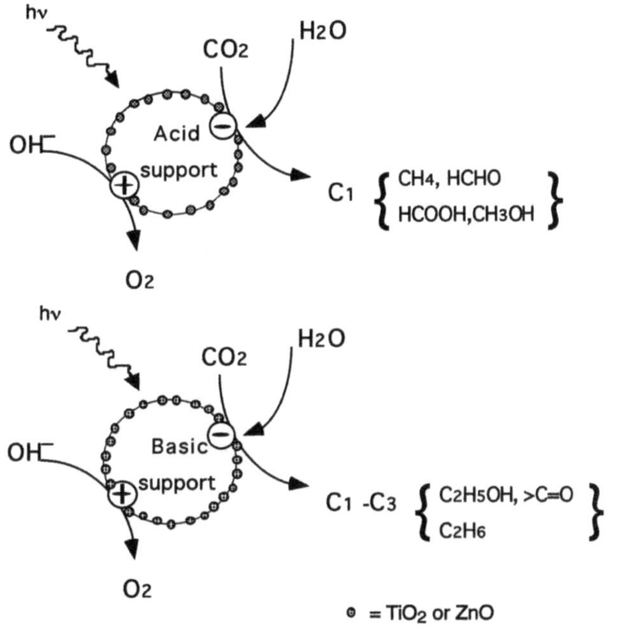

Fig. 3.6 Schematic diagram showing the photoinduced process on acid and basic oxide supported catalysts. Original diagram adapted from Subrahmanyam et al. (1999)

Nguyen and Wu 2008a). Hydrocarbon compounds (C1–C3) are preferentially formed in the photocatalytic reactions catalyzed by basic photocatalysts such as MgO (Srinivas et al. 2011). In contrast, short chain hydrocarbon (C1) compounds, such as methane and carbon monoxide, are commonly obtained from reaction processes catalyzed by acidic metal oxide photocatalysts such as Pd, Al_2O_3 and SiO_2. Figure 3.6 shows the schematic of the photoinduced process on both acid and basic oxide catalysts (Subrahmanyam et al. 1999; Srinivas et al. 2011).

In addition to the basicity and acidity of the metal oxides, another important aspect that manipulates the product distribution is the formation of anion CO_3^{2-} on the metal oxide photocatalysts. Liu et al. (2009) has proposed this important concept in their recent research on the photocatalytic reduction of CO_2 using monoclinic bismuth vanadate, $BiVO_4$. They explained that during the photoreaction process, with the nature of $BiVO_4$, it is difficult for the photoexcited electrons to capture protons in the water to form hydrogen. Hence, the reaction proceeds in an alternative manner where CO_2 is converted into anion CO_3^{2-} using the photoexcited electrons e^- and protons from water, H^+. Subsequently, anion CO_3^{2-} is likely to adopt the photogenerated electrons and is dimerized into methanol or ethanol. Therefore, oxygenated compounds are commonly derived when photocatalysts with such natural behaviour is utilized (Liu et al. 2009).

Addition of metal dopant on the photocatalyst can also affect the product selectivity. In a research work reported on the photoreduction of CO_2 using Fe/Cu-TiO_2 photocatalysts, the reaction products were compared with single metal doped catalysts, i.e. Cu/TiO_2 and Fe/TiO_2. The presence of the metal dopant was found to affect the product selectivity among each other. The presence of Fe as co-dopant in Cu/TiO_2 was found to favour the formation of ethylene (C2 hydrocarbon) and depress the formation of methane (C1 hydrocarbon), while methane was favourably produced in reactions catalyzed by Cu/TiO_2 and Fe/TiO_2 (Nguyen and Wu 2008a).

Yamashita and co-workers (2002 and 2004) demonstrated improved selectivity for methanol formation with the addition of fluorine in titanium catalyst supported on mesoporous silica. Fluorination of Ti-silica photocatalysts could induce the hydrophobicity of the photocatalyst surface, resulting in changes in product selectivity. Hydrophobic Ti-photocatalyst was found to exhibit higher selectivity for the production of longer chain hydrocarbons in the process. The hydrophobicity of the photocatalysts is in fact contributed by the mesoporous silica support instead of the TiO_2 particles. On the other hand, in a hydrophilic photocatalyst, water molecules can easily access to the tetrahedral coordinates of TiO_2, thus resulting in higher reactivity compared to photocatalysts with high hydrophobicity. It was reported that hydrophilic Ti-Beta(OH) zeolites showed higher selectivity towards the formation of methane, whereas hydrophobic Ti-Beta(F) zeolites demonstrated higher activity for the production of methanol (Yamashita et al. 1998, 2004; Ikeue et al. 2001a, b).

Overall, since the photocatalytic process is a chain reaction which involves a combination of radicals, electrons and ions, a huge variety of compounds can be formed. Examples of some possible compounds and the corresponding equations for their formation are shown as following (Usubharatana et al. 2006; Centi and Perathoner 2009):

$$\textbf{Carbon monoxide}: CO_2 + 2H^+ + 2e^- \rightarrow CO + H_2O \tag{3.7}$$

$$\textbf{Formic acid}: CO_2 + 2H^+ + 2e^- \rightarrow HCO_2H \tag{3.8}$$

$$\textbf{Formaldehyde}: CO_2 + 4H^+ + 4e^- \rightarrow HCHO + H_2O \tag{3.9}$$

$$\textbf{Methanol}: CO_2 + 6H^+ + 6e^- \rightarrow CH_3OH + H_2O \tag{3.10}$$

$$\textbf{Methane}: CO_2 + 8H^+ + 8e^- \rightarrow CH_4 + 2H_2O \tag{3.11}$$

3.4 Key Parameters in Enhancing Photocatalyst Efficiency of CO_2 Reduction

3.4.1 Influence of Reaction Temperature

It is well-known that the rate of photocatalytic reaction increases progressively at high temperatures due to an increase in diffusion rate and collision frequency to overcome the activation energy. The effect of reaction temperature for CO_2 photoreduction has been studied by several researchers (Tan et al. 2006a; Zhang et al. 2009). Anpo et al. (1995) investigated the influence of reaction temperature on the photocatalytic process of CO_2 and H_2O and found that the yields of reaction products such as CH_4, CH_3OH and CO were greater at 323 K than that at 275 K under the illumination of UV light.

Similar trend of results were reported in the studies of Slamet et al. (2005) and Zhang et al. (2009). Slamet et al. (2005) studied the temperature dependence of the CO_2 photoreduction ranging from 43 to 100 °C. They discovered that the yield of methanol after 6 h of light irradiation was higher at 100 °C than that at 43 °C. At low temperatures, the surface is covered with products and hence they do not easily desorb. On the other hand, the adsorption of reactants becomes more crucial owing to scarcely covered surface and also products desorbs readily at high temperatures. Furthermore, they indicated that the product desorption is always the rate determining step as observed from the activation energy values in a typical photocatalytic reaction.

In a recent research by Zhang et al. (2009), they claimed that as the photocatalytic temperature increased from 323 to 343 K, the CH_4 yield using Pt/TiO_2 catalysts was significantly enhanced. This is due to the fact that the increase in temperature favors the desorption process of the reaction products and also increases the frequency of effective collisions between the excited state of TiO_2 $[Ti^{3+}–O^{-}]^{*}$ and the adsorbed CO_2 molecule.

3.4.2 Influence of H_2O/CO_2 Molar Ratio

The overall chemical equation to represent the reduction of CO_2 to CH_4 is shown as follows:

$$CO_2 + 2H_2O \xrightarrow{hv} CH_4 + 2O_2 \tag{3.12}$$

From this stoichiometric equation, the amount of water required has to be twice as much as CO_2 in order to totally complete the CO_2 reduction process. Therefore, insufficient water in the photocatalytic system will cease the reduction reaction leading to remarkably low yield of products formed.

Undeniably, the molar ratio of H_2O/CO_2 plays an extremely important factor in the CO_2 photoconversion system. The amount of moisture in the CO_2 gas flowing through the catalysts is estimated based on the temperature of the water in the flask (Tan et al. 2006a). It is noteworthy that when the TiO_2 is irradiated by light, the TiO_2 surface becomes super hydrophilic whereby water molecules will cover most of the TiO_2 surface (Wu et al. 2005). With an ample water content, physisorbed H_2O on the surface of photocatalysts as well as abundant –OH groups possessed by anatase TiO_2 will further promote the production of hydroxyl radicals (•OH) and hydrogen ions (H^+) (Mori et al. 2012). Evidently, the yield of CH_4 was close to zero in the CO_2 reaction without significant water content but it is considerably enhanced with increasing H_2O content.

Zhang et al. (2009) examined the effect of H_2O/CO_2 molar ratio in the range between 0 and 1 and claimed that the CH_4 yield was gradually improved for the Pt/TiO_2 nanotube photocatalysts with increasing H_2O/CO_2 ratio. Wu et al. (2005) found out that when the H_2O/CO_2 ratio was less than 0.02, less water coverage causes the decline in the reaction rate. Therefore, absolute humidity around the photocatalysts in the heterogeneous system is an essential factor to determine if redox reactions are able to proceed photocatalytically at the solid-gas interface.

3.4.3 Influence of Metal Loading

There exists an optimum metal loading to be doped with TiO_2 particles. According to Slamet et al. (2005), an optimum amount of Cu loading of 3 wt% loaded on TiO_2 gave the highest yield of methanol which was about 3.3 times as compared to the conventional TiO_2 (Degussa P25). The yield of methanol was found to increase with Cu loading until it reached the optimum limit. This is because more Cu loading will act as the active sites for the CO_2 reduction process. Not only that, the amount of surface –OH groups on TiO_2 increased with increasing Cu loading due to the introduction of much more oxygen vacancies in the TiO_2 crystal lattice (Wu et al. 2005; Han et al. 2009).As a result, this causes a substantial enhancement of photocatalytic activity due to greater adsorption of the CO_2 onto the photocatalysts. Beyond the optimum Cu loading (3 wt%), the methanol yield was significantly reduced due to the masking and shading effects of Cu_2O clusters on the TiO_2 surface (Tseng et al. 2002). Consequently, excess Cu_2O clusters on the TiO_2 surface will cause less photon energy to be absorbed by the TiO_2 contributing to a significant decline in the photoexcitation of electron-hole pairs.

More importantly, Slamet et al. (2005) quoted that the activation energy for 3 wt% CuO/TiO_2 and Degussa P−25 was calculated to be 12 and 26 kJ/mol, respectively. Thus, with the apparent lower activation energy of 3 wt% CuO/TiO_2 catalyst, this implies that Cu serves as an active species for the CH_3OH products enhancing the CO_2 photocatalytic performance. A compromise between metal loading and the CO_2 photocatalytic efficiency is required to estimate the optimal metal dopant.

In addition to Cu doping, Kočí et al. (2010a) studied the effect of silver (Ag) doping on the TiO_2 powder which was prepared by the sol-gel process controlled in the reverse micellar environment. They claimed that Ag contents above 5 % were no longer randomly located in TiO_2 crystal but formed Ag clusters inside TiO_2 crystal instead. This causes the formation of Schottky barrier at the metal-semiconductor contact region leading to the decrease in recombination rate of electron-hole pairs. Consequently, this improves the overall photocatalytic activity. The highest yields for the CO_2 reduction were obtained by employing 7 % Ag/TiO_2.

3.4.4 Influence of Particle Size of Photocatalysts

Particle size of the photocatalysts also plays an influential role in affecting the photocatalytic efficiency of CO_2 reduction. Smaller particle size promotes larger surface area per unit volume of photocatalyst which can greatly enhance the photocatalytic performance. More importantly, when the particle size is reduced, the probability of electron-hole recombination decreases because the distance for the photoexcited electrons and holes to travel to the surface reaction sites becomes shorter (Leary and Westwood 2011). As a result, the lifetimes of the photogenerated electrons and holes are markedly enhanced leading to higher yields of reaction products. Koči et al. (2009) reported that higher yields of methane and methanol were obtained over the TiO_2 particles when the particle size was gradually reduced to the size of 14 nm.

However, ultra-fine particles (below the optimal particle size of 14 nm) can have undesired effects such as fast electron-hole recombination and rapid agglomeration of particles which in turn reduces the availability of active surface sites (Lin et al. 2006). On top of that, the band gap of the semiconductor becomes larger with decreasing particle size since the valence band energy is shifted to lower energy whereas the conduction band energy is moved to higher energy (Koči et al. 2009). Consequently, the photocatalytic reduction is not effective under the visible light spectrum inhibiting the red shift of the absorption spectra. On the other hand, the oxidizing and reduction power become stronger when there is an enlargement of the band gap of photocatalyst (Li et al. 2011). Overall, an optimal particle size of the semiconductor is to be achieved to maximize the photocatalytic performance.

3.4.5 *Influence of Illumination Time*

Effect of irradiation time for the CO_2 photoconversion was investigated by several researchers (Tseng et al. 2002; Li et al. 2010; Zhao et al. 2012). Li et al. (2010) reported that the production rate of CO and CH_4 increased with the irradiation time up to a maximum rate at 4 h. After 4 h, the production rate did not remain constant but subsequently the product yields were gradually reduced (Li et al. 2010). The decline in the production rate can be attributed to the deactivation of catalysts resulting from the saturation of the adsorption sites with intermediate products or the recombination of proton and the radicals with the deposited products on the surface of photocatalysts (Adachi et al. 1994). Similar trend of results have been reported by Sasirekha et al. (2006). However, some other studies revealed contradicting results by claiming that the production rate remained steady state after irradiation of light for more than 10 h. Tseng et al. (2002) showed that a steady state methanol yield of approximately 250 µmol/g-catalyst was obtained using 2 wt% Cu/TiO_2 under the illumination of UV light for 20 h. Very recently, Zhao et al. (2012) reported the synthesis of nitrogen-doped TiO_2 nanotubes via a hydrothermal method. They investigated the effect of irradiation duration for periods of 0–48 h under visible light irradiation and discovered that the yield of formic acid (3543.6 µmol/g-catalyst) remained constant after 12 h of light irradiation.

3.4.6 *Influence of Emission Power/Intensity of the Light Source*

Last but not least, the irradiating power emitted by the light source has a significant effect on the overall photocatalytic reaction. According to Tan et al. (2006a), they examined the total emission power of UVC lamps (4.8 W) and UVA lamps (3.0 W). They claimed that the product yields under the illumination of UVC lamps were higher than that of UVA lamps owing to greater photonic energy. At higher light incident intensity, excessive photons contribute to the generation of more reactive species (Laoufi et al. 2008) and hence more easily to degrade CO_2 to valuable fuels. Similar results have been published by Wu et al. (2005). They reported that photoactivity would be directly proportional to the light intensity. For 2 wt% Cu/TiO_2, the yield of methanol under 16 W/cm^2 was found to be higher ($\approx$0.42 μmol/g-catalyst) as compared to the intensity of 1 W/cm^2 ($\approx$0.12 μmol/g-catalyst).

Besides utilizing high intensities of light, the increase in CO_2 photoactivity notably depends on the absorption capacity of the semiconductor. Overall, if both emission power by the light source and absorption capacity by the photocatalysts are large, the product yields will be remarkably enhanced.

3.5 Concluding Remarks

Conversion of CO_2 into useful hydrocarbon products can be a potential solution to control the CO_2 emission into the environment. CO_2 is a highly chemical stable molecule that requires high activation energy to break the strong C–O bonding. Therefore, the conventional approach for the conversion of CO_2 mainly involves the hydrogenation of CO_2 such as reverse-water-gas-shift reaction or Fischer-Tropsch reaction, which is conducted at high temperatures to ensure sufficient energy supply for the reaction process to take place.

The photocatalytic reduction of CO_2 utilizes light source as an energy resource to fulfil the high activation energy requirement to break the strong C–O chemical bonding in the CO_2 molecule. The energy also renders the photoreduction of CO_2 possible with reasonable conversions and product yield. Derivation of hydrocarbon products from the photocatalytic reduction of CO_2 can be more viable than the conventional approaches in view of lower energy consumption. In this work, the important parameters that control the photocatalytic reduction of CO_2 have been vividly discussed. The types of photocatalyst, reaction configuration, and reducing agent used for the reaction process are concluded as important parameters that can control the photocatalytic reduction of CO_2. Nevertheless, the types of hydrocarbon derived from the photoreduction reaction are highly dependent on any of these three parameters. The proper selection of parameters must be done in order to achieve the desired hydrocarbon products.

Undeniably, visible light is the utmost desired light source for the photocatalytic reduction of CO_2 seeing as it is abundantly available for free from natural sunlight. However, the energy level that can be supplied by visible light source is much lower than UV light. Therefore, a highly photoreactive catalyst is required to achieve reasonable conversions if a visible light source is to be employed. Effort has to be focused on broadening and extending the spectrum to visible light range. Currently, most of the photocatalysts being used for visible light activated photoreduction of

CO_2 are mainly derived from ruthenium Ru(II) and rhenium Re(I) metal complexes or metal oxides added with dye that act as photosensitizers to absorb visible light. In addition, the absorption of visible light can also be achieved by doping semiconductor (TiO_2) with other elements such as metals and non-metals to reduce the overall band gap energy.

To date, photocatalytic reduction of CO_2 is still in progress by leaps and bounds and a number of questions remain to be solved by researchers: enhancement of photocatalysts by utilizing visible light instead of UV light and also improvement in the photocatalytic efficiency by considering all reaction parameters (temperature, H_2O/CO_2 molar ratio, illumination duration and light intensity) in order to commercialize this photocatalysis process for large scale applications.

Acknowledgement The authors would like to thank the Ministry of Higher Education, Malaysia for the Long-term Research Grant Scheme (account number: 2110226-113-00) for the financial support given and Chemical and Sustainable Process Engineering (CSPE) research group, Monash University Sunway Campus, for the equipment and facilities support.

References

Abou Asi M, He C, Su M, Xia D, Lin L, Deng H, Xiong Y, Qiu R, Li XZ (2011) Photocatalytic reduction of CO_2 to hydrocarbons using $AgBr/TiO_2$ nanocomposites under visible light. Catal Today 175(1):256–263

Adachi K, Ohta K, Mizuno T (1994) Photocatalytic reduction of carbon dioxide to hydrocarbon using copper-loaded titanium dioxide. Sol Energy 53(2):187–190

Ahmed N, Shibata Y, Taniguchi T, Izumi Y (2011) Photocatalytic conversion of carbon dioxide into methanol using zinc–copper–M(III) (M = aluminum, gallium) layered double hydroxides. J Catal 279(1):123–135

Anpo M, Yamashita H, Ichihashi Y, Ehara S (1995) Photocatalytic reduction of CO_2 with H_2O on various titanium oxide catalysts. J Electroanal Chem 396(1–2):21–26

Centi G, Perathoner S (2004). Heterogeneous catalytic reactions with CO_2: Status and perspectives. In: Sang-Eon Park J-SC, Kyu-Wan L (eds.) Stud Surf Sci Catal. Elsevier, Amsterdam, The Netherlands

Centi G, Perathoner S (2009) Opportunities and prospects in the chemical recycling of carbon dioxide to fuels. Catal Today 148(3–4):191–205

Centi G, Perathoner S (2010) Towards solar fuels from water and CO_2. ChemSusChem 3(2):195–208

Centi G, Santen RA (2007). Catalysis for renewables: from feedstock to energy production. Wiley-VCH Verlag GmbH & Co. KGaA, Weinheim, Germany

Chen L, Graham ME, Li G, Gentner DR, Dimitrijevic NM, Gray KA (2009) Photoreduction of CO_2 by TiO_2 nanocomposites synthesized through reactive direct current magnetron sputter deposition. Thin Solid Films 517(19):5641–5645

Cheung K-C, Guo P, So M-H, Lee LYS, Ho K-P, Wong W-L, Lee K-H, Wong W-T, Zhou Z-Y, Wong K-Y (2009) Electrocatalytic reduction of carbon dioxide by a polymeric film of rhenium tricarbonyl dipyridylamine. J Organomet Chem 694(17):2842–2845

Chunshan S (2006) Global challenges and strategies for control, conversion and utilization of CO_2 for sustainable development involving energy, catalysis, adsorption and chemical processing. Catal Today 115(1–4):2–32

Chunshan S, Pan W (2004) Tri-reforming of methane: a novel concept for synthesis of industrially useful synthesis gas with desired H_2/CO ratios using CO_2 in flue gas of power plants without CO_2 separation. Prepr Pap—Am Chem Soc. Div Fuel Chem 49:128–131

Davison J, Thambimuthu K (2005). Technologies for capture of carbon dioxide. In: Rubin ES, Keith DW, Gilboy CF, Wilson M, Morris T, Gale J, K. Thambimuthu ESR, D. W. Keith C. F. Gilboy M. Wilson T. Morris J. Gale (eds.) Greenhouse Gas Control Technologies 7. Elsevier Science Ltd, Oxford

Dey GR, Belapurkar AD, Kishore K (2004) Photo-catalytic reduction of carbon dioxide to methane using TiO2 as suspension in water. J Photochem Photobiol A 163(3):503–508

Feng X, Sloppy JD, Latempa TJ, Paulose M, Komarneni S, Bao N, Grimes CA (2011) Synthesis and deposition of ultrafine Pt nanoparticles within high aspect ratio TiO_2 nanotube arrays: application to the photocatalytic reduction of carbon dioxide. J Mater Chem 21(35):13429–13433

Fu Y, Sun D, Chen Y, Huang R, Ding Z, Fu X, Li Z (2012) An amine-functionalized titanium metal–organic framework photocatalyst with visible-light-induced activity for CO_2 reduction. Angew Chem Int Ed 51(14):3364–3367

Gholamkhass B, Mametsuka H, Koike K, Tanabe T, Furue M, Ishitani O (2005) Architecture of supramolecular metal complexes for photocatalytic CO_2 reduction: ruthenium–rhenium bi- and tetranuclear complexes. Inorg Chem 44(7):2326–2336

Grodkowski J, Neta P (2000) Cobalt corrin catalyzed photoreduction of CO_2. J Phys Chem A 104(9):1848–1853

Grodkowski J, Neta P, Fujita E, Mahammed A, Simkhovich L, Gross Z (2002) Reduction of cobalt and iron corroles and catalyzed reduction of CO_2. J Phys Chem A 106(18):4772–4778

Guan G, Kida T, Harada T, Isayama M, Yoshida A (2003a) Photoreduction of carbon dioxide with water over $K_2Ti_6O_{13}$ photocatalyst combined with Cu/ZnO catalyst under concentrated sunlight. Appl Catal A 249(1):11–18

Guan G, Kida T, Yoshida A (2003b) Reduction of carbon dioxide with water under concentrated sunlight using photocatalyst combined with Fe-based catalyst. Appl Catal B 41(4):387–396

Han F, Kambala VSR, Srinivasan M, Rajarathnam D, Naidu R (2009) Tailored titanium dioxide photocatalysts for the degradation of organic dyes in wastewater treatment: a review. Appl Catal A 359(1–2):25–40

Hirose T, Maeno Y, Himeda Y (2003) Photocatalytic carbon dioxide photoreduction by Co(bpy) 32 + sensitized by Ru(bpy) 32 + fixed to cation exchange polymer. J Mol Catal A: Chem 193:27–32

Hori H, Koike K, Suzuki Y, Ishizuka M, Tanaka J, Takeuchi K, Sasaki Y (2002) High-pressure photocatalytic reduction of carbon dioxide using [fac-Re(bpy)$(CO)_3$P$(OiPr)_3$] + (bpy = 2,2′-bipyridine). J Mol Catal A: Chem 179(1–2):1–9

Hurum DC, Agrios AG, Gray KA, Rajh T, Thurnauer MC (2003) Explaining the enhanced photocatalytic activity of Degussa P25 mixed-phase TiO_2 using EPR. J Phys Chem B 107(19):4545–4549

Hussain ST, Khan K, Hussain R (2009) Size control synthesis of sulfur doped titanium dioxide (anatase) nanoparticles, its optical property and its photo catalytic reactivity for CO_2 + H_2O conversion and phenol degradation. J Nat Gas Chem 18(4):383–391

Iea (2011) *CO_2 emissions from fuel combustion highlights* [Online]. http://www.iea.org/co2highlights/CO2highlights.pdf. Accessed 18 July 2012

Ikeue K, Nozaki S, Ogawa M, Anpo M (2002) Characterization of self-standing Ti-containing porous silica thin films and their reactivity for the photocatalytic reduction of CO_2 with H_2O. Catal Today 74(3–4):241–248

Ikeue K, Yamashita H, Anpo M, Takewaki T (2001a) Photocatalytic reduction of CO_2 with H_2O on Ti–β zeolite photocatalysts: effect of the hydrophobic and hydrophilic properties. J Phys Chem B 105(35):8350–8355

Ikeue K, Yamashita H, Takewaki T, Davis ME, Anpo M (2001b) Characterization of Ti-Beta zeolites and their reactivity for the photocatalytic reduction of CO_2 with H_2O. J Synchorotron Radiat 8(2):602–604

Jansen A, Pitter S (2004) Homogeneously catalysed reduction of carbon dioxide with silanes: a study on solvent and ligand effects and catalyst recycling. J Mol Catal A: Chem 217(1–2):41–45

Jia L, Li J, Fang W (2009) Enhanced visible-light active C and Fe co-doped $LaCoO_3$ for reduction of carbon dioxide. Catal Comm 11(2):87–90

Jiang Z, Xiao T, Kuznetsov VL, Edwards PP (2010) Turning carbon dioxide into fuel. Philos Trans R Soc London, Ser A 368(1923):3343–3364

Kaneco S, Kurimoto H, Shimizu Y, Ohta K, Mizuno T (1999) Photocatalytic reduction of CO_2 using TiO_2 powders in supercritical fluid CO_2. Energy 24(1):21–30

Kočí K, Obalová L, Matějová L, Plachá D, Lacný Z, Jirkovský J, Šolcová O (2009) Effect of TiO_2 particle size on the photocatalytic reduction of CO_2. Appl Catal B 89(3–4):494–502

Kočí K, Matějů K, Obalová L, Krejčíková S, Lacný Z, Plachá D, Čapek L, Hospodková A, Šolcová O (2010) Effect of silver doping on the TiO_2 for photocatalytic reduction of CO_2. Appl Catal B 96(3–4):239–244

Kočí K, Reli M, Kozák O, Lacný Z, Plachá D, Praus P, Obalová L (2011a) Influence of reactor geometry on the yield of CO_2 photocatalytic reduction. Catal Today 176(1):212–214

Kočí K, Zatloukalová K, Obalová L, Krejčíková S, Lacný Z, Čapek L, Hospodková A, Šolcová O (2011b) Wavelength effect on photocatalytic reduction of CO_2 by Ag/TiO_2 catalyst. Chin J Chem Eng 32(5):812–815

Kohno Y, Tanaka T, Funabiki T, Yoshida S (2000) Photoreduction of CO_2 with H_2 over ZrO_2. A study on interaction of hydrogen with photoexcited CO_2. PCCP 2(11):2635–2639

Laitar DS, Müller P, Sadighi JP (2005) Efficient homogeneous catalysis in the reduction of CO_2 to CO. J Am Chem Soc 127(49):17196–17197

Laoufi NA, Tassalit D, Bentahar F (2008) The degradation of phenol in water solution by TiO_2 photocatalysis in a helical reactor. Global NEST J 10:404–418

Leary R, Westwood, A (2011) Carbonaceous nanomaterials for the enhancement of TiO2 photocatalysis. Carbon 49(3):741–772

Li H, Lei Y, Huang Y, Fang Y, Xu Y, Zhu L, Li X (2011) Photocatalytic reduction of carbon dioxide to methanol by Cu_2O/SiC nanocrystallite under visible light irradiation. J Nat Gas Chem 20(2):145–150

Li Y, Wang W-N, Zhan Z, Woo M-H, Wu C-Y, Biswas P (2010) Photocatalytic reduction of CO_2 with H_2O on mesoporous silica supported Cu/TiO_2 catalysts. Appl Catal B 100(1–2):386–392

Liang YT, Vijayan BK, Gray KA, Hersam MC (2011) Minimizing graphene defects enhances titania nanocomposite-based photocatalytic reduction of CO_2 for improved solar fuel production. Nano Lett 11(7):2865–2870

Lin H, Huang CP, Li W, Ni C, Shah SI, Tseng Y-H (2006) Size dependency of nanocrystalline TiO_2 on its optical property and photocatalytic reactivity exemplified by 2-chlorophenol. Appl Catal B 68(1–2):1–11

Liu Y, Huang B, Dai Y, Zhang X, Qin X, Jiang M, Whangbo M-H (2009) Selective ethanol formation from photocatalytic reduction of carbon dioxide in water with $BiVO_4$ photocatalyst. Catal Comm 11(3):210–213

Lo CC, Hung CH, Yuan CS, Wu JF (2007) Photoreduction of carbon dioxide with H_2 and H_2O over TiO_2 and ZrO_2 in a circulated photocatalytic reactor. Sol Energy Mater Sol Cells 91(19):1765–1774

Luo D, Bi Y, Kan W, Zhang N, Hong S (2011) Copper and cerium co-doped titanium dioxide on catalytic photo reduction of carbon dioxide with water: Experimental and theoretical studies. J Mol Struct 994(1–3):325–331

Ma J, Sun N, Zhang X, Zhao N, Xiao F, Wei W, Sun Y (2009) A short review of catalysis for CO_2 conversion. Catal Today 148(3–4):221–231

Macfarlane JW, Scott TB (2012) Reduction of carbon dioxide on jet spray formed titanium dioxide surfaces. J Hazard Mater 211–212:247–254

Mei B, Wiktor C, Turner S, Pougin A, Tendeloo G, Fisher RA, Muhler M, Strunk J (2013) Evidence for metal–support interactions in Au modified TiOx/SBA-15 materials prepared by photodeposition. ACS Catal 3(12):3041–3049

Morris AJ, Meyer GJ, Fujita E (2009) Molecular approaches to the photocatalytic reduction of carbon dioxide for solar fuels. Acc Chem Res 42(12):1983–1994

Nguyen T-V, Wu JCS (2008a) Photoreduction of CO_2 in an optical-fiber photoreactor: Effects of metals addition and catalyst carrier. Appl Catal A 335(1):112–120

Nguyen T-V, Wu JCS (2008b) Photoreduction of CO_2 to fuels under sunlight using optical-fiber reactor. Sol Energy Mater Sol Cells 92(8):864–872

Nguyen T-V, Wu JCS, Chiou C-H (2008) Photoreduction of CO_2 over Ruthenium dye-sensitized TiO_2-based catalysts under concentrated natural sunlight. Catal Comm 9(10):2073–2076

Olah GA, Goeppert A, Prakash GKS (2008) Chemical recycling of carbon dioxide to methanol and dimethyl ether: From greenhouse gas to renewable, environmentally carbon neutral fuels and synthetic hydrocarbons. J Org Chem 74(2):487–498

Olah GA, Prakash GKS, Goeppert A (2011) Anthropogenic chemical carbon cycle for a sustainable future. J Am Chem Soc 133(33):12881–12898

Ozcan O, Yukruk F, Akkaya E, Uner D (2007) Dye sensitized CO_2 reduction over pure and platinized TiO_2. Top Catal 44(4):523–528

Pan P-W, Chen Y-W (2007) Photocatalytic reduction of carbon dioxide on NiO/$InTaO_4$ under visible light irradiation. Catal Comm 8(10):1546–1549

Pathak P, Meziani MJ, Li Y, Cureton LT, Sun Y-P (2004) Improving photoreduction of CO_2 with homogeneously dispersed nanoscale TiO_2 catalysts. Chem Commun 10:1234–1235

Qin S, Xin F, Liu Y, Yin X, Ma W (2011) Photocatalytic reduction of CO_2 in methanol to methyl formate over CuO–TiO_2 composite catalysts. J Colloid Interface Sci 356(1):257–261

Romero-Gomez P, Borras A, Barranco A, Espinos JP, Gonzalez-Elipe AR (2011) Enhanced photoactivity in bilayer films with buried rutile–anatase heterojunctions. ChemPhysChem 12(1):191–196

Sasirekha N, Basha SJS, Shanthi K (2006) Photocatalytic performance of Ru doped anatase mounted on silica for reduction of carbon dioxide. Appl Catal B 62(1–2):169–180

Sato S, Koike K, Inoue H, Ishitani O (2007) Highly efficient supramolecular photocatalysts for CO_2 reduction using visible light. Photochem Photobiol 6(4):454–461

Sato S, Morikawa T, Saeki S, Kajino T, Motohiro T (2010) Visible-light-induced selective CO_2 reduction utilizing a ruthenium complex electrocatalyst linked to a p-type nitrogen-doped Ta_2O_5 semiconductor. Angew Chem Int Ed 49(30):5101–5105

Schaub G, Unruh D, Rohde M (2004). Synthetic hydrocarbon fuels and CO_2 utilization. In: Sang-Eon Park J-SC, Kyu-Wan L (eds.) Stud Surf Sci Catal. Elsevier, Amsterdam, The Netherlands

Shi D, Feng Y, Zhong S (2004) Photocatalytic conversion of CH_4 and CO_2 to oxygenated compounds over Cu/CdS–TiO_2/SiO_2 catalyst. Catal Today 98(4):505–509

Shioya Y, Ikeue K, Ogawa M, Anpo M (2003) Synthesis of transparent Ti-containing mesoporous silica thin film materials and their unique photocatalytic activity for the reduction of CO_2 with H_2O. Appl Catal A 254(2):251–259

Slamet, Nasution HW, Purnama E, Kosela S, Gunlazuardi J (2005) Photocatalytic reduction of CO_2 on copper-doped Titania catalysts prepared by improved-impregnation method. Catal Comm 6(5):313–319

Srinivas B, Shubhamangala B, Lalitha K, Reddy PaK, Kumari VD, Subrahmanyam M, Ranjan B (2011) Photocatalytic reduction of CO_2 over Cu-TiO_2/molecular sieve 5A composite. Photochem Photobiol 87:995–1001

Subrahmanyam M, Kaneco S, Alonso-Vante N (1999) A screening for the photo reduction of carbon dioxide supported on metal oxide catalysts for C1–C3 selectivity. Appl Catal B 23(2–3):169–174

Takeda H, Koike K, Inoue H, Ishitani O (2008) Development of an efficient photocatalytic system for CO_2 reduction using rhenium(I) complexes based on mechanistic studies. J Am Chem Soc 130(6):2023–2031

Tan SS, Zou L, Hu E (2006a) Photocatalytic production of methane and hydrogen through reduction of carbon dioxide with water using titania pellets. Int J Green Energy 3(3):283–290

Tan SS, Zou L, Hu E (2006b) Photocatalytic reduction of carbon dioxide into gaseous hydrocarbon using TiO_2 pellets. Catal Today 115(1–4):269–273

Tan SS, Zou L, Hu E (2007) Photosynthesis of hydrogen and methane as key components for clean energy system. Sci Tech Adv Mater 8(1–2):89

Teramura K, Tsuneoka H, Shishido T, Tanaka T (2008) Effect of H_2 gas as a reductant on photoreduction of CO_2 over a Ga_2O_3 photocatalyst. Chem Phys Lett 467(1–3):191–194

Teramura K, Okuoka S-I, Tsuneoka H, Shishido T, Tanaka T (2010) Photocatalytic reduction of CO_2 using H_2 as reductant over $ATaO_3$ photocatalysts (A = Li, Na, K). Appl Catal B 96(3–4):565–568

Truong QD, Liu J-Y, Chung C-C, Ling Y-C (2012) Photocatalytic reduction of CO_2 on $FeTiO_3$/TiO_2 photocatalyst. Catal Comm 19:85–89

Tseng IH, Chang W-C, Wu JCS (2002) Photoreduction of CO_2 using sol–gel derived titania and titania-supported copper catalysts. Appl Catal B 37(1):37–48

Tseng IH, Wu JCS, Chou H-Y (2004) Effects of sol–gel procedures on the photocatalysis of Cu/TiO_2 in CO_2 photoreduction. J Catal 221(2):432–440

Tsuneoka H, Teramura K, Shishido T, Tanaka T (2010) Adsorbed Species of CO_2 and H_2 on Ga_2O_3 for the Photocatalytic Reduction of CO_2. J Phys Chem C 114(19):8892–8898

Ulagappan N, Frei H (2000) Mechanistic study of CO_2 photoreduction in Ti silicalite molecular sieve by FT-IR spectroscopy. J Phys Chem A 104(33):7834–7839

Uner D, Oymak MM (2012) On the mechanism of photocatalytic CO_2 reduction with water in the gas phase. Catal Today 181(1):82–88

Usubharatana P, Mcmartin D, Veawab A, Tontiwachwuthikul P (2006) Photocatalytic process for CO_2 emission reduction from industrial flue gas streams. Ind Eng Chem Res 45(8):2558–2568

Varghese OK, Paulose M, Latempa TJ, Grimes CA (2009) High-rate solar photocatalytic conversion of CO_2 and water vapor to hydrocarbon fuels. Nano Lett 9(2):731–737

Wang C, Thompson RL, Baltrus J, Matranga C (2010a) Visible light photoreduction of CO_2 using CdSe/Pt/TiO_2 heterostructured catalysts. J Phys Chem Lett 1(1):48–53

Wang X, He Z, Zhong S, Xiao X (2007) Photocatalytic synthesis of hydrocarbon oxygenates from C_2H_6 and CO_2 over Pd-MoO_3/SiO_2 catalyst. J Nat Gas Chem 16(2):173–178

Wang Z-Y, Chou H-C, Wu JCS, Tsai DP, Mul G (2010b) CO_2 photoreduction using NiO/InTaO4 in optical-fiber reactor for renewable energy. Appl Catal A 380(1–2):172–177

Woolerton TW, Sheard S, Reisner E, Pierce E, Ragsdale SW, Armstrong FA (2010) Efficient and clean photoreduction of CO_2 to CO by enzyme-modified TiO_2 nanoparticles using visible light. J Am Chem Soc 132(7):2132–2133

Wu J (2009) Photocatalytic reduction of greenhouse gas CO_2 to fuel. Catal Surv Asia 13(1):30–40

Wu JCS, Lin H-M (2005) Photo reduction of CO_2 to methanol via TiO_2 photocatalyst. Int J Photoenergy 7(3):115–119

Wu JCS, Lin H-M, Lai C-L (2005) Photo reduction of CO_2 to methanol using optical-fiber photoreactor. Appl Catal A 296(2):194–200

Xia X-H, Jia Z-J, Yu Y, Liang Y, Wang Z, Ma L-L (2007) Preparation of multi-walled carbon nanotube supported TiO_2 and its photocatalytic activity in the reduction of CO_2 with H_2O. Carbon 45(4):717–721

Yamashita H, Fujii Y, Ichihashi Y, Zhang SG, Ikeue K, Park DR, Koyano K, Tatsumi T, Anpo M (1998) Selective formation of CH_3OH in the photocatalytic reduction of CO_2 with H_2O on titanium oxides highly dispersed within zeolites and mesoporous molecular sieves. Catal Today 45(1–4):221–227

Yamashita H, Ikeue K, Takewaki T, Anpo M (2002) In situ XAFS studies on the effects of the hydrophobic-hydrophilic properties of Ti-Beta zeolites in the photocatalytic reduction of CO_2 with H_2O. Top Catal 18(1–2):95–100

Yamashita H, Okazaki M, Ikeue K, Anpo M (2004). Photocatalytic reduction of CO_2 with H_2O on Ti-containing mesoporous silica hydrophobically modified using fluoride ions. In: Sang-Eon Park J-SC, Kyu-Wan L (eds.) Stud Surf Sci Catal. Elsevier, Amsterdam, The Netherlands

Yan SC, Ouyang SX, Gao J, Yang M, Feng JY, Fan XX, Wan LJ, Li ZS, Ye JH, Zhou Y, Zou ZG (2010) A room-temperature reactive-template route to mesoporous $ZnGa_2O_4$ with improved photocatalytic activity in reduction of CO_2. Angew Chem 122(36):6544–6548

Yang C-C, Yu Y-H, Van Der Linden B, Wu JCS, Mul G (2010) Artificial photosynthesis over crystalline TiO_2-based catalysts: Fact or fiction? J Am Chem Soc 132(24):8398–8406

Yang H-C, Lin H-Y, Chien Y-S, Wu J, Wu H-H (2009) Mesoporous TiO_2/SBA-15, and Cu/TiO_2/SBA-15 composite photocatalysts for photoreduction of CO_2 to methanol. Catal Lett 131(3):381–387

Yang X, Xiao T, Edwards PP (2011) The use of products from CO_2 photoreduction for improvement of hydrogen evolution in water splitting. Int J Hydrogen Energy 36(11):6546–6552

Ying-Min YU, Jin-Hua FEI, Yi-Ping Z, Xiao-Ming Z (2006) Carbon dioxide hydrogenation catalyzed by ruthenium complexes immobilized on MCM-41. J Fuel Chem Tech 34(6):700–705

Yoshida H, Maeda K (2010). Preparation of gallium oxide photocatalysts for reduction of carbon dioxide. In: E.M. Gaigneaux MD, S. Hermans, P. A. Jacobs, J. A. Martens, P. Ruiz (eds.) Stud Surf Sci Catal. Elsevier, Amsterdam, The Netherlands

Zhang Q-H, Han W-D, Hong Y-J, Yu J-G (2009) Photocatalytic reduction of CO_2 with H_2O on Pt-loaded TiO_2 catalyst. Catal Today 148(3–4):335–340

Zhang Q, Li Y, Ackerman EA, Gajdardziska-Josifovska M, Li H (2011) Visible light responsive iodine-doped TiO_2 for photocatalytic reduction of CO_2 to fuels. Appl Catal A 400(1–2):195–202

Zhao Z-H, Fan J-M, Wang Z-Z (2007) Photo-catalytic CO_2 reduction using sol–gel derived titania-supported zinc-phthalocyanine. J Cleaner Prod 15(18):1894–1897

Zhao Z, Fan J, Liu S, Wang Z (2009a) Optimal design and preparation of titania-supported CoPc using sol–gel for the photo-reduction of CO_2. Chem Eng J 151(1–3):134–140

Zhao Z, Fan J, Xie M, Wang Z (2009b) Photo-catalytic reduction of carbon dioxide with in-situ synthesized CoPc/TiO_2 under visible light irradiation. J Cleaner Prod 17(11):1025–1029

Zhao Z, Fan J, Wang J, Li R (2012) Effect of heating temperature on photocatalytic reduction of CO_2 by N–TiO_2 nanotube catalyst. Catal Comm 21:32–37

Chapter 4
Fluoride in Drinking Water: Health Effects and Remediation

Meththika Vithanage and Prosun Bhattacharya

Contents

Abstract Fluoride at low concentration is an essential element for dental health. However groundwater in many countries has exceeding concentrations of fluoride, which poses a health threat to millions of people around the world. It has been estimated that more than 200 million people from among 25 nations are suffering from fluorosis due to the consumption of fluoride-rich drinking water. Fluoride contamination is mostly geogenic however, in some cases anthropogenic industrial inputs may cause a threat. Many techniques have been developed for defluoridation. However a solution is still to be found especially for the household and community supply. This

M. Vithanage (✉)
Chemical and Environmental Systems Modeling Research Group, Institute of Fundamental Studies, Hantana Road, Kandy 20000, Sri Lanka
e-mail: meththikavithanage@gmail.com

P. Bhattacharya
KTH-International Groundwater Arsenic Research Group, Department of Land and Water Resources Engineering, Royal Institute of Technology, SE-100 44, Stockholm, Sweden

© Springer International Publishing Switzerland 2015
E. Lichtfouse et al. (eds.), *CO_2 Sequestration, Biofuels and Depollution*,
Environmental Chemistry for a Sustainable World 5, DOI 10.1007/978-3-319-11906-9_4

article presents a global overview of the distribution of fluoride rich groundwaters and defluoridation techniques. Focus is given on efficient and cost-effective defluoridation for rural communities for both household filtration units and community water supplies. Rocks containing fluoride-rich minerals are the main sources of high fluoride in groundwater especially in the developing world. Other sources are volcanic and industrial. Gneisses and granites sometimes contain very high concentrations of fluoride reaching more than 5000 mg/kg. Weathering and release of fluorides in rocks, secondary minerals and soils are ruled by alkalinity, specific conductivity and evaporation. Exposure to high fluoride levels in drinking water cause endemic dental and skeletal fluorosis, and crippling. Those diseases are common in India, Sri Lanka, Pakistan, Ethiopia, Ghana, Kenya and China. The most common defluoridation techniques are coagulation and adsorption, although those methods are not the best for fluoride removal. Membrane processes are quite efficient but not economical for the developing communities. Eletrocoagulation may be a solution but requires high energy and, as a consequence, is less available to poor communities. To conclude, most defluoridation techniques are not achieving social acceptance and sucessful implementation. Therefore, a better technology is lacking for fluoride removal.

Keywords Fluorosis · Groundwater · Adsorption · Ion exchange · Coagulation · Co-precipitation

4.1 Introduction

Arsenic and fluoride are the two major elements that are in interest in the recent world whereas recognized as the most serious inorganic contaminants in drinking water on a global basis. Fluoride rich drinking water is a widespread problem which can be seen all over the world. Fluorosis is endemic in at least 25 countries around the world, and is most prevalent in India, China, and parts of Africa. It is not known how many people are currently afflicted with the disease, but conservative estimates are in the tens of millions of people (WHO 2004). Fluorine, is the most electronegative and reactive element in the periodic table, occurring primarily as the fluoride ion (F^-). In the natural environment fluoride occurs as the fluoride ion, F^-. Fluorine is an ubiquitous element in the environment. Most environmental fluoride problems are due to the fluoride mobilization in natural conditions. Fluorine, the element of fluoride, associates with many mineral deposits containing fluoride bearing minerals and weathering, dissolution and other pedogenic processes can release fluoride into groundwater. Even though fluoride is considered as an essential element for human health, especially for the strengthening of tooth enamel, excessive doses can be harmful. While fluoride is present in air, water, and food, the most common way it enters the food chain is via drinking water (Fawell et al. 2006).

According to the World Health Organization (WHO), at concentrations above 1.5 mg/L, fluoride is considered as dangerous to human health. Excessive fluoride can lead to dental and skeletal fluorosis (Fig. 4.1), a disease that can cause mottling of the teeth, calcification of ligaments (Fawell et al. 2006; Kowalski 1999). Long

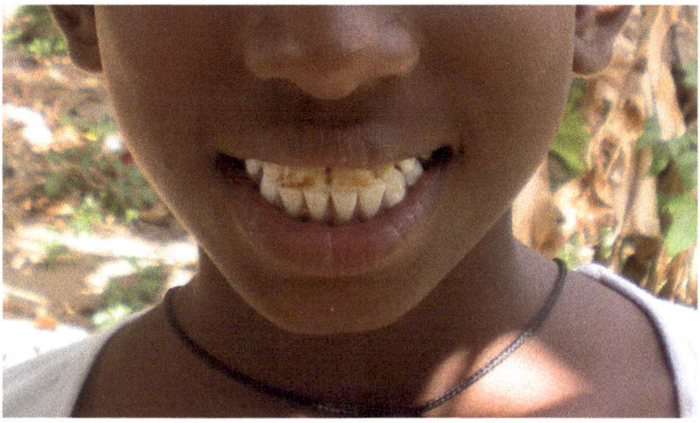

Fig. 4.1 Dental fluorosis affected child in Anuradhapura, Sri Lanka

term ingestion of fluoride rich drinking water may show the way to crippling bone deformities, cancer (Kowalski 1999; Yiamouyiannis 1993) decreased cognitive ability, lower Intelligence Quotient, and developmental issues in children (Li et al. 1994). The populations in tropical belt are having close contacts with their surrounding environment and thereby the geochemical anomalies play a role in the people's health (Dissanayake and Chandrajith 2009). One argument is that the fluoride toxicity increases with the amount of drinking water consumed especially in the humid tropics, thus the maximum permissible fluoride concentration in drinking water has to be lowered from the existing WHO limit (Dissanayake 2005; WHO 1996, 2004).

Although arsenic in water have received an enormous attention in wider aspects and have been well documented in reviews (Bhattacharya et al. 2002; Bundschuh et al. 2004; Kar et al. 2010; Mandal and Suzuki 2002; Mukherjee et al. 2009; Stollenwerk 2002; Wang and Mulligan 2008) however, the attention on fluoride mainly is on its removal (Ayoob et al. 2008b; Mohapatra et al. 2009; Veressinina et al. 2001). Some baseline studies reported fluoride as a groundwater contaminant in tropical world (Dissanayake 1991; Jacks et al. 1993). Once, Ayoob and Gupta (2006) paid their attention on reviewing the sources, geochemistry, distribution and health impacts and have reported in detail. The presence of fluoride in water does not impart any color, odor or taste. Hence, it acts as an invisible poison such as arsenic in groundwater. Unless otherwise tested, one cannot reveal the high concentrations of fluoride in their waters. Records have increased with time on fluoride rich drinking waters similar to that of arsenic. This review therefore focuses on the sources, distribution and treatment methods of fluoride available in literature. Furthermore, we discuss the problems of the processes involved in the removal methods which have been proposed as well as future perspectives of fluoride research.

4.2 Sources of Fluoride

The possible sources of fluoride in the environment are schematically shown in Fig. 4.2. The largest fluoride reserve is considered as the natural input from the rocks and minerals containing fluoride in there composition.

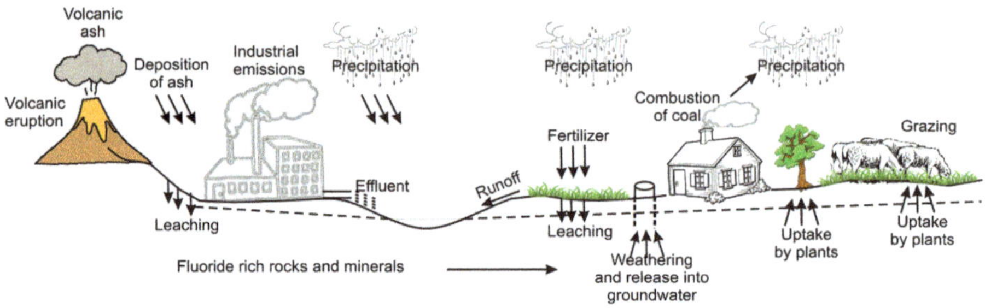

Fig. 4.2 The schematic diagram showing the fluoride existing sources in the environment

4.2.1 *Rocks and Minerals*

Fluoride is one of the most abundant trace elements in the Earth's crust, with an average concentration of 625 mg/kg in different rock types (Edmunds and Smedley 2005; Tavener and Clark 2006). The largest fluoride reserve is the rocks containing fluorine rich minerals (WHO 1984). Fluoride concentrations of rocks typically range from 100 mg/kg in ultramafic rocks and some limestones to 1000 mg/kg in alkaline igneous rocks and 1300 mg/kg in marine shales (Faure 1991; Hem 1985; IGRAC 2003b), and in some cases values as high as 2000 mg/kg in volcanic rocks produced at a subduction zone boundary (Anazawa 2006). Table 4.1 provides data for fluorine in fresh rocks, weathered material and soils from Coimbatore district in Tamil Nadu, India.

The highest fluoride levels are associated with syenites, granites, quartz monzonites, granodiorites, felsic and biotite gneisses, and alkaline volcanics (Apambire et al. 1997b; Chae et al. 2006a, 2007; Dissanayake 1991; Handa 1975; Hyndman 1985; Jones et al. 1977; Moore 2004; Nanyaro et al. 1984a; Ozsvath 2006; Robinson and Kapo 2003; Rosi et al. 2003; Stormer and Carmichael 1970; Taylor and Fallick 1997). Although these rock types can contain a variety of fluoride-rich accessory minerals, laboratory experiments and field studies have shown that the presence of biotite alone is sufficient to produce dissolved fluoride concentrations above 4 mg/L (Chae et al. 2006a, b, 2007).

Among all fluoride rich minerals fluorite (CaF_2), fluoroapatite ($Ca_5(PO_4)_3F$), micas, amphiboles, cryolite (Na_3AlF_6), villiaumite (NaF) and topaz ($Al_2(SiO_4)F_2$) are considered as the most abundant mineral which occurs in most rocks and sediments (Apambire et al. 1997b; Chae et al. 2007; Cronin et al. 2000; Edmunds and Smedley 2005; Handa 1975; Hem 1985; Saxena and Ahmed 2003). The granitic rocks containing fluoride minerals such as amphibolites, pegmatites, hornblende, muscovite and biotite micas supply fluoride to soils and groundwater by weathering and soil forming processes. Among the few laboratory model studies, it has been observed that granites, acid volcanic rocks, basic dikes, hornblende in gneisses contribute to fluoride rich soils and waters in the surrounding (Saxena and Ahmed 2001).

In Coimbatore district in Tamil Nadu, rocks are found to contain 180–2600 mg/kg F (Table 4.2). The dark mineral fraction of gneisses (separated in a high intensity magnetic field) contains high concentrations of F (Jacks et al. 2005). Calcrete and

Table 4.1 Fluorine in fresh rocks, weathered material and soils from Coimbatore district in Tamil Nadu, India. (Modified from Jacks et al. 2005)

Sample/site	Fraction	Weight %	F concentration, mg/kg	Fluoride in groundwater, mg/L	Fluoride in calcrete, mg/kg
Gneiss/Nallur	Light	74	1520	5.9	800–900
	Dark	26	5700		
	Fresh rock		2500		
	Weathered		1880		
	Soil (<2 mm)		2000		
Gneiss/Ponnan-dakavundanur	Light	70	1270	4.0	510–1300
	Dark	30	5400		
Gneiss/Kodavadi	Light	65	195	5.2	1600–2500
	Dark	38	2700		
	Fresh rock	1070			
	Weathered	900			
	Soil (<2 mm)	690			
Granite/ Karunagarapuri	Light	95	360	1.5	
	Dark	5	2100		
Vedasandur	Fresh rock	380			1080
	Weathered	360			
	Soil (<2 mm)	470			

dolocrete have also found as sinks for fluoride (Jacks et al. 1993). Fractions of soil in a high fluoride area in Rajasthan have been found to contain from about 10 (sand) to 130 (clay) mg/kg (Madhavan and Subramanian 2002).

4.2.1.1 Geochemical Processes in Fluoride Release from Minerals

Wall rock interaction is believed to be the foremost process on fluoride release to groundwater (Abdelgawad et al. 2009; Handa 1975; Saxena and Ahmed 2003). High-F groundwaters are common in the dry parts of the world. The reason behind is that the F originates mainly from hydroxypositions in biotite and hornblende and is concentrated through evapotranspiration in soil and groundwater exhibiting residual alkalinity. Such waters are common in areas with generally alkaline soils. Along the flow paths of the water from hill tops to valley bottoms calcite, dolomite and fluorite seem to precipitate, in that order (Jacks et al. 2005).

The high fluoride groundwater typically has high pH value, high HCO_3^- content, and high Na^+ content (Guo et al. 2007; Handa 1975; Jacks et al. 1993, 2005). Guo et al. (2007) indicated that the fluoride concentration is positively correlated with HCO_3^- and Na^+ contents. The alkaline water can mobilize fluoride from soils and weathered rocks. Alkaline conditions, moderate specific conductivity and their

Table 4.2 Fluoride levels reported in different countries

Country	Location	Water source	Fluoride concentration or range (mg/L)	Reference
Ghana	Upper regions	Shallow and deep groundwater	0.11–4.6	Apambire et al. (1997a)
	Nathenje and Lilongwe	Shallow and deep groundwater	0.5–7.02	Msonda et al. (2007)
Pakistan	Naranji	Shallow groundwater	1.08–1.38	Tahir Shah and Danishwar (2003)
	Faisalabad	Groundwater	0.38–1.15	Kausar et al. (2003)
	Kalalanwala	Shallow groundwater	2.47–21.1	Farooqi et al. (2007a, b)
	Lahor	Shallow and deep groundwater	ND–8.46	Naeem et al. (2007)
	Sialkot	Shallow groundwater	0.41–0.99	Ullah et al. (2009)
	Nagar Parkar	Groundwater	1.13–7.85	Naseem et al. (2010)
Canada	Gaspe, Quebec	Shallow, intermediate and deep groundwater	0.05–10.9	Boyle and Chagnon (1995)
India	Nalgonda	Shallow groundwater	0.1–8.8	Brindha et al. (2011)
	Majhiaon	Shallow and deep groundwater	–	Avishek et al. (2010)
	Guntur, Andhra Pradesh	Shallow groundwater	0.6–2.5	Rao (2009)
	Anantapur, Andhra Pradesh	Shallow groundwater	0.56–5.8	Rao and Devadas (2003)
	Hydrabad, Andhra Pradesh	Shallow and intermediate groundwater	0.38–4.0	Sreedevi et al. (2006)
	Ranga Reddy, Andhra Pradesh	Groundwater	0.4–4.8	Sujatha (2003)
	Karbi Anglong, Assam	Groundwater	0.4–20.6	Chakraborti et al. (2000)
	Bihar	Shallow groundwater	0.1–2.5	Ray et al. (2000)
	Delhi	Groundwater	0.2–32.5	Raju et al. (2009)
	Gujarat	Groundwater	0.1–40	Raju et al. (2009) and reference therein
	Bellary, Karnataka	Groundwater and surface water	0.33–7.8	Wodeyar and Sreenivasan (1996)
	Karnataka	Shallow groundwater	1–7.4	Suma Latha et al. (1999)

Table 4.2 (continued)

Country	Location	Water source	Fluoride concentration or range (mg/L)	Reference
	Palghat, Kerala	Shallow, intermediate and deep groundwater	0.2–5.75	Shaji et al. (2007)
	Chandidongri, Madhya Pradesh	Shallow groundwater	1.5–4.0	Chatterjee and Mohabey (1998)
	Shivpuri, Madhya Pradesh	Groundwater	0.2–6.4	Ayoob and Gupta (2006)
	Orissa	Groundwater	0.1–10.1	Kundu et al. (2001)
	Churu/Dungapur, Rajasthan	Groundwater	0.1–14	Muralidharan et al. (2002); Choubisa (2001)
	Kacheepurum, Tamil Nadu	Shallow–deep groundwater	1–3.24	Dar et al. (2011)
	Tamil Nadu	Shallow–deep groundwater	0.5–4.0	Handa (1975); Raju et al. (2009) and references therein
	Varanasi, Uttar Pradesh	Groundwater	0.2–2.1	Raju et al. (2009) and references therein
	Agra, Uttar Pradesh	Shallow–deep groundwater	0.1–17.5	Gupta et al. (1999)
	Mathura, Uttar Pradesh	Shallow–deep groundwater	0.6–2.5	Misra et al. (2006)
	Sonbhadra, Uttar Pradesh	Shallow–deep groundwater	0.48–6.7	Raju et al. (2009)
	Cambay, North Gujarat	Deep groundwater	0–10	Gupta et al. (2005)
Sri Lanka	Dry Zone	Shallow–deep groundwater	0.02–5.30	Chandrajith et al. (2011)
	Udawalawe	Shallow groundwater	0.09–5.9	Van der Hoek et al. (2003)
Manitoba	Lake Saint Martin	Groundwater	0–15.1	Desbarats (2009)
Camaroon	Mayo Tsanaga	Shallow groundwater	0.19–15.2	Fantong et al. (2010)
Yemen	Hidhran & Alburayhi Basin and Taiz City	Groundwater	1.08–10	Al–Amry (2009)
Ethiopia		Shallow–deep groundwater	0–204	Ayenew et al. (2008)
Iran	Posht-e-Kooh-e-Dashtestan	Shallow groundwater	0.7–6.6	Battaleb-Looie and Moore (2010)
	Maku area	Groundwater	0.46–5.96	Moghaddam and Fijani (2008)

Table 4.2 (continued)

Country	Location	Water source	Fluoride concentration or range (mg/L)	Reference
South Korea		Thermal groundwater	0–40.8	Chae et al. (2007)
	Gimcheon	Deep groundwater	0.04–2.15	Kim et al. (2011)
China	Yung–Chen Basin	Shallow–deep groundwater	0–3.3	Currell et al. (2011)
	Zhuiger Basin, Kuitun area	Groundwater	0–21.5	Wang et al. (2011)
	Taiyuan Basin	Groundwater	0.4–3.32	Guo et al. (2007)
	Taiyuan Basin	Shallow groundwater	0.4–2.4	Li et al. (2011b)
Turkey		Groundwater	0.51–33.0	Oruc (2008)
Germany	Muenster Region	Groundwater	0.01–8.8	Queste et al. (2001)
Mexico	San Luis Potosi Basin	Groundwater	0–3.7	Carrilo–Rivera et al. (2002)
	Hermesillo city, Sonara	Shallow groundwater	0–7.59	Valenzuela-Vasquez et al. (2006)

ratios play a significant role in F dissolution from rocks (Saxena and Ahmed 2001; Saxena and Ahmed 2003). For the dissolution of fluorite in groundwater with high HCO_3^- contents, the reaction is as follows:

$$CaF_2 + 2HCO_3^- \Leftrightarrow CaCO_3 + 2F^- + H_2O + CO_2$$

Moreover, groundwaters with high HCO_3^- and Na^+ content are usually alkaline and have relative high OH^- content, so the OH^- can replace the exchangeable F^- of fluoride-bearing minerals, increasing the F^- content in groundwater. The reactions are basically as follows:

Muscovite:

$$KAl_2\left[AlSi_3O_{10}\right]F_2 + 2OH^- \Leftrightarrow KAl_2\left[AlSi_3O_{10}\right]OH_2 + 2F^-$$

Biotite:

$$KMg_3\left[AlSi_3O_{10}\right]F_2 + 2OH^- \Leftrightarrow KMg_3\left[AlSi_3O_{10}\right]OH_2 + 2F^-$$

The occurrence of groundwater with high HCO_3^- and Na^+ contents and high pH value under the control of above water–rock interactions is the important reason for fluoride release from the aquifer matrix into groundwater (Guo et al. 2007; Handa 1975; Jacks et al. 1993, 2005). Evaporation is another important factor resulting in the occurrence of high fluoride groundwater (Datta et al. 1999). Calcium ions precipitate out as $CaCO_3$, due to high evaporation resulting reducing Ca^{2+} concentration of the groundwater, and consequently the solubility control of CaF_2 on fluoride enrichment in the aqueous phase becomes weaker. This has been described in detail by Handa (1975).

In more arid climates the dissolution of fluoride may happen in a different sequence due to the precipitation of the secondary minerals such as sepiolite and palygorskite (Jacks et al. 1993, 2005) as they act as Mg sinks. Also, soil pH plays an important role in the dissolution of fluoride where it has been described high in alkaline pHs (Jacks et al. 2005). Excessive irrigation may be responsible for this effect.

4.2.1.2 Fluoride Mobility in Soils and Groundwater

Although the fluorine content of most rocks ranges from 100 to 1300 mg/kg (Faure 1991), soil concentrations typically vary between 20 and 500 mg/kg (Edmunds and Smedley 2005). In some cases, higher concentrations have been reported in soils affected by anthropogenic inputs, such as phosphate fertilizers (Kabata-Pendias and Pendias 2001), sewage sludge (Rea 1979) and industrial pollution (Cronin et al. 2000).

Fluoride mobility in soil is highly dependent on the soil's sorption capacity, which varies with pH, the types of sorbents present, and soil salinity (Cronin et al. 2000; Fuhong and Shuqin 1988; Pickering 1985). In general, fluorine found in soils occurs within minerals or is adsorbed to clays and oxyhydroxides, with only a few percent or less dissolved in the soil solution (Cronin et al. 2000; Pickering 1985). The minimum fluoride mobility occurs in fine-grained soils at a pH of 6.0–6.5 (Gilpin and Johnson 1980; Larsen et al. 2002; Wenzel and Blum 1992). As pH rises above that range, colloid surfaces release adsorbed F^- as it is displaced by increasing OH^- concentrations (Larsen et al. 2002).

Fluoride is commonly found in soils ranging from 200–300 mg/kg (WHO 2002), however, strong association of fluoride with the soil components it does not readily released from soil (Ayoob and Gupta (2006) and references therein). The chemical speciation, soil chemistry and climate are the factors influencing the fluoride release from soils. Below pH 6, in acidic soils fluoride form complexes with Fe and Al in the soil, and the adsorption is significantly high in low pHs (below pH 4.0) and decreases above pH 6.5 (WHO 2002). Alkalinisation of soils by application of fertilizers under irrigation may increase the fluoride release from soils to the groundwater.

Fluoride occurs in almost all groundwaters at least in minute concentrations. The primary controls on dissolved fluoride concentrations include the types of minerals, residence time, and climate as well as the water quality (e.g., pH, hardness and ionic strength) also has an important role through its influence on mineral solubility, complexation, and sorption/exchange reactions (Apambire et al. 1997b). The maximum concentration of fluoride in groundwater is usually controlled by the solubility of fluorite (Apambire et al. 1997b; Chae et al. 2007; Cronin et al. 2000; Edmunds and Smedley 2005; Handa 1975; Nordstrom and Jenne 1977; Pekdeger et al. 1992; Reardon and Wang 2000; Saxena and Ahmed 2003). It has been reported that once the solubility limit for fluorite (CaF_2) is reached, an inverse relationship will exist between fluoride and calcium concentrations (Ozsvath 2009). Many studies have found a strong association between high fluoride and soft, alkaline (i.e., sodium-bicarbonate) groundwater that is depleted in calcium (Bårdsen et al. 1996; Chae et al. 2006a, 2007; Conrad et al. 1999; Dhiman and Keshari 2006; Earle and Krogh 2004; Gupta et al. 1999; Handa 1975; IGRAC 2003a; Kohut et al. 2001; Pekdeger et al. 1992; Robertson 1986; Whittemore et al. 1993). Other than these, pH and the development of secondary minerals are responsible in high mobility of fluoride in soil (Jacks et al. 2005).

4.2.2 Other Sources

Several studies conducted in India have reported different sources of fluoride for soil and groundwater other than weathering.

Other common sources of fluoride in soil and groundwater are the following (Datta et al. 1996):

1. Hydro-geothermal sources
2. Volcanic ash
3. Wet and dry depositions of gaseous (e.g., HF, SiF_4), particulate fluorides (e.g., AlF_3, $NaAlF_6$, CaF_2) emissions from steel, aluminium, glass, phosphate fertilizer, brick and tile industries, soil dust and crustal material. Burning of coal and fly-ash deposition (Pickering 1985; Skjelkvåle 1994).
4. Phosphate fertilizers, fumigants, rodenticide, insecticides and herbicides containing fluoride as impurity or constituent (Poovaiah 1988; Ware 1975), e.g., cryolite, barium fluorosilicate, sodium silicofluoride, sulfuryl fluoride, trifluralin.

4.2.2.1 Fluorine in Volcanic and Hydrothermal Regions

Fluorine has been found to have a higher affinity for silicate melts than solid phases (Anazawa 2006; Sawyer and Oppenheimer 2006; Xiaolin et al. 1998) and is progressively enriched in magmas and hydrothermal solutions through time (Dolejš and Baker 2004; Fuge 1977; Hyndman 1985; Li et al. 2003b; Scaillet and Macdonald 2004; Taylor and Fallick 1997). It has been found that the hydrothermal vein deposits and rocks that crystallize from highly evolved magmas contain fluorite, fluorapatite, and fluoride-enriched micas and/or amphiboles (Dolejš and Baker 2004; Hyndman 1985; Stormer and Carmichael 1970; Taylor and Fallick 1997).

High fluoride concentrations sometimes exceeding 1000 mg/L found in closed-basin alkaline lakes (Jones et al. 1977) as a result of rift valleys with hyperalkaline volcanic rocks and geothermal activity. Acidic geothermal waters can also contain high fluoride concentrations (Ellis and Mahon 1977; Nordstrom and Jenne 1977) under low pH conditions. High sodium and bicarbonate concentrations, near-neutral to alkaline pH values, and under saturation with respect to fluorite (Gizaw 1996) favorable to high fluoride concentrations, which are further increased through evaporation. Similarly, the fluoride solubility is enhanced by the formation of complexes with hydrogen and aluminum which prevent fluorite formation.

4.2.2.2 Atmospheric Sources

The natural atmospheric sources of fluorine in precipitation include marine aerosols, volcanic gas emissions, and air-borne soil dust (De Angelis and Legrand 1994; Saether et al. 1995; Tavener and Clark 2006). Volcanic eruptions are common in Iceland and fluorosis poisoning in livestock and humans was identified long ago in 1978 from the Laki eruption (Brindha and Elango 2011). The fluoride content in ash from Hekla eruption in 2010 was 23–35 mg/kg (Matvaelastofnun 2010). Volcanic ash is readily soluble and thus the risk of fluoride contamination in groundwater is very high. These volcanic sources have also been found to cause fluoride contamination in groundwater of Kenya (Gaciri and Davies 1993). There have been records of acute and chronic fluorosis in animals that grazed on pastures covered by ash and other particulates from recent volcanic eruptions (Cronin et al. 2000; Sawyer and Oppenheimer 2006) or by dust derived from phosphate rocks (Cronin et al. 2000). The industrial aerosols from brickworks, aluminum smelters, iron and steel production, fossil fuel burning, ceramic industries and phosphate fertilizers plants are the primary anthropogenic sources of fluorine (Bonvicini 2006; Cronin et al. 2000; Feng et al. 2003; Fuge 1977; Tavener and Clark 2006; Walna et al. 2007). Both gaseous (e.g., HF, SiF_4, F_2, and H_2SiF_4) and particulate forms (e.g., CaF_2, NaF, and Na_2SiF_6) of fluoride is being released by the industrial sources. It has been reported that the rainfall contaminated by such industrial emissions may contain fluoride concentrations exceeding 1 mg/L (Feng et al. 2003; Neal 1989; Saether and Andreassen 1989; Walna et al. 2007), and these concentrations can persist up to 2 km from the source (Mirlean and Roisenberg 2007).

4.3 Exposure to Fluoride

The major sources of exposure to fluoride are drinking water, food, dental products, and pesticides (NRC 2006). The biggest contributor to exposure for fluoride in the world is drinking water. Based on the report from the Environmental Protection Agency (NRC 2006). in the United States of America, it was estimated that in 1992 approximately 1.4 million people in the United States had drinking water with natural fluoride concentrations of 2.0–3.9 mg/L, and just over 200,000 people had concentrations equal to or exceeding 4 mg/L while in the year 2000, approximately 162 million people had artificially fluoridated water (0.7–1.2 mg/L). In India, about 62 million people at risk of developing fluorosis due to the ingestion of water with high concentrations of fluoride (Andezhath et al. 1999). Highly exposed subpopulations include individuals who have high concentrations of fluoride in drinking water, who drink unusually large volumes of water, or who are exposed to other important sources of fluoride. Some subpopulations consume much greater quantities of water than the 2 L per day that EPA assumes for adults, including outdoor workers, athletes, and people with certain medical conditions, such as diabetes insipidus. On a per-body-weight basis, infants and young children have approximately 3–4 times greater exposure than do adults.

Other than the water consumption beverages such as tea and food stuff contribute most estimated fluoride intake (Cao et al. 1998; Chandrajith et al. 2007; Rao and Mahajan 1990). Dry tea leaves also have significantly high levels of fluoride of up to 400 mg/kg, however due to the ingestion of tea the fluoride exposure ends up ranging from 0.04 to 2.7 mg/person/day (Murray 1986). The other sources of non-dietary fluoride intake are dental products, primarily toothpastes, background air, from certain pesticide residues, certain pharmaceuticals and consumer products (NRC 2006). A detail study showed that the 32 of the 98 of analyzed food items from an area of endemic skeletal fluorosis were ranged from 0.2 to 11 mg/kg (Rao and Mahajan 1990). Furthermore, they calculated the combined fluoride intake from water and food as 0.05–0.32 mg/kg of body weight. Tea plants (*Camellia sinensis*) are found as hyperaccumulator for fluoride, and hence it is released into the tea infusion (Cao et al. 2004; Fung et al. 1999; Lung et al. 2003; Malde et al. 2006; Shu et al. 2003). The fluoride content in tea leaves are estimated about 1000 times of the soluble fluoride in the soils (Fung et al. 1999). Tea infusions are considered as the most popular beverage consumed by human society worldwide, second only to water (Graham 1992). It provides an appreciable proportion of the daily dietary intake of fluoride (Duckworth and Duckworth 1978; Walters et al. 1983). Tea prepared in distilled water showed fluoride concentrations varies from 0.32 to 3.3 mg/L (Chandrajith et al. 2007; Whyte et al. 2005). However, this may vary with the type, fertilizer and the geographical differences in tea plantations. Dental-care products are also a special consideration for children, because many tend to use more toothpaste than is advised, their swallowing control is not as well developed as that of adults, and many children under the care of a dentist undergo fluoride treatments.

Other exposures include aluminum smelters, industrial plants manufacturing hydrofluoric acid, phosphate fertilizer plants, enamel, glass, brick and tile builders, dye and plastic factories, coal power plants etc (Ayoob and Gupta 2006). Cigarette smokers may expose to high fluoride intake due to the presence of increased fluoride concentrations in cigarettes (Okamura and Matsuhisa 1965).

4.3.1 Medical Geology of Fluoride

Much of the input into the human body is from drinking water and hence the geochemistry of fluoride in groundwater is therefore of particular significance in the aetiology of various diseases. The majority of the people affected by high fluoride concentration in groundwater live in the tropical countries where the per capita consumption of water is more because of the prevailing climate. In such areas, people consume 3–4 L of water which is higher than the WHO estimate of 2 L/adult/day (Apambire et al. 1997b). The risk of fluorosis is higher in these places. However, incidence of fluorosis in people living in other parts of the world has also been reported. Table 4.3 shows the impact of fluoride on health based on the WHO recommendation (Dissanayake 1991), however, many scientists have shown that the optimum dosage of fluoride can be vary with the geography and climate as well as with the volume of daily water intake (NRC 2006; WHO 2006).

Table 4.3 Effects of prolonged fluoride intake for human health. (Dissanayake 1991)

F concentration, mg/L	Health outcome
<0.5	Dental caries
0.5–1.5	Optimum dental health
1.5–4.0	Dental fluorosis
4.0–10	Dental and skeletal fluorosis
>10.0	Crippling fluorosis

The most salient feature of fluoride geochemistry is its ability to exchange with the OH^- ion and that is due to the similar ionic radius of both ions. Fluorapatite $Ca_5(PO_4)_3F$ and hydroxylapatite $Ca_5(PO_4)_3$ OH are isomorphic end members in the apatite solid-solution series $Ca_5(PO_4)_3$ OH F. Human and other animal teeth are composed mainly of hydroxylapatite. The substitution of OH^- by F^- ion results in the replacement of hydroxylapatite in teeth and bones by fluoroapatite.

4.4 Global Distribution of Fluoride

Fluoride content in drinking water varies around the world depending on the geographical location. Fluoride contamination is widely reported in groundwater in different parts of the world especially from the humid tropics. These areas include Africa, China, South Asia and Middle East (Ayoob and Gupta 2006). It is estimated that more than 200 million people worldwide rely on drinking water with fluoride concentrations that exceed the present WHO guideline of 1.5 mg/L (WHO 2004). In Mexico, it has been estimated about more than 5 million people are affected by fluoride in groundwater (Ayoob and Gupta (2006) and references therein). Dangerous levels of fluoride that are increasingly found in groundwater in South and Southeastern Asia are of growing concern, along with infectious or other toxic substances (WHO 2000). A detailed description on the concentration of fluoride in groundwater and its sources in various regions of the world based on literature (Ayoob and Gupta 2006; Brindha and Elango 2011) are given in Table 4.2 and Fig. 4.1.

4.4.1 Asian Scenario

The two most populated nations in the world, China and India, stand the top in the list of worst hit nations with high fluoride groundwater. In Taiyuan basin of China, interaction between recharge area and fluoride containing minerals were the sources for high fluoride whereas in discharge area evaporation and mixing of karst water contributed to high fluoride (Guo et al. 2007). Groundwater studies on fluoride in South Korea show that the concentration of fluoride depends on the residence time (Kim and Jeong 2005) due to geogenic source of fluoride (Chae et al. 2007; Kim et al. 2011). People living in Ikeno district of Japan were exposed to drinking water containing 7.8 mg/L fluoride for 12 years (Ish and Suckling 1991). Ash from volcanic explosion of Sakurajima volcano, Japan was found to contain average fluoride concentration of 788.1 mg/kg (Nogami 2006). Oversaturation of fluoride in groundwater in Mizunami area, Japan is due to weathering and alteration of granitic rocks (Abdelgawad et al. 2009).

Of the 85 million t of fluoride deposits on the earth's crust, 12 million t are found in India (Teotia and Teotia 1994). Hence, fluoride contamination is widespread, intensive and alarming in India. About 50 % of the groundwater in Delhi exceeds the maximum permissible limit for fluoride in drinking water (Datta et al. 1996). Jacks et al. (2005) observed that high fluoride in groundwater in many parts of India was due to evapotranspiration of groundwater with residual alkalinity. Fluoride content was higher in deeper aquifers of Maharashtra (Madhnure et al. 2007) which was due to long residence time than shallow groundwater. The rocks in southern India are rich with fluoride which forms the major reason for fluoride contamination in groundwater. It is a well established fact that groundwater in Nalgonda district, Andhra Pradesh, has high fluoride due to the inherent fluoride rich granitic rocks with 325–3200 mg/kg. The mean fluoride content in Hyderabad granites is 910 mg/kg (Ramamohana Rao et al. 1993). In Kurmapalli watershed, rocks are enriched in fluoride from 460 to 1706 mg/kg (Mondal et al. 2009). Co-precipitation and/or adsorption of fluoride by calcrete deposits in Wailapalli watershed had resulted in high fluoride in groundwater (Reddy et al. 2010).

Considerably higher concentrations, around 0.1–0.3 mg/L have been reported from two sites in Uttar Pradesh (Satsangi et al. 1998) and Madhya Pradesh in India (Das 1981; Singh 2001). The reason behind this was predicted as the deposition of soil dust. Jain et al. (2000) found wet deposition of crustal material in Haryana south of New Delhi. About 0.05–0.22 mg/L fluoride concentrations have reported by Chandrawanshi and Patel (1999) from eastern Madhya Pradesh comprising 13 sites and this area is close to an industrial Al plant. Recently published measurements of dry deposition near Agra, Satsangi et al. (2002) indicate even larger amounts of F derived by atmospheric deposition. Thus, as several authors claim that the atmospheric deposition is largely from a crustal source.

Fertilizer containing leachable fluoride ranging from 53 to 255 mg/kg and coal containing fluoride ranging from 5 to 20 mg/kg were reported to pollute groundwater with high fluoride in east Punjab, Pakistan by Farooqi et al. (2007a, b) where 2 million people are at risk of being exposed to high fluoride. The granitic rocks with average fluoride concentration of 1939 mg/kg in Nagar Parkar area, Pakistan, contain fluoride in kaolin deposits between 468 and 1722 mg/kg and secondary kaolin deposits have 270 mg/kg which are the source of fluoride up to 7.85 mg/L in groundwater in this area (Naseem et al. 2010). Studies on fluoride in groundwater in Sri Lanka carried out by Dissanayake (1991) and Young et al. (2010) shows that the condition has not changed even after about two decades with fluoride above 4 mg/L in groundwater. It was found that high fluoride areas lie within low plains. It may be due to the contact time with the geological material was longer in the plains and there exists slow groundwater movement compared to highlands (Dharmagunawardhane and Disanayake 1993).

The public drinking water supply system in Isparta, Turkey draws water from lakes and springs discharged from volcanic rocks, Golcuk pyroclastic and Miocene clastic rocks contained fluoride between 0.39–5.62 mg/L respectively. Moghaddam and Fijani (2008) found that groundwater occurring almost everywhere in basaltic rocks in north western Iran contain fluoride beyond the suitable range. High concentrations of fluoride up to 2.3 mg/L have found in groundwater in Algeria (Messaïtfa 2008). It has been estimated that about 70 % of the fluoride intake for the people of this region is through groundwater used for drinking. Apart from these, dates and tea contribute to 10 and 20 % of fluoride intake respectively. Thus the daily intake of fluoride ingested by an adult exceeds the threshold limit of 0.05–0.07 mg of fluoride/kg/day (Burt 1992).

4.4.2 European Scenario

The concentration of fluoride in spring and stream waters used to determine the occurrences of fluorite in Osor district, Spain (Schwartz and Friedrich 1973). In Poland, fluoride concentration of 1.38 mg/L was detected around a phosphate industry waste disposal site (Czarnowski et al. 1996). The fluoride concentrations of about 7 mg/L occur naturally in western Estonia which is due to Silurian-Ordovician aquifer system (Indermitte et al. 2009). Alumina production plants had increased the fluoride concentration in nearby soils (0.3–9.2 mg/L) in Greece (Haidouti 1991).

4.4.3 African Scenario

An example from around the world with volcanic activity leading to high fluoride concentration in the waters is Tanzania and the area surrounding the East African Rift system. Many of the lakes in this area have fluoride concentrations reaching up to 1640 and 2800 mg/L (IPCS 2002). Fluoride contents in some rivers (12–26 mg/L), springs (15–63 mg/L) and alkaline ponds and lakes (60–690 mg/L) were found to be very high in Tanzania (Nanyaro et al. 1984b). Gaciri and Davies (1993) noticed that in natural waters of Kenya, fluoride concentration was greater in lake water than groundwater and springs which was greater than river water. Evaporation would have been a major cause to increase the concentration of fluoride in lakes of this region.

4.4.4 America's Scenario

High fluoride concentrations in groundwater are reported from United States of America in the industrial facility wells in Pennsylvania having 3.2 and 6.5 ppm, deep aquifers of Western US with 5–15 ppm and Southern California Lakeland having 3.6–5.3 ppm (Cohen and Conrad 1998). The prevalence of fluorosis in different states of the USA such as Arizona, Arkansas, California, Colorado, Idaho, Illinois, Iowa, Kansas, Minnesota, Nevada, New Mexico, North Carolina, North Dakota, Oklahoma, Oregon, South Carolina, South Dakota, Texas, Utah and Virginia. In Mexico, 5 million people (about 6% of the population) are affected by fluoride in groundwater. Throughout Canada, there are a number of communities whose sources of drinking-water contain elevated levels of fluoride (as high as 4.3 ppm) from natural sources. However, in most of the cases the fluoride contamination reported by Canada and USA are due to industrial emissions (Rose and Marier 1977). Some parts of Argentina consist of groundwaters with fluoride levels about 5 mg/L (Kruse and Ainchil 2003).

4.5 Defluoridation

Many defluoridation techniques already exist, but there is still no one method that has been found effective, safe, and cheap enough to implement widely. An in-depth analysis through the annals of defluoridation research reveals that very few proven sustainable solutions have been developed so far (Ayoob et al. 2008a; Jagtap et al. 2012). Furthermore, the coagulation and adsorption/ion-exchange processes are still the most widely used fluoride removal techniques practiced in endemic areas of the developing world. Many countries like India and Tanzania use both the domestic and community based such defluoridation techniques in different levels. Of late, a paradigm shift has occurred in the perception of people in India and Sri Lanka, toward community based water supply treatment systems, using activated alumina like sorbents and electro-coagulation respectively. However, some of these defluoridation systems are not affordable to the bulk of population in the fluoride-endemic rural areas. Similarly, other techniques as reverse osmosis, electrodialysis, and nanofiltration assure good quality water, however, very high cost and high technical competence is a must, which limits the use of these techniques in the developing community.

The defluoridation techniques can be broadly classified into four categories, namely coagulation and precipitation, adsorption and/or ion exchange, electrochemical and membrane techniques.

4.5.1 Coagulation

Chemical coagulation is a treatment process commonly used for surface waters in order to remove broad range of impurities in water such as colloids and dissolve organic matter (DOC). In this process, the chemical coagulant are placed in the raw water under specific dosages and conditions to form a solid flocculent or flakes that may be easily filtered from the water (Fawell et al. 2006). The precipitated floc removes the dissolved fluoride contaminant by charge neutralization, adsorption and entrapment. Hence, this is a combined process of coagulation and precipitation.

This process is also known as the Nalgonda process that was developed for low-income Indian and African households (Fawell et al. 2006). Lime, other calcium salts, alum and magnesium oxides are the cheapest and most commonly used coagulants for this process (Ayoob et al. 2008b). This process will remove fluoride up to 50% and possibly more depending on the nature and degree of the fluoride content in the water (Fawell et al. 2006). Addition of lime, calcium salts and magnesium oxides allow fluoride to precipitate. However, in the case of lime, the process is mainly co-precipitation. Co-precipitation is the method used in Nalgonda process of fluoride removal.

4.5.2 Adsorption and Ion Exchange

Adsorption is defined as the accumulation of compounds, ions or gases on surface layer of a solid in comparison with the bulk phase related to unit surface area. The process can occur at an interface between any two phases, such as liquid–liquid, gas–liquid, gas–solid, or liquid–solid interfaces. Adsorption occurring as a result of physical (van der Waals forces) or chemical bonding. Physisorption usually predominates at low temperatures and is characterized by a relatively low energy of adsorption, and hence adsorption is weak. If the adsorbate undergoes chemical interaction with the adsorbent the phenomenon is referred to as chemisorption. Chemisorption exhibits high energies of adsorption and is favored at higher temperatures forming strong interaction between adsorbate and adsorbent. Ion exchange happens when ions of one substance concentrate at a surface as a result of electrostatic attraction to charged sites at the surface. Because of the limited applications of fluoride treatment techniques, adsorption/ion exchange is the most frequently used (Table 4.4). Alumina, bone charcoal and many clays and soils have been tested and used for defluoridation since 1930s (Ayoob et al. 2008b; Boruff 1934; Harmon and Kalichman 1965; Jagtap et al. 2012; Mohapatra et al. 2009).

Activated alumina, made of aluminum oxide and, has a very high surface area to weight ratio allowing it to have many small pores that run through it (Fawell et al. 2006). This process will have a success rate of up to 80% removal of fluoride with less that 1 mg/L of fluoride content left in the water (Fawell et al. 2006). Similarly, many aluminum based sorbents have shown better adsorption capacity (Table 4.4) however, activation has shown an increase in surface area and thereby increasing fluoride adsorption (Shimelis et al. 2006). Defluoridation has also been tested using aluminum based adsorbents together with calcium, iron and manganese oxides and minerals (Table 4.4). However, all these have shown different defluoridation capacities depending on dosage of fluoride, adsorbate, temperature, reaction time and etc. Only few studies have concentrated on the mechanisms behind the adsorptive removal of fluoride by different sorbents. In reality, defluoridation using these sorbents can even show rather different results due to the differences in scale, the presence and influence of other ions and differences in other environmental conditions such as pH, EC, hardness, alkalinity etc.

Table 4.4 Fluoride removal studies, maximum adsorption capacities and other information based on the previous studies

Adsorbent	Adsorption capacity	Concentration range	Contact time	pH	Reference
Granulated Fe–Al–Ce trimetal hydroxide	51.3 mg/g	10–250 mg/L	24 or 48 h	7.0 ± 0.2	Zhao et al. (2012)
Zr(IV)–ethylenediamine (ZrEDA)	> 99 % (37.03 mg/g)	10 mg/L	14 min	7	Swain et al. (2012)
Mechanochemically activated kaolinites	134–114 mg/kg	2–10 mg/L	30 min	3–7	Meenakshi et al. (2004)
Activated red mud	77 %	21 mg/L	2 h	5.5	Tor et al. (2009)
Heavily-weathered tertiary soil	146 μg/g	10 mg/L	1.5 h		Wang and Reardon (2001)
Quartz sand	220.6–1110.4 mg/kg	0.5–45 mg L	48 h	2.94–5.79	Usunoff et al. (2009)
Calcite	90–99 %	3–2100 mg/L	24 h	Neutral pH	Turner et al. (2005)
Kaolinte	< 5 % 50 %	1 mM < 0.1 mM	24 h	5 7	Weerasooriya et al. (1998)
Activated alumina	1–10 mg/g	10–100 mg/L	10 h	5–10.5	Tang et al. (2009a)
Goethite		2.7–25.6 mg/L	24 h	3–11	Tang et al. (2010)
Stilbite zeolite modified with Fe(III)	2.31 mg/g	10 mg/L	2 h	6.94	Sun et al. (2011)
Bauxite	5.16 mg/g	4–24 mg L	120 min	6 ± 0.1	Sujana and Anand (2011)
Bentonite Charfines Kaolinite clay Nirmali seeds and lignite	46 % 38 % 18.2 % 6–8 %	5 mg/L	5 h	2.8 2.8 – –	Srimurali (1998)
Manganese dioxide	1198–1888 mg/kg	2 mg/L	35 min	5–7	Sivasankar et al. (2011)
Modified attapulgite ($MgCl_2.6H_2O:AlCl_3.2H_2O$) = 2:1:2 = 4:3:3 = 2:2:1	19.1–44.0 mg/g 17.1–40.1 mg/g 11.1–27.6 mg/g	20–200 mg/L	48 h	7	Zhang et al. (2009)
Limestone	Below 4 mg/g	10–109 mg/L	2 h		Reardon and Wang (2000)

Table 4.4 (continued)

Adsorbent	Adsorption capacity	Concentration range	Contact time	pH	Reference
Serpentine-Green Yellow Black	4 mg/g 5 mg/g 6 mg/g	10 mg/L	24 h	9.4 9.2 8.0	Rao et al. (1975)
Sulphonated carbonaceous materials from coconut shell Carbion Tulsion Zeocarb 225	780 mg/kg 820 mg/kg 960 mg/kg 1650 mg/kg	5 mg/L		4.0 –7.0	Rao and Bhaskaran (1988)
Brushite	6.373–6.591 mg/g	25–30 mg/L	10–180 min		Mourabet et al. (2011)
Bio-char (pine wood) pine bark	2.11–4.08 mg/g	1–100 mg/L	48 h	2.0	Mohan et al. (2011)
Fired clay chips	200 mg/kg	5–20 mg/L	120–150 h	5–7	Moges (1996)
Chemical treated laterite	36.3–39.1 mg/g	3–50 mg/L	3–24 h	3–5	Maiti (2011)
Kaolin clay	1.65–3.48 mg/g	10–250 mg/L	24 h	6–7.25	Kau et al. (1997)
Regenerated bone char	0.75 mg/g	21.26 mg/L	2–180 min		Kaseva (2006)
Modified fly ash		100 mg/L		4–6	Goswami and Das (2006)
Granular ferric hydroxide	1–10 mg/g	10 to 105 mg/L	24 h	3–6.5	Tang et al. (2009b)
Kaolinite Bentonite	2.65–10.65 mg/g 1.51–64.6 mg/g	50–2000 mg/L	1 h—28 days	6–7	Kau et al. (1998)
Acid activated kaolinite clay		3 mg/L	150 min	4	Gogoi and Baruah (2008)
Aluminum hydroxide coated rice husk ash	9 and 10 mg/g	10–60 mg/L	1 h	7±0.2	Ganvir and Das (2011)
Quick lime	16.67 mg/g	10 mg/L and 50 mg/L	75 min	12.55–12.77	Islam and Patel (2007)
Calcium phosphate systems	0.6 mg/g	10.4 mg/L	Over 24 h	6.5–8.5	He and Beijing (1996)
Hydrous aluminium oxides Gibbsite or alumina	9 mol/kg 10–20 mmol/kg	0.1–1 mM		5.5–6.5 5–7.5	Farrah et al. (1987)

Table 4.4 (continued)

Adsorbent	Adsorption capacity	Concentration range	Contact time	pH	Reference
Hydrous iron oxide	84–44 g/kg	2–100 mg/L		4–6	Farrah and Pickering (1986)
Mn–Ce oxide- powdered Mn–Ce oxide -granular	79.5 mg/g 45.5 mg/g	10 mg/L	1–3 h	6	Deng et al. (2011)
Kanuma mud	0.7376–2.4 mg/g	5 mg/L	20 min- 2 h	5.0–8.0.	Chen et al. (2010a)
Alumina cement granules	2–4 mg/g	8.65 mg/L	10–180 min	6.9±0.4 to 11.7±0.4	Ayoob et al. (2008b)
Cynodon dactylon based thermally activated carbon	4.702 mg/g	2.0–10.0 mg/L	105 min	7.0±0.2	Alagumuthu et al. (2010)
Zirconium (IV)–Impregnated Ground-nut shell carbon	1.26 mg/g	3 mg/L	180 min	3	Alagumuthu and Rajan (2010)
Acid treated spent bleaching earth	7.752 mg/g	5–45 mg/dm^3	30 min	3.5	Mahramanlioglu et al. (2002)
Hydroxyapatite Fluorspar Activated quartz Calcite Quartz	4.54 mg/g 1.79 mg/g 1.16 mg/g 0.39 mg/g	2.5×10^{-5} to 6.34×10^{-2} mg/L		6.0	Fan et al. (2003)
Basic oxygen furnace slag	4.58–8.07 mg/g	1–50 mg/L	35 min	7.0	Islam and Patel (2011)
Alum sludge	5.394 mg/g	5–35 mg/L	240 min	6.0	Sujana et al. (1998)
Activated alumina (γAl_2O_3)	16.34 mg/g	15–100 mg/L	16–24 h	5.0–6.0	Ku and Chiou (2002)
Activated alumina (Grade OA −25)	14.5 mg/g	2.5–14 mg/L		7.0	Ghorai and Pant (2005)
Metallurgical grade Alumina	12.21 mg/g			5.0–6.0	Pietrell (2005)
Aluminium hydroxide (hydrated alumina)	23.7 and 7.0 mg/g	5.0–30 mg/L	24 h	7.0±0.3	Shimelis et al. (2006)
La(III) impregnated on alumina	0.350 mM/g	2 mM/L	20 h	5.7–8.0	Puri and Balani (2000)
Alum-impregnated activated alumina	40.68 mg/g	1–35 mg/L	3 h	6.5	Tripathy et al. (2006)
Manganese oxide coated alumina	2.851 mg/g	2.5–30 mg/L	3 h	7.0±0.2	Maliyekkal et al. (2006)

Table 4.4 (continued)

Adsorbent	Adsorption capacity	Concentration range	Contact time	pH	Reference
Hydrous manganeseoxide-coated alumina	7.09 mg/g	10–70 mg/L	4 h	5.2±0.05	Teng et al. (2009)
Copper oxide coated alumina (COCA)	7.770 mg/g	10 mg/L	24 h		Bansiwal et al. (2010)
Magnesia amended activated alumina	10.12 mg/g	5–150 mg/L	3 h	6.5–7.0	Maliyekkal et al. (2008)
Calcium oxide-modified activated alumina and Manganese oxide modified activated alumina	101.01 mg/g and 10.18 mg/g	1–1000 mg/L	48 h	5.5	Camacho et al. (2010)
Alkoxide origin Alumina	2.0 mg/g	0–25 mg/L	24 h	7.0	Kamble et al. (2010)
Quick lime	16.67 mg/g	10–50 mg/L	75 min	–	Islam and Patel (2007)
Lime stone (LS) and aluminium hydroxide Impregnated lime stone (AlLS)	43.10 mg/g and 84.03 mg/g	0–100 mg/L	5 h	8.0	Jain and Jayaram (2009)
Schwertmannite	50.2–55.3 mg/g	10–90 mg/L	24 h	3.8	Eskandarpour (2008)
Granular ferric hydroxide (GFH)	7.0 mg/g	1–100 mg/L	24 h	6.0–7.0	Kumar et al. (2009)
Nano-geothite	59 mg/g	5–150 mg/L	2 h	5.75	Mohapatra et al. (2009)
Synthetic Siderite	1.775 mg/g	3–20 mg/L	8–12 h	4.0–9.0	Liu et al. (2010)
Magnesia amended silicon dioxide granules	12.6 mg/g		60 min	3.0	Zhu et al. (2009)
Iron(III)-tin(IV) mixed oxide	10.47 mg/g	10–50 mg/L	2 h	6.4±0.2	Biswas et al. (2009)
Hydrated iron(III)–aluminium(III)–chromium(III) ternary mixed oxide	31.89 mg/g	10–80 mg/dm3	1.5 h	5.6±0.2	Biswas et al. (2010a)
Activated cerium(IV) oxide/SiMCM-41 adsorbent	6.02 mmol/g		24 h	6.0	Liu et al. (2010)
Al–Ce hybrid adsorbent	91.4 mg/g	2–15 mg/L	24 h	6.0	Liu et al. (2010)
Fe–Al–Ce trimetal oxide	178 mg/g	2–110 mg/L	24 h	7.0±0.1	Wu et al. (2007)

Table 4.4 (continued)

Adsorbent	Adsorption capacity	Concentration range	Contact time	pH	Reference
Fe–Al mixed hydroxide (Molar ratio:1)	91.7 mg/g	10–90 mg/L	2 h	4.0	Sujana et al. (2009b)
	16.51–34.25 mg/g	10–30 mg/L	2 h	4.0	Sujana and Anand (2010)
Aluminum titanate (AT)	3.01 mg/g	2–10 mg/L	40 min	7.0	Karthikeyan and Elango (2009)
Bismuth aluminate (BA)	7.09 mg/g				Karthikeyan and Elango (2009)
Aluminium impregnated hierarchal web of carbon fibers	17 mg/g	0.1–50 mg/L			Gupta et al. (2009)
Manganese oxide-coated granular activated carbon		3–35 mg/L	3 h	5.2±0.2	Ma et al. (2009)
Graphite	0.16–3.13 mg/g	2–10 mg/L	60 min	7.0	Karthikeyan and Elango (2008)
Lignite (LN)	6.9–7.44 mg/g	3–90 mg/dm^3	10 h	5–10	Sivasamy et al. (2001)
Fired clay chips	72–90%	5–20 mg/L		5–7	Moges et al. (1996)
Montmorillonite clay	1.485 mg/g	3.0 mg/L	50 min	Neutral pH	Karthikeyan et al. (2005)
Montmorillonite	0.263 mg/g	2–120 mg/L	180 min	6.0	Tor (2006)
Clay minerals	69.44–93.45 mg/g	23.6–2360.0 mg/L	3–4 days	3.0	Hamdi and Srasra (2007)
Laterite	0.8461 mg/g	10–50 mg/dm^3	195 min	7.5	Sarkar et al. (2006)
Laterite	0.1854–0.3586 mg/g	10–50 mg dm^{-3} 20–50 mg dm^{-3}	195 min	6.8	Sarkar et al. (2007)
Acid–base treated raw laterite (TL)	11.8 mg/g	3–50 mg/L	24 h	5.0	Maiti et al. (2011)
Acid activated kaolinite clay	0.0450 mg/g	3.0 mg/L	100 min	Low pH	Gogoi and Baruah (2008)
Chemically modified bentonite clay (10% La-bentonite)	4.24 mg/g		24 h	7.0	Kamble et al. (2009)

Table 4.4 (continued)

Adsorbent	Adsorption capacity	Concentration range	Contact time	pH	Reference
Magnesium incorporated bentonite clay	2.26 mg/g	5.0 mg/L	12 h	3–10	Thakre et al. (2010b)
Algerian clay (montmorillonite) (with and without calcium)	1.013 mg/g and 1.324 mg/g	1–6 mg/L	30 min	4.0	Ramdani et al. (2010)
Zeolite F-9	28–41 mg/g	10–80 mg/L	24 h		Onyango et al. (2004)
Al^{3+} pretreated low-silica synthetic zeolites	28.21–41.35 mg/g	5–80 mg/L	2 days	4.0–8.0	Onyango et al. (2006)
Metal ion loaded natural zeolite	2.04–4.13 mg/g	1–20 mg/L	24 h		Samatya et al. (2007)
Neodymium modified chitosan	22.380 mg/g	10–100 mg/L	24 h	7.0	Yao et al. (2009)
Chitosan coated silica	44.4 mg/g	10–20 mg/L	180 min	4.0	Vijaya and Krishnaiah (2009)
La(III) incorporated carboxylated chitosan beads	4711 mg/kg	11–19 mg/L	60 min	Neutral pH	Viswanathan and Meenakshi (2008a)
Fe(III) loaded carboxylated chitosan beads	4230 mg/kg	11–19 mg/L	40 min	Neutral pH	Viswanathan and Meenakshi (2008b)
Magnesia	2175 mg/kg	10–23 mg/L	60 min	10.1–10.4	Sairam Sundaram et al. (2009a)
Magnesia/chitosan composite	4440 mg/kg		30 min		Sairam Sundaram et al. (2009a)
Nano-hydroxyapatite	1296 mg/kg	10 mg/L	30 min	7.0	Sairam Sundaram et al. (2009b)
Nano-hydroxyapatite/chitin composite	2840 mg/kg				Sairam Sundaram et al. (2009b)
Hydrocalcite	1030 mg/kg	9–15 mg/L	30 min	Acidic pH	Viswanathan and Meenakshi (2010a)
Hydrocalcite/chitosn composites	1255 mg/kg				Viswanathan and Meenakshi (2010a)

Table 4.4 (continued)

Adsorbent	Adsorption capacity	Concentration range	Contact time	pH	Reference
Zirconium (IV) tungstophosphate/chitosan composite	2025–2142 mg/kg	10 mg/L	30 min	7.0, 3.0	Viswanathan and Meenakshi (2010b)
Lanthanum incorporated chitosan beads	4.7 mg/g	5.34 mg/L	24 h	5.0	Bansiwal et al. (2009)
Aluminium impregnated chitosan	1.73 mg/g	10 mg/L	60 min	6.5	Swain et al. (2009)
La-incorporated chitosan beads (10 wt% La)	4.7 mg/g	5.34 mg/L	24 h		Thakre et al. (2010a)
Commercial activated alumina	1.73 mg/g				Thakre et al. (2010a)
Magnetic chitosan particles	22.49 mg/g	5–140 mg/L	150 min	7.0±0.2	Ma et al. (2007)
Titanium/chitosan	7.20 mg/g		24 h	7.0	Jagtap et al. (2009)
Chitosan based mesoporous Ti–Al binary metal oxide supported beads	2.22 mg/g		24 h	3.0–9.0	Thakre et al. (2010a)
Spirogyra IO_2	1.272 mg/g	5.0 mg/L	120 min	7.0	Mohan and Pittman (2007)
Fungal biomass (Pleurotus ostreatus 1804)	1.272 mg/g	5.0 mg/L	240 min	7.0	Ramanaiah (2007)
Eichhornia crassipes biomass and its carbonized form	0.523–1.54 mg/g	2–25 mg/L	24 h	5.5	Sinha et al. (2003)
Zirconium(IV)- impregnated collagen fiber	2.18 mmol/g	1–5 mmol/L	24 h	5.0–8.0	Liao and Shi (2005)
Glutaraldehyde-crosslinked calcium alginate (GCA)	73.6 mg/g	10–25 mg/L	90–120 min	8.0	Vijaya et al. (2011)
Zirconium impregnated coconut shell carbon	6.41 mg/g		6 h	4.0	Sathish et al. (2007)
Zirconium ion impregnated coconut fiber carbon	40.016 mg/g		6 h	4.0	Sathish et al. (2008)
Waste carbon slurry	4.306 mg/g	1–11 mg/L	1 h	7.58	Gupta et al. (2007)

Table 4.4 (continued)

Adsorbent	Adsorption capacity	Concentration range	Contact time	pH	Reference
Original waste mud	4.2 mg/g	5.4–914 mg/L	1 h	5.0	Kemer et al. (2009)
Acid-activated mud	2.8 mg/g				Kemer et al. (2009)
Precipitated mud	27.2 mg/g				Kemer et al. (2009)
Activated titanium rich bauxite	3.70–4.13 mg/g	2–50 mg/L	1–1.5 h	6	Das et al. (2005)
Calcined Mg-Al-CO_3 layered double hydroxides	213.2 mg/g	50 mg/L	5 h	6.0	Lv et al. (2006)
Calcined Zn/Al hydrotalcite-like compound (HTlc)	13.43 mg/g	10 mg/L	4 h	6.0	Das et al. (2003)
MgCHT	1.185 mg/g	2–10 mg/L	10 h	7.5–10.5	Jimenez-Nu˜nez et al. (2007)
NiCHT	1.202 mg/g			7.0–7.6	Jimenez-Nu˜nez et al. (2007)
CoCHT	0.842 mg/g			6.9–7.6	Jimenez-Nu˜nez et al. (2007)
Synthetic nanohydroxy apatite	4.575 mg/g	3–80 mg/L	100 min	5.0–6.0	Gao et al. (2009a)
Biogenic apatite	4.99 mg/g				Gao et al. (2009a)
Treated biogenic apatite	6.849 mg/g				Gao et al. (2009a)
Geogenic apatite	0.014 mg/g				Gao et al. (2009a)
Synthetic hydroxyapatite	0.295–0.489 mg/g	3–80 mg/L	120 min	5.0	Gao et al. (2009b)
Light weight concrete	5.15 mg/g		60 min	6.9	Oguz (2007)
Hydrated cement	2.6788 mg/g		24 h	6.7	Kagne et al. (2008)
Hardened alumina cement granules (ALC)	34.36 mg/g	2.5–100 mg/L	3 h		Ayoob and Gupta (2009)
Carbon nanotubes (Al_2O_3/CNTs)	28.7 mg/g	50 mg/L	12 h	6.0	Li et al. (2001)
Aligned carbon nanotubes (CNTs)	4.5 mg/g	15 mg/L	180 min	7.0	Li et al. (2003a)
Nano-AlOOH	3259 mg/kg	3–35 mg/L	6 h	5.2±0.2	Wang et al. (2009)
CaO nanoparticles	163.3 mg/g	10–100 mg/L	30 min	2.0–8.0	Patel et al. (2009)

Table 4.4 (continued)

Adsorbent	Adsorption capacity	Concentration range	Contact time	pH	Reference
Fe–Al–Ce nano adsorbent	2.22 mg/g	0.001 M	36 h	7.0	Chen et al. (2009)
Aluminum-type superparamagnetic adsorbents (MASG)	38 g/kg			6.0	Chang et al. (2006)
Aluminum-type superparamagnetic adsorbents (MAHP)	8 g/kg				Chang et al. (2006)
Fe_3O_4@$Al(OH)_3$ magnetic nanoparticles	88.48 mg/g	0–160 mg/L	240 min	6.5	Zhao et al. (2010)
Nanomagnesia	267.82 mg/g	5–200 mg/L	90–120 min	7.0	Maliyekkal et al. (2010)
Nanoalumina	14.0 mg/g	1–100 mg/L	24 h	6.15	Kumar et al. (2011)
Geomaterials	12.3–15.17 mg/g	10–50 mg/L	4 h	5.0	Sujana et al. (2009a)
Bleaching powder	0.1308 mg/g	5.0 mg/L	1 h	6.7	Kagne et al. (2009)
Conducting polypyrrole	6.37 mg/g	2–10 mg/L	30 min		Karthikeyan et al. (2009)
Granular ceramic	12.12 mg/g	5–50 mg/L	72 h	6.9 ± 0.1	Chen et al. (2010b)
Magnesialoaded fly ash cenospheres (MLC)	6.0 mg/g	100 mg/L	1440 min	3.0	Xu et al. (2011)
Al and Fe dispersed in porous granular ceramics	1.79 mg/g	10 mg/L	48 h	6.0	Chen et al. (2011b)
Polypyrrole/Fe_3O_4 magnetic nanocomposite	17.6–22.3 mg/g	5–100 mg/L	24 h	6.5	Bhaumik et al. (2011)
Glass derived hydroxyapatite (G-HAP)	17.34 mg/g	100 mg/L	12 h	6.72	Liang et al. (2011)
Chitosan based mesoporous alumina	8.264 mg/g	5.0 mg/L	24 h		Jagtap et al. (2011)
$KMnO_4$ modified carbon	15.9 mg/g	20 mg/L	3 h	2.0	Daifullah et al. (2007)
Alumina/chitosan composite	3809 mg/kg	10 mg/L	30 min	Neutral pH	Viswanathan and Meenakshi (2010c)
Zirconium-iron	9.80 mg/g	10 mg/L	10 h	7.0	Dou et al. (2011)

Table 4.4 (continued)

Adsorbent	Adsorption capacity	Concentration range	Contact time	pH	Reference
Calcium chloride modified natural zeolite	1.766 mg/g	25–100 mg/L	6 h	6.0	Zhang et al. (2011)
Serpentinite-derived soil Quartz schist- derived soil Amphibolite-derived soils	1100 mg/kg 989 mg/kg 1783 mg/kg	0–200 mg/L	23 h		Gago et al. (2012)
Ceramic adsorbent	2.16 mg/g	20–100 mg/L	0–48 h	2–12	Chen et al. (2010c)
Cuttlefish bones	80 %	5 mg/L	1 h	7.2	Ben Nasr et al. (2011)
Nanosized hydroxyapatite	1.438 mg/g	50 mg/L	120 min		Wang et al. (2011)
Al (III) modified calicium hydroxyapatite	32.57 mg/g	0–50 mg/L		7	Nie et al. (2012)
Mesoporous alumina	14.26 mg/g	20–250 mg/L	24 h	6	Lee et al. (2010)
Granular acid-treated bentonite	0.094 mg/g	2.85–20 mg/L	40 min	4.95	Ma et al. (2011)
Mixed-phase nano iron oxides	53.19 mg/g	10–100 mg/L	8 h	5.75	Mohapatra et al. (2011)
Disposed earthenware with 0.025 % of manganese dioxide	1888 mg/kg	2 mg/L	35 min	7	Sivasankar et al. (2011)
Industrial waste (regenerated spent bleaching earth)	0.6 mg/g	2.5–8 mg	180 min	7	Malakootian et al. (2011)
Functionalize pumice stone	41 mg/g	1.5–20 mg/L	1440 min	6	Asgari et al. (2012)
Iron-impregnated granular ceramics (GC) GC ($FeSO_4 \cdot 7H_2O$) GC (Fe_2O_3)	2.157 mg/g 1.699 mg/g	5–50 mg/L	48 h	6.9	Chen et al. (2011a)
Synthetic iron(III)–aluminum (III)–chromium(III) ternary mixed oxide	31.889 mg/g	10–80 mg/L	1.5 h	5.±0.2	Biswas et al. (2010b)
Bone char	5.19 mg/g	1–20 mg/L		7	Leyva-Ramos et al. (2010)
Mg-doped nano ferrihydrite	64 mg/g	10–150 mg/L	5 h	5.75	Mohapatra et al. (2012)

Table 4.4 (continued)

Adsorbent	Adsorption capacity	Concentration range	Contact time	pH	Reference
Orange waste loaded with multi-valent metal ions (SOJR) Al-SOJR La-SOJR Ti-SOJR Sn-SOJR	1.01 mol/kg 1.07 mol/kg 0.91 mol/kg 1.18 mol/kg	0.5–8 mM	24 h	6 4 3 3	Paudyal et al. (2012)
Fe(III) modified montmorillonite	0.90 μ mol m^{-2} 0.76 μ mol m^{-2}	5–70 mg/L	24 h	4.5 7	Bia et al. (2012)
Granular ceramic	0.929 mg/g	5–50 mg/L	72 h	6.90±0.10	Chen et al. (2010b)
Natural stilbite zeolite modified with Fe(III)	2.31 mg/g	5.0–40 mg/L	2 h	6.94	Sun et al. (2011)
Highly ordered mesoporous aluminas	450 mg/g	2–1000 mg/L	12 h	6.5	Li et al. (2011a)
Fe-Ti oxide nano-adsorbent	47.0 mg/g		12 h		Chen et al. (2012)
Meso-structured zirconium phosphate	4.268 mg/g	1–10 mg/L	60 min	6.0	Swain et al. (2011)
Superparamagnetic zirconia material	14.7 mg/g	10–105 mg/L		4.0	Chang et al. (2011)
Zirconium(IV)-Impregnated Collagen Fiber	2.18 mmol/g	1–5 mg/L	24 h	5.5	Liao and Shi (2005)

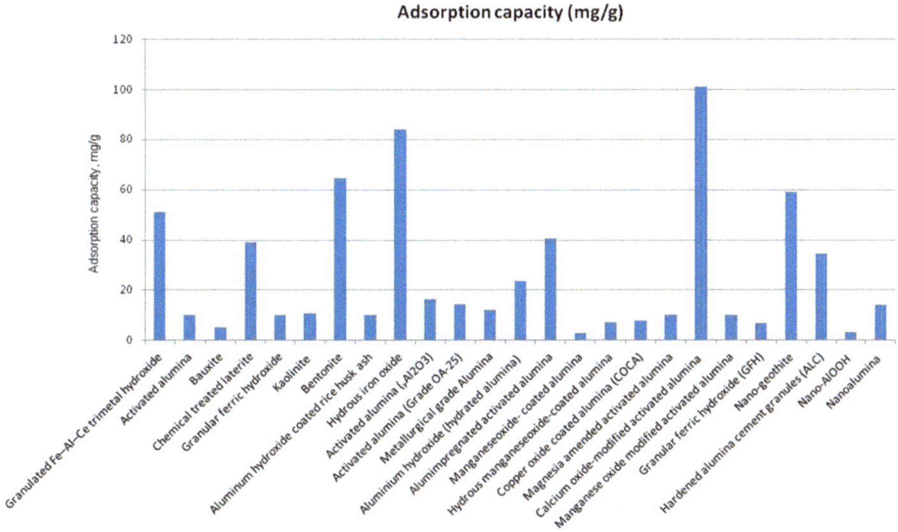

Fig. 4.3 Maximum adsorption capacities of fluoride sorption into different materials

Activated carbon also has a long history for defluoridation (McKee and Johnston 1934). Many carbon types, bone char, wood charcoal, black carbon, petroleum waste and carbon nanotubes, have been tested similar to that of aluminum based sorbents (Abe et al. 2004; Bhargava and Killedar 1992; Daifullah et al. 2007; Gupta et al. 2007; Li et al. 2003b). Also many different materials have been tested for efficient fluoride removal (Table 4.4). Dey et al. (2004) describe Hydrous Ferric Oxide (HFO) as a "scavenger" for fluoride and show that HFO is over three times more effective at fluoride removal than activated alumina, currently the most widely used adsorbent in field studies. In their study, Dey et al. (2004) clearly demonstrate that the quantity of fluoride removed from water by adsorption to HFO is dependent on the HFO surface area present in solution.

4.5.3 *Electrochemical Methods*

Electrocoagulation (EC) utilizes an electrolytic process to generate a coagulant in situ by oxidation of an appropriate anodic material and the coagulant ions then react with the target pollutant ions, initiating coagulation (Ayoob et al. 2008b). The defluoridation of water by EC using aluminum electrodes was introduced by Ming et al. (1983). With the electric current passing through the aluminum electrodes, an anodic reaction releases Al(III) ions, which then react with hydroxide ions produced at the cathode and with fluoride ions in solution (Fig. 4.3). This method has been successfully applied by a group of scientists and engineers (Padmasiri et al. 2012) in the rural villages in Sri Lanka where people are suffering from Chronic Kidney

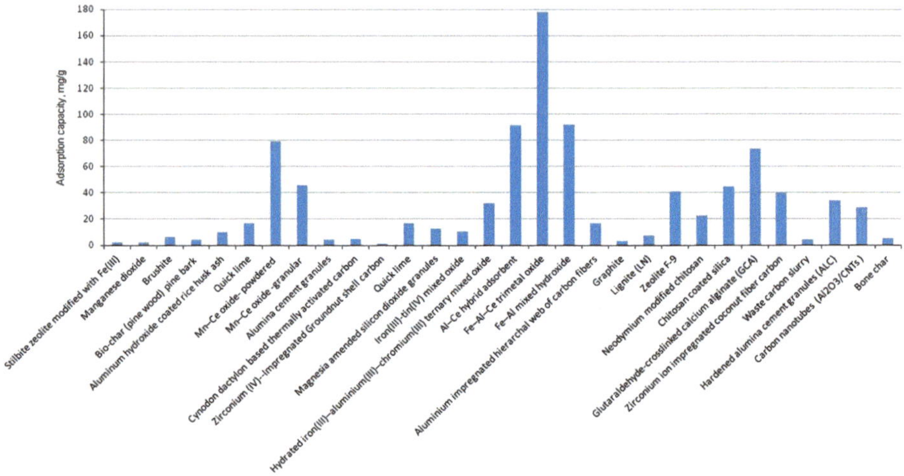

Fig. 4.4 Maximum adsorption capacities of fluoride sorption into different aluminum and iron based materials

Fig. 4.5 Electrocoagulation plant implemented in rural Sri Lanka (Picture by Mr. J.P. Padmasiri, Institute of Fundamental Studies, Kandy, Sri Lanka)

Disease of unknown etiology (CKDu) to provide drinking water with less fluoride and hardness (Fig. 4.4) (Fig. 4.5).

4.5.4 Membrane Process

The most significant processes in water treatment for membrane processes include reverse osmosis, ultra-filtration, micro-filtration, and nano-filtration (Fawell et al. 2006). These processes are now recently being applied to the treatment of drinking water. Membrane operations generally utilize artificial membranes to separate the mixtures and capture the undesired material. This process is successful in fluoride removal from drinking water up to 80% or more, leaving the water with a fluoride content of less than 1 mg/L (Fawell et al. 2006).

An ion exchange fibrous adsorbent in the form of linear microporous fiber (diameter ~30 μm) has been tested for removing fluoride through the cross-linking reaction of hydrazine-modified polyacrylonitrile fiber (PANF) (Ruixia et al. 2002). They have reported a rapid and high removal of fluoride about 90.4% at pH 3, which decreases slowly with an increase of pH. Sol-gel method was recently applied to develop a novel ion exchanger based on double hydrous oxides of aluminum and iron ($Fe_2O_3 \cdot Al_2O_3 \cdot xH_2O$) and high fluoride sorption capacity has been reported with 88 mg/g (Chubar et al. 2005).

Recently, Reverse Osmosis (RO) modules of spiral-wound polyamide membrane were applied successfully for the treatment of electronic industry wastewater rich in fluoride (400 mg/L) (Ndiaye et al. 2005). Application of commercially available cellulose acetate membrane showed 93.3% removal (Ayoob et al. 2008a). Similarly, about 59–67% removal of fluoride using a RO unit, employing spiral wound cellulose acetate membranes, after adjusting the pH from 8.2 to 6.4 has been reported (Schneiter and Middlebrooks 1983). Also, a successful implementation of a pilot scale RO plant for has been reported for fluoride removal in Lakeland, southern California (Cohen and Conrad 1998).

Nanofiltration (NF) was used in commercial level in Finland with a capacity of 380–600 m^3/day which was intended for the removal of fluoride and aluminum from groundwater (Kettunen and Keskitalo 2000). Selective defluoridation was observed for the first time using an NF membrane that has mass transfer properties very similar to RO membranes (Pontie et al. 2003). Electrodialysis (ED) is an excellent technique for simultaneous defluoridation and desalination of brackish water (Adhikary et al. 1989). Fluoride removal by Donnan Dialysis (DD) was investigated by many researchers (Dieye et al. 1998; Garmes et al. 2002; Hichour et al. 1999a, b).

Problems and perspectives

Many defluoridation techniques have been examined since the 1930s, when the danger of excess fluoride in drinking water was first identified (Boruff 1934). More than 70 years since the problem was recognized, however, the attempts to develop a method of defluoridation that can be sustained under differing social, financial, environmental and technical constraints have not been successful. Although coagulation methods are generally effective and are in use commonly in defluoridation, the major limitation is that the process is unsuccessful in bringing fluoride to desired concentration levels. Ion exchange and/or adsorption are widely accepted technologies utilized on a full-scale basis however, the regeneration and disposal of the material has made the process quite questionable. Though a number of adsorbents with very high potential have had been developed, only activated alumina and bone char were reported successful at the implementation level (Ayoob et al. 2008b). For many methods the most critical factor in the sorption process is pH as it dictates the entire process chemistry of defluoridation. Membrane methods are relatively expensive to install and operate and prone to fouling, scaling, or membrane degradation as well as these need skillful operational capability which will be a limitation for the developing world where the excessive fluoride problem is mostly reported. Similarly, the electrochemical techniques are with high cost factor, during both installation and maintenance. Especially for developing countries, the high cost of technology, cost of power supply, expensive chemicals, operational and maintenance costs, regeneration cost, use of sophisticated accessories act as major constraints for implementation. In the developing world, people always are in contact with the immediate environment and they prefer to use drinking water directly from the source or tap. Such household filter systems not succeed due to this reason. Most of the techniques are not achieving the social acceptance and the implementation fails. Very limited studies have been carried out on bioremediation of fluoride (Evans-Tokaryk 2011; Ramanaiah 2010). It may be important to focus on the bioremediation potential for the defluoridation. Although defluoridation research has made significant advancement, still no sustainable solution to this salient crisis. Hence, still need is there for a best available technology for defluoridation of drinking water.

References

Abdelgawad AM, Watanabe K, Takeuchi S, Mizuno T (2009) The origin of fluoride-rich groundwater in Mizunami area, Japan-mineralogy and geochemistry implications. Eng Geol 108:76–85

Abe I, Iwasaki S, Tokimoto T, Kawasaki N, Nakamura T, Tanada S (2004) Adsorption of fluoride ions onto carbonaceous materials. J Colloid Interface Sci 275:35–39

Adhikary SK, Tipnis UK, Harkare WP, Govindan KP (1989) Defluoridation during desalination of brackish water by electrodialysis. Desalination 71:301–312

Al-Amry AS (2009) Hydrogeochemistry and origin of fluoride in groundwater of Hidhran & Alburayhi Basin, Northwest Taiz City, Yemen. Delta J Sci 33:10–20

Alagumuthu G, Rajan M (2010) Kinetic and equilibrium studies on fluoride removal by zirconium (IV): impregnated groundnut shell carbon. Hemijska industrija 64:295–304

Alagumuthu G, Veeraputhiran V, Venkataraman R (2010) Adsorption isotherms on fluoride removal: batch techniques. Arch Appl Sci Res 2:170–185

Anazawa K (2006) Fluorine and coexisting volatiles in the geosphere: the role in Japanese volcanic rocks. In: Tressaud A (ed) Fluorine and the environment: atmospheric chemistry, emissions, and lithosphere. Elsevier, Amsterdam

Andezhath SK, Susheela AK, Ghosh G (1999) Fluorosis management in India: the impact due to networking between health and rural drinking water supply agencies. IAHS-AISH Publication, Wallingford, ROYAUME–UNI

Apambire WB, Boyle DR, Michel FA (1997a) Geochemistry, genesis, and health implications of fluoriferous groundwaters in the upper regions of Ghana. Environ Geol 33:13–24. doi:10.1007/s002540050221

Apambire WB, Boyle DR, Michel FA (1997b) Geochemistry, genesis, and health implications of fluoriferous groundwater in the upper regions of Ghana. Environ Geol 33:13–24

Asgari G, Roshani B, Ghanizadeh G (2012) The investigation of kinetic and isotherm of fluoride adsorption onto functionalize pumice stone. J Hazard Mater 217–218:123–132

Avishek K, Pathak G, Nathawat M, Jha U, Kumari N (2010) Water quality assessment of Majhiaon block of Garwa district in Jharkhand with special focus on fluoride analysis. Environ Monit Assess 167:617–623. doi:10.1007/s10661-009-1077-4

Ayenew T, Demlie M, Wohnlich S (2008) Hydrogeological framework and occurrence of groundwater in the Ethiopian aquifers. J Afr Earth Sci 52:97–113

Ayoob S, Gupta AK (2006) Fluoride in drinking water: a review on the status and stress effects. Crit Rev Environ Sci Technol 36:433–487. doi:10.1080/10643380600678112

Ayoob S, Gupta AK (2009) Performance evaluation of alumina cement granules in removing fluoride from natural and synthetic waters. Chem Eng J 150:485–491

Ayoob S, Gupta AK, Bhat VT (2008a) A conceptual overview on sustainable technologies for the defluoridation of drinking water. Crit Rev Environ Sci Technol 38:401–470. doi:10.1080/10643380701413310

Ayoob S, Gupta AK, Bhakat PB, Bhat VT (2008b) Investigations on the kinetics and mechanisms of sorptive removal of fluoride from water using alumina cement granules. Chem Eng J 140:6–14

Bansiwal A, Thakre D, Labhshetwar N, Meshram S, Rayalu S (2009) Fluoride removal using lanthanum incorporated chitosan beads. Colloids Surf B Biointerfaces 74:216–224

Bansiwal A, Pillewan P, Biniwale RB, Rayalu SS (2010) Copper oxide incorporated mesoporous alumina for defluoridation of drinking water. Micropor Mesopor Mater 129:54–61

Bårdsen A, Bjorvatn K, Selvig KA (1996) Variability in fluoride content of subsurface water reservoirs. Acta Odontol Scand 54:343–347. doi:10.3109/00016359609003549

Battaleb-Looie S, Moore F (2010) A study of fluoride groundwater occurrence in Posht-e-Kooh-e-Dashtestan, South of Iran. World App Sci J 8:1317–1321

Ben Nasr A, Walha K, Charcosset C, Ben Amar R (2011) Removal of fluoride ions using cuttlefish bones. J Fluor Chem 132:57–62

Bhargava DS, Killedar DJ (1992) Effect of variation in sorbent fluoride and sorbent concentration on fluoride removal by fishbone charcoal in a moving fluidized reactor. J Instn Engrs 73:11–14

Bhattacharya P, Jacks G, Ahmed KM, Routh J, Khan AA (2002) Arsenic in groundwater of the Bengal Delta plain aquifers in Bangladesh. Bull Environ Contam Toxicol 69:538–545. doi:10.1007/s00128-002-0095-5

Bhaumik M, Leswifi TY, Maity A, Srinivasu VV, Onyango MS (2011) Removal of fluoride from aqueous solution by polypyrrole/Fe3O4 magnetic nanocomposite. J Hazard Mater 186:150–159

Bia G, De Pauli CP, Borgnino L (2012) The role of Fe(III) modified montmorillonite on fluoride mobility: adsorption experiments and competition with phosphate. J Environ Manage 100:1–9

Biswas K, Gupta K, Ghosh UC (2009) Adsorption of fluoride by hydrous iron(III)-tin(IV) bimetal mixed oxide from the aqueous solutions. Chem Eng J 149:196–206

Biswas K, Debnath S, Ghosh UC (2010a) Physicochemical aspects on fluoride adsorption for removal from water by synthetic hydrous iron(III)-chromium(III) mixed oxide. Sep Sci Technol 45:472–485. doi:10.1080/01496390903526667

Biswas K, Gupta K, Goswami A, Ghosh UC (2010b) Fluoride removal efficiency from aqueous solution by synthetic iron(III)-aluminum(III)-chromium(III) ternary mixed oxide. Desalination 255:44–51

Bonvicini G, Fregni A, Palmonari C (2006) Fluorine compounds from industrial sources: the case of ceramic industries. In: Tressaud A (ed) Fluorine and the environment: atmospheric chemistry, emissions, and lithosphere. Elsevier, Amsterdam, pp 225–249

Boruff CS (1934) Removal of fluorides from drinking waters. Ind Eng Chem 26:69–71. doi:10.1021/ie50289a016

Boyle DR, Chagnon M (1995) An incidence of skeletal fluorosis associated with groundwaters of the maritime carboniferous basin, Gaspé region, Quebec, Canada. Environ Geochem Health 17:5–12. doi:10.1007/bf00188625

Brindha K, Elango L (2011) Fluoride in groundwater: causes, implications and mitigation measures. In: Monroy SD (ed) Fluoride properties, applications and environmental management. Nova Publishers, pp 111–136

Brindha K, Rajesh R, Murugan R, Elango L (2011) Fluoride contamination in groundwater in parts of Nalgonda district, Andhra Pradesh, India. Environ Monit Assess 172:481–492. doi:10.1007/s10661-010-1348-0

Bundschuh J, Farias B, Martin R, Storniolo A, Bhattacharya P, Cortes J, Bonorino G, Albouy R (2004) Groundwater arsenic in the Chaco-Pampean Plain, Argentina: case study from Robles county, Santiago del Estero province. Appl Geochem 19:231–243. doi:http://dx.doi.org/10.1016/j.apgeochem.2003.09.009
Burt BA (1992) The changing patterns of systemic fluoride intake. J Dent Res 71:1228–1237
Camacho LM, Torres A, Saha D, Deng S (2010) Adsorption equilibrium and kinetics of fluoride on sol-gel-derived activated alumina adsorbents. J Colloid Interface Sci 349:307–313
Cao J, Zhao Y, Liu JW (1998) Safety evaluation and fluorine concentration of pu'er brick tea and bianxiao brick tea. Food Chem Toxicol 36:1061–1063
Cao J, Fei Luo S, Wei Liu J, Li Y (2004) Safety evaluation on fluoride content in black tea. Food Chem 88:233–236
Carrillo-Rivera JJ, Cardona A, Edmunds WM (2002) Use of abstraction regime and knowledge of hydrogeological conditions to control high-fluoride concentration in abstracted groundwater: San Luis Potosí basin, Mexico. J Hydrol 261:24–47
Chae G-T, Yun S-T, Kim K, Mayer B (2006a) Hydrogeochemistry of sodium-bicarbonate type bedrock groundwater in the Pocheon spa area, South Korea: water-rock interaction and hydrologic mixing. J Hydrol 321:326–343
Chae GT, Yun ST, Kwon MJ, Kim YS, Mayer B (2006b) Batch dissolution of granite and biotite in water: implication for fluorine geochemistry in groundwater. Geochem J 40:95–102
Chae G-T, Yun S-T, Mayer B, Kim K-H, Kim S-Y, Kwon J-S, Kim K, Koh Y-K (2007) Fluorine geochemistry in bedrock groundwater of South Korea. Sci Total Environ 385:272–283
Chakraborti D, Chanda CR, Samanta G, Chowdhury UK, Mukherjee SC, Pal AB, Sharma B, Mahanta KJ, Ahmed HA, Sing B (2000) Fluorosis in Assam India. Curr Sci 78:1421–1423
Chandrajith R, Abeypala U, Dissanayake C, Tobschall H (2007) Fluoride in Ceylon tea and its implications to dental health. Environ Geochem Health 29:429–434. doi:10.1007/s10653-007-9087-z
Chandrajith R, Dissanayake CB, Ariyarathna T, Herath HMJMK, Padmasiri JP (2011) Dose-dependent Na and Ca in fluoride-rich drinking water -Another major cause of chronic renal failure in tropical arid regions. Sci Total Environ 409:671–675
Chandrawanshi CK, Patel KS (1999) Fluoride deposition in central India. Environ Monit Assess 55:251–265
Chang CF, Lin PH, Höll W (2006) Aluminum-type superparamagnetic adsorbents: synthesis and application on fluoride removal. Colloids Surf A 280:194–202
Chang C-F, Chang C-Y, Hsu T-L (2011) Removal of fluoride from aqueous solution with the superparamagnetic zirconia material. Desalination 279:375–382
Chatterjee MK, Mohabey NK (1998) Potential fluorosis problems around Chandidongri, Madhya Pradesh, India. Environ Geochem Health 20:1–4. doi:10.1023/a:1006529925395
Chen L, Wu H-X, Wang T-J, Jin Y, Zhang Y, Dou X-M (2009) Granulation of Fe-Al-Ce nano-adsorbent for fluoride removal from drinking water by spray coating on sand in a fluidized bed. Powder Technol 193:59–64
Chen N, Zhang Z, Feng C, Li M, Chen R, Sugiura N (2010a) Removal of fluoride from aqueous solution by adsorption onto Kanuma mud. Water Sci Technol 62:1888–1897
Chen N, Zhang Z, Feng C, Sugiura N, Li M, Chen R (2010b) Fluoride removal from water by granular ceramic adsorption. J Colloid Interface Sci 348:579–584
Chen N, Zhang Z, Feng C, Li M, Zhu D, Chen R, Sugiura N (2010c) An excellent fluoride sorption behavior of ceramic adsorbent. J Hazard Mater 183:460–465
Chen N, Zhang Z, Feng C, Li M, Zhu D, Sugiura N (2011a) Studies on fluoride adsorption of iron-impregnated granular ceramics from aqueous solution. Mater Chem Phys 125:293–298
Chen N, Zhang Z, Feng C, Zhu D, Yang Y, Sugiura N (2011b) Preparation and characterization of porous granular ceramic containing dispersed aluminum and iron oxides as adsorbents for fluoride removal from aqueous solution. J Hazard Mater 186:863–868
Chen L, He B-Y, He S, Wang T-J, Su C-L, Jin Y (2012) Fe—Ti oxide nano-adsorbent synthesized by co-precipitation for fluoride removal from drinking water and its adsorption mechanism. Powder Technol 227:3–8

Choubisa SL (2001) Endemic fluorosis in Southern Rajasthan, India. Fluoride 34:61–70
Chubar NI, Samanidou VF, Kouts VS, Gallios GG, Kanibolotsky VA, Strelko VV, Zhuravlev IZ (2005) Adsorption of fluoride, chloride, bromide, and bromate ions on a novel ion exchanger. J Colloid Interface Sci 291:67–74
Cohen D, Conrad HM (1998) 65,000 GPD fluoride removal membrane system in Lakeland, California, USA. Desalination 117:19–35. doi:http://dx.doi.org/10.1016/S0011-9164(98)00063-0
Conrad P, Carey DI, Webb JS, Dinger JS, Fisher RS, McCourt MJ (1999) Groundwater quality in Kentucky. Kent Geol Survey Info, p 4
Cronin SJ, Manoharan V, Hedley MJ, Loganathan P (2000) Fluoride: a review of its fate, bioavailability, and risks of fluorosis in grazed-pasture systems in New Zealand. New Zeal J Agric Res 43:295–321. doi:10.1080/00288233.2000.9513430
Currell M, Cartwright I, Raveggi M, Han D (2011) Controls on elevated fluoride and arsenic concentrations in groundwater from the Yuncheng Basin, China. Appl Geochem 26:540–552
Czarnowski W, Wrześniowska K, Krechniak J (1996) Fluoride in drinking water and human urine in Northern and Central Poland. Sci Total Environ 191:177–184
Daifullah AAM, Yakout SM, Elreefy SA (2007) Adsorption of fluoride in aqueous solutions using KMnO4-modified activated carbon derived from steam pyrolysis of rice straw. J Hazard Mater 147:633–643
Dar MA, Sankar K, DarImran A (2011) Fluorine contamination in groundwater: a major challenge. Environ Monit Assess 173:955–968
Das DK, Burman GK, Kidwai AL (1981) Chemical composition of monsoon rainwater over Bhopal, Madhya Pradesh during 1977 and 1978. Mausam 32:221–228
Das DP, Das J, Parida K (2003) Physicochemical characterization and adsorption behavior of calcined Zn/Al hydrotalcite-like compound (HTlc) towards removal of fluoride from aqueous solution. J Colloid Interface Sci 261:213–220
Das N, Pattanaik P, Das R (2005) Defluoridation of drinking water using activated titanium rich bauxite. J Colloid Interface Sci 292:1–10
Datta PS, Deb DL, Tyagi SK (1996) Stable isotope (18O) investigations on the processes controlling fluoride contamination of groundwater. J Contam Hydrol 24:85–96
Datta PS, Tyagi SK, Mookerjee P, Bhattacharya SK, Gupta N, Bhatnagar PD (1999) Groundwater NO3 and F contamination processes in Pushkar Valley, Rajasthan as reflected from 18O isotopic signature and 3H recharge studies. Environ Monit Assess 56:209–219. doi:10.1023/a:1005903619718
De Angelis M, Legrand M (1994) Origins and variations of fluoride in Greenland precipitation. J Geophys Res 99:1157–1172. doi:10.1029/93jd02660
Deng S, Liu H, Zhou W, Huang J, Yu G (2011) Mn-Ce oxide as a high-capacity adsorbent for fluoride removal from water. J Hazard Mater 186:1360–1366
Desbarats A (2009) On elevated fluoride and boron concentrations in groundwaters associated with the Lake Saint-Martin impact structure, Manitoba. Appl Geochem 24:915–927. doi:citeulike-article-id:4450394
Dey S, Goswami S, Ghosh UC (2004) Hydrous ferric oxide (HFO)-a scavenger for fluoride from contaminated water. Water Air Soil Pollut 158:311–323
Dharmagunawardhane HA, Disanayake CB (1993) Fluoride problemsin Sri Lanka. Environ Manage Health 4:9–16
Dhiman SD, Keshari AK (2006) Hydrogeochemical evaluation of high-fluoride groundwaters: a case study from Mehsana district, Gujarat, India. Hydrol Sci J 51:1149–1162. doi:10.1623/hysj.51.6.1149
Dieye A, Larchet C, Auclair B, Mar-Diop C (1998) Elimination des fluorures par la dialyse ionique croisée. Eur Polym J 34:67–75
Dissanayake CB (1991) The fluoride problem in the groundwater of Sri Lanka-environmental management and health. Int J Environ Stud 38:137–156
Dissanayake CB (2005) Water quality in the dry zone of Sri Lanka-some interesting health aspects. J Nat Sci Found Sri Lanka 33:161–168

Dissanayake CB, Chandrajith R (2009) Introduction to medical geology-focus on tropical environment. Springer, Heidelberg
Dolejš D, Baker D (2004) Thermodynamic analysis of the system Na2O-K2O-CaO-Al2O3-SiO2-H2O-F2O-1: stability of fluorine-bearing minerals in felsic igneous suites. Contrib Mineral Petrol 146:762–778. doi:10.1007/s00410-003-0533-3
Dou X, Zhang Y, Wang H, Wang T, Wang Y (2011) Performance of granular zirconium-iron oxide in the removal of fluoride from drinking water. Water Res 45:3571–3578
Duckworth CS, Duckworth R (1978) The ingestion of fluoride in tea. Br Dent J 145:368–370
Earle S, Krogh E (2004) Geochemistry of Gabriola's groundwater. Shale J Gabriola His Mus Soc 7:35–42
Edmunds WM, Smedley PL (2005) Fluoride in natural waters. In: Selinus O (ed) Essentials of medical geology. Elsevier Academic Press, London, pp 301–329
Ellis AJ, Mahon WAJ (1977) Chemistry and geothermal systems. Academic, New York
Eskandarpour A, Onyango MS, Ochieng A, Asai S (2008) Removal of fluoride ions from aqueous solution at low pH using schwertmannite. J Hazard Mater 152:571–579
Evans-Tokaryk K (2011) Biogeochemical defluoridation. Department of Geology, University of Toronto
Fan X, Parker DJ, Smith MD (2003) Adsorption kinetics of fluoride on low cost materials. Water Res 37:4929–4937
Fantong W, Satake H, Ayonghe S, Suh E, Adelana S, Fantong E, Banseka H, Gwanfogbe C, Woincham L, Uehara Y, Zhang J (2010) Geochemical provenance and spatial distribution of fluoride in groundwater of Mayo Tsanaga River Basin, Far North Region, Cameroon: implications for incidence of fluorosis and optimal consumption dose. Environ Geochem Health 32:147–163. doi:10.1007/s10653-009-9271-4
Farooqi A, Masuda H, Firdous N (2007a) Toxic fluoride and arsenic contaminated groundwater in the Lahore and Kasur districts, Punjab, Pakistan and possible contaminant sources. Environ Pollut 145:839–849
Farooqi A, Masuda H, Kusakabe M, Naseem M, Firdous N (2007b) Distribution of highly arsenic and fluoride contaminated groundwater from east Punjab, Pakistan, and thecontrolling role of anthropogenic pollutants in the natural hydrological cycle. Geochem J 41:213–234
Farrah H, Pickering W (1986) Interaction of dilute fluoride solutions with hydrous iron oxides. Soil Res 24:201–208. doi:http://dx.doi.org/10.1071/SR9860201
Farrah H, Slavek J, Pickering W (1987) Fluoride interactions with hydrous aluminum oxides and alumina. Soil Res 25:55–69. doi:http://dx.doi.org/10.1071/SR9870055
Faure G (1991) Principles and applications of inorganic geochemistry. Macmillan, New York, pp 626
Fawell J, Bailey K, Chilton J, Dahi E, Fewtrell L, Magara Y (2006) Fluoride in drinking-water. IWA Publishing, London
Feng YW, Ogura N, Feng ZW, Zhang F, Shimizu H (2003) The concentrations and sources of fluoride in atmospheric depositions in Beijing, China. Water Air Soil Pollut 145:95–107
Fuge R (1977) On the behaviour of fluorine and chlorine during magmatic differentiation. Contrib Mineral Petrol 61:245–249. doi:10.1007/bf00376699
Fuhong R, Shuqin J (1988) Distribution and formation of high-fluorine groundwater in China. Environ Geol 12:3–10. doi:10.1007/bf02574820
Fung KF, Zhang ZQ, Wong JWC, Wong MH (1999) Fluoride contents in tea and soil from tea plantations and the release of fluoride into tea liquor during infusion. Environ Pollut 104:197–205
Gaciri SJ, Davies TC (1993) The occurrence and geochemistry of fluoride in some natural waters of Kenya. J Hydrol 143:395–412
Gago C, Romar A, Fernández-Marcos ML, Álvarez E (2012) Fluorine sorption by soils developed from various parent materials in Galicia (NW Spain). J Colloid Interface Sci 374:232–236
Ganvir V, Das K (2011) Removal of fluoride from drinking water using aluminum hydroxide coated rice husk ash. J Hazard Mater 185:1287–1294
Gao S, Cui J, Wei Z (2009a) Study on the fluoride adsorption of various apatite materials in aqueous solution. J Fluor Chem 130:1035–1041

Gao S, Sun R, Wei Z, Zhao H, Li H, Hu F (2009b) Size-dependent defluoridation properties of synthetic hydroxyapatite. J Fluor Chem 130:550–556
Garmes H, Persin F, Sandeaux J, Pourcelly G, Mountadar M (2002) Defluoridation of groundwater by a hybrid process combining adsorption and Donnan dialysis. Desalination 145:287–291
Ghorai S, Pant KK (2005) Equilibrium, kinetics and breakthrough studies for adsorption of fluoride on activated alumina. Sep Purif Technol 42:265–271
Gilpin L, Johnson AH (1980) Fluorine in agricultural soils of southeastern Pennsylvania. Soil Sci Soc Am J 44:255–258
Gizaw B (1996) The origin of high bicarbonate and fluoride concentrations in waters of the main Ethiopian Rift Valley, East African Rift system. J Afr Earth Sci 22:391–402
Gogoi PK, Baruah R (2008) Fluoride removal from water by adsorption on acid activated kaolinite clay. Indian J Chem Technol 15:500–503
Goswami D, Das AK (2006) Removal of fluoride from drinking water using a modified fly ash adsorbent. J Sci Ind Res 65:77–79
Graham HN (1992) Green tea composition, consumption, and polyphenol chemistry. Prev Med 21:334–350
Guo Q, Wang Y, Ma T, Ma R (2007) Geochemical processes controlling the elevated fluoride concentrations in groundwaters of the Taiyuan Basin, Northern China. J Geochem Explor 93:1–12
Gupta MK, Singh V, Rajwanshi P, Agarwal M, Rai K, Srivastava S, Shrivastav R, Dass S (1999) Groundwater quality assessment of Tehsil Kheragarh, Agra (India) with special reference to fluoride. Environ Monit Assess 59:275–285. doi:10.1023/a:1006117604763
Gupta SK, Deshpande RD, Agarwal M, Raval BR (2005) Origin of high fluoride in groundwater in the North Gujarat-Cambay region, India. Hydrogeol J 13:596–605. doi:10.1007/s10040-004-0389-2
Gupta VK, Ali I, Saini VK (2007) Defluoridation of wastewaters using waste carbon slurry. Water Res 41:3307–3316
Gupta AK, Deva D, Sharma A, Verma N (2009) Adsorptive removal of fluoride by micro-nano hierarchical web of activated carbon fibers. Ind Eng Chem Res 48:9697–9707
Haidouti C (1991) Fluoride distribution in soils in the vicinity of a point emission source in Greece. Geoderma 49:129–138
Hamdi N, Srasra E (2007) Removal of fluoride from acidic wastewater by clay mineral: effect of solid-liquid ratios. Desalination 206:238–244
Handa BK (1975) Geochemistry and genesis of fluoride-containing ground waters in India. Ground Water 13:275–281. doi:10.1111/j.1745-6584.1975.tb03086.x
Harmon JA, Kalichman SC (1965) Defluoridation of drinking water in Southern California. J Am Water Work Assoc 57:245–251
He GL, Cao SR (1996) Assessment of fluoride removal from drinking water by calcium phosphate systems. Fluoride 29:212–216
Hem JD (1985) The study and interpretation of the chemical characteristics of natural water, U.S. Geological Survey Water-Supply
Hichour M, Persin F, Sandeaux J, Gavach C (1999a) Fluoride removal from waters by Donnan dialysis. Sep Purif Technol 18:1–11
Hichour M, Persin F, Molénat J, Sandeaux J, Gavach C (1999b) Fluoride removal from diluted solutions by Donnan dialysis with anion-exchange membranes. Desalination 122:53–62
Hyndman DW (1985) Petrology of igneous rocks, 2 edn. McGraw-Hill, New York
IGRAC (2003a) Fluoride in groundwater. International Groundwater Resources Assessment Centre
IGRAC (2003b) Fluoride in groundwater. International Groundwater Resources Assessment Centre
Indermitte E, Saava A, Karro E (2009) Exposure to high fluoride drinking water and risk of dental fluorosis in Estonia. Int J Environ Res Public Health 6:710–721
IPCS (2002) Fluorides. World Health Organization International Programme on Chemical Safety (Environmental Health Criteria 227), Geneva

Ish T, Suckling G (1991) The severity of dental fluorosis in children exposed to water with a high fluoride content for various periods of time. J Dent Res 70:952–956. doi:10.1177/002203459 10700060801

Islam M, Patel RK (2007) Evaluation of removal efficiency of fluoride from aqueous solution using quick lime. J Hazard Mater 143:303–310

Islam M, Patel R (2011) Thermal activation of basic oxygen furnace slag and evaluation of its fluoride removal efficiency. Chem Eng J 169:68–77

Jacks G, Rajagopalan K, Alveteg T, Jönsson M (1993) Genesis of high-F groundwaters, Southern India. Appl Geochem 8(Suppl 2):241–244

Jacks G, Bhattacharya P, Chaudhary V, Singh KP (2005) Controls on the genesis of some high-fluoride groundwaters in India. Appl Geochem 20:221–228

Jagtap S, Thakre D, Wanjari S, Kamble S, Labhsetwar N, Rayalu S (2009) New modified chitosan-based adsorbent for defluoridation of water. J Colloid Interface Sci 332:280–290

Jagtap S, Yenkie MKN, Labhsetwar N, Rayalu S (2011) Defluoridation of drinking water using chitosan based mesoporous alumina. Micropor Mesopor Mater 142:454–463

Jagtap S, Yenkie MK, Labhsetwar N, Rayalu S (2012) Fluoride in drinking water and defluoridation of water. Chem Rev 112:2454–2466. doi:10.1021/cr2002855

Jain S, Jayaram RV (2009) Removal of fluoride from contaminated drinking water using unmodified and aluminium hydroxide impregnated blue lime stone waste. Sep Sci Technol 44:1436–1451. doi:10.1080/01496390902766074

Jain M, Kulshrestha UC, Sarkar AK, Parashar DC (2000) Influence of crustal aerosols on wet deposition at urban and rural sites in India. Atmos Environ 34:5129–5137

Jiménez-Núñez ML, Olguín MT, Solache-Ríos M (2007) Fluoride removal from aqueous solutions by magnesium, nickel, and cobalt calcined hydrotalcite-like compounds. Sep Sci Technol 42:3623–3639. doi:10.1080/01496390701626669

Jones BF, Eugster HP, Rettig SL (1977) Hydrochemistry of the Lake Magadi Basin, Kenya. Geochim Cosmochim Acta 41:53–72

Kabata-Pendias A, Pendias H (2001) Trace elements in soils and plants, 3rd edn. CRC, Boca Raton

Kagne S, Jagtap S, Dhawade P, Kamble SP, Devotta S, Rayalu SS (2008) Hydrated cement: a promising adsorbent for the removal of fluoride from aqueous solution. J Hazard Mater 154:88–95

Kagne S, Jagtap S, Thakare D, Devotta S, Rayalu SS (2009) Bleaching powder: a versatile adsorbent for the removal of fluoride from aqueous solution. Desalination 243:22–31

Kamble SP, Dixit P, Rayalu SS, Labhsetwar NK (2009) Defluoridation of drinking water using chemically modified bentonite clay. Desalination 249:687–693

Kamble SP, Deshpande G, Barve PP, Rayalu S, Labhsetwar NK, Malyshew A, Kulkarni BD (2010) Adsorption of fluoride from aqueous solution by alumina of alkoxide nature: Batch and continuous operation. Desalination 264:15–23

Kar S, Maity JP, Jean J-S, Liu C-C, Nath B, Yang H-J, Bundschuh J (2010) Arsenic-enriched aquifers: occurrences and mobilization of arsenic in groundwater of Ganges Delta Plain, Barasat, West Bengal, India. Appl Geochem 25:1805–1814. doi:http://dx.doi.org/10.1016/j.apgeochem.2010.09.007

Karthikeyan M, Elango KP (2008) Removal of fluoride from aqueous solution using graphite: a kinetic and thermodynamic study. Indian J Chem Technol 15:525–532

Karthikeyan M, Elango KP (2009) Removal of fluoride from water using aluminium containing compounds. J Environ Sci 21:1513–1518

Karthikeyan G, Pius A, Alagumuthu G (2005) Fluoride adsorption studies of montmorillonite clay. Indian J Chem Technol 12:263–272

Karthikeyan M, Satheeshkumar KK, Elango KP (2009) Removal of fluoride ions from aqueous solution by conducting polypyrrole. J Hazard Mater 167:300–305

Kaseva ME (2006) Optimization of regenerated bone char for fluoride removal in drinking water: a case study in Tanzania. J Water Health 4:139–147

Kau PMH, Smith DW, Binning P (1997) Fluoride retention by kaolin clay. J Contam Hydrol 28:267–288

Kau PMH, Smith DW, Binning P (1998) Experimental sorption of fluoride by kaolinite and bentonite. Geoderma 84:89–108

Kausar R, Ahmad S, Rehman KL, Ahmad R (2003) Fluoride status of underground water of Faisalabad City area. Int J Agric Biol 5:536–539

Kemer B, Ozdes D, Gundogdu A, Bulut VN, Duran C, Soylak M (2009) Removal of fluoride ions from aqueous solution by waste mud. J Hazard Mater 168:888–894

Kettunen R, Keskitalo P (2000) Combination of membrane technology and limestone filtration to control drinking water quality. Desalination 131:271–283

Kim K, Jeong GY (2005) Factors influencing natural occurrence of fluoride-rich groundwaters: a case study in the southeastern part of the Korean Peninsula. Chemosphere 58:1399–1408

Kim Y, Kim J-Y, Kim K (2011) Geochemical characteristics of fluoride in groundwater of Gimcheon, Korea: lithogenic and agricultural origins. Environ Earth Sci 63:1139–1148. doi:10.1007/s12665-010-0789-7

Kohut AP, Foweraker J, Hodge W (2001) Ground water resources of the basins, lowlands, and plains: gulf islands, ground water resources of British Columbia. British Columbia Ministry of Environment

Kowalski F (1999) Fluoridation. J Am Water Works Assoc 91:4–4

Kruse EK, Ainchil JA (2003) Fluoride variations in groundwater of an area in Buenos Aires province, Argentina. Environ Geol 44:86–89. doi:10.1007/s00254-002-0702-0

Ku Y, Chiou H-M (2002) The adsorption of fluoride ion from aqueous solution by activated alumina. Water Air Soil Pollut 133:349–361. doi:10.1023/a:1012929900113

Kumar E, Bhatnagar A, Ji M, Jung W, Lee S-H, Kim S-J, Lee G, Song H, Choi J-Y, Yang J-S, Jeon B-H (2009) Defluoridation from aqueous solutions by granular ferric hydroxide (GFH). Water Res 43:490–498

Kumar E, Bhatnagar A, Kumar U, Sillanpää M (2011) Defluoridation from aqueous solutions by nano-alumina: characterization and sorption studies. J Hazard Mater 186:1042–1049

Kundu N, Panigrahi MK, Tripathy S, Munshi, Powell MA, Hart BR (2001) Geochemical appraisal of fluoride contamination of groundwater in the Nayagarh district of Orissa, India. Environ Geol 41:451–460

Larsen PR, Davies TF, Schlumberger MJ, Hay ID (2002) Thyroid physiology and diagnostic evaluation of patients with thyroid disorders. In: Larsen PR et al (eds) Williams textbook of endocrinology. Saunders, Philadelphia, pp 331–373

Lee G, Chen C, Yang S-T, Ahn W-S (2010) Enhanced adsorptive removal of fluoride using mesoporous alumina. Micropor Mesopor Mater 127:152–156

Leyva-Ramos R, Rivera-Utrilla J, Medellin-Castillo NA, Sanchez-Polo M (2010) Kinetic modeling of fluoride adsorption from aqueous solution onto bone char. Chem Eng J 158:458–467

Li XS, Zhi JL, Gao RO (1994) Effect of fluoride exposure on intelligence in children. Fluoride 28:189–192

Li Y-H, Wang S, Cao A, Zhao D, Zhang X, Xu C, Luan Z, Ruan D, Liang J, Wu D, Wei B (2001) Adsorption of fluoride from water by amorphous alumina supported on carbon nanotubes. Chem Phys Lett 350:412–416

Li YH, Wang S, Zhang X, Wei J, Xu C, Luan Z, Wu D (2003a) Adsorption of fluoride from water by aligned carbon nanotubes. Mater Res Bull 38:469–476

Li Z, Tainosho Y, Shiraishi K, Owanda M (2003b) Chemical characteristics of fluorine-bearing biotite of early paleozoic plutonic rocks from the Sor Rondane mountains, East Antarctica. Geochem J 37:145–161

Li W, Cao C-Y, Wu L-Y, Ge M-F, Song W-G (2011a) Superb fluoride and arsenic removal performance of highly ordered mesoporous aluminas. J Hazard Mat 198:143–150

Li X, Hou X, Zhou Z, Liu L (2011b) Geochemical provenance and spatial distribution of fluoride in groundwater of Taiyuan basin, China. Environ Earth Sci 62:1635–1642. doi:10.1007/s12665-010-0648-6

Liang W, Zhan L, Piao L, Rüssel C (2011) Fluoride removal performance of glass derived hydroxyapatite. Mater Res Bull 46:205–209

Liao X-P, Shi B (2005) Adsorption of fluoride on zirconium(IV)-impregnated collagen fiber. Environ Sci Technol 39:4628–4632. doi:10.1021/es0479944

Liu Q, Guo H, Shan Y (2010) Adsorption of fluoride on synthetic siderite from aqueous solution. J Fluor Chem 131:635–641

Lung S-CC, Hsiao P-K, Chiang K-M (2003) Fluoride concentrations in three types of commercially packed tea drinks in Taiwan. J Expo Anal Environ Epidemiol 13:66–73

Lv L, He J, Wei M, Evans DG, Duan X (2006) Factors influencing the removal of fluoride from aqueous solution by calcined Mg-Al-CO3 layered double hydroxides. J Hazard Mater 133:119–128

Ma W, Ya F-Q, Han M, Wang R (2007) Characteristics of equilibrium, kinetics studies for adsorption of fluoride on magnetic-chitosan particle. J Hazard Mater 143:296–302

Ma Y, Wang S-G, Fan M, Gong W-X, Gao B-Y (2009) Characteristics and defluoridation performance of granular activated carbons coated with manganese oxides. J Hazard Mater 168:1140–1146

Ma Y, Shi F, Zheng X, Ma J, Gao C (2011) Removal of fluoride from aqueous solution using granular acid-treated bentonite (GHB): batch and column studies. J Hazard Mater 185:1073–1080

Madhavan N, Subramanian V (2002) Fluoride in fractionated soil samples of Ajmer district, Rajasthan. J Environ Monitor 4:821–822

Madhnure P, Sirsikar DY, Tiwari AN, Ranjan B, Malpe DB (2007) Occurence of fluoride in the groundwaters Pandharkawada area, Yawatmal district, Maharastra, India. Curr Sci 92:675–679

Mahramanlioglu M, Kizilcikli I, Bicer IO (2002) Adsorption of fluoride from aqueous solution by acid treated spent bleaching earth. J Fluor Chem 115:41–47

Maiti A, Basu JK, De S (2011) Chemical treated laterite as promising fluoride adsorbent for aqueous system and kinetic modeling. Desalination 265:28–36

Malakootian M, Fatehizadeh A, Yousefi N, Ahmadian M, Moosazadeh M (2011) Fluoride removal using Regenerated Spent Bleaching Earth (RSBE) from groundwater: case study on Kuhbonan water. Desalination 277:244–249

Malde M, Greiner-Simonsen R, Julshamn K, Bjorvatn K (2006) Tealeaves may release or absorb fluoride, depending on the fluoride content of water. Sci Total Environ 366:915–917

Maliyekkal SM, Sharma AK, Philip L (2006) Manganese-oxide-coated alumina: a promising sorbent for defluoridation of water. Water Res 40:3497–3506

Maliyekkal SM, Shukla S, Philip L, Nambi IM (2008) Enhanced fluoride removal from drinking water by magnesia-amended activated alumina granules. Chem Eng J 140:183–192

Maliyekkal SM, Anshup, Antony KR, Pradeep T (2010) High yield combustion synthesis of nanomagnesia and its application for fluoride removal. Sci Total Environ 408:2273–2282

Mandal BK, Suzuki KT (2002) Arsenic round the world: a review. Talanta 58:201–235

Matvaelastofnun (2010) Volcanic activity in Iceland. Possible effects on food safety, animal health and animal welfare, MAST-Icelandic Food and Veterinary Authority. http://www.mast.is/Uploads/document/leidbeiningar/Enska/volcaniceruptioninfofoodsafetyanimalhealth.pdf. Accessed on 15 Feb 2013

McKee RH, Johnston WS (1934) Removal of fluorides from drinking water. Ind Eng Chem 26:849–851

Meenakshi, Garg VK, Kavita, Renuka, Malik A (2004) Groundwater quality in some villages of Haryana, India: focus on fluoride and fluorosis. J Hazard Mater 106:85–97

Messaïtfa A (2008) Fluoride contents in groundwaters and the main consumed foods (dates and tea) in Southern Algeria region. Environ Geol 55:377–383. doi:10.1007/s00254-007-0983-4

Ming L, Yi SR, Hua ZJ, Lei BYW, Ping L, Fuwa KC (1983) Elimination of excess fluoride in potable water with coacervation by electrolysis using aluminium anode. Fluoride 20:54–63

Mirlean N, Roisenberg A (2007) Fluoride distribution in the environment along the gradient of a phosphate-fertilizer production emission (Southern Brazil). Environ Geochem Health 29:179–187. doi:10.1007/s10653-006-9061-1

Misra AK, Mishra A, Premraj (2006) Escalation of groundwater fluoride in the Ganga Alluvial plain of India. Fluoride 39:35–38

Moges G, Zewge F, Socher M (1996) Preliminary investigations on the defluoridation of water using fired clay chips. J Afr Earth Sci 22:479–482
Moghaddam AA, Fijani E (2008) Distribution of fluoride in groundwater of Maku area, Northwest of Iran. Environ Geol 56:281–287. doi:10.1007/s00254-007-1163-2
Mohan SV, Ramanaiah SV, Rajkumar B, Sarma PN (2007) Biosorption of Fuoride from Aqueous Phase onto Algal Spirogyra IO1 and Evaluation of Adsorption Kinetics. Bioresour Technol 98:1006
Mohan D, Sharma R, Singh VK, Steele P, Pittman CU (2011) Fluoride removal from water using bio-char, a green waste, low-cost adsorbent: equilibrium uptake and sorption dynamics modeling. Ind Eng Chem Res 51:900–914. doi:10.1021/ie202189v
Mohapatra M, Anand S, Mishra BK, Giles DE, Singh P (2009) Review of fluoride removal from drinking water. J Environ Manage 91:67–77
Mohapatra M, Rout K, Singh P, Anand S, Layek S, Verma HC, Mishra BK (2011) Fluoride adsorption studies on mixed-phase nano iron oxides prepared by surfactant mediation-precipitation technique. J Hazard Mater 186:1751–1757
Mohapatra M, Hariprasad D, Mohapatra L, Anand S, Mishra BK (2012) Mg-doped nano ferrihydrite-A new adsorbent for fluoride removal from aqueous solutions. Appl Surf Sci 258:4228–4236
Mondal NC, Prasad RK, Saxena VK, Singh Y, Singh VS (2009) Appraisal of highly fluoride zones in groundwater of Kurmapalli watershed, Nalgonda district, Andhra Pradesh (India). Environ Earth Sci 59:63–73. doi:10.1007/s12665-009-0004-x
Moore RB (2004) Quality of water in the fractured-bedrock aquifer of New Hampshire, U.S. Geological Survey Scientific Investigation, pp 2004–5093
Mourabet M, El Boujaady H, El Rhilassi A, Ramdane H, Bennani-Ziatni M, El Hamri R, Taitai A (2011) Defluoridation of water using Brushite: equilibrium, kinetic and thermodynamic studies. Desalination 278:1–9
Msonda KWM, Masamba WRL, Fabiano E (2007) A study of fluoride groundwater occurrence in Nathenje, Lilongwe, Malawi. Phys Chem Earth Parts A/B/C 32:1178–1184
Mukherjee A, Bhattacharya P, Shi F, Fryar AE, Mukherjee AB, Xie ZM, Jacks G, Bundschuh J (2009) Chemical evolution in the high arsenic groundwater of the Huhhot basin (Inner Mongolia, PR China) and its difference from the western Bengal basin (India). Appl Geochem 24:1835–1851. doi:http://dx.doi.org/10.1016/j.apgeochem.2009.06.005
Muralidharan D, Nair AP, Sathyanarayana U (2002) Fluoride in shallow aquifers in Rajgarh Tehsil of Churu district, Rajasthan-an arid environment. Curr Sci 83:699–702
Murray JJ (1986) (Ed.) Appropriate use of fluorides for human health. World Health Organization, Geneva
Naeem M, Khan K, Rehman S, Iqbal J (2007) Environmental assessment of ground water quality of Lahore area, Punjab, Pakistan. J Appl Sci 7:41–46
Nanyaro JT, Aswathanarayana U, Mugure JS (1984a) A geochemical model for the abnormal fluoride concentration in waters in parts of Northern Tanzania. J Afr Earth Sci 2:129–140
Nanyaro JT, Aswathanarayana U, Mungure JS, Laiiermo PW (1984b) A geochemicalmodel for the abnormal fluoride concentrations in waters in parts of Northern Tanzania. J Afr Earth Sci 2:129–140
Naseem S, Rafique T, Bashir E, Bhanger MI, Laghari A, Usmani TH (2010) Lithological influences on occurrence of high-fluoride groundwater in Nagar Parkar area, Thar Desert, Pakistan. Chemosphere 78:1313–1321
Ndiaye PI, Moulin P, Dominguez L, Millet JC, Charbit F (2005) Removal of fluoride from electronic industrial effluentby RO membrane separation. Desalination 173:25–32. doi:citeulike-article-id:9855464
Neal C (1989) Fluorine variations in welsh streams and soil waters. Sci Total Environ 80:213–223
Nie Y, Hu C, Kong C (2012) Enhanced fluoride adsorption using Al (III) modified calcium hydroxyapatite. J Hazard Mater 233–234:194–199

Nogami K, Iguchi M, Ishihara K, Hirabayashi J, Miki D (2006) Behavior of fluorineand chlorine in volcanic ash of Sakurajima volcano, Japan in the sequence of its eruptiveactivity. Earth Planets Space 58:595–600

Nordstrom DK, Jenne EA (1977) Fluorite solubility equilibria in selected geothermal waters. Geochim Cosmochim Acta 41:175–188

NRC (2006) Fluoride in drinking water: a scientific review of EPA's standards. The National Academies, Washington, DC

Oguz E (2007) Equilibrium isotherms and kinetics studies for the sorption of fluoride on light weight concrete materials. Colloids Surf A 295:258–263

Okamura T, Matsuhisa T (1965) The content of fluorine in cigarettes. J Food Hyg Soc Jpn 6:382–385

Onyango MS, Kojima Y, Aoyi O, Bernardo EC, Matsuda H (2004) Adsorption equilibrium modeling and solution chemistry dependence of fluoride removal from water by trivalent-cation-exchanged zeolite F-9. J Colloid Interface Sci 279:341–350

Onyango MS, Kojima Y, Kumar A, Kuchar D, Kubota M, Matsuda H (2006) Uptake of fluoride by Al3+ pretreated low-silica synthetic zeolites: adsorption equilibrium and rate studies. Sep Sci Technol 41:683–704. doi:10.1080/01496390500527019

Oruc N (2008) Occurrence and problems of high fluoride waters in Turkey: an overview. Environ Geochem Health 30:315–323. doi:10.1007/s10653-008-9160-2

Ozsvath D (2006) Fluoride concentrations in a crystalline bedrock aquifer Marathon County, Wisconsin. Environ Geol 50:132–138. doi:10.1007/s00254-006-0192-6

Ozsvath D (2009) Fluoride and environmental health: a review. Rev Environ Sci Biotechnol 8:59–79. doi:10.1007/s11157-008-9136-9

Padmasiri JP, Jayawardhene WM, Dissanayake CB (2012) Water quality improvements in the dry zone areas in Sri Lanka. In: Ileperuma OA, Priyantha N, Navaratne A, Yatigammana SK, Weragoda SK (eds) Symposium proceedings, International symposium on water quality and human health: challenges ahead, PGIS, Peradeniya, Sri Lanka

Patel G, Pal U, Menon S (2009) Removal of fluoride from aqueous solution by CaO nanoparticles. Sep Sci Technol 44:2806–2826

Paudyal H, Pangeni B, Nath Ghimire K, Inoue K, Ohto K, Kawakita H, Alam S (2012) Adsorption behavior of orange waste gel for some rare earth ions and its application to the removal of fluoride from water. Chem Eng J 195–196:289–296

Pekdeger A, Ozgur N, Schneider HJ (1992) Hydrogeochemistry of fluorine in shallow aqueous systems of the Golcuk area, southwest turkey. In: Kharaka YK, Maest AS (eds) Seventh international symposium on water rock interactions. Balkema, Rotterdam

Pickering WF (1985) The mobility of soluble fluoride in soils. Environ Pollut Ser B Chem Phys 9:281–308

Pietrelli L (2005) Fluoride wastewater treatment by adsorption onto metallurgical grade alumina. Ann Chim (Rome) 95:303–312. doi:10.1002/adic.200590035

Pontie M, Buisson H, Diawara CK, Essis-Tome H (2003) Studies of halide ions mass transfer in nanofiltration - application to selective defluorination of brackish drinking water. Desalination 157:127–134. doi:http://dx.doi.org/10.1016/S0011-9164(03)00391-6

Poovaiah BW (1988) Senescence and ageing of plants. In: Nooden L, Leopold AC (eds). Academic, New York 369–389

Puri BK, Balani S (2000) Trace determination of fluoride using lanthanum hydroxide supported on alumina. J Environ Sci Health Part A 35:109–121. doi:10.1080/10934520009376957

Queste A, Lacombe M, Hellmeier W (2001) High concentrations of fluoride and boron in drinking water wells in the Muenster region – Results of a preliminary investigation. International Journal of Hygiene and Environmental Health 203:221–224

Raju JN, Dey S, Das K (2009) Fluoride contamination in groundwaters of Sonbhadra district, Uttar Pradesh, India. Curr Sci 96:979–985

Ramamohana Rao NV, Rao N, Surya Prakash Rao K, Schuiling RD (1993) Fluorine distribution in waters of Nalgonda district, Andhra Pradesh, India. Environ Geol 21:84–89. doi:10.1007/bf00775055

Ramanaiah SV (2010) Bioremediation of fluoride by biosorption. VDM Verlag
Ramanaiah SV, Venkata Mohan S, Sarma PN (2007) Adsorptive removal of fluoride from aqueous phase using waste fungus (Pleurotus ostreatus 1804) biosorbent: kinetics evaluation. Ecol Eng 31:47–56
Ramdani A, Taleb S, Benghalem A, Ghaffour N (2010) Removal of excess fluoride ions from Saharan brackish water by adsorption on natural materials. Desalination 250:408–413
Rao N (2009) Fluoride in groundwater, Varaha River Basin, Visakhapatnam district, Andhra Pradesh, India. Environ Monit Assess 152:47–60. doi:10.1007/s10661-008-0295-5
Rao NVRM, Bhaskaran CS (1988) Studies on defluoridation of water. J Fluor Chem 41:17–24
Rao KV, Mahajan CL (1990) Fluoride content of some common South Indian foods and their contribution to fluorosis. J Sci Food Agric 51:275–279. doi:10.1002/jsfa.2740510215
Rao KV, Purushottam D, Vaidyanadham D (1975) Uptake of fluoride by serpentine. Geochim Cosmochim Acta 39:1403–1411
Ray D, Ravindar Rao R, Bhoi AV, Biswas AK, Ganguly AK, Sanyal PB (2000) Physico-chemical quality of drinking water in Rohtas district of Bihar. Environ Monit Assess 61:387–398. doi:10.1023/a:1006165615097
Rea RE (1979) A rapid method for the determination of fluoride in sewage sludges. Water Pollut Control (Maidstone) 78:139–142
Reardon EJ, Wang Y (2000) A limestone reactor for fluoride removal from wastewaters. Environ Sci Technol 34:3247–3253. doi:10.1021/es990542k
Reddy A, Reddy D, Rao P, Prasad K (2010) Hydrogeochemical characterization of fluoride rich groundwater of Wailpalli watershed, Nalgonda district, Andhra Pradesh, India. Environ Monit Assess 171:561–577. doi:10.1007/s10661-009-1300-3
Robertson FN (1986) Occurrence and solubility controls of trace elements in groundwater in alluvial basins of Arizona. Am Water Res Assoc Monog Series 7:69–80
Robinson GR, Kapo KE (2003) Generalized lithology and lithogeochemical character of near-surface bedrock in the New England Region, U.S. Geological Survey
Rose D, Marier JR (1977) Environmental fluoride NRC associate committee on scientific criteria for environmental quality, National Research Council of Canada
Rosi M, Papale P, Lupi L, Stoppato M (2003) Volcanoes. Firefly Books Ltd, Spain
Ruixia L, Jinlong G, Hongxiao T (2002) Adsorption of fluoride, phosphate, and arsenate ions on a new type pf ion exchange fiber. J Colloid Interface Sci 248:268–274
Saether OM, Andreassen BT (1989) Fluoride in precipitation in Southern Norway: amounts and sources, Geol Survey Norway
Saether OM, Andreassen BT, Semb A (1995) Amounts and sources of fluoride in precipitation over southern Norway. Atmos Environ 29:1785–1793
Sairam Sundaram C, Viswanathan N, Meenakshi S (2009a) Defluoridation of water using magnesia/chitosan composite. J Hazard Mater 163:618–624
Sairam Sundaram C, Viswanathan N, Meenakshi S (2009b) Fluoride sorption by nano-hydroxyapatite/chitin composite. J Hazard Mater 172:147–151
Samatya S, Yüksel Ü, Yüksel M, Kabay N (2007) Removal of fluoride from water by metal ions (Al3+, La3+ and ZrO2+) loaded natural zeolite. Sep Sci Technol 42:2033–2047. doi:10.1080/01496390701310421
Sarkar M, Banerjee A, Pramanick PP, Sarkar AR (2006) Use of laterite for the removal of fluoride from contaminated drinking water. J Colloid Interface Sci 302:432–441
Sarkar M, Banerjee A, Pramanick PP, Sarkar AR (2007) Design and operation of fixed bed laterite column for the removal of fluoride from water. Chem Eng J 131:329–335. doi:10.1016/j.cej.2006.12.016
Sathish RS, Raju NSR, Raju GS, Rao GN, Kumar KA, Janardhana C (2007) Equilibrium and kinetic studies for fluoride adsorption from water on zirconium impregnated coconut shell carbon. Sep Sci Technol 42:769–788
Sathish RS, Sairam S, Raja VG, Rao GN, Janardhana C (2008) Defluoridation of water using zirconium impregnated coconut fiber carbon. Sep Sci Technol 43:3676–3694

Satsangi GS, Lakhani A, Khare P, Singh SP, Kumari KM, Srivastava SS (1998) Composition of rain water at a semi-arid rural site in India-Inter-relation of acid base components. Atmos Environ 32:3783–3793
Satsangi GS, Lakhani A, Khare P, Singh SP, Kumari KM, Srivastava SS (2002) Measurements of major ion concentration in settled coarse particles and aerosols at a semiarid rural site in India. Environ Int 28:1–7
Sawyer GM, Oppenheimer C (2006) Volcanic fluorine emissions: observations by Fourier transform infrared spectroscopy. In: Tressaud A (ed) Fluorine and the environment: atmospheric chemistry, emissions, and lithosphere. Elsevier, Amsterdam
Saxena VS, Ahmed SA (2001) Dissolution of fluoride in groundwater: a water-rock interaction study. Environ Geol 40:1084–1087. doi:10.1007/s002540100290
Saxena VS, Ahmed S (2003) Inferring the chemical parameters for the dissolution of fluoride in groundwater. Environ Geol 43:731–736. doi:10.1007/s00254-002-0672-2
Scaillet B, Macdonald R (2004) Fluorite stability in silicic magmas. Contrib Mineral Petrol 147:319–329. doi:10.1007/s00410-004-0559-1
Schneiter RW, Middlebrooks EJ (1983) Arsenic and fluoride removal from groundwater by reverse osmosis. Environ Int 9:289–291. doi:http://dx.doi.org/10.1016/0160-4120(83)90087-9
Schwartz MO, Friedrich GH (1973) Secondary dispersion patterns of fluoride in the osor area, province of Gerona, Spain. J Geochem Explor 2:103–114
Shaji E, Bindu, Viju E, Thambi DS (2007) High fluoride in groundwater of Palghat district, Kerala. Curr Sci 92:240–245
Shimelis B, Zewge F, Chandravanshi BS (2006) Removal of excess fluoride from water by aluminum hydroxide. Bull Chem Soc Ethiop 20:17–34
Shu WS, Zhang ZQ, Lan CY, Wong MH (2003) Fluoride and aluminium concentrations of tea plants and tea products from Sichuan province, PR China. Chemosphere 52:1475–1482
Singh SP, Khare P, Satsangi GS, Lakhani A, Kumari SS, Srivastava SS (2001) Rainwater composition at a regional representative site of a semi-arid region of India. Water Air Soil Pollut 127:93–108
Sinha S, Pandey K, Mohan D, Singh KP (2003) Removal of fluoride from aqueous solutions by Eichhornia crassipes biomass and its carbonized form. Ind Eng Chem Res 42:6911–6918. doi:10.1021/ie030544k
Sivasamy A, Singh KP, Mohan D, Maruthamuthu M (2001) Studies on defluoridation of water by coal-based sorbents. J Chem Technol Biotechnol 76:717–722. doi:10.1002/jctb.440
Sivasankar V, Ramachandramoorthy T, Darchen A (2011) Manganese dioxide improves the efficiency of earthenware in fluoride removal from drinking water. Desalination 272:179–186
Skjelkvåle BL (1994) Factors influencing fluoride concentrations in Norwegian lakes. Water Air Soil Pollut 77:151–167. doi:10.1007/bf00483055
Sreedevi P, Ahmed S, Made B, Ledoux E, Gandolfi J (2006) Association of hydrogeological factors in temporal variations of fluoride concentration in a crystalline aquifer in India. Environ Geol 50:1–11. doi:10.1007/s00254-005-0167-z
Srimurali M, Pragathi A, Karthikeyan J (1998) A study on removal of fluorides from drinking water by adsorption onto low-cost materials. Environ Pollut 99:285–289
Stollenwerk KG (2002) Geochemical processes controlling transport of arsenic in groundwater: a review of adsorption. In: Welch AH, Stollenwerk KG (eds) Arsenic in groundwater: geochemistry and occurrence. Springer, New York
Stormer JC, Carmichael ISE (1970) Villiaumite and the occurrence of fluoride minerals in igneous rocks. Am Miner 55:126–134
Subba Rao N, John Devadas D (2003) Fluoride incidence in groundwater in an area of Peninsular India. Environ Geol 45:243–251. doi:10.1007/s00254-003-0873-3
Sujana MG, Anand S (2010) Iron and aluminium based mixed hydroxides: a novel sorbent for fluoride removal from aqueous solutions. Appl Surf Sci 256:6956–6962
Sujana MG, Anand S (2011) Fluoride removal studies from contaminated ground water by using bauxite. Desalination 267:222–227

Sujana MG, Thakur RS, Rao SB (1998) Removal of fluoride from aqueous solution by using alum sludge. J Colloid Interface Sci 206:94–101
Sujana MG, Pradhan HK, Anand S (2009a) Studies on sorption of some geomaterials for fluoride removal from aqueous solutions. J Hazard Mater 161:120–125
Sujana MG, Soma G, Vasumathi N, Anand S (2009b) Studies on fluoride adsorption capacities of amorphous Fe/Al mixed hydroxides from aqueous solutions. J Fluor Chem 130:749–754
Sujatha D (2003) Fluoride levels in the groundwater of the south-eastern part of Ranga Reddy district, Andhra Pradesh, India. Environ Geol 44:587–591. doi:10.1007/s00254-003-0795-0
Suma Latha S., Ambika SR, Prasad SJ (1999) Fluoride contamination status of groundwater in Karnataka. Current Science 76:730–734
Sun Y, Fang Q, Dong J, Cheng X, Xu J (2011) Removal of fluoride from drinking water by natural stilbite zeolite modified with Fe(III). Desalination 277:121–127
Swain SK, Dey RK, slam MI, Patel RK, Jha U, Patnaik T, Airoldi C (2009) Removal of fluoride from aqueous solution using aluminum-impregnated chitosan biopolymer. Sep Sci Technol 44:2096–2116
Swain SK, Patnaik T, Singh VK, Jha U, Patel RK, Dey RK (2011) Kinetics, equilibrium and thermodynamic aspects of removal of fluoride from drinking water using meso-structured zirconium phosphate. Chem Eng J 171:1218–1226
Swain SK, Mishra S, Patnaik T, Patel RK, Jha U, Dey RK (2012) Fluoride removal performance of a new hybrid sorbent of Zr(IV)-ethylenediamine. Chem Eng J 184:72–81
Tahir Shah M, Danishwar S (2003) Potential fluoride contamination in the drinking water of Naranji area, Northwest frontier province, Pakistan. Environ Geochem Health 25:475–481. doi:10.1023/B:EGAH.0000004579.54947.e7
Tang Y, Guan X, Su T, Gao N, Wang J (2009a) Fluoride adsorption onto activated alumina: Modeling the effects of pH and some competing ions. Colloids Surf A 337:33–38
Tang Y, Guan X, Wang J, Gao N, McPhail MR, Chusuei CC (2009b) Fluoride adsorption onto granular ferric hydroxide: effects of ionic strength, pH, surface loading, and major co-existing anions. J Hazard Mat 171:774–779
Tang Y, Wang J, Gao N (2010) Characteristics and model studies for fluoride and arsenic adsorption on goethite. J Environ Sci 22:1689–1694
Tavener SJ, Clark JH (2006) Fluorine: friend or foe? A green chemist's perspective. In: Tressaud A (ed) Fluorine and the environment: agrochemicals, archaeology, green chemistry and water. Elsevier, Amsterdam
Taylor RP, Fallick AE (1997) The evolution of fluorine-rich felsic magmas: source dichotomy, magmatic convergence and the origins of topaz granite. Terra Nova 9:105–108
Teng S-X, Wang S-G, Gong W-X, Liu X-W, Gao B-Y (2009) Removal of fluoride by hydrous manganese oxide-coated alumina: Performance and mechanism. J Hazard Mater 168:1004–1011
Teotia SPS, Teotia M (1994) Dental caries: a disorder of high fluoride and low dietary calcium interactions (30 years of personal research). Fluoride 27:59–66
Thakre D, Jagtap S, Bansiwal A, Labhsetwar N, Rayalu S (2010a) Synthesis of La-incorporated chitosan beads for fluoride removal from water. J Fluor Chem 131:373–377
Thakre D, Rayalu S, Kawade R, Meshram S, Subrt J, Labhsetwar N (2010b) Magnesium incorporated bentonite clay for defluoridation of drinking water. J Hazard Mater 180:122–130
Tor A (2006) Removal of fluoride from an aqueous solution by using montmorillonite. Desalination 201:267–276
Tor A, Danaoglu N, Arslan G, Cengeloglu Y (2009) Removal of fluoride from water by using granular red mud: batch and column studies. J Hazard Mater 164:271–278
Tripathy SS, Bersillon J-L, Gopal K (2006) Removal of fluoride from drinking water by adsorption onto alum-impregnated activated alumina. Sep Purif Technol 50:310–317
Turner BD, Binning P, Stipp SLS (2005) Fluoride removal by calcite: evidence for fluorite precipitation and surface adsorption. Environ Sci Technol 39:9561–9568. doi:10.1021/es0505090
Ullah R, Malik RN, Qadir A (2009) Assessment of groundwater contamination in an industrial City, Sialkot, Pakistan. Afr J Environ Sci Technol 3:429–446

Usunoff E, Winzettel P, Dietrich S (2009) Sorption of fluoride by quartz sand: batch tests. Geoacta 34:9–17

Valenzuela-Vásquez L, Ramírez-Hernández J, Reyes-López J, Sol-Uribe A, Lázaro-Mancilla O (2006) The origin of fluoride in groundwater supply to Hermosillo City, Sonora, México. Environ Geol 51:17–27. doi:10.1007/s00254-006-0300-7

Van Der Hoek W, Ekanayake L, Rajasooriyar L, Karunaratne R (2003) Source of drinking water and other risk factors for dental fluorosis in Sri Lanka. Int J Environ Health Res 13:285–293. doi:10.1080/0960312031000122433

Veressinina Y, Trapido M, Ahelik V, Munter R (2001) Fluoride in drinking water: the problem and its possible solutions. Proc Estonian Acad Sci 50:81–88

Vijaya Y, Krishnaiah A (2009) Sorptive response profile of chitosan coated silica in the defluoridation of aqueous solution. E-J Chem 6:713–724

Vijaya Y, Popuri SR, Reddy AS, Krishnaiah A (2011) Synthesis and characterization of glutaraldehyde-crosslinked calcium alginate for fluoride removal from aqueous solutions. J Appl Polym Sci 120:3443–3452. doi:10.1002/app.33375

Viswanathan N, Meenakshi S (2008a) Enhanced fluoride sorption using La(III) incorporated carboxylated chitosan beads. J Colloid Interface Sci 322:375–383

Viswanathan N, Meenakshi S (2008b) Selective sorption of fluoride using Fe(III) loaded carboxylated chitosan beads. J Fluor Chem 129:503–509

Viswanathan N, Meenakshi S (2010a) Selective fluoride adsorption by a hydrotalcite/chitosan composite. Appl Clay Sci 48:607–611

Viswanathan N, Meenakshi S (2010b) Development of chitosan supported zirconium(IV) tungstophosphate composite for fluoride removal. J Hazard Mater 176:459–465

Viswanathan N, Meenakshi S (2010c) Enriched fluoride sorption using alumina/chitosan composite. J Hazard Mater 178:226–232

Walna B, Kurzyca I, Siepak J (2007) Variations in fluoride level in precipitation in a region of human impact. Water Air Soil Pollut Focus 7:33–40

Walters CB, Sherlock JC, Evans WH, Read JI (1983) Dietary intake of fluoride in the United Kingdom and fluoride Content of Some foodstuffs. J Sci Food Agric 34:523–528. doi:10.1002/jsfa.2740340514

Wang S, Mulligan CN (2008) Speciation and surface structure of inorganic arsenic in solid phases: a review. Environ Int 34:867–879

Wang Y, Reardon EJ (2001) Activation and regeneration of a soil sorbent for defluoridation of drinking water. Appl Geochem 16:531–539

Wang S-G, Ma Y, Shi Y-J, Gong W-X (2009) Defluoridation performance and mechanism of nanoscale aluminum oxide hydroxide in aqueous solution. J Chem Technol Biotechnol 84:1043–1050. doi:10.1002/jctb.2131

Wang Y, Chen N, Wei W, Cui J, Wei Z (2011) Enhanced adsorption of fluoride from aqueous solution onto nanosized hydroxyapatite by low-molecular-weight organic acids. Desalination 276:161–168

Ware GW (1975) Pesticides, Theory and Application. W.H. Freeman and Company, San Francisco, CA

Weerasooriya R, Wickramarathne HUS, Dharmagunawardhane HA (1998) Surface complexation modeling of fluoride adsorption onto kaolinite. Colloids Surf A 144:267–273

Wenzel WW, Blum WEH (1992) Fluorine speciation and mobility in f-contaminated soils. Soil Sci 153:357–364

Whittemore DO, MacFarlane PA, Doveton JH, Butler JJ, Chu T, Bassler R, Smith M, Mitchell J, Wade A (1993) The Dakota aquifer program annual report, FY 92, Kansas Geological Survey open-file report 93-1

WHO (1984) Fluorine and fluorides, environmental health criteria. World Health Organization, Geneva

WHO (1996) Trace elements in human nutrition and health. World health Organization, Geneva

WHO (2000) Chapter 6.5, Fluorides. World Health Organization, Regional Office for Europe, Copenhagen

WHO (2002) Fluorides, environmental health criteria number. World Health Organization, Geneva

WHO (2004) Guidelines for drinking-water quality. Volume 1 recommendations. World Health Organization, Geneva

WHO (2006) Fluoride in drinking-water. In: Fawell J et al (eds) London. ISBN 92-4-156319-2

Whyte MP, Essmyer K, Gannon FH, Reinus WR (2005) Skeletal fluorosis and instant tea. Am J Med 118:78–82

Wodeyar BK, Sreenivasan G (1996) The occurrence and movement of ground water Bellary district, Karnataka-A preliminary study. Curr Sci 70:71–74

Wu X, Zhang Y, Dou X, Yang M (2007) Fluoride removal performance of a novel Fe-Al-Ce trimetal oxide adsorbent. Chemosphere 69:1758–1764

Xiaolin X, Zhenhua Z, Jinchu Z, Bing R, Mingyuan L (1998) Partitioning of F between aqueous fluids and albite granite melt and its petrogenetic and metallogenetic significance. Chin J Geochem 17:303–310. doi:10.1007/bf02837981

Xu X, Li Q, Cui H, Pang J, Sun L, An H, Zhai J (2011) Adsorption of fluoride from aqueous solution on magnesia-loaded fly ash cenospheres. Desalination 272:233–239

Yao R, Meng F, Zhang L, Ma D, Wang M (2009) Defluoridation of water using neodymium-modified chitosan. J Hazard Mater 165:454–460

Yiamouyiannis JA (1993) Fluoridation and cancer. The biology and epidemiology of bone and oral cancer related to fluoridation. Fluoride 26:83–96

Young SM, Pitawala A, Ishiga H (2010) Factors controlling fluoride contents of groundwater in north-central and northwestern Sri Lanka. Environmental Earth Sciences. DOI 10.1007/s12665-010-0804-z

Zhang J, Xie S, Ho Y-S (2009) Removal of fluoride ions from aqueous solution using modified attapulgite as adsorbent. J Hazard Mater 165:218–222

Zhang Z, Tan Y, Zhong M (2011) Defluorination of wastewater by calcium chloride modified natural zeolite. Desalination 276:246–252

Zhao X, Wang J, Wu F, Wang T, Cai Y, Shi Y, Jiang G (2010) Removal of fluoride from aqueous media by Fe3O4@Al(OH)3 magnetic nanoparticles. J Hazard Mater 173:102–109

Zhao B, Zhang Y, Dou X, Wu X, Yang M (2012) Granulation of Fe-Al-Ce trimetal hydroxide as a fluoride adsorbent using the extrusion method. Chem Eng J 185-186:211–218

Zhu P, Wang H, Sun B, Deng P, Hou S, Yu Y (2009) Adsorption of fluoride from aqueous solution by magnesia-amended silicon dioxide granules. J Chem Technol Biotechnol 84:1449–1455. doi:10.1002/jctb.2197

Chapter 5
Selenium in Agriculture: Water, Air, Soil, Plants, Food, Animals and Nanoselenium

Hassan R. El-Ramady, Éva Domokos-Szabolcsy, Tarek A. Shalaby, József Prokisch and Miklós Fári

Contents

H. R. El-Ramady (✉)
Soil and Water Sciences Department, Faculty of Agriculture, Kafrelsheikh University, 33516 Kafr El-Sheikh, Egypt
e-mail: ramady2000@gmail.com

É. Domokos-Szabolcsy · M. Fári
Plant Biotechnology Department, Debrecen University, AGTC Böszörményi Útca. 138, 4032 Debrecen, Hungary
e-mail: domokosszabolcsy@gmail.com

M. Fári
e-mail: fari@agr.unideb.hu

T. A. Shalaby
Horticulture Department, Faculty of Agriculture, Kafrelsheikh University, 33516 Kafr El-Sheikh, Egypt
e-mail: tashalaby@yahoo.com

J. Prokisch
Institute of Bio- and Environmental Enegetics, Debrecen University, Böszörményi Útca. 138, 4032 Debrecen, Hungary
e-mail: jprokisch@agr.unideb.hu

© Springer International Publishing Switzerland 2015

E. Lichtfouse et al. (eds.), *CO_2 Sequestration, Biofuels and Depollution*,
Environmental Chemistry for a Sustainable World 5, DOI 10.1007/978-3-319-11906-9_5

"ظَهَرَ الْفَسَادُ فِي الْبَرِّ وَالْبَحْرِ بِمَا كَسَبَتْ أَيْدِي النَّاسِ لِيُذِيقَهُمْ بَعْضَ الَّذِي عَمِلُوا لَعَلَّهُمْ يَرْجِعُونَ" الروم الآية 41

Corruption has appeared throughout the land and sea by [reason of] what the hands of people have earned so He may let them taste part of [the consequence of] what they have done that perhaps they will return [to righteousness].
*(**Holy Quran**: Ar-Room, verse 41)*

Abstract Selenium (Se) is an example of an essential element becoming more and more insufficient in food crops as a result of intensive plant production in many countries. Se is an essential biological trace element. It is an essential constituent of several enzymes in which it is present in the form of the unusual amino acid selenocysteine (SeCys). Se was first recognized as an essential nutrient in the late 1950s when it was found to replace vitamin E in the diets of rats and chicks for the prevention of vascular, muscular and/or hepatic lesions. Until that time, Se had been thought of only as a toxicant, being associated with "alkali disease" in grazing livestock in the northern Great Plains of the United States. Since that time, Se has become the subject of investigations in many parts of the world. Se enters soils primarily as a result of the weathering of Se-containing rocks, although volcanic activity, dusts such as in the vicinity of coal burning, Se-containing fertilizers, and some waters can also be sources. Se cycles through the food system, being removed from soils by plants and soil microorganisms, which can take up the element into their tissue proteins and metabolize some of it to volatile forms e.g., dimethylselenide. The latter enter the atmosphere to be brought down with precipitation and airborne particulates. This chapter reviews the present knowledge of the Se in agroecosystem. The occurrence of selenium in the environment from soil to food systems is discussed. The most promising and important nanotechnology applications in agriculture; and nano-selenium particles production, agricultural nanotechnology and its use in sustainable development will also be highlighted.

Keywords Selenium · Agroecosystem · Nanoselenium · Elemental selenium · Nanotechnology · Nanoparticles · Agrifoods

Abbreviations

ATSDR	Agency for Toxic Substances and Disease Registry
BSA	Bovine serum albumin
ETA-AAS	Electrothermal atomization—Atomic absorption spectroscopy
F-AAS	Fluorescence—Atomic absorption spectroscopy
FAO/WHO	Food and Agriculture Organization/World Health Organization
FDA	Food and Drug Administration
FW	Fresh weight
GSH	Glutathione
ICP MS	Inductively coupled plasma mass spectrometry

ICP-OES	Inductively coupled plasma optical emission spectroscopy
IUPAC	International Union of Pure and Applied Chemistry
HNO_3	Nitric acid
LC-ICP-MS	Liquid Chromatography—Inductively Coupled Plasma Mass Spectrometry
MAD	Multi-wavelength anomalous diffraction (or dispersion)
MAK	Maximum concentration of a chemical substance on air at the workplace (in German)
RDAs	Recommended dietary allowances
SeCys	Selenocysteine
SeGSH	Selenoglutathione
SeMet	Selenomethionine
SeMC	Selenomethylocysteine
SCENIHR	Scientific Committee on Emerging and Newly Identified Health Risks
US EPA	U.S. Environmental Protection Agency
WHO	World Health Organization

5.1 Selenium in Agriculture

5.1.1 Introduction

Selenium (Se) is a contradictory nutrient. It has been called *the essential poison*—too much of it in the diet can be toxic; too little can result in chronic, and sometimes fatal, deficiency. Even health authorities have at times been confused. Although today in most countries, selenium appears among the trace elements for which recommended dietary allowances (RDAs) have been established, it was at one time declared by the Food and Drug Administration (FDA) to be a carcinogen and banned as an additive in food. Undeterred by reports of their possible toxic effects, today millions of people worldwide consume selenium supplements. They are encouraged to do so, not only by articles in the popular media but also by the results of investigations by reputable scientists which indicate that selenium has a vital role to play in human health, not least in the prevention of cancer. Their findings indicate that selenium is a key player in cellular metabolism, is an essential component of enzymes that protect the body against oxidative damage, and has important roles in thyroid metabolism, human fertility, and many other vital functions (Reilly 2006).

Soil is not homogenous system, and the microscale heterogenity creates a real problem in representative sampling. Also the variability in sampling procedure from plants and other organisms is a subject of concern and this has made the assessment and evaluation of some data almost impossible. Therefore, quantitative comparisons of analytical data for soils, plants, and human/animal tissues have often been difficult. Different chemical preparations of samples—such as HNO_3 microwave decomposition, ashing with aqua regia, and total digestion—as well as different instrumental methods used for the determination of elements (e.g., ICP MS, ICP-OES, F-AAS, ETA-AAS) have an influence on final results. Luckily, analytical quality assurance and the use of reference materials have decreased uncertainties of analytical data. Therefore, each measurement of trace elements builds up a database and contributes to a better understanding of their overall distribution and behavior in given media and in the total environment (Kabata-Pendias 2011).

Common trace cation descriptors are "*trace metals*" and/or "*heavy metals*." The trace metalloids are simple "*trace elements*." The other terms: "*micronutrients*," "*essential elements*," and "*toxic elements*" are related to their physiological functions and are rather confusing since their effects on organisms and health depend upon concentrations. All these terms are inadequate, and a great deal of confusion has occurred in the literature where authors have been imprecise in their use of these terms. Especially the term "heavy metals" has recently become a subject of a broad discussion that emphasizes its nonprecise definition. Duffus (2002) has written: "Over the past two decades, the term 'heavy metals' has been widely used … and related to chemical hazards. It is often used as a group name for metals and semimetals or metalloids that have been associated with contamination and potential toxicity or ecotoxicity." This term is based on various criteria i.e., atomic weight, atomic number, density, chemical properties, etc. Thus, the inconsistent use of the term "heavy metals" reflects inconsistency in the scientific literature. This term has never been defined by any authoritative body, such as the International Union of Pure and Applied Chemistry (IUPAC) (Kabata-Pendias 2011).

Therefore, the specific objectives of this chapter are mainly to follow the distribution and occurance of Se in the different agricultural environments from soils, water, air, plants to humans and animanls. The nano-selenium, from agricultural nanotechnology and sustainable development, nanotechnology in agriculture and food, recent developments, risks and regulation of nanotechnology, and finally synthesis methods of nanoparticles will be also highlighted.

5.1.2 Discovery of the Essential Poison

The essential poison or selenium began its formal scientific life in controversy, in a laboratory in Uppsala, during an era when Swedish scientists were world leaders in the rapidly developing field of chemistry. Preeminent among them was Jöns Jakob Berzelius, an outstanding theorist and skilled experimenter. The electrochemical theory and the use of chemical symbols to represent the elements are among the valuable contributions he made to science. His laboratory skills, under conditions that may be hard for modern researchers to visualize, enabled him to determine atomic and molecular weights with such accuracy that many of his calculations are very close to those in use today. He also discovered, or isolated for the first time, several new elements, including Si, cerium (Ce), thorium (Th), vanadium (V), zirconium (Zr), and, in 1817, selenium (Hurd and Kipling 1964).

A few years earlier, a German chemist, M. H. Klaproth, who had come to work in Sweden with Berzelius had examined a reddish material found in the residue after copper pyrites had been roasted in a lead chamber during the manufacture of sulfuric acid. The residue had an unpleasant smell, described as rotten radishes, and blistered the skin of those who handled it. Klaproth, who had already made a name for himself as the discoverer of several new elements, including titanium, uranium, and zirconium, concluded that the deposit contained tellurium, another of his discoveries. This had also been found as a residue of copper ore roasting and, like the new material, had an unpleasant smell. But his friend Berzelius disagreed. In collaboration with another Swedish chemist, J. G. Gahn, who had discovered manganese, Berzelius reexamined the residue. As he explained in a letter to the French chemist Claude-Louis Berthollet, he found no evidence of the presence of tellurium, but rather of "an unknown substance with properties very like those of tellurium. For this reason I gave it the name *selenium* from the Greek word Σεληνη (Selēnē), which signifies the moon, while *tellus* is the name of our own planet" (Berzelius 1818). The letter, which was published in the French scientific journal, *Annales de Chimie et de Physique* in 1818, was the first of more than 100,000 papers on selenium to appear in the scientific press over the following 2 centuries (Table 5.1; Reilly 2006).

While Berzelius was the first to isolate and chemically characterize selenium, he was probably not its original discoverer. That achievement should, perhaps, be attributed to the thirteenth century Italian scholar, Arnold of Villanova, who, in his book *Rosarium Philosophorum*, described a "red sulfur" sometimes found on the walls of chambers in which native yellow sulfur was heated to make "flowers of sulfur." This "red sulfur," it has been argued by Hoefer (1842), could have been the same type of deposit investigated by Berzelius 500 years later. However, even of it were, Villanova was in no position to isolate and study the element (Crombie 1959). Another thirteenth century Italian, the Venetian traveler Marco Polo, may also have stumbled across selenium, or at least evidence of its less desirable qualities, at about the same time as Villanova (Fig. 5.1; Reilly 2006).

Therefore, Jöns Jakob Berzelius is the first to isolate and chemically characterize selenium, he was probably not its original discoverer. That achievement should, perhaps, be attributed to the thirteenth century Italian scholar, Arnold of Villanova and anothe Italian, the Venetian traveler Marco Polo, may also have stumbled across selenium, or at least evidence of its less desirable qualities, at about the same time as Villanova.

5.1.3 Selenium in Agroecosystem

It is well known that trace elements mean elements present at low concentrations (mg kg^{-1} or less) in agroecosystems. Some trace elements, including copper, zinc, manganese, iron, molybdenum, and boron are essential to plant growth and are called micronutrients. Except for B, these elements are also heavy metals, and are

Table 5.1 Discovery and origin and selected properties of Se comparing with some micronutrients and beneficial mineral elements (from Shehata and E1-Ramdy 2012)

	Discovery (year)	Name origin	Ran. No.	Concentration in ocean (ppm)	Elec. Neg.	Most important minerals
B	Gay- Lussac and Thénard (1808)	Arabic word *buraq or* Persian *burah* (name of borax)	37	4.44	2.04	Borax, tincal $Na_2B4O_7 \times 10H_2O$ Colemanite $Ca_2B_6O_{11} \times 5H_2O$ Kernite $Na_2B_4O_7 \times 4H_2O$ Ulexite $NaCaB_5O_9 \times 8H_2O$
Cl	C. W. Scheele (1774)	From the Greek word *khlôros* (green)	19	1.94×10^4	3.16	Halite, rock salt NaCl Sylvite, sylvine KC1 Sylvinite NaCl (KCl)
Co	Georg Brandt (1735)	Germon world *kobalt* or *kobold* evil spirit	30	2×10^{-5}	1.88	Skutterudite (Ni, Co) As_3Cobaltite (CoAsS) Linnaeite $(Co, Ni)_2S_2$
Cu	Known 5000 BC	Latin *Cuprum*	26	2.5×10^{-4}	1.90	Chalcopyrite $CuFeS_2$ Malachite $Cu_2(OH)_2(CO)_2$ Cuprite Cu_2O
Fe	Known ancient times	From the Latin word *ferrum* (iron)	4	0.002	1.83	Hematite Fe_2O_3 Magnetite $FeCO_4$ Siderite $FeCO_3$
Mn	J. Gahn (1774)	From *Latin word mangnes,* magnet	12	2×10^{-4}	1.55	Pyrolusite MnO_2 manganite MnO(OH)
Mo	P. Hjelm (1781)	From Gr. word *molubdos* (lead)	58–59	0.01	2.16	Molybdenite (MoS_2) Molibdite (MoO_3)
Ni	Alex F. Cronstedt (1751)	*kupfernickel* (false copper)	23	5.6×10^{-4}	1.91	Pentlandite $(Ni,Fe)_9S_2$ Gersdorffite NiAsS Garnierite $(Ni,Mg)_3(OH)_2Si_4O_{10}$
Se	Berzelius (1817)	From Gr.word selēnē (Moon)	69	2×10^{-4}	2.55	Berzelianite Cu_2 Se Ferroselenite Fe Se_2
Si	J. J Berzelius (1824)	word *silex* (flint)	2	2.2	1.90	Quartz, SiO_2 Kaolinite $Al_2(OH)_4Si_2O_5$ Serpentine $Mg_2(OH)_4Si_2O_5$
Zn	Marggraf (1746)	German word *zin* (meaning tin)	24	0.0049	1.65	Sphalerite, zinc blende (Zn Fe)S Smithsonite $ZnCO_3$

Ran. No Ranking in of abundance in earth crust
Elec.Neg Electronegativity, is a chemical property that describes the tendency of an atom a functionalgroup to attract electrons (or electron density) towards itself (Pauling 1932)
Conc.in ocean Mean content in oceans in ppm or g ton^{-1} (Data from Enghag 2004)
Name Origin from http://www.chemicaleelements.com/index.html

Black and red allotropes of selenium (www. http://en.wikipedia.org/wiki/Selenium/1.2.2013)

Grinded selenium rich yoghurt powder, before drying Dried selenium rich yoghurt

Fig. 5.1 Black and red allotropes of selenium and production of selenium rich yoghurt. Under very high Se concentration, LactoBacteria defend themselves against toxic selenium by converting toxic Se to non-toxic elemental Se nanospheres inside themselves. It could be used Se enriched yoghurt powder for food and feed supplements. The commercial product called LactoMicroSel, where the spheres Se are not separated from the bacteria, and can use milk as bacterium media. (Photos by Nano Food Lab, Debrecen Uni., Hungary)

toxic to plants at high concentrations. Some trace elements, such as cobalt and selenium, are not essential to plant growth but are required by animals and human beings. Other trace elements such as cadmium, lead, chromium, nickel, mercury, and arsenic have toxic effects on living organisms and are often considered as contaminants. Trace elements in an agroecosystem are either inherited from soil parent materials or inputs through human activities. Soil contamination with heavy metals and toxic elements due to parent materials or point sources often occurs in a limited area and is easy to identify. Repeated use of metal-enriched chemicals, fertilizers, and organic amendments such as sewage sludge as well as wastewater may cause contamination at a large scale. A good example is the increased concentration of Cu and Zn in soils under long-term production of citrus and other fruit crops (He et al. 2005).

Many chemical processes are involved in the transformation of trace elements in soils, but precipitation–dissolution, adsorption–desorption, and complexation are the most important processes controlling bioavailability and mobility of trace elements in soils. Both deficiency and toxicity of trace elements occur in agroecosystems. Application of trace elements in fertilizers is effective in correcting micronutrient deficiencies for crop production, whereas remediation of soils contaminated with metals is still costly and difficult although phytoremediation appears promising as a costeffective approach. Soil microorganisms are the first living organisms subjected to the impacts of metal contamination. Being responsive and sensitive, changes in microbial biomass, activity, and community structure as a result of increased metal concentration in soil may be used as indicators of soil contamination or soil environmental quality. Future research needs to focus on the balance of trace elements in an agroecosystem, elaboration of soil chemical and biochemical parameters that can be used to diagnose soil contamination with or deficiency in trace elements, and quantification of trace metal transport from an agroecosystem to the environment (He et al. 2005).

The most important aim of studying trace elements in agroecosystems is to know their benefit and harmfulness to human health, and how to control them, because most agro-productions, especially from crops, are the main food resources for human beings whether in direct or indirect (Lam et al. 2004). Basically, the effects of trace elements on crops consist of two aspects. One is element deficiency like Fe, Mn, Mo, Zn, Cu, and so on, while the other is toxicity like heavy-metal contaminants Hg, Pb, Cd, As, Cr, Ni, and so on. However, more attention is paid to agro-production qualities caused by these trace elements. The reason partly lies in the fact that element deficiency in food is not more dangerous than those of element toxicity. Thus, the harmfulness of some elements to crops, along with implication to human health, should be considered (Wei and Zhou 2008).

It is well documented that the normal abundance of an element in earth material is commonly referred to by the geochemist as background, and for any particular element this value, or range of values is likely to vary according to the nature of the materials (Thornton 1981). Trace elements in soil are derived from parent materials and anthropogenic inputs. In remote or mountain areas where impacts of human activity are relatively small, trace elements in soil are mainly inherited from parent materials, whereas in urban areas or agricultural land with a long history of crop production, the concentrations of trace elements in soil can be higher than those found in the parent materials. For instance, Cu concentrations in some citrus grove soils in Florida have been found to be as high as several hundreds mg kg^{-1}, or 10–20 times greater than the background level, due to repeated use of Cu-containing fungicides/pesticides/herbicides for sustaining citrus production (Alva 1992).

Increased anthropogenic inputs of trace elements in soils have received considerable attention, since transport of the elements may result in an increased content of trace elements in the groundwater or surface water (Moore et al. 1998). trace elements inputs include those from commercial fertilizers, liming materials, and agrochemicals, sewage sludges and other wastes used as soil amendments, irrigation waters and atmospheric deposition (Senesi et al. 1999). Soils receiving repeated applications of organic manures, fungicides, and pesticides exhibited high concentrations of extractable metals and subsequently resulted in increased trace elements concentrations in runoff (He et al. 2005).

Selenium is one of the elements playing a most important role in human and animal health and is essential to all other organisms including bacteria and algae.

Most plants contain rather low foliar Se, around 25 μg kg^{-1} and rarely exceed 100 μg kg^{-1}. However, some plants exhibit a great capability to accumulate Se and they may concentrate Se to extremely high levels over 1000 mg kg^{-1}. However, some plants reveal a great capability to accumulate Se and they may concentrate Se to extremely high levels that may be toxic to humans and animals. Although Se is not an essential element for plants, with some exceptions, it is being added to soil to ensure that both food and feed products contain adequate amounts for the dietary needs. It should be emphasized that the margin of safety of Se concentrations is rather narrow (Kabata-Pendias 2011).

Selenium and its compounds have been recognized as inorganic carcinogens of concern in the spectrum of identified environmental pollutants, based on observation of proved toxic effects and relative accessibility. From a public health standpoint, selenium holds a rather unique place among the elements because of small differences between concentrations which are essential and those that are toxic to animals and human beings. Movement of these toxic elements through the geocycle and their biological methylation in the environment to volatile selenium species further complicates the problem. As selenium is of interest as a potential environmental toxicant, lots of attention has been focused on the detailed understanding of biomethylation and transformation of selenium. This methylation of toxic elements is an important transformation because it often leads to a change in both mobility and toxicity of the element. With selenium, biomethylation represents the conversion of a nonvolatile precursor to volatile products such as dimethyl selenide and dimethyl diselenide, which are less toxic than other forms of nonvolatile selenium species (Spallholz 1994). Furthermore, the formation of volatile compounds may play an important role in the cycling of this element in the biosphere (Jiang 1994).

Selenium is considered by some to be a serious hazard to the environment and to animal health. Selenium-contaminated water has brought deformity and death to wildlife in nature reserves in western USA. There is even concern that because of selenium contamination of soil, crops supplied to the great cities of California could become unfit for human consumption. In large areas of China, endemic selenium toxicity is a hazard for locals who depend on crops grown on selenium rich soil. Yet, in the UK, and in other parts of Europe, fears are expressed that soil selenium levels are inadequate. There are demands that the example of Finland should be followed and soil selenium levels increased by the addition of selenium to fertilizers. There may be controversy among the experts and health authorities about selenium, but this has not deterred the general public from deciding that the element has an important role to play in health. In New Zealand, when the use of selenium was first permitted to prevent deficiency in farm animals, but was still not approved as a supplement for humans, people took the matter into their own hands. Veterinary preparations containing selenium were used by those who believed that what was good for animals must also be good for humans (Reilly 2006).

Selenium is an essential nutrient for many organisms, but also toxic at higher levels. While certain algae require Se to make selenoproteins, no such requirement has been shown for higher plants. Still, plants readily take up and assimilate Se

Table 5.2 Selected properties of Se comparing with some micronutrients and beneficial mineral elements. (Source: compiled from Enghag 2004, Kabata-Pendias and Mukherjee 2007 and Kabata-Pendias 2011)

Element	Atomic No.	Atomic mass	Atomic radius[a] (pm)	Density (20° C, gcm^{-3}	Valence[b]	Melting point (°C)
Boron (**B**)	5	10.81	117	2.34	+3	2079
Chlorine (**Cl**)	17	35.42	97	3.21	−1	−100.9
Cobalt (**Co**)	27	58.93	167	8.9	+2, +3, +4	1495
Copper (**Cu**)	29	63.54	157	8.96	+1, +2	1083
Iron (**Fe**)	26	55.8	172	7.87	+2, +3, +4, +6	1535
Manganese (**Mn**)	25	54.9	179	7.44	+2	1244
Molyb-demm (**Mo**)	42	95.9	201	10.2	+2, +3, +4, +5, +6	2617
Nickel (**Ni**)	28	58.69	162	8.90	+2, +1, +3, +4	1454
Selenium (**Se**)	34	78.96	122	4.28	−2, +2, +4, +6	217
Silicon (**Si**)	14	28.08	170	2.33	+2, −4, +4	1410
Zinc (**Zn**)	30	65.38	153	7.13	+2	419.6

[a]Approximately average values for the main oxidation states
[b]Valences value in bold are for the main oxidation states. PM=picometers=10^{-12} m

using sulfur (S) transporters and biochemical pathways, and can also volatilize methylated Se. Some plants can even hyperaccumulate Se to levels around 1% of plant dry weight, in the form of methyl-selenocysteine, probably as a defense mechanism. Plants may be used both to provide dietary Se in areas of Se deficiency, and to clean up Se pollution from seleniferous areas. These applications benefit from better insight into the genetic and biochemical mechanisms that control plant Se tolerance and accumulation (Table 5.2; Pilon-Smits and Quinn 2010).

Therefore, trace elements like selenium, in the agroecosystem, are either inherited from soil parent materials or inputs through human activities. Soil contamination with these elements and toxic elements due to parent materials or point sources often occurs in a limited area and is easy to identify. Repeated use of metal-enriched chemicals, fertilizers, and organic amendments such as sewage sludge as well as wastewater may cause contamination at a large scale. Selenium is considered by some to be a serious hazard to the environment and to animal health. For example, selenium-contaminated water has brought deformity and death to wildlife in nature reserves in western USA.

5.1.4 Selenium Characterization

Selenium is quite unique as trace element, because it is a component of an amino acid, selenocysteine, typical of selenoproteins and therefore involved in very specific biological roles. Enzymes depending on selenium perform very important roles in the cells (Rayman 2000). The main roles are in protection against oxidative damages, defenses against infection, and modulation of growth and development. The main exposure to selenium occurs through food, and its distribution in the natural environment has a marked effect on its content in soils, crops, and the human body (WHO 1987). In most instances, a reduced intake of selenium can be connected to several effects: loss of immunocompetence, increased virulence of viral diseases including HIV, early pregnancy loss, depression and other negative mood states, hypothyroidism, cardiovascular diseases, inflammation states, and cancer incidence. Many of these effects can be linked to roles of the selenoproteins, and especially of glutathione peroxidases active as antioxidants and cell protecting agents. Deficiency of selenium in the diet causes diseases which often are endemic to specific regions (Marmiroli and Maestri 2008).

Selenium is one of the rarest elements. It is about 70th in abundance among the 88 elements that naturally occur in the earth's crust (Shriver and Atkins 1999). Yet, in spite of its scarcity, selenium plays a key role in all animal life. It is an essential component of the human diet, though only in minute amounts. If this intake is exceeded by relatively little, disastrous consequences can follow. Selenium is a two-faced element. Like the moon, after which it is named, it has both a dark and a bright side. This duality has, right from the time the element was first isolated, presented science with a dilemma: how to reconcile its apparently contradictory properties and roles? Now, nearly 200 years later, in spite of thousands of hours of research and the publication of great numbers of scientific studies, the dilemma has not yet been fully resolved. Nevertheless, the gaps in our understanding of selenium are rapidly being filled by the efforts of an extraordinary array of investigators, working in a range of disciplines, aided by powerful new research tools and techniques (Reilly 2006).

Selenium has an atomic weight of 78.96 and its atomic number is 34. It occurs along with oxygen, sulfur, tellurium, and polonium in Group 16/VIA, and between arsenic and bromine in Period 4 of the Periodic Table of the elements. This location accounts for many of its biological interactions with sulfur, as well as with arsenic and its neighbor phosphorus, and, as was noted by an eminent biochemist more than 3 decades ago (Frost 1972), places selenium in "a frontier that will challenge advances in biochemistry." Subsequent research has confirmed the accuracy of that prediction. The outer electronic configuration of the element is $3d^{10}4s^24p^4$, with three completely filled inner shells. Its chemical properties are intermediate between those of sulfur and tellurium, and its compounds resemble the corresponding sulfur and tellurium compounds in behavior. Its electronic configuration and position in the Periodic Table place selenium in the important group of half metals, or metalloids, elements that are neither fully metals nor nonmetals, but share chemical and physical properties of both (Reilly 2006).

Selenium, like sulfur, has several allotropes. They include monoclinic or red selenium, an amorphous powder that exists in two forms, one of which is comparable to crystalline "flowers of sulfur." There is also a black amorphous form. A vitreous form that changes to gray selenium on heating also occurs. Gray, also known as metallic selenium is stable at ordinary temperatures and is the most common allotrope. Due to close crystallochemical behaviors of Se and S, much of the Se geochemistry is close to the more abundant S. It reveals complex behavior in geochemical processes that resulted from relatively easy changes of its oxidation states. The easy methylation of Se, mainly due to biological processes, yields the formation of volatile Se forms and has a significant role in the biogeochemical cycling of this element. Selenium exists in four valence states, of which the −2 state predominates in organic Se compounds. Commonly occurring species, selenites (Se^{4+}) and selenates (Se^{6+}), in geochemical environments do not form stable compounds and are preferably absorbed by minerals, particularly clay minerals, and Fe and Mn oxides and hydroxides. Approximately 50 Se minerals are known, of which the relatively common ones are: klockmanite, CuSe; berzelianite, $Cu2-_{x}Se$; clausthalite, PbSe; tiemannite, HgSe; ferroselenite, $FeSe_2$ and crookesite, $(Cu, Tl, Ag)_2Se$. The association of Se with host minerals, such as pyrite, chalcopyrite, and sphalerite, is relatively common (Table 5.2). The great affi nity of Se to different organic substances resulted in a large number of organic compounds that are analogous to those of S organic compounds and are easily accumulated in some biolithes. Its concentrations (in mg kg^{-1}) may be up to 10.7 in coal, and up to 1.4 in crude oils and bituminous shales. Extremely high Se concentrations, up to 6500 mg kg^{-1}, is in Chinese stone coal (Plant et al. 2004). Se concentrations of as much as about 0.1 % are not uncommon in the immediate vicinity of some oxidized sandstone-type uranium deposits. Atmospheric deposition of Se in some regions might be of an environmental concern (Kabata-Pendias 2011).

In the environment, however, there are concerns with both deficiency and toxicity. Since 1983 there has been dynamic growth in both studies and understanding of Se cycling and importance in human and animal health. Since that time an explosion in scientific papers and review articles has been observed. The average content of selenium in the Earth's crust is estimated as 0.05 mg kg^{-1}; however, a higher value, up to 0.5 mg kg^{-1}, is also given. It is slightly more concentrated in mafic rocks, but rarely exceeds 0.1 mg kg^{-1} (Table 5.3). In sedimentary rocks, Se is associated with the clay fraction and thus its abundance in argillaceous sediments (0.3–0.6 mg kg^{-1}) is higher than in sandstones and limestones (0.01–0.1 mg kg^{-1}). Enriched Se concentration in Cretaceous rocks (up to above 100 mg kg^{-1}) was derived from volcanic gases and dust brought down by rain into the Cretaceous sea; thus sediments of that period are likely to be enriched with Se. Some sedimentary rocks formed in nonvolcanic periods also may be enriched with Se, probably in a normal course of weathering (Kabata-Pendias 2011).

Selenium has unique electrical properties. Its conductivity, which is low in the dark, is increased several 100-fold on exposure to light which also generates a small electrical current in the element. It is, in addition, a semiconductor, possessing what

Table 5.3 Abundance of selenium in the environment comparing with some micronutrients and beneficial mineral elements. (Source: compiled from Kabata-Pendias and Mukherjee 2007 and Kabata-Pendias 2011)

Element	Earth crust	Igneous rock, acid	Sedimentary rocks			Soils, mg kg^{-1}	Water (μg l^{-1})[a]	Air[b] (ng m^{-3})
			Argill.	Sand.	Calcare-ous			
Concentration (mg kg^{-1}) in different environmental compartments								
B	10	10–30	120–130	30–35	20–30	15–35	10–100	–
Cl	145	300–850	500–800	50–270	50–350	300	50–2700	1–7
Co	10–12	1–15	14–20	0.3–10	0.1–3.0	8.0	0.15	0.05
Cu	26	5–30	40–60	5–30	2–10	20	0.27–35	150–1600
Mn	950	350–1200	400–850	100–500	200–1000	50–2000	0.2–130	2.8–4.5
Mo	1.2	1–2	2–2.5	0.2–0.8	0.2–0.4	1.8	0.1	<0.2
Ni	20	5–20	40–90	5–20	5–20	19–22	0.8	0.9
Se	0.05	0.01–0.05	0.3–0.6	0.01–0.08	0.03–0.10	0.33	0.07	0.2
Zn	52–80	40–100	80–120	15–30	10–25	63	3.5–10.3	18–41
Concentration (%) in different environmental compartments							(gl^{-1})	(ng m^{-3})
Fe	5.6	1.4–4.7	3.3–4.7	10–30	0.4–1.0	3.5	66	166–171
Si	26–29	31–34	25–28	30–40	5–30	54	1.3	20–100

Argill. Argillaceous and sands., Sandstones
[a]Water of river
[b]greenland

is known as asymmetrical conductivity, able to conduct more easily in one direction than in the other. These properties account for the element's exceptional usefulness to the electrical and electronic industries. Elemental selenium boils at the relatively low temperature of 684°C. As a consequence, atmospheric pollution can be caused by industrial processes that involve heating the element or its compounds (Crystal 1973). However, elemental selenium itself is very stable and highly insoluble. These properties are important from an environmental point of view since, under reducing conditions, selenates and other soluble compounds of selenium that occur in certain soils can be converted into elemental selenium and thus become unavailable for absorption by plants. The process can also remove selenium from active recycling and thus reduce the possibility of environmental pollution (Reilly 2006).

Compounds of Se are used in photoelectric cells of broad utility from photometers to photovoltaic-batteries. They are used as pigment (maroon and orange colors), mainly in glass and plastic production (approximately 50% of the total Se production) and serves as vulcanizing and galvanizing agents. Its addition to alloys increases the machinability of stainless steel. Selenide compounds (e.g., WSe_2) are used in lubricants for metals. So, the largest use of Se, as a pigment, is in glass and ceramic manufacture. It is used (with Bi) in brasses to replace more toxic Pb. The Se is also used to improve abrasion resistance in vulcanized rubbers. Its use in toning of photographic prints, as well as in various photoelectric cells, is relatively common. In agriculture, Se is used as an addition (mainly as sodium selenite, Na_2SeO_3) to insecticides, fertilizers, and foliar sprays. Selenium is widely used, in small doses, in vitamins and other dietary supplements. Also, some livestock feeds are fortified with this element. It is a relatively common component of various cosmetics and medications, as a therapeutic agent (e.g., in cardiology as an antioxidant). It is one of the elements that play the most important role in human and animal health. Recently, numerous studies have been curried out on this element and this has resulted in a huge number of publications (Kabata-Pendias 2011).

A considerable number of selenium analogs of organosulfur compounds are known. Many have been isolated from biological materials and their properties investigated. Although some aspects of the metabolism of organosulfur compounds resemble those of their sulfur analogs, their metabolic pathways diverge considerably (Levander 1976). Many have also been synthesized and their possible uses, industrially and medicinally, have been investigated (Klayman and Gunther 1973). Of particular interest from the nutrition point of view are the selenoaminocarboxylic acids, selenium-containing peptides, and selenium derivatives of nucleic acids that occur naturally in body cells and tissues (Reilly 2006).

Therefore, selenium is quite unique as trace element, because it is a component of an amino acid, selenocysteine, typical of selenoproteins and therefore involved in very specific biological roles. Enzymes depending on selenium perform very important roles in the cells. Selenium is extracted from selenide minerals in many sulfide ores, such as those of copper, silver, or lead. It finds application as a catalyst in many chemical reactions and in various industrial and laboratory synthesis. It is also widely used in structure determination of proteins and nucleic acids by X-ray crystallography for the incorporation of one or more Se atoms that help with MAD phasing. The use of selenium is principally associated to glass and ceramic productions.

5.1.5 Production, Sources and Uses of Selenium

There are no mines in the world that specifically extract Se; instead it is a by-product of the production of other metals such as refining of Pb and Cu, or recovered from the sludge accumulated in H_2SO_4 plants (Johnson et al. 2010). The supply of selenium is directly affected by the supply of the materials from which it is a by product—copper, and to a lesser extent, nickel. Estimated domestic selenium production was slightly higher in 2012 compared with that of 2011 owing to a slight increase in copper production (1980 and 2000 t in 2011 and 2012, respectively; USGS 2013). Most of the world's selenium is produced in the USA, Japan, and Canada, with smaller amounts coming from China, Russia, Belgium, Finland, Australia, Peru, Zambia, and other countries with a copper refining industry. Production has been increasing in recent years as new uses for the element, especially in the electronic and related industries, have been found. Although ultra-pure selenium of, it is claimed, 99.99 % purity is available, most marketed refined grades contain >99.5 %

Table 5.4 Main sources of selenium in the environment (adapted from Fordyce 2005)

Main sources	Comments
Natural sources	
Volcanic activity	Important source
Weathering of rocks	Important source
Sea spray	Concentrations in ocean water are only an order of magnitude lower than those in rocks
Atmospheric flux	From the ocean surface to the atmosphere
Volatilization and recycling from biota	For example, in the UK annual selenium deposition
Aerial deposition	=2.2–6.5 g ha^{-1}
Man-made sources	
Selenium-based industries	
Metal processing industries	Important source
Burning of fossil fuels	Important source
Disposal of sewage sludge to land	Typical selenium contents 1–17 mg kg^{-1}
Agricultural use of pesticides	Potassium ammonium sulfide ($[K(NH_4)S]_9Se$)
Agricultural use of lime	Typical selenium contents 0.08 mg kg^{-1}
Agricultural use of manure	Typical selenium contents 2.4 mg kg^{-1}
Agricultural use of phosphate fertilizers	Typical selenium contents 0.08–25 mg kg^{-1}

selenium with, as impurities, up to 0.2 % tellurium and lesser amounts of iron, lead, and copper. Several compounds of selenium are commercially available, including ferro- and nickel-selenide, selenium dioxide, cadmium sulfoselenide, and selenium diethyldithiocarbamate, as well as sodium selenite and selenate (Reilly 2006).

At the global scale, selenium is constantly recycled in the environment via the atmospheric, marine, and terrestrial systems. Estimates of selenium flux indicate that anthropogenic activity (76,000–88,000 ton per year) is a major source of selenium release in the cycle, whereas the marine system (38,250 ton per year) constitutes the main natural pathway (Haygarth 1994). Selenium cycling through the atmosphere (15,300 ton per year) is significant because of the rapidity of transport, but the terrestrial system (15,380 ton per year) is most important in terms of animal and human health because of the direct links with agricultural activities and the food chain. Although the element is derived from both natural and man-made sources, an understanding of the links between environmental geochemistry and health is particularly important for selenium as rocks are the primary source of the element in the terrestrial system (Table 5.4).

Selenium is dispersed from the rocks through the food chain via complex biogeochemical cycling processes including weathering to form soils, rock-water interactions, and biological activity. As a result, selenium is not distributed evenly across the planet, rather concentrations differ markedly depending on local conditions and an understanding of these variations is essential to aid the amelioration of health problems associated with selenium deficiency and toxicity. The following sections of this chapter provide a brief summary of anthropogenic sources of the element before going on to discuss the important aspects of selenium in the natural biogeochemical cycle and impacts on health (Fordyce 2005).

Selenium forms natural compounds with several other elements and is a constituent of more than 60 mineral species, chiefly sulfides. However, selenium minerals, the most common of which is clausthalite (PbSe), are finely disseminated and do not form economically exploitable deposits. Primary commercial sources are sulfide deposits of copper and other base metals with which small amounts of selenium are associated. The element is recovered as a by-product from slimes produced during electrolytic refining, particularly of copper ores, along with a variety of precious metals, including silver, gold, platinum, and germanium. In some ores in which it is present as the double selenide CuAgSe, selenium may make up more than 50 % of the total metal content of the slime (Oldfield 1990). Other lesser sources are sludges and dusts produced during the manufacture of sulfuric acid. This was the source from which selenium was first isolated by Berzelius. Treatment of such slimes and dusts requires that the selenium be converted into a water-soluble form, followed by reduction to the elemental state. This can be achieved by a number of different processes such as heating with soda ash or sulfuric acid as well as by direct oxidation. A growing, but still relatively small amount of selenium is recovered by recycling from discarded electronic equipment and other scrap machinery. It can be recovered mechanically by milling, sandblasting, by use of high-pressure water jets, or by solution in aqueous sodium sulfite, fused caustic soda, or other such solvents (Reilly 2006).

As long ago as 1873 it was discovered by two British engineers, W. Smith and J. May, that on exposure to light the electrical resistance of selenium was decreased (van der Krogt 2004). A selenium photovoltaic cell was constructed in 1883 and the application of the element in rectifiers described in 1909. Industrial exploitation of the element's photoelectric and semiconducting properties began in the 1920s when the first commercially available selenium rectifier was produced. Today, the electrical and electronic industries are major users of selenium, taking about a third of the world's production. Until recently, large amounts of selenium were also used on photoreceptor drums of plain paper copiers, in the process known as "xerography" or dry photocopying (Zingaro and Cooper 1974). However, this use has decreased, mainly for health and economic reasons, as selenium is replaced by more environment-friendly photosensitive materials. Selenium is also used in laser printers, solar photovoltaic cells, and X-ray machines. Significant amounts are used by the glass industry, both to decolorize and to color glass. Selenium ruby glass is one of the most brilliant reds known and is used in airfield and other warning lights and in decorative stained glass. Selenium compounds added to the glass mix can also produce other colors as well as the bronze and smoky plate glass used in curtain walls of many modern buildings to block solar heat transmission. Selenium compounds, such as cadmium sulfide-selenide, are used in a range of pigments in ceramics, paints, and plastics (Reilly 2006).

Selenium has a number of important agricultural and horticultural applications. These include the use of sodium selenite and selenate as additives and dietary supplements in animal feeds. Soil deficiencies are corrected by adding selenium compounds to fertilizers and top dressings. Potassium ammonium sulfoselenide has considerable pesticidal properties and was one of the first systemic insecticides to be marketed in the 1930s. It is still in use, but is restricted to nonfood crops because of its toxicity. Sodium selenate has also been used for a similar purpose in commercial greenhouses growing flowers for cutting. The selenate is added to irrigation water and is taken up through the roots of the plants. It is converted in the leaves into volatile selenide, which is released by the plant to repel red spiders, aphids, and similar pests (Reilly 2006).

Therefore, it could be concluded that there are no mines in the world that specifically extract Se; instead it is a by-product of the production of other metals such as refining of Pb and Cu, or recovered from the sludge accumulated in H_2SO_4 plants. Estimated domestic selenium production was slightly higher in 2012 about 2000 t. Most of the world's selenium is produced in the USA, Japan, and Canada, with smaller amounts coming from China, Russia, Belgium, Finland, Australia, Peru, Zambia, and other countries with a copper refining industry. Selenium has a number of important agricultural and horticultural applications. These include the use of sodium selenite and selenate as additives and dietary supplements in animal feeds. Sodium selenate has also been used for a similar purpose in commercial greenhouses growing flowers for cutting. The selenate is added to irrigation water and is taken up through the roots of the plants. It is converted in the leaves into volatile selenide, which is released by the plant to repel red spiders, aphids, and similar pests.

5.1.6 Occurrence of Selenium in the Environment

The naturally occurring element Se is rarely recovered in a free state. Its chemistry is related to sulfur and tellurium and occurs in the environment, especially as inorganic forms. Selenium oxidation states are +6 (VI), +4 (IV), 0, and −2. Selenate Se(VI) and selenite Se(IV) are the oxidized water-soluble forms recovered in soil solutions and in natural waters. The highly stable elemental or metallic selenium (Se°) is also recovered in soil, but not in water solution because it is insoluble. Metallic selenium exists in several allotropic forms, and some of them have been identified: trigonal gray selenium (containing Se_n helical chain polymers), rhombohedral selenium (containing Se_6 molecules), three deep-red monoclinic forms, α, β-, and γ-selenium (containing Se_8 molecules), amorphous red selenium, and black vitreous selenium. The most thermodynamically stable form is the gray (trigonal) selenium, which contains countless helical chains of selenium atoms. Grey selenium is the only allotropic form that conducts electricity. The red and black amorphous allotropes are the forms that are most likely to occur in soils. At temperature greater than 30 °C, red amorphous selenium gradually reverts to the black amorphous form. The latter form is then slowly transformed into the much more stable gray hexagonal allotrope or, depending on the redox conditions and the pH of the soils, it is reoxidized (Di Gregorio 2008).

Se is an element essential to human and animal health in trace amounts but is harmful in excess. It is the toxic effects of excess Se that first brought attention to the health impacts of this element (Smith et al. 1936), i.e. selenosis in human subjects and blind staggers and alkali disease in livestock. Even before the symptoms were attributed to Se poisoning there are historical references to the toxic effects of Se on livestock. Pedro Simon, a sixteenth century missionary, recorded that in areas of Columbia farm animals (1560 in South Dakota, USA) suffered from hair loss and other abnormalities (Reilly 2006). Approximately 400 years later the cause was attributed to poisoning by Se taken up by plants in seleniferous soils (Rosenfeld and Beath 1964). Even earlier in the thirteenth century there are apocryphal tales of Marco Polo's horses suffering from lost hair and hooves in ancient Suzhou of western China, again attributed to Se poisoning. However, recent research concludes that whilst suffering symptoms similar to Se poisoning, the ailment of the horses recorded by Marco Polo in 1295 might not have been selenosis (Shao and Zheng 2008). Literature on the role of Se as an essential trace element was published in the 1950s (Scharz and Foltz 1957), although it is really in the last few decades that there has been a substantial increase in research into the health impacts of Se deficiency. This research has been catalysed by certain key medical investigations such as the discovery in 1971 that glutathione peroxidise is a selenoenzyme and the landmark clinical trial that appears to show that Se can reduce the risk of cancer (Johnson et al. 2010).

The interest in Se as an essential trace element in the human diet, as well as its potential to be toxic in Se-rich areas, has increased the interest in Se geochemistry and this position is reflected in the many texts that discuss Se in the environment such as McNeal and Balistrieri (1989), Haygarth (1994), Plant et al. (2004) and Fordyce (2005). A better knowledge of the distribution of Se in the surface environment facilitates an understanding of its possible impact on human and animal nutrition and health. In particular, an understanding of the physico-chemical properties that control the movement of Se in the food chain such as soil pH, organic content and speciation, enable a better prediction of potential risks from Se deficiency and toxicity. Since its discovery in 1817 by Jöns Jakob Berzelius Se has been an element that has attracted little attention and study by geochemists. As its concentrations in most geological materials are very low it has been difficult to study its distribution at the Earth's surface. It was Se toxicity in the environment that first attracted substantial attention to Se geochemistry and at about the same time, in the 1930s, industrial uses for Se began to develop such as invention of the photocopier and Se rectifiers, driving a demand for Se as a commodity (Johnson et al. 2010).

Selenium presents a complex chemical behaviour that allows it to combine with a variety of elements in nature. This makes selenium compounds widespread over all the Earth compartments: rocks, soils, waters and air. However, selenium is considered a trace element in the Earth crust, with average concentrations ranging from 0.05 to 0.09 mg kg^{-1} (Neal and Sposito 1989). A recent exhaustive compilation of published data of inorganic selenium–metal solubility constants, selenium acid-base equilibrium constants, standard potentials of selenium redox couples and some complex dissociation constants for the soluble species of selenium can be

Table 5.5 Main selenium aqueous or gaseous chemical species present in the environment or biologically relevant. (Adapted from Fernández-Martnez and Charlet 2009)

Species	Chemical formula	Comment
Inorganic species		
Selenate [Se(VI)]	$H_2SeO4°$, $HSeO_4^-$, SeO_4^{2-}	Predominant species in soils, sediments and waters
Selenite, Selenium dioxide [Se(IV)]	$H_2SeO_3°$, $HSeO_3^-$ SeO_3^{2-}, SeO_2	Selenite presents in mildly oxidizing, acidic environments (e.g., oil refinery waste waters)
		Selenium dioxide, gas present in volcanic eruptions and combustion processes
Elemental selenium [Se(°)	Se°	Precipitated after microbial reduction process and, to a less extent, by inorganic processes
Selenide [Se(-II)]	H_2Se	Volatile compound formed upon microbial processes
Organic species		
Dimethylselenide (DMSe)	$(CH_3)_2Se$	Volatile compound formed upon bacterial methylation processes
Dimethyldiselenide (DMDSe)	$(CH_3)_2Se_2$	Volatile compound formed upon bacterial methylation processes
Dimethylselenium-sulfide	$(CH_3)_2SeS$	Product of microbial methylation processes
Dimethylselenium-disulfide	$(CH_3)_2SeS_2$	Product of microbial methylationprocesses
Selenodi-glutathione	GSSeSG	Formed during in vitro experiments mimicking biochemical processes
Selenocysteine(SeC)	$HSeCH_2CHNH_2COOH$	Main selenium specie in organic tissues
Selenomethionine (SeM)	$CH_3Se(CH_2)_2CHNH_2$–COOH	Predominant specie in plants
Trimethylselenonium	$(CH_3)_3Se^-$	Urinary metabolite
Selenocyanate	$SeCN^-$	Present in Wastewaters from petroleum refineries
Selenoproteins		Various proteins and enzymes (1.e., GPX, Selenoprotein P, TR.....)

found in Seby et al. (2001) or in Olin et al. (2005). Besides inorganic forms of selenium, non-volatile organic selenides such as seleno amino-acids and volatile methylated forms of selenides, principally dimethylselenide and dimethyldiselenide, can occur in surface water, groundwater and waste water (Lenz et al. 2006). Table 5.5 shows the main selenium-containing organic compounds found in nature (Fernández-Martánez and Charlet 2009).

Selenium is recovered in rocks combined with sulfide minerals with silver, copper, lead, and nickel. The metalloid is present also in coal and actually in significant amounts; in fact, a coal refinery is a severe source of selenium environmental contamination (Lawson and Macy 1995). Examples of selenide minerals are as follows: (1) crookesite $(Cu,Tl,Ag)_2Se$, a copper selenide mineral containing also thallium and silver; it is formed by precipitation from hydrothermal fluids; (2) tilleite ZnSe, a zinc selenide found only as microscopic gray crystals associated with other selenides; (3) tiemannite HgSe, a mercury selenide that occurs in hydrothermal veins associated with other selenides, or other mercury minerals such as cinnabar, and often with calcite; (4) umangite CuSe, a copper selenide that occurs in small grains or fine granular aggregates with other copper minerals of the sulfide group (Kabayta-Pendias 2011).

Selenium is distributed in the environment by processes such as volcanic activity and hot springs, combustion of fossil fuels, weathering of rocks and soils, soil leaching, sea salt spray, forest wildfires, groundwater movements, soil adsorption and desorption, chemical and biological redox reactions, and mineral formation, but also incineration of municipal waste, copper/nickel production, lead and zinc smelting, iron and steel production, crop growth and irrigation practices, and plant and animal uptake and release (Nriagu 1989). Estimated Se fluxes indicate that the natural sources of Se emission are as important as anthropogenic emission (Masscheleyn et al. 1990). Most soils contain between 0.1 and 2 mg Se kg^{-1} (Mayland et al. 1989). However, elevated concentrations of this metalloid are associated with various environments, notably those of marine sedimentary parent material and soils impacted by industrial activity (Haygarth 1994). Thus, as a general rule, Se concentration in soil or ground and fresh water depends upon the parent material, climate, topography, age of the soil, and agricultural or industrial utilization. Under acidic, reducing conditions in soils that may be waterlogged and rich in organic matter, elemental Se and selenides (Se^{2-}) are the predominant species (McNeal and Balilestri 1989). From pH 4 to 8, stable adsorption complexes or co-precipitates with sesquioxides are prevalent. At moderate redox potentials in soil solution, Se(IV) is the predominant form, while at high redox potential in well-aerated, alkaline soils the predominant form is Se(VI) (Elrashidi et al. 1989), which does not form stable adsorption complexes or co-precipitates with sesqioxides. Most natural waters have low concentrations of Se, ranging from 0.1 to 100 mg L^{-1}. However, some evaporation ponds in the California San Joaquin Region, USA, have reached levels in excess of 1000 mg L^{-1} (Thompson-Eagle et al. 1989). Soils in this region were derived from seleniferous cretaceous sediments and contain high levels of Se that is mobilized by infiltrating irrigation water that is then discharged to surface water by subsurface drainage or to groundwater by deep percolation (Di Gregerio 2008).

Therefore, it is a matter of the fact that, besides the positive effect of the metalloid on biological systems, uncontrolled release of selenium in the environment may be trouble some to living organisms. Because Se undergoes microbial and plant transformations, their application may be potentially useful to the development of bioremediation strategies. The naturally occurring element Se is rarely recovered in a free state. Its chemistry is related to sulfur and tellurium and occurs in the environment, especially as inorganic forms. The highly stable elemental or metallic selenium (Se°) is also recovered in soil, but not in water solution because it is insoluble.

Table 5.6 Soil factors affecting the mobility of selenium. (Adapted from Kabata-Pendias 2011)

Soil factors	Se form	Mobility
Soil acidity (pH)		
High (alkaline)	Selenates (Se^{6+})	High
Medium (neutral)	Selenites (Se^{4+})	Medium
Low (acid)	Selenides (Se^{2-})	Low
Redox potential (Eh)		
High (high oxidation, >400 mv)	Selenates (SeO_4^{2-})	High
Moderate (200–400 mv)	Selenites (SeO_3^{2-})	Medium
Low (oxidation, <200 mv)	Selenides (Hse^-)	Low
Hydroxides (Fe, Mn)		
High content	Absorbed all forms of Se	Low
Low content	Slight absorption	High
Organic Matter[a]		
Undecayed	Absorbed	Low
Decayed (e.g., peat)	Complexed	High
Enhanced biomethylation	Volatilized	High
Clays[b]		
High content	Absorbed all Se forms	Low
Low content	Not fixed or soluble all Se forms	High
Interaction with		
S, P and N	Antagonistic effects	Rather low

[a]Variable impact of organic matter depends on its Kind

[b]Absorption by clay minerals decreases with increasing pH values and, at pH 8, is almost negligible

5.1.6.1 Selenium in Soils

The Se content in soils is inherited from parent material and its distribution strongly reflects soil-forming processes and atmospheric deposition. The lowest amounts of Se are in sandy soils developed under humid climate, particularly in podzols, with the highest contents occurring most often in organic and calcareous soils (Table 5.6). Selenium contents of worldwide soils have been calculated at an average value of 0.33 mg kg^{-1}, but the range of its concentrations is very broad, from 0.005 to 3.5 mg kg^{-1}. Soils from the Se-deficient endemic areas e.g., in China contain very low watersoluble Se, in the range from 0.2 to 2 μg kg^{-1}, whereas soils from non-endemic regions contain easily soluble Se from 1 to 11 μg kg^{-1}. The total Se range in topsoils of New Zealand is between 0.1 and 4.0 mg kg^{-1} and the common range is 0.3–0.9 mg kg^{-1}. Higher contents of Se are observed in surface layer of forest soils, organic rich soils, calcareous soils, and volcanic soils. Behavior of Se in soil has been studied extensively and summarized in several publications, such as Frankenberger and Endberg (1998) and Kabata-Pendias and Pendias (2001). It can be generalized that the main factors controlling Se forms and behavior in soils are Eh and pH, however several other parameters like organic ligands, clays, and hydroxides also play very significant roles (Table 5.6). Inorganic species of Se associated with defined soil parameters reveal variable properties:

- Selenates (Se^{6+}) are mobile in inorganic forms, especially in neutral and alkaline soils, and are not absorbed on hydrous sesquioxides in particular $Fe_2O_3 \times H_2O$.
- Selenites (Se^{4+}) are slightly mobile in neutral and acid soils of humid temperate regions, and are easily absorbed on hydrous sesquioxides and organic matter.
- Selenides (Se^{2-}) are rather immobile in acid soils due to the formation of stable mineral and organic compounds (Kabata-Pendias 2011).

Forms and concentrations of Se in soil solution are governed by various physicoal, chemical and biological factors, and common inorganic anions are: SeO_3^{2-}, SeO_4^{2-}, $HSeO_4^-$, $HSeO_3^-$, and $H_2SeO_4^-$. Under oxidizing conditions, SeO_4^{2-} is the favored form, while in mild reducing conditions, SeO_3^{2-} is likely to dominate. This anion is strongly sorbed on oxides and precipitates such as $Fe(SeO_3)_3$, whereas SeO_4^{2-} is very weakly sorbed, especially at high pH. Hence, mobile and easily phytoavailable Se occurs in alkaline well aerated soils, most commonly in arid and semi arid regions. Organic matter has a strong tendency to form organometallic complexes which remove Se from soil solution. High Se mobility can be expected in soils of high pH and Eh, and conversely low mobility in soils with high contents of hydroxides, organic matter and granulometric clay fractions. In acid soils Se is likely to occur as Se^{4+}, strongly adsorbed by Fe oxides to form ferric selenite, $Fe_2(OH)_4SeO_3$, and iron selenide, FeSe. Maximum Se adsorption occurs between pH 3 and pH 5 and decreases as the pH rises. In alkaline Se-rich soils, the predominated species is Se^{6+} which is very weakly adsorbed. Hence, selenates occur in soluble forms in soil of arid and semi-arid regions (Kabata-Pendias and Mukherjee 2007).

The phytoavailability of different Se species in soils decreases in the following order: selenate > selenomethionine > selenocysteine > selenite > elemental selenium > selenide. A close relationship between Se and organic C has been observed in most soils. Microbial processes in both the formation and mineralization of organic Se such as selenomethionine and selenocysteine play a crucial role in Se cycling and especially in its volatilization from Se-contaminated soils. These processes are important, especially in the reduction of Se, principally through the reduction of Se^{4+} and Se^{6+}. The activity of microbiota reveals variable sensitivity to increased Se contents in soil. However, a 5 mM concentration of selenic acid inhibited the activity of all soil enzymes (Nowak et al. 2002). In poorly drained soils, an accumulation of insoluble Se^{2-} compounds is likely to occur. Due to methylation processes under anaerobic condition, Se may volatilize in the form of $(CH_3)_2Se$, as well as in forms of several other methane and sulfide Se compounds. A number of bacteria and fungi species are involved in Se volatilization processes (Frankenberger and Karlson 1994). They reported that, organic amendments may significantly increase the rate of Se volatilization from soils (Kabata-Pendias and Mukherjee 2007).

Although parent geology is the primary long-term determinant of selenium in soils, significant inputs of selenium to soils occur following deposition of selenium from natural i.e., volcanoes, sea spray, volatilization/recycling via biotic cycling and anthropogenic e.g., fossil fuel combustion, sewage, and agricultural inputs such as fertilizers and lime sources (Johnson et al. 2010). Annually, fluxes of selenium to soils from anthropogenic activities are greater than those from all natural sources combined. The effect can be seen in longterm agricultural experiments, where fossil fuel combustion practices correlate with selenium deposition to crops and soils. Crop selenium uptake is influenced greatly by the availability and chemical species of selenium in soils. Inorganic selenium occurs in three soil-phases—fixed, adsorbed, and soluble—and only adsorbed/soluble forms of selenium are thought to be available for plant uptake. In addition, availability of selenate (+6 oxidation state) and selenite (+4) forms to plants varies markedly, with selenate taken up much more rapidly than selenite under most soil conditions. Until recently, it was possible to quantify selenium species in different soil phases only from soils with high adsorbed/soluble selenium loads (50–9000 μg soluble selenium per kg soil) using Hydride Generation Atomic Absorption Spectroscopy techniques (Stroud et al. 2010). However, anion-exchange liquid chromatography (LC) coupled to inductively coupled plasma mass spectrometry (ICP-MS) have enabled selenium species to be quantified in soils of low selenium concentrations i.e., <20 μg soluble seleniumper kg soil (Fairweather-Tait et al. 2011).

Therefore, it could be concluded that, Se content in soils is inherited from parent material and its distribution strongly reflects soil-forming processes and atmospheric deposition. The lowest amounts of Se are in sandy soils developed under humid climate. Selenium contents of worldwide soils have been calculated at an average value of 0.33 mg kg^{-1}, but the range of its concentrations is very broad, from 0.005 to 3.5 mg kg^{-1}. Soils from the Se-deficient endemic areas e.g., in China contain very low watersoluble Se, in the range from 0.2 to 2 μg kg^{-1}, whereas soils from non-endemic regions contain easily soluble Se from 1 to 11 μg kg^{-1}. The total Se range in topsoils of New Zealand is between 0.1 and 4.0 mg kg^{-1} and the common range is 0.3–0.9 mg kg^{-1}. Higher contents of Se are observed in surface layer of forest soils, organic rich soils, calcareous soils, and volcanic soils. It can be generalized that the main factors controlling Se forms and behavior in soils are Eh and pH, however several other parameters like organic ligands, clays, and hydroxides also play very significant roles.

5.1.6.2 Selenium in Waters

As in soils, in aquatic environment, under most pH and redox conditions, the two oxyanions are dominant with several forms of Se^{2-} also being present (Cutter and Bruland 1984). In these ecosystems, selenium can be absorbed by organisms, can form stable adsorption complexes with particulate/colloidal matter and sediments, or can be dissolved in solution. Most transport processes are governed by movement into and out of the top layer of sediment through biogeochemical processes. In contaminated aquifers showing high biological activity, a locally oxidative environment may occur. These environmental conditions oxidize and solubilize the reduced selenium forms that enter in the food web, and Se levels in biota can remain high for years after inputs have ceased (Lemly 1997). Also selenium accumulation in plants and algae from water can be considered a mobilization process, because the Se species are concentrated in a potentially biologically available form that can accumulate through the food chain (Denina et al. 2005). On the contrary, processes that immobilize/sequester Se include chemical and microbial reduction of oxidized forms to Se° (Schlekat et al. 2000) as well as adsorption of selenate and selenite to clay, minerals, particularly iron, and dissolved organic carbon (Di Gregorio 2008).

Natural waters usually contain Se at a level of < 1 µg l^{-1}. The Se concentrations in seawater commonly vary between 0.1 and 0.35 µg l^{-1}. The median Se concentration in oceans is 0.2 µg l^{-1}. According to another sources of data, the average Se level in oceans is 0.09 µg l^{-1}, and in the North Pacific is 0.1 µg kg^{-1} (Nozaki 2005). It has been estimated that 7.7–8.0 kt of Se is introduced into the sea annually (Schrauzer 2004). Organic selenide, mainly dimethyl selenide, $(CH_3)_2Se$, makes up around 80 % of total dissolved Se in ocean surface water and its outgassing may be an important removal mechanism for dissolved Se from aquatic systems. The global Se average in river-waters is given at a value of 0.07 µg l^{-1}, with a range of 0.02–0.5 µg l^{-1} (Gaillardet et al. 2003). However, some rivers, e.g., Colorado River, contain Se in the range of 1–4 µg l^{-1}, although much higher values, up to 400 µg l^{-1} are also reported (ATSDR 2002). Much of this Se is thought to have been derived from industrial sources, e.g., oil refineries contribute up to 75 % of the total Se load entering the San Francisco Bay (Plant et al. 2003). The world average riverine flux of Se is given as 2.6 kt yr^{-1} (Kabata-Pendias 2011).

The levels of selenium in groundwater and surface water range from 0.06 µg l^{-1} to about 400 µg l^{-1}. In some areas, selenium levels in groundwater may approach 6000 µg l^{-1}. Concentrations increase at high and low pH as a result of conversion into compounds of greater solubility in water. Levels of selenium in tap water samples from public water supplies around the world are usually much less than 10 µg l^{-1} but may exceed 50 µg l^{-1}. Drinking-water from a high soil selenium area in China was reported to contain 50–160 µg l^{-1} (WHO 2011).

Usually, the labile Se in most soils and the Se deposited atmospherically onto soils are rapidly leached into groundwater. Wang et al. (1994) reported that Se levels in stream and river waters of Finland increased up to an average of 180 µg l^{-1} and in bottom sediments up to around 4 mg kg^{-1}, after the Se-fertilizing program. The threshold value for Se in drinking water has been established by the WHO as 10 µg l^{-1} whereas in the USA it ranges between 10 and 45 µg l^{-1}. The maximum critical level value for the Se concentration in waters of all states of the USA is 50 µg l^{-1}. The limit level for Se in water used for irrigation is 20 µg l^{-1}. In most countries the maximum permisibale concentration for Se are (µg l^{-1}) for: drinking water for humans 10, drinking water for livestock 50, and irrigation water 20 (Kabata-Pendias and Mukherjee 2007).

The concentrations and chemical forms of selenium in soils or drainage water are governed by various physical and chemical factors, including pH, chemical and mineralogical composition, adsorbing surface, and oxidation–reduction status

(Dhillon and Dhillon 1999). These processes include as well weathering of elemental selenium or metallic selenides (selenium-bearing FeS_2, $FeSe_2$), and oxidation of Se(-II) to selenite (under acidic conditions) or selenate (under alkaline conditions). The general trend for the occurrence of the different species of selenium in soils is deduced directly from their physico–chemical properties: the oxyanions selenite and selenate exist predominantly in the aqueous solutions of aerated alkaline soils, while insoluble selenides and elemental forms of selenium are mostly present in poorly aerated, acid, organic-rich soils of strong reducing conditions (Fordyce 2007). However, this is a very general rule that is not always fulfilled (Fernández-Martánez and Charlet 2009).

Therefore, it could be summarized that the levels of selenium in groundwater and surface water range from 0.06 $\mu g\ l^{-1}$ to about 400 $\mu g\ l^{-1}$. In some areas, selenium levels in groundwater may approach 6000 $\mu g\ l^{-1}$. Concentrations increase at high and low pH as a result of conversion into compounds of greater solubility in water. Natural waters usually contain Se at a level of $< 1\ \mu g\ l^{-1}$. The Se concentrations in seawater commonly vary between 0.1 and 0.35 $\mu g\ l^{-1}$. The median Se concentration in oceans is 0.2 $\mu g\ l^{-1}$. In aquatic environment, under most pH and redox conditions, the two oxyanions are dominant with several forms of Se^{2-} also being present. Selenium can be absorbed by organisms, can form stable adsorption complexes with particulate/colloidal matter and sediments, or can be dissolved in solution. Most transport processes are governed by movement into and out of the top layer of sediment through biogeochemical processes. In contaminated aquifers showing high biological activity, a locally oxidative environment may occur.

5.1.6.3 Selenium in Air

Concentrations of Se in the atmosphere are highly variable due to differentiated sources: (1) evaporation from ocean and sea surface, (2) volcanic eruption, and (3) industrial emissions. The Se concentration in air above the South Pole is 0.06 $ng\ m^{-3}$ and the average value for worldwide air from remote regions is 0.2 $ng\ m^{-3}$ whereas the median for polluted areas is 4.0 $ng\ m^{-3}$ (Reimann and Caritat 1998). There is evidence that the ocean is a significant source of Se to coastal areas. Significant enrichment of Se in marine aerosols results from the formation of volatile organoselenium compounds, mainly dimethyl selenide, $(CH_3)_2Se$. Increased Se levels in mosses ($> 1\ mg\ kg^{-1}$) and in peat ($> 2\ mg\ kg^{-1}$) in the marine regions clearly indicate the impact of Se volatilization from the surface of seawaters (Steinnes 2003). Selenium is released into the air as hydrogen selenide, produced metabolically by plants, and as elemental selenium, selenites and selenates in particulate form. The level of selenium in most urban air ranges from 0.1 to 10 $ng\ m^{-3}$, but higher levels may be found in certain areas, such as in the vicinity of copper smelters (WHO 2011).

Inhalation exposure limits have been established by the US EPA as follows (in $\mu g\ m^{-3}$): 12,700 for hydrogen selenide, 400 for Se-hexafluoride, and 200 for other Se-compounds (Fordyce 2005). According to the guidelines presented by ATSDR (2002), the concentration of Se in air may vary from 160 to 1000 $\mu g\ m^{-3}$. The recommended threshold limit value for Se in workplace is 200 $\mu g\ m^{-3}$, whereas in

Germany, the MAK value at the workplace has been established at 50 μg m^{-3}, and in Russia at 100 μg m^{-3} (Schrauzer 2004). Selenium is released during fossil fuel combustion. Its annual emission from all sources in Europe (only) was 420 t in 1979 (Schrauzer 2004). Its global emission is >6 kt yr^{-1}, in both small particles and volatile compounds which make around 40 % of the total aerial abundance (Kapata-Pendias and Mukherjee 2007).

Therefore, it could be concluded that concentrations of Se in the atmosphere are highly variable due to differentiated the following three sources: evaporation from ocean and sea surface, volcanic eruption, and industrial emissions. There is evidence that the ocean is a significant source of Se to coastal areas. Significant enrichment of Se in marine aerosols results from the formation of volatile organoselenium compounds, mainly dimethyl selenide, $(CH_3)_2Se$. Finally, selenium is released into the air as hydrogen selenide, produced metabolically by plants, and as elemental selenium, selenites and selenates in particulate form. The level of selenium in most urban air ranges from 0.1 to 10 ng m^{-3}, but higher levels may be found in certain areas, such as in the vicinity of copper smelters.

5.1.6.4 Selenium in Humans and Animals

Selenium is an essential trace element also for humans and animals, however, even apparently low concentrations of selenium, in the order of few ppm, can provoke health disturbances (Vinceti et al. 2001). Humans and animals require selenium for the function of a number of selenium-dependent enzymes. Also in humans and animals, Se-Cys is incorporated into a very specific location in the amino acid sequence of selenoproteins, and in humans and animals the uncontrolled S substitution of Se may cause toxic effect. At least 11 selenoproteins have been characterized in animal systems. However, there is evidence that additional selenoproteins may be recovered. The first to be characterized was the glutathione peroxidases; more precisely, four selenium-containing glutathione peroxidases (GPx) have been identified: cellular or classical GPx, plasma or extracellular GPx, phospholipid hydroperoxide GPx, and gastrointestinal GPx (Holben and Smith 1999). These enzymes reduce damaging reactive oxygen species (ROS) oxidizing glutathione. A well characterized example of this enzyme family is the sperm mitochondrial capsule selenoprotein, a phospholipid hydroperoxide GPx; this is an antioxidant enzyme that protects developing sperm from oxidative damage and is responsble for a structural protein required by mature sperm. Se-Cys is present also in the active site of the thioredoxin reductase that, in conjunction with the compound thioredoxin, participates in the regeneration of several antioxidant systems in animal cells, possibly including vitamin C (Mustacich and Powis 2000). Moreover, maintenance of thioredoxin in a reduced form by thioredoxin reductase is important for regulating cell growth and viability (Di Gregorio 2008).

Selenium occurs in mammalian tissues in the range from 0.7 in heart tissue to 2.5 mg kg^{-1} in muscles. The average Se content in human soft tissues is estimated as 0.11 mg kg^{-1} (Li 2000). Concentrations of Se in kidneys of humans from other

European countries are reported by Zduôska et al. (1994) as follows (in mg kg^{-1} FW): Bulgaria, 2.5; Germany 0.7; and Italy, 1.9. In human fluids, mean Se concentrations are (in μg l^{-1}): blood, 107; serum, 80; urine, 22; and milk, 13 (Li 2000). The range of Se in milk of women in Poland varies from <9 to >11 μg l^{-1} (Zachara and Pileexi 2000).

In Finland, the Se level in serum was between 0.63 and 0.76 μmol l^{-1}, and after the Se supplementation of fertilizers, it increased to the range of 1.2–1.4 μmol l^{-1} (Hartikainen 2005). Deficiency symptoms of improper Se supply to humans could be summarized as follows: muscle weakness and pain, inflammation of muscles, fragile red blood cells, degeneration of pancreas, abnormal skin coloration, heart muscle dysfunction, prolonged illness condition, susceptibility to cancer, Keshan disease (KD) and Kashin-Beck disease (KBD). Whereas, the toxicity symptoms could be summarized as follows: liver and kidneys damage, blood clotting, necrosis of heart and liver, skin lesions, hair and nail loss and Nausea and vomiting (Kabata-Pendias and Mukherjee 2007).

Furthermore, selenoprotein P has been identified in plasma; its function has not been clearly delineated, and it has been suggested to function as a transport protein and also as an antioxidant capable of protecting endothelial cells from damage by a reactive nitrogen species (RNS). On the other hand, a selenoprotein W has been found in muscle, whose function is presently unknown but thought to be related to muscle metabolism (Holben and Smith 1999). On the other hand, selenium appears to stimulate the immune response (Roy et al. 1994) in humans, playing a role in regulating the expression of cell signaling molecules called cytokines, which orchestrate the immune response (Baum et al. 2000). Many studies suggest that selenium supplementation at high levels reduces the incidence of cancer in animals and that the methylated forms of selenium are the active species against tumors (Combs and Gray 1998). Results of epidemiological studies of cancer incidence in groups of humans with variable selenium intakes show a trend for individuals with lower blood selenium levels to have a higher incidence of several different types of cancer. The trend is less pronounced in women. Several mechanisms have been proposed for the cancer prevention effects of selenium: (1) maximizing the activity of antioxidant selenoenzymes and improving antioxidant status, (2) improving immune system function, (3) affecting the metabolism of carcinogens, and (4) increasing the levels of selenium metabolites that inhibit tumor cell growth, an interesting developing aspect of research in selenium reviewed in the next paragraph (Di Gregorio 2008).

Finally, selenium recommended dietary allowance changes with human life stage. In fact, although selenium is required for health, high doses can be toxic. Acute and fatal toxicities have occurred with accidental or suicidal ingestion of gram quantities of selenium. Chronic selenium toxicity (selenosis) may occur with smaller doses of selenium over long periods of time. As mentioned before, the most frequently reported symptoms of selenosis are hair and nail brittleness and loss, gastrointestinal disturbances, skin rashes, a garlic breath odor, fatigue, irritability, and nervous system abnormalities (Di Gregorio 2008).

Therefore, it coule be concluded that, selenium is an essential trace element also for humans and animals, however, even apparently low concentrations of selenium, in the order of few ppm, can provoke health disturbances. Humans and animals require selenium for the function of a number of selenium-dependent enzymes. Also in humans and animals, Se-Cys is incorporated into a very specific location in the amino acid sequence of selenoproteins, and in humans and animals the uncontrolled S substitution of Se may cause toxic effect. At least 11 selenoproteins have been characterized in animal systems. However, there is evidence that additional selenoproteins may be recovered. The first to be characterized was the glutathione peroxidases.

5.1.6.5 Selenium in Plants

It has been demonstrated that also higher plants metabolize Se via the sulfur assimilation pathway (Zayed et al. 2000), synthesizing Se analogues of various S metabolites. This involves the fact that the nonspecific incorporation of Se into selenoamino acids and proteins as well as Se volatilization occurs, when the metalloid is supplied to plants in excess of any potential, but not demonstrated, requirement. Although there are evidences that Se is required for the growth of algae (Yokota et al. 1988), the question of the essentiality of Se as a micronutrient in higher plants is controversial. Although Se is not an essential element for plants, with some exceptions, it is being added to soil to ensure that both food and feed products contain adequate amounts for the dietary needs. It should be emphasized that the margin of safety of Se concentrations is rather narrow. The Se content of crops received recently much attention because of its importance in the food chain. Thus, most data that are available are for food and fodder plants. Especially, cereal grains, as the most common source of Se in diets, have been broadly analyzed. In general, mean concentrations of Se in grains are higher in countries from arid climates than in countries from humid climates (Kabata-Pendias 2011).

The function of Se in plants has been investigated in many studies and there is still little evidence that Se is essential for all plants. As mentioned before, there is some evidence that Se is required for the growth of algae, but its essentiality to higher plants is controversial and yet unresolved. However, there are some indications that this element may be required for Se-accumulating plants. It is well established that some grasses and vegetables provide indications that at proper Se addition the growth rate of these plants may be enhanced (Hartikainen 2005). It is also cited that there are several naturally occurring organic Se species: (1) selenocysteine, (2) methylselenocysteine, (3) selenomethionine, (4) selenotaurine, (5) seleniobetaine, (6) seleniocholine, (7) dimethylselenide, (8) dimethyldiselenide, and (9) trimethylselenonium. Although the essentiality of selenoproteins in higher plants has not been documented, syntheses of selenoproteins in some plants such as sugar beet, have been reported (Terry et al. 2000). Several selenoamino acids, SeMet (selenomethionine), SeCys (selenocysteine), and SeMC (selenomethylocysteine) in association with glutathione peroxidases have been found in both bacteria and higher plants (Kabata-Pendias 2011).

The range of Se in cereals at the worldwide level is estimated as 100–800 μg kg^{-1} FW (Fordyce 2005). The range of mean Se varies from 142 to 970 μg kg^{-1} for countries with high Se levels in grains, and from 14 to 90 μg kg^{-1} for countries with low Se levels in grains. The anticarcinogenic effectiveness of various Se compounds in plants has been recently broadly investigated. Plants differ in their ability to accumulate Se in their tissues. Actually, certain native plants are able to hyperaccumulate Se in their shoots when they grow on seleniferous soil. These species are called Se hyperaccumulators and include a number of species of *Astragalus*, *Stanleya*, *Morinda*, *Nepturia*, *Oonopsis*, and *Xylorhiza* (Brown and Shrift 1982). On the other hand, most forage and crop plants, as well as grasses, contain less than 25 mg Se kg^{-1} dry weight and do not accumulate Se much above 100 mg Se kg^{-1} dry weight when grown on seleniferous soils. These plants are referred to as Se nonaccumulators (White et al. 2004). A third category of plants, known as secondary Se accumulators (Brown and Shrift 1982), grow on soil of low-to-medium Se content and accumulate up to 1000 mg Se kg^{-1} dry weight. Examples of plants in this group are species of *Aster*, *Astragalus*, *Atriplex*, *Castilleja*, *Comandra*, *Grayia*, *Grindelia*, *Gutierrezia*, and *Machaeran hera* (Parker and Page 1994). Indian mustard (*Brassica juncea*) and canola (Brassica napus), are also secondary Se accumulator plant species with a typical Se concentration of several hundred micrograms of Se g^{-1} dry weight in their shoots when grown on soils containing moderate levels of Se (Banuelos et al. 1997).

The uptake of Se by plants depends on several factors, such as climate, soil parameters, and plant capacity to accumulate. When present in soluble forms, Se is readily absorbed by plants, although differences between plant species are very pronounced. Primarily, Se is taken up from the soil as selenate, SeO_4^{2-} or selenite, SeO_3^{2-} (Ellis and Salt 2003). In most cases there is a positive linear correlation between Se in plant tissues and Se contents of soils. However, the complex impact of variable factors on Se uptake by plants can significantly alter the relation between Se in plants and soils. Finland is an exception, having implemented the program of Se addition to fertilizers. Food plants of the USA contain fairly similar amounts of Se that does not exceed 100 μg kg^{-1} FW. Its mean concentration is higher in roots and tubers (13 μg kg^{-1} in potato and 17 μg kg^{-1} in carrot) than in fruits (from 1 μg kg^{-1} in oranges to 4 μg kg^{-1} in apples). Se-enriched garlic has often been reported as an important dietary supplement (Kabata-Pendias and Mukherjee 2007).

There is a great variation in plants' capability to absorb Se from soils, especially from seleniferous ones. While there are indications that Se may be required for Se-accumulating plants (Broyer et al. 1972), which are actually endemic on seleniferous soils, there is no evidence for a Se requirement in nonaccumulators (Shrift 1969). In order to investigate the essentiality of Se in higher plants, attempts have been made to establish whether plants contain essential selenoproteins, such as those discovered for bacteria and animals. However, on the basis of all the available information published to date, no essential higher plant selenoprotein has been clearly identified by either protein or DNA sequences analysis (Terry et al. 2000). Obscure is also the ecological significance of selenium hyperaccumulators except the hypothesis that protection from insect herbivore is a major ecological advantage of Se hyperaccumulation. The hypothesis is corroborated by the evidence that accumulation in leaves of two very different forms of Se, inorganic selenate in nonaccumulator plants and organic Se-methylcysteine in hyperaccumulators, had similar protective effects against herbivores. Additionally, an intermediate Se concentration caused an intermediate degree of herbivore protection (Freeman et al. 2007). These results shed light on the possible selection pressures that have driven the evolution of Se hyperaccumulation (Di Gregrio 2008).

A notable problem with phytoextraction is that plants with elevated Se content may become available to wildlife potentially causing toxicity, therefore they should be harvested, removed and utilized elsewhere. Volatilization of Se, both directly from the soil or by plants, may have a practical application in specific ecosystems. Microorganisms (bacteria and fungi) and several higher plants can biomethylate inorganic Se species and exhibit a great capability to remove Se from soil (Frankenberger and Karlson 1994). The main Se species that volatilize from soils are those that also volatilize from plants (dimethylselenides) as well as other methane and sulfide-methane compounds. As Lin et al. (1999) calculated, the phytovolatilization of Se during the growing season of *Salicornia* species was 34.6 mg Se m^{-2} and exceeded about two times the amount of Se removed by phytoextraction. Phytovolatilization seems to be promising methods since it removes Se from soils directly into the atmosphere (Kabata-Pendias and Mukherjee 2007).

Therefore, it could be concluded that, although there are evidences that Se is required for the growth of algae, the question of the essentiality of Se as a micronutrient in higher plants is controversial. Although Se is not an essential element for plants, with some exceptions, it is being added to soil to ensure that both food and feed products contain adequate amounts for the dietary needs. It should be emphasized that the margin of safety of Se concentrations is rather narrow. The Se content of crops received recently much attention because of its importance in the food chain. Thus, most data that are available are for food and fodder plants. Especially, cereal grains, as the most common source of Se in diets, have been broadly analyzed. In general, mean concentrations of Se in grains are higher in countries from arid climates than in countries from humid climates.

5.1.6.6 Selenium in Food Systems

Most people obtain virtually all of their selenium from the foods they eat. In plant and animal tissues, selenium is found mostly bound to proteins. Therefore, the most important food sources of selenium are meats and seafood (0.3–0.5 mg kg^{-1}), because of their high protein contents, and cereals (0.1–10 mg kg^{-1}), because they tend to be consumed in large amounts. In contrast, foods with relatively low protein levels, such as vegetables and fruits, tend to have relatively low selenium contents (<0.01 mg kg^{-1}). In all cases, the selenium content of foods reflects the available selenium content of the soils used to produce those foods (and the feedstuffs used to produce livestock). FAO/WHO (1998) noted that global selenium intakes vary significantly; average intakes were relatively high in North America (85–150 μg/day), moderate in Europe (40–90 μg/day) and low in parts of China (10–20 μg/day) (WHO 2011).

It is well documted that the amount of selenium in the diet largely depends on where crops are grown and cultivated, the soil/fodder to which animals are exposed, and the actual foods consumed. The effect of selenium species on bioavailability has been reviewed recently and data on the selenium content of foods are available (Fairweather-Tait et al. 2010). The main food groups providing selenium in the diet or contribution of each food group to total population dietary exposure in the UK are bread and cereals (26 %), meat (26 %), milk/dairy products (21 %), fish (10 %), vegtables and fruits (7 %) and eggs (4 %). Some Brazil nuts are a particularly rich source, with selenium concentrations ranging from ~ 0.03–512 mg kg^{-1} fresh weight (Rayman et al. 2008).

The selenium content of bread and cereals can vary widely from ~0.01–30 mg kg^{-1} (Rayman et al. 2008). On average, bread and cereals provide a quarter of the selenium intake in the UK. The predominant species of selenium in wheat and bread are selenomethionine (usually ~55–85 %), selenocysteine (~4–12 %), and selenate/selenite (~12–19 %) (Whanger 2002). The selenium content of meat depends on many factors. Offal contains relatively high levels of selenium, in particular liver and kidneys; the selenium concentrations of kidney, liver, and heart tissue from beef were 4.5, 0.93, and 0.55 mg kg^{-1}, respectively, whereas muscle was in the region of 0.2 mg kg^{-1}. Supplementation of cattle with selenium-enriched yeast increased muscle selenium concentration to ~0.6 mg kg^{-1} (Juniper et al. 2008). In the United States, the average selenium content of chicken is ~0.2 mg kg^{-1} and beef ~0.25–0.3 mg kg^{-1} (Fairweather-Tait et al. 2011).

Meat generally provides a relatively large proportion of the selenium intake in omnivorous populations, and in the UK, it provides one quarter of the total estimated intake. The predominant species of selenium in edible portions of meat may be selenomethionine (~50–60 % of total extractable selenium species) and selenocysteine (20–31 and ~50 % of total extractable selenium species in chicken and lamb, respectively). However, the total content and species depends mainly on the animals' diet. The selenium content in fish is between 0.1 and ~5.0 mg kg^{-1} (Fairweather-Tait et al. 2010); some marine fish are relatively high in selenium; for example, the selenium content of cod, shark, and canned tuna is ~1.5, 2.0, and 5.6 mg kg^{-1}, respectively (Reyes et al. 2009). In the UK, the average selenium content of fish is ~0.42 mg kg^{-1}. The main selenium species in fish are selenomethionine (29–70 %) and selenite/selenate (12–45 %) with the species profile differing between fish species and the total selenium content. Hens' eggs contain from ~3 to ~25 mg selenium per whole egg (Lipiec et al. 2010). Selenium supplementation of the hen's diet may increase the selenium content of eggs to 0.34–0.58 mg kg^{-1}; selenium-enriched eggs are widely produced around the world (Fisinin et al. 2009). The main selenium species in eggs are selenocysteine, selenomethionine, and possibly selenite, with selenomethionine and selenocysteine as the predominant species (> 50 %) in egg white and egg yolk, respectively (Lipiec et al. 2010).

The selenium content of milk and dairy products varies widely; in the UK, milk and dairy products contain ~0.01–0.03 mg kg^{-1} selenium. The predominant selenium species in cows' milk are selenocysteine and selenite. Supplementation of dairy cows with selenium-enriched yeast alters the species profile in the milk and the major species after supplementation are selenocysteine, selenomethionine, and selenite (Muniz-Naveiro et al. 2007). On the other land, fruit and vegetables typically contain relatively small amounts of selenium. In unenriched vegetables with low levels of selenium, the major species may be, for example, selenate in onions or selenomethionine (53 %), γ-glutamyl-Se-methylselenocysteine (31 %), Se-methylselenocysteine (12 %), and selenate (4 %) in garlic with natural selenium content of <0.5 mg kg^{-1}. However, certain vegetables, such as onions, garlic, and broccoli when grown on selenium-rich soil can accumulate selenium, resulting in selenium-enrichment from <0.5 mg kg^{-1} up to 140–300 mg kg^{-1}. The main selenium species in Se-enriched food such as onions is γ-glutamyl-Se-methylselenocysteine, accounting for ~63 % of the species, with a relatively smaller proportion of ~10 % selenate and 5 % selenomethionine, plus other species (Hurst et al. 2010). In Se-enriched garlic, similar to Se-onions, γ-glutamyl-Se-methylselenocysteine may be the predominant species (~73 %) with also ~13 % selenomethionine, 4 % g-glutamyl-selenomethionine, 3 % Se-methylselenocysteine, and 2 % selenate. Selenium-enriched broccoli sprouts may contain predominantly Se-methylselenocysteine (~45 %) with smaller amounts (~12–20 %) of selenate and selenomethionine, plus other species of selenium such as adenosylselenohomocysteine (Finley et al. 2001). In summary, in vegetables such as broccoli, onions, and garlic the selenium species profile is variable depending on the total level of selenium enrichment, the forms of selenium used for enrichment, and the type of vegetable; predominant species in selenium-enriched vegetables analyzed to date are Se-methylselenocysteine or γ-glutamyl-Se-methylselenocysteine; these forms of selenium in foods have received attention due to purported protection against cancer in animal models when compared with other forms of selenium (Fairweather-Tait et al. 2011).

Therefore, it could concluded that, the amount of selenium in the diet largely depends on where crops are grown and cultivated, the soil/fodder to which animals are exposed, and the actual foods consumed. The effect of selenium species on bioavailability has been reviewed recently and data on the selenium content of foods are available. The main food groups providing selenium in the diet or contribution of each food group to total population dietary exposure in the UK are bread and cereals (26 %), meat (26 %), milk/dairy products (21 %), fish (10 %), vegtables and fruits (7 %) and eggs (4 %). Some Brazil nuts are a particularly rich source, with selenium concentrations ranging from ~0.03–512 mg kg^{-1} fresh weight.

5.1.7 Fundamental Importance of Soil Selenium

It is well established that the significance of selenium in the nutrition of human subjects has grown rapidly during the past 20 years. Demonstrations of its essentiality to rats and farm animals were followed by appreciation that the development of selenium-responsive diseases often reflected the distribution of geochemical variables which restricted the entry of the element from soils into food chains. Such findings were the stimulus to in-depth investigations of the regional relevance of selenium in human nutrition. These studies have now yielded an increased understanding of the complex metabolic role of this trace nutrient. Selenium has been implicated in the protection of body tissues against oxidative stress, maintenance of defences against infection, and modulation of growth and development (FAO/WHO 2004).

It is well documented that acidic and reducing conditions reduce inorganic selenites to elemental selenium, whereas alkaline and oxidizing conditions favour the formation of selenates. Because selenites and selenates are soluble in water, selenium is leached from well-aerated alkaline soils that favour its oxidation. In contrast, elemental selenium and selenides are insoluble in water; therefore, selenium tends to be retained in wet, poorly aerated soils, the reducing conditions of which favour those forms. Thus, selenium in alkaline soils is available for uptake by plants, whereas the availability of selenium in acidic soils tends to be limited by the adsorption of selenites and selenates to iron and aluminium oxide sols (WHO 2011).

Selenium enters soils primarily as a result of the weathering of Se-containing rocks, although volcanic activity, dusts (e.g., in the vicinity of coal burning), Se-containing fertilizers, and some waters can also be sources. Some parts of the world such as Denmark, eastern Finland, New Zealand, eastern and central Siberia, and a long belt extending from northeast to south-central China including parts of Heilongjiang, Jilin, Liaoning, Hebei, Shanxi, Shaanxi, Sichuan and Zhejiang Provinces and Inner Mongolia are notable for having very low amounts of Se in their soils and, therefore, their food systems. Other areas such as the Great Plains of the USA and Canada; Enshi County, Hubei Province, China; and parts of Ireland, Colombia and Venezuela, in contrast, are seleniferous. For example, soils derived from the Se-rich Niobara and Pierre shales of North Dakota contain as much as 90 mg Se kg^{-1} soil, while most non-seleniferous soils based on low-Se granites and metamorphic sandstone contain appreciably less than 2 mg Se kg^{-1} soil. The biogeochemical mapping of Se has only been accomplished for the United States and parts of Canada and China, as well as for parts of Europe, the former Soviet Union, New Zealand and Australia (Combs 2005).

Selenium is an essential element for humans and animals, and therefore various national and international organizations have established recommended daily intakes of selenium. The joint World Health Organization (WHO)/Food and Agriculture Organization of the United Nations (FAO) consultation on preparation and use of food-based dietary guidelines (FAO/WHO 1998) listed recommended intakes of 6–21 μg of selenium per day for infants and children, according to age, 26 and 30 μg of selenium per day for adolescent females and males, respectively, and 26 and 35 μg of selenium per day for adult females and males, respectively. In 2000, the United States National Academy of Sciences Panel on Dietary Oxidants and Related Compounds revised the recommended intake of selenium to 55 μg/day for both men and women and 70 μg/day for women during pregnancy and lactation. Recommended selenium intakes for children are between 15 μg/day for infants 0–6 months of age and 30 μg/day for children 4–8 years old. The United Kingdom Expert Group on Vitamins and Minerals recommended selenium intakes of 60 μg/day for women and 70 μg/day for men. However, it is clear that the position with regard to selenium requirements is more complex than these recommendations would suggest, because some groups, such as New Zealanders and Swedish vegans, have very low intakes, comparable to those in selenium-deficient parts of China, with no apparent adverse effects. Therefore, other aspects of the diet would appear to be important in mitigating the effects of low selenium intakes (FAO/WHO 1998). Because of concern about the adverse effects resulting from exposure to excessive levels of selenium, various national and international organizations have established upper limits of exposure for selenium. The United States National Academy of Sciences Panel on Dietary Oxidants and Related Compounds set an upper tolerable limit for selenium at 400 μg/day. This level was also recommended by FAO/WHO (1998) and the United Kingdom Expert Group on Vitamins and Minerals (WHO 2011).

Much of tissue selenium is found in proteins as selenoanalogues of sulfur amino acids; other metabolically active forms include selenotrisulphides and other acid-labile selenium compounds. At least 15 selenoproteins have now been characterized. Examples are given in Table 5.7. Functionally, there appear to be at least two distinct families of seleniumcontaining enzymes. The first includes the glutathione peroxidases and thioredoxin reductase, which are involved in controlling tissue concentrations of highly reactive oxygen-containing metabolites. These metabolites are essential at low concentrations for maintaining cell-mediated immunity against infections but highly toxic if produced in excess (FAO/WHO 2004).

In the early 1970s, Se was found to be an essential component of the enzyme glutathione peroxidase (GPx). As that enzyme was known to participate in the antioxidant protection of cells by reducing hydroperoxides, this finding was taken to explain the nutritional "sparing" by Se of vitamin E, a known lipid-soluble antioxidant. Discoveries over the last 15 years have revealed that several Se-enzymes are recognized: at least five GPx isoforms, three iodothyronine 5′-deiodinases (DIs), three thioredoxin reductases (TRs), and selenophosphate synthetase. Other proteins are recognized as specifically incorporating Se, although their metabolic functions remain unclear: plasma selenoprotein P, muscle selenoprotein W, and selenoproteins in prostate and placenta. Each of these contains Se as the amino acid selenocysteine (SeCys). Selenium is incorporated into SeCys by the co-translational modification of tRNA-bound serinyl residues at certain loci encoded by specific TGA codons containing SeCys-insertion sequences in the 3′-untranslated regions of their respective mRNAs. Thus, TGA is decoded as SeCys rather than as a stop signal. It is likely that more SeCys-proteins remain to be discovered, as other Se-containing proteins have been identified (Combs 2005).

Therefore, it could concluded that, the significance of selenium in the nutrition of human subjects has grown rapidly during the past two decades. Demonstrations of its essentiality to rats and farm animals were followed by appreciation that the development of selenium-responsive diseases often reflected the distribution of geochemical variables which restricted the entry of the element from soils into food chains. Such findings were the stimulus to in-depth investigations of the regional relevance of selenium in human nutrition. These studies have now yielded an

Table 5.7 The known selenocysteine—proteins or SeCys-proteins. (Adapted from Combs 2005)

Protein	Sub unit Mass kDa	SeCys content moles/mole	Enzymatic function	Tissue distribution
GPx-1	22	4 (tetramer)	GSH-dependent reduction of H_2O_2 or lipoperoxides	Cytosol and mitochondrial matrix space; ubiquitous
GPx-2	22	4 (tetramer)	GSH-dependent reduction of H_2O_2 or lipoperoxides	Gastrointestinal tract
GPx-3	23*	4 (tetramer)	GSH-dependent reduction of H_2O_2 or lipoperoxides	Plasma
GPx-4	19	1	Phospholipid hydroperoxide removal	Ubiquitous
DI-1	28	1	Conversion of T_4 to T_3	Thyroid, liver, kidney, brain, pituitary
DI-2	30 5	1	Conversion of T_4 to T_3	Pituitary, thyroid, heart, muscles
DI-3	31 5	1	Conversion of T_4 to r T_3	Placenta, brain, skin
TR-1	55	2 (dimer)	NAPDH-dependent reduction of oxidized thi oredoxin	Cytosol; ubiquitous
TR-2	53	2 (dimer)	NAPDH-dependent reduction of oxidized thi oredoxin	Mitochondria; kidney, liver, adrenal, heart
TR-3	57	2 (dimer)	NAPDH-dependent reduction of oxidized thi oredoxin	Ubiquitous
Seleno-phosphate synthetase	48	1	ATP-dependent formation of seleno-phosphate from selenide	Liver, testes, kidney, thymus, spleen
SeP	57[a](43net)	Up to 10 (monomer)	Unknown	Plasma
SeW	10	1	Unknown	Muscle
Prostate Se-protein	15	1	Unknown	Prostate
Placental Se-protein	15	1	Unknown	Placenta

GPx glutathione peroxidase
TRs thioredoxin reductases
DIs iodothyronine 5'-deiodinases
GSH glutathione
NADP Nicotinamide adenine dinucleotide phosphate
SeP plasma selenoprotein
SeW muscle selenoprotein
kDa kilo Dalton
[a]glycosylated

increased understanding of the complex metabolic role of this trace nutrient. Selenium has been implicated in the protection of body tissues against oxidative stress, maintenance of defences against infection, and modulation of growth and development. Selenium is an essential element for humans and animals, and therefore various national and international organizations have established recommended daily intakes of selenium.

5.1.8 Global Variation in Selenium Status

Selenium was first recognized as an essential nutrient in the late 1950s when it was found to replace vitamin E in the diets of rats and chicks for the prevention of vascular, muscular and/or hepatic lesions. Until that time, Se had been thought of only as a toxicant, being associated with "alkali disease" in grazing livestock in the northern Great Plains of the United States. Since that time, Se has become the subject of investigations in many parts of the world (Combs 2005).

The intake of Se varies considerably among populations in different parts of the world, and it is clear that the diets of millions of people do not provide enough of the element to support maximal expressions of the SeCys-enzymes. The best-described Se-deficient areas are New Zealand, Finland prior to 1984, and a long belt of mountainous terrain extending from the northeast to south-central portions of mainland China. Evidence of low Se intakes has also been reported in parts of Eastern Europe, and parts of Russia and Africa. In each case, low amounts of available Se in soils results in a generalized deficiency of the element throughout the food system. Typical Se intakes have been estimated for several counties on the basis of published Se contents of local foods (Table 5.8). These indicate significant inter-regional differences in food system Se contents, which are manifest as differences in nutritional Se status. Such data indicate that millions of people appear to be deficient in the element. Using the criterion of a serum/plasma Se concentration of 80 $\mu g\ ml^{-1}$ as the minimum level associated with maximal expression of plasma GPX, sub-clinical Se deficiency appears to be highly prevalent in Austria, Bulgaria, Chile, China, Cuba, the Czech Republic, Estonia, Germany Greece, Hungary, Jamaica, Nigeria, Poland, Spain, Zambia, and parts of Venezuela. Selenium deficiency may affect substantial numbers in Australia, Belgium, Denmark, Egypt, England, Finland, India, Ireland, Italy, Mexico, Portugal, Saudi Arabia, Sweden, Switzerland, Turkey, and parts of the USA. Among the countries for which data are available, Se deficiency would not appear likely on a wide scale only in Canada, Japan, Norway and the USA. This classification must be considered provisional, as the data base admittedly sparse for most of the countries listed and includes little or no information for several large parts of the world such as most of Africa, South American, central and south Asia (Combs 2005).

Therefore, it could be concluded that, the intake of Se varies considerably among populations in different parts of the world, and it is clear that the diets of millions of people do not provide enough of the element to support maximal expressions of the SeCys-enzymes. Evidence of low Se intakes has also been reported in parts

Table 5.8 Mean measures of nutritional selenium status reported for healthy adults worldwide. (Adapted from Combs 2005)

Country	Serum/plasma Se ($\mu g\ ml^{-1}$)	Country	Serum/plasma Se ($\mu g\ ml^{-1}$)
Countries under minimal level (<80 µg ml⁻¹)		*Countries upper minimal level (>80 µg ml⁻¹)*	
Zambia	40	Switzerland	84
Bulgaria	45	Egypt	85
Nigeria	50	Italia	85
Estonia	55	Spain	85
Hungary	61	Jamaica	86
Czech Republic	62	Demark	93.
Greece	65	India	94
Chile	66	China (eastern urban)	95
Austria	67	Australia	96
Cuba	69	Portugal	97
Poland	70	Mexico	100
Germany	71	Finland (after 1984)	100
Sweden	76	Netherlands	101
Finland (before 1984)	77	Ireland	103
Niger	78	France	103
		Saudi Arabia	103
		Turkey	109
		England	110
		USA (Eastern States)	113
		USA (Western States)	130
		Norway	138
		Japan	141
		Belgium	165
		Canada	165
		Venezuela	197

of Eastern Europe, and parts of Russia and Africa. In each case, low amounts of available Se in soils results in a generalized deficiency of the element throughout the food system. Typical Se intakes have been estimated for several counties on the basis of published Se contents of local foods.

5.1.9 *Nano-selenium and the Environment*

In fact, *nano* refers to any parameter when it is expressed as a measure of 10^{-9} times of SI units. Until recent past, the very existence of nanoparticles and their applications remained undetected. Nanotechnology was first proposed to have applications in the field of electronics for the miniaturization of the electronic devices. In fact, the term "Nanotechnology" has been coined by Norino Taniguchi, a researcher at the University of Tokyo, Japan (Taniguchi 1974). This slowly expanded to various fields. Even when various scientists reported the remediation of various heavy metals by microorganisms, the remediated nanosized zero valent metal crystal remained unnoticed (Deepak et al. 2011).

There is no doubt that nanomaterials will play a key role in many technologies of the future. One key aspect of nanotechnology concerns the development of reliable experimental protocols for the synthesis of nanomaterials over a range of chemical compositions, sizes and high monodispersity. In the context of the current drive to develop green technologies in materials synthesis, this aspect of nanotechnology assumes considerable importance. An attractive possibility is to use micro-organisms in the synthesis of nanoparticles. It is worth to mention that the field of nanotechnology is an immensely developing field as a result of its wide-ranging applications in different areas of science and technology. The term nanotechnology is defined as the creation, exploitation and synthesis of materials at a scale smaller than 1 mm. The word "nano" is derived from a Greek word meaning dwarf or extremely small (Rai et al. 2008). The concept of nanotechnology was given by physicist Professor Richard Feynman in his historic talk "*there's plenty of room at the bottom*" (Feynman 1959), though the term nanotechnology was introduced by Tokyo Science University Professor Norio Taniguchi. Nanobiotechnology is a multidisciplinary field and involves research and development of technology in different fields of science like biotechnology, nanotechnology, physics, chemistry, and material science (Huang et al. 2007). Nanoparticles are metal particles with size 1–100 nm and exhibit different shapes like spherical, triangular, rod, etc. Research on synthesis of nanoparticles is the current area of interest due to the unique visible properties (chemical, physical, optical, etc.) of nanoparticles compared with the bulk material (Rai et al. 2009; Rai et al. 2011).

There is tremendous current excitement in the study of nanoscale matter (matter having nanometre dimensions, $1\ nm = 10^{-7}$ cm) with respect to their fundamental properties, organization to form superstructures and applications. The unusual physicochemical and optoelectronic properties of nanoparticles arise primarily due to confinement of electrons within particles of dimensions smaller than the bulk electron delocalization length, this process being termed quantum confinement. The exotic properties of nanoparticles have been considered in applications such as optoelectronics, catalysis, reprography, single-electron transistors (SETs) and light emitters, nonlinear optical devices and photoelectrochemical applications. Recognizing the importance of nanomaterials in key future technologies, many countries have launched major initiatives into the development of a strong fundamental and applied knowledge base in the area of nanotechnology. It certainly does appear that 'there's plenty of room at the bottom' (Feynman 1959) in this fascinating area (Sastry et al. 2003).

Progress in the field of nanotechnology has been rapid and with the development of innovative synthesis protocols and characterization techniques (Sharma et al. 2009). But most of the synthesis methods are limited to synthesis of nanoparticles in small quantities and poor morphology (Sau and Rogach 2010). Chemical and physical synthesis methods often result in synthesis of a mixture of nanoparticles with poor morphology, and these methods also prove to be toxic to the environment due to the use of toxic chemicals and also of elevated temperatures for synthesis process (Birla et al. 2009). Biogenic synthesis of nanoparticles with controlled morphology needs more attention, as the biogenic synthesis of nanoparticles is carried out by using biological means like bacteria (Shahverdi et al. 2007, 2009), fungi (Kumar et al. 2007a, b; Parikh et al. 2008; Gajbhiye et al. 2009; Govender et al. 2009), actinomycetes (Ahmad et al. 2003a, b), lichens (Shahi and Patra 2003), algae (Singaravelu et al. 2007; Chakraborty et al. 2009), etc. The biogenic entities are found to secrete large amount of proteins which are found to be responsible for metal–ion reduction and morphology control (Thakkar et al. 2010). The microbial cultures are easy to handle and also the downstream processing of biomass is simpler as compared to the synthetic methods (Ingle et al. 2008). Biogenic nanoparticles are toward a greener approach and environment friendly, as no toxic chemical is involved in synthesis, and also the synthesis process takes place at ambient temperature and pressure conditions (Gade et al. 2008). Hence, a number of researchers are focusing toward the synthesis of biogenic nanoparticles compared with the chemically or physically synthesized nanoparticles (Ingle et al. 2008; Kumar and Yadav 2009; Rai et al. 2011).

Therefore, it could be included that there is no doubt that nanomaterials will play a key role in many technologies of the future. One key aspect of nanotechnology concerns the development of reliable experimental protocols for the synthesis of nanomaterials over a range of chemical compositions, sizes and high monodispersity. In the context of the current drive to develop green technologies in materials synthesis, this aspect of nanotechnology assumes considerable importance. An attractive possibility is to use micro-organisms in the synthesis of nanoparticles. Progress in the field of nanotechnology has been rapid and with the development of innovative synthesis protocols and characterization techniques.

5.1.9.1 Introduction

Nanoscale science and nanotechnology have been demonstrated to have great potential in providing novel and improved solutions to many grand challenges facing agriculture and society today and in the future. Presently, the agricultural sector is facing various global challenges: climate change, urbanization, sustainable use of resources, and environmental issues such as run-off and accumulation of pesticides and fertilizers. These situations are further exacerbated by the growing food demand that will be needed to sustain an estimated population growth from the current level of about 6 billion to 9 billion by 2050. In addition, considering the world's diminishing petroleum resources, agricultural products and materials will soon be viewed again as the foundation of commerce and manufacturing, hence additional demands on agricultural production. At the same time there are new opportunities emerging. For example, the use of agricultural waste for the generation of energy and electricity could be a viable solution pending workable economics and encouraging policy. This aforementioned scenario of rapidly evolving and yet complex agriculture system is, and will pose even greater challenges to developing countries. The agricultural sector and commodity production in developing regions are the backbone of the national economy where multitude critical issues such as lack of new arable soil, reduction of the current agricultural land due to competing economic development activities, commodity dependence, poverty and malnutrition are closely intertwined (Chen and Yada 2011).

A variety of microorganisms, bacteria, yeast, fungi and algae, can adsorb and accumulate metals but only a few groups can selectively reduce metal ions to produce nano-scale mineral phases (Oremland et al. 2004). These organisms have the unique ability to produce inorganic phases of constant chemical composition and size (Pearce et al. 2008). The majority of studies on the biogenesis of nano-Se particles have concentrated on anaerobic systems that have certain limitations, such as culture conditions and isolate characteristics that make optimization and scale-up in bio-manufacturing processes challenging. Selenium tolerant aerobic organisms, however, provide the opportunity to overcome these limitations in the biosynthetic process. The isolate tolerates selenium oxyanions and generates selenium nanoparticles, thus combining the detoxification of oxidized seleniferous environments with the biotechnological production of nanomaterials (Prakash et al. 2010).

Owing to the greater surface area of nanoparticles per mass unit, they are expected to be more biologically active than larger sized particles of the same chemical composition. This offers several perspectives for food applications. Nanoparticles can, for instance, be used as bioactive compounds in functional foods (Chau et al. 2007). Bioactive compounds that can be found naturally in certain foods have physiological benefits and might help to reduce the risk of certain diseases, including cancer. By reducing particle size, nanotechnology can contribute to improve the properties of bioactive compounds, such as delivery properties, solubility, prolonged residence time in the gastrointestinal tract and efficient absorption through cells (Chen et al. 2006b). Omega 3 and omega 6 fatty acids, probiotics, prebiotics, vitamins and minerals have found their applications in food nanotechnology as bioactive compounds (Watanabe et al. 2005). In the food industry, several novel applications of nanotechnologies have become apparent, including the use of nanoparticles, such as micelles, liposomes, nanoemulsions, biopolymeric nanoparticles and cubosomes, as well as the development of nanosensors, which are aimed at ensuring food safety (Sozer and Kokini 2009).

Selenium has been recognized as an essential dietary nutrient. It is common practice to supplement broiler diets with Se. The Se supplement that primarily has been used in animal diets is the inorganic form, sodium selenite, which has a very narrow margin between its nutritional dosage and its toxicity (Wolffram et al. 1986). Nano elemental selenium (Nano-Se), which is bright red, highly stable, soluble and of nano defined size in the redox state of zero (Se°), has been manufactured for use in nutritional supplements and developed for applications in medical therapy (Gao et al. 2002). It has been reported that Nano-Se have a higher efficiency in upregulating selenoenzymes and exhibit less toxicity than selenite (Wang et al. 2007). Nanomaterials exhibit novel properties, such as great specific surface area, high surface

Table 5.9 Size of DNA comparing with range of sizes of nanomaterials in the food sector. (Adapted from Sozer and Kokini 2009)

Structures	Diameter or Length (nm)
DNA (Deoxyribonucleic acid)	12
Glucose	21–75
Liposome	30–10,000
LDH (Layered double hydroxides)	40–300
Amylopectin	44–200
Casein micelle	60–100
Polylactic acid (PLA) nanosphere	100–300
Zein	200
Cubosome	500
Nanosensors	<1000

activity, a lot of surface active centers and high catalytic efficiency (Table 5.9; Gao and Hiroshi 2005). Due to the advantage of size effect and high surface reactivity, nanoparticle has been already used in pharmaceutical applications to increasing the bioavailability of drugs and targeting therapeutic agents to particular organs (Davda and Labhasetwar 2002). It has been reported that nanoparticle showed new characteristics of transport and uptake and exhibited higher absorption efficiencies (Zha et al. 2008; Liao et al. 2010). However, there is little data on intestinal absorption and Se retention of Nano-Se (Hu et al. 2012).

This review highlights some of the most promising and important nanotechnology applications in agriculture; and nano-selenium particles production. Agricultural nanotechnology and its use in sustainable development will be also highlighted.

5.1.9.2 Elemental Selenium

Studies have reported Se-induced toxicity at a cellular level, the effects of which include cellular apoptosis, DNA damage and inhibition of enzyme activity (Spallholz and Hoffman 2002). However, bacteria play an important role in the biogeochemical cycle of Se in nature (Haudin et al. 2007). During bacterial metabolism, Se is transformed by diverse processes (oxidation, reduction and/or methylation), and Se-tolerant bacteria (STB) have shown a great potential for use in environmental sciences (bioremediation and phytoremediation) and technology (glassware manufacturing, electronic devices) (Fesharaki et al. 2010; Prakash et al. 2010; Narayanan and Sakthivel 2010). Despite its potential toxicity, Se is also a recognized micronutrient with antioxidant properties, and dietary deficiencies of Se in humans can affect cancer suppression, HIV treatment, free radical-induced diseases and protection from toxic heavy metals (Fairweather-Tait et al. 2010). Thus, agronomic biofortification with Se-supplemented fertilizers is a common practice in cereal crops to increase the Se content and nutritional quality of grains (Banuelos et al. 2005).

However, the transformation of Se by bacteria and the effect of these bacteria on the Se availability to plants are poorly understood (Acuña et al. 2013).

It is the great potential of nano-scale Se° phases in electronic, optical, catalytic and medical application that has led to extensive investigations of the production and post-preparative modification of these materials by various inorganic routes (Pearce et al. 2008). These include, solid-solution-solid transformation from amorphous Se (*a*-Se) colloids to *t*-Se nanowires (Gates et al. 2002a) and sono-chemistry based synthesis and transformation of *a*-Se to *t*-Se nanowires (Gates et al. 2002b). The potential of the environment to yield organisms that can produce functional bionanominerals is demonstrated by selenium-tolerant, aerobic bacteria isolated from a seleniferous rhizosphere soil. An isolate, NS3, was identified by Prakash et al. (2010) as a *Bacillus* species (EU573774.1) based on morphological and 16 S rRNA characterization. This strain reduced Se(IV) under aerobic conditions to produce amorphous a Se° nanospheres. A room-temperature washing treatment was then employed to remove the biomass and resulted in the production of clusters of hexagonal Se° nanorods. The Se° nanominerals were analyzed using electron microscopy and X-ray diffraction techniques. This Bacillus isolate has the potential to be used both in the neutralizing of toxic Se(IV) anions in the environment and in the environmentally friendly manufacture of nanomaterials (Prakash et al. 2010).

It is well known that selenium possesses many complex chemical and biochemical properties. It has four different possible chemical oxidation states, Se(IV) and Se(VI), present as SeO_3^{2-} and SeO_4^{2-} in the aqueous phase, elemental Se° present in solid phase or in a colloidal form and Se(−II), occurring as selenide in minerals under reducing conditions or in organic and biochemical compounds. With the later chemical valence, many chemical compounds can be formed and the great majority of them exist in organic forms or in proteins (Kyriakopoulos and Behne 2002). Se° is a relatively biochemically inert species, being much less bioavailable compared to Se(IV), Se(VI) and some other forms of organic compounds. The formation of Se in natural environments is usually through a biotic process involving the reduction of selenate (Zhang and Frankenberger 2005) or selenite (Di Gregorio et al. 2005) by bacteria. The abiotic reduction has also been reported. Possible transformations of Se° include its oxidation to Se(IV) or Se(VI) or its incorporation into iron sulfides or selenides (Belzile et al. 2000). As Se° sets between the higher chemical valence of Se(IV) and Se(VI), and the lower valence of Se(−II), it could be transformed to either direction depending on the redox potential of the environment and/or the presence of biological activity. In non-toxic surface soil, Se concentration varies from 0.01 to 2.0 μg g^{-1} in many parts of the world (Chen et al. 2006c).

Elemental Se is rare, occurring mostly in sedimentary rocks (White et al. 2004). From three allotropes of elemental Se the gray and the black one are biologically inert, which may due to their insolubility (Huang et al. 2003). The red allotrope has been produced by several kinds of bacteria from selenite such as Hunter and Manter 2008, Prokisch et al. 2008 and Prokisch and Zommara (2008). It was found that the red elemental Se particles of nano-size scale have good free radical scavenging effects on different free radicals in vitro (Huang et al. 2003). The nano-sized

Table 5.10 Summary of mineralogical, physical and chemical properties of elemental selenium. (Adapted from Chen et al. 2006c)

Allotropic from	Common name	Solubility	Density (g cm^{23})
Crystalline, hexagonal	β-Se, or grey-Se or black-Se, or metallic Se	The most stable from Soluble: ether, chloroform; CS_2: 2 mg 100 mL^{-1}	4.82
Crystalline, monoclinic red	Crystalline red two forms: α-monoclinic, β-monoclinic	Soluble in CS_2	α-monoclinic 4.46 β-monoclinic 4.50
Amorphous	May exist as black or red amorphous, or colloidal	Se Soluble in CS_2, CH_2I_2, benzene or quinoline	

red elemental Se (abbreviated as nanoSe) was shown lower acute toxicity as compared with selenite in mice; however bioavailability to selenite was similar in terms of inducing seleno-enzymes in cultured cells and in Se-deficient rats (Zhang et al. 2001). Based on chemoprevention-related responses in mice the nanoSe could also be served as a potential chemopreventive agent with significantly reduced risk of toxicity compared with that of Se-methylselenocysteine (Zhang et al. 2008; Domokos-Szabolcsy et al. 2012).

The accurate determination of Se° is a key step in order to understand any process whether it is geological, environmental, or biological. However, finding an appropriate method to identify and measure Se° in natural systems such as soil or sediment is a difficult task due to its low concentration and the complexity of the matrix. An extraction method requires selectivity in the dissolution of Se° from a complex sample and accuracy in its determination. One of the most difficult problems analysts are facing in quantitative speciation of environmental samples is the lack of appropriate standard reference materials (SRM). It is particularly problematic for Se° because the exact chemical and mineralogical properties of this species in nature are unknown. Elemental selenium exists in several allotropic forms of remarkably different mineralogical, physical and chemical properties. Unstable amorphous elemental selenium is also susceptible to a phase transition to possess a more stable mineral structure. Some information of allotropic elemental Se is summarized in Table 5.10 (Chen et al. 2006c).

It should be kept in mind that though this information can be found in few literatures, the original work on identifying or synthesizing those forms of selenium is very difficult to find and often lacks of details. For instance, the solubility of selenium in various solvents given in those sources is impossible to verify due to the unavailability of those specific selenium forms, which often rises important discrepancies between different literatures. Various solvents, including potassium thiocyanide, methyl iodide, cyclohexene, quinoline and carbon disulphide, were tested to dissolve the purchased red elemental selenium standard (PF-Se) sample. Only carbon disulphide could partially dissolve PF-Se, which resulted in a remarkable yellowish color in the solution. It was also noted that only a limited fraction of PF-Se could be dissolved in CS_2. However, all the PF-Se was dissolved using the Na_2SO_3 method. The huge solubility difference in both solvents and the apparent low solubility of PF-Se in CS_2 stimulated our interests in further investigating on the nature of the PF-Se sample and in the comparison of these two techniques of Se° extraction in sediment samples (Chen et al. 2006c).

It could be generally assumed that Se° freshly formed in natural sediments is likely present in an amorphous form. Therefore, the chemical and mineralogical properties of red or black amorphous Se° would be the closest to elemental Se formed in natural sediments when compared to other types of well crystallized Se° which are synthesized under rather unnatural laboratory conditions. The analytical method for the determination of Se° in the aquatic environment and agriculture soil are still scarce. Some promising techniques based on X-rays such as absorption near edge structure (XANES) and extended X-ray absorption fine structure spectroscopy (EXAFS) have been used to study the redox transformations of Se in sediments and soils (Myneni et al. 1997), but these methods provide only qualitative or descriptive information. Only a few papers have dealt with the quantitative determination of Se° in natural systems. The technique more frequently used is the one proposed by Velinsky and Cutter (1990), in which an extraction with sodium sulphite is applied at pH 7 to form a water soluble selenosulfate ($Na–Se–SO_3$). However, it seems that the proportion of Se° in samples extracted with this method are often suspiciously high, ranging from more than 30% to as high as 90% of the total selenium present in the sediments (Zhang and Frankenberger 2003). Uchida et al. (1980) have proposed to determine Se (−II+0) in water samples by subtracting Se(IV) from the measured values after oxidation by a 3% (v/v) bromine solution. It was believed that with bromine oxidation, Se (−II) plus Se° could be selectively determined. With this method, no distinction between Se° and Se(−II) could be made. In their paper they mentioned to use carbon disulfide to dissolve elemental selenium. Unfortunately they did not give any detail on the characteristics of the elemental selenium used in the experiment. Yamada et al. (1999) and Wright et al. (2003) mentioned that the sodium sulphite extraction might not be specific for elemental selenium; the later authors speculated a possible overestimation of the percentage of Se° by large values due to the solubilization of organic and iron selenide in sodium sulphite. Yamada et al. (1999) used carbon disulfide to extract elemental selenium from soil, but their work seriously lacks of necessary detailed studies. In addition the purpose of back extraction of Se° into an acetate buffer, through a reaction with KCN to form SeCN− ion, is unclear. Because CS_2 is a non-polar solvent, it is unlikely that any other charged or polar bonded Se species could be extracted into this solvent. Besides, potassium cyanide is an extremely dangerous chemical, therefore, its use should be avoided as much as possible (Chen et al. 2006c).

Therefore, it could be concluded that, it is the great potential of nano-scale Se° phases in electronic, optical, catalytic and medical application that has led to extensive investigations of the production and post-preparative modification of these materials by various inorganic routes. Elemental Se is rare, occurring mostly in sedimentary rocks. From three allotropes of elemental Se the gray and the black one are biologically inert, which may due to their insolubility. The red allotrope has been produced by several kinds of bacteria from selenite. It was found that the red elemental Se particles of nano-size scale have good free radical scavenging effects on different free radicals in vitro. The nano-sized red elemental Se was shown lower acute toxicity as compared with selenite in mice; however bioavailability to selenite was similar in terms of inducing seleno-enzymes in cultured cells and in Se-deficient rats. Based on chemoprevention-related responses in mice the nanoSe could also be served as a potential chemopreventive agent with significantly reduced risk of toxicity compared with that of Se-methylselenocysteine.

5.1.9.3 Nanotechnology in Agriculture and Food

Science and technology are at the core of human endeavor, and the process of creating new tools and products has been accelerated by reaching at the basic building blocks at the nanoscale. In today's world of laptop computers, cell phones, regenerative medicine, targeted drugs, fuel cells, environmentally friendly technologies and carbon sequestration, it is natural to imagine that technology can take us even further. Nanotechnology is an enabler and catalyst of current and future possibilities. It can help us realize a wide spectrum of applications not only in engineered materials, nanomanufacturing, electronics, and communication, but also in energy, environment, biomedicine, food and agricultural systems. As long ago as 3.5 billion years, cells developed the ability to perform photosynthesis, mainly inside chloroplast. Chloroplast may have been one of the first nanoscale machines, a nanoscale biomachine in this case. In fact, all matter as we know it has nanostructure naturally; nanotechnology aims to harness nanoscale properties by controlling and manufacturing matter at that scale. The control and restructuring at the nanoscale is new (Chen and Roco 2009).

As mentioned before, it is well documented that nanotechnology is the understanding and control of matter at dimensions of roughly 1–100 nm, where unique phenomena enable novel applications. The application of nanotechnology to the agricultural and food industries was first addressed by the United States Department of Agriculture in its roadmap published in September 2003. It is now emerging as a rapidly evolving field with a potential to revolutionize agriculture and food systems, across the entire agricultural value chain (Opara 2004). Nanotechnology is beginning to be seen as an important option for enhancing agricultural productivity, along with other emerging technologies such as biotechnology, to complement conventional agricultural technologies (Kalpana Sastry et al. 2007). However, to make an impact on the rural economy, it is important to recognize that this new technology needs to be extended beyond the farm to all the links across the entire agricultural value chain. The key role of this technology is also envisaged in agri-biotechnology in the areas of gene delivery, gene expressions, gene sequencing, gene therapy, gene regulation, DNA targeting, DNA extraction, DNA hybridization, fingerprints for

DNA and RNA detection, cell probes, specific targeting, cell sorting and bioimaging, single-cell-based assay, drug delivery, tissue engineering, proteomics and nanobiogenomics (Kalpana Sastry et al. 2010).

That means nanotechnology is the control and restructuring of matter at the intermediate dimensions, between the sizes of one atom to about hundred molecules shoulder to shoulder (about 100 nm), where new phenomena enable new applications. Nanoscale science and engineering operates at the first level of organization of atoms and molecules for both living and anthropogenic systems. A nanomaterial is defined as an "*insoluble or biopersistent and intentionally manufactured material with one or more external dimensions, or an internal structure, on the scale from 1 to 100 nm*" as detailed in the recent EC Cosmetics Regulation (EC No. 1223/2009). Efforts are underway to establish a more comprehensive definition for nanomaterials. Hence, this is a provisional definition until a uniform, European and international definition is made available (Mildau and Huber 2010). The nanoscale is a natural threshold between discrete behavior of single atoms or single molecules with given properties, on one side, and collective behavior of assemblies of atoms and molecules where the properties are a function of size, structure and composition, on the other side. It is where the fundamental properties and functions of all materials and systems are defined, and where novel properties can be exploited and changed. Such fundamental control promises a broad and revolutionary technology platform for industry, biomedicine, environmental engineering, safety and security, shared resources such as food, water, energy, nanoinformatics, and countless other areas (Chen and Roco 2009).

Food security is the state achieved when food systems operate such that "all people, at all times, have physical and economic access to sufficient, safe and nutritious food to meet their dietary needs and food preferences for an active and healthy life" (FAO 1996). Food systems encompass three components: (1) food availability (production, distribution and exchange) (2) food access (affordability, allocation and preference) and (3) food utilization (nutritional value, social value and food safety) (Gregory et al. 2005). Food security is diminished or a state of food insecurity occurs when any one of the three components of the food systems are diminished (Tables 5.11 and 5.12; Kalpana Sastry et al. 2011).

Advances in science and technology could offer potential solutions for developing countries to innovate and add value to their current commodities production systems. Many technologies being developed have the potential not only to increase farm productivity but also to reduce the environmental and resource costs often associated with agricultural production. These include technologies that conserve land and water by increasing yields with the same or fewer inputs and technologies that protect environmental quality. It will be crucial, however, to support these applications even though they may not be commercially lucrative while avoiding the risk that some advances in science and technology may increase the disparity between developed and developing countries. Therefore, serious consideration of the social and ethical implications on new agriculture technologies will be necessary. It should also be recognized that while new agrifood technologies may deliver efficiencies in some areas, they may not necessarily solve existing problems of global food production

Table 5.11 Nanoresearch areas and applications in agri-food sector which can contribute to enhanced food security: I Enhancing productivity. (Adapted from Kalpana Sastry et al. 2011)

Agri-food thematic areas	Nanoresearch area and indicative applications
Plant/animal disease diagnostics	Nanobarcodes as ID tags for multiplexed analysis for gene expression and intracellular histopathology
	Quantum dots as fluorescence marker coupled with immunomagnetic separation for detection of *E. coli*
	Label-free sensor chip assembled from peptide nanotubes for electrical detection of low detection limit viruses
	Sensor array containing six non-covalent gold nanoparticles for detection and qualification for protein targets
Delivery mechanisms in plant, soil and animal systems	Mesoporous silica nanoparticles for delivery of DNA and chemicals intoplants
	Smart magnetic silica core for specific targeting, cell sorting and bioimaging
	Nanocontainers for delivery of drugs to organs or tissues
	Organically modified silica nanoparticles as DNA carriers, for gene delivery and promoters of transgene expression
	Carbon nanofibers for gene therapy of plants
	Oligonucleotide-loaded nanoparticles for enhancing the expression of rice a-galactosidase gene in yeast cells
	Micro/nanofluidic device-single-cell-based assay
	Carbon nanotubes as molecular transporters
	Tin oxide nanowires for water vapor detection
Developing new genetic types/breeds/cultivars and crop production	Atomically modified rice by drilling a nanosized hole through the wall and membrane of a rice cell for inserting a nitrogen atom
	Functionalized cow pea mosaic virus (CPMV) nanostructures for use in sensing applications
	Magnetic nanoparticles coated with tetramethylammonium hydroxide enhancing the growth of Zea mays plants in early ontogenetic stages
	Carbon nanotubes for experiments on artificial photosynthesis
Livestock breeding and livestock management	Blue shift of CdSe/ZnS nanocrystal-labels upon DNA-hybridization
	Nanoparticles, nanocapsules and nanospheres in veterinary medicines
	Nanodevices implanted in an animal for detecting the presence of disease and notifying the farmer and veterinarian to targeted treatment delivery system

and distribution. In this regard, it is essential for developing countries to actively participate in research and development while respecting their needs and capacity to utilize these new technologies. Therefore, critical to the innovation, capacity building is the establishment of relevant, complementary and synergistic partnerships between developing countries and more advanced countries (Chen and Yada 2011).

Table 5.12 Nanoresearch areas and applications in agri-food sector which can contribute to enhanced food security: II. Improving soil health, raising water use efficiency and Food products on shelf. (Adapted from Kalpana Sastry et al. 2011)

Determinants of food security	Agri-food thematic areas	Nanoresearch area and indicative applications
Improving soil health	Natural resource management–efficient use of soil resources	Nanoparticles for soil in situ remediation
		Sorption and release of contaminants in the soil onto the surface of engineered nanoparticles
		Nanoscale iron particles for rapid destruction of chlorinated hydrocarbons in soil and ground water
		Nanosensors for continuous monitoring of heavy metals
Determinants of food security	Agri-food thematic areas	Nanoresearch area and indicative applications
Raising water use efficiency	Natural resources management–efficient use of water resources	Nanotechnology for desalination and water purification
		Nanoporous membranes for filtration of viruses
		Nanosponges to absorbs toxic metals
		Ozone nanobubbles to sterilize water
		Nanowire immunosensors array-for detection of microbial pathogens
		Ultra sensitive pathogen quantification in drinking water using high piezoelectric PMN-PT micro cantilevers
Food products On helf	Food processing	Quick detection of food borne pathogens using bioconjugated nanomaterials, biosensors, nano-cantilevers, carbon nanotubes, nanowires, BioMEMS
		Nanosensors for enhancing flavor and taste of the processed products
		Edible nanosensors for detection of bacterial contamination in the packaged foods
	Food packaging	Natural biopolymer-based nanocomposite films used for food packaging for safe storage
		Nanoscale titanium dioxide particles as blocking agent of UV light in plastic packaging

Since food systems encompass food availability, access and utilization, the scope of applications of nanotechnology for enhancing food security must encompass entire agricultural production–consumption systems. Further, in a rapidly globalizing economy, increasing access to food and its utilization in rural areas will be determined primarily by increase in rural incomes. The primary source of increasing rural incomes has been recognized as value addition across the different links in the agricultural production–consumption chain. These links include farm inputs, farm production systems, post harvest management and processing and finally markets and consumers. From the food security perspective, it is therefore necessary that application of nanotechnology be not limited to the farm production level, but be extended across all the links of the agricultural value chain to increase agricultural productivities, product quality, consumer acceptance and resource use efficiencies. This will help to reduce farm costs, raise the value of production, increase rural incomes and enhance the quality of the natural resource base of agricultural production systems (Kalpana Sastry et al. 2007).

In doing so, it is important to view nanotechnology as an enabling technology that can complement conventional technologies and biotechnology (Salerno et al. 2008). Considering the concerns on biosafety and consumer acceptance emerging after agribiotechnology basedproducts have entered the market place during last two decades, it is also essential that integrating and deploying new technologies like nanotechnology in agricultural and food systems be made after understanding the various societal and environmental implications (Kalpana Sastry et al. 2011).

Nanoscience and nanotechnology have already been applied in various fields, such as computer electronics, communication, energy production, medicine and the food industry. The nanoscale devices are often manufactured with the view to imitate the nanodevices found in nature and include proteins, DNA, membranes and other natural biomolecules (Sanguansri and Augustin 2006). In today's world, food materials are often considered not only a source of nutrients but also as having to contribute to the health of consumers. Most of the nanoparticles used traditionally belong to the group of colloids i.e., emulsions, micelles, mono- and bi-layers. One of the first colloidal gold dispersions was prepared by Michael Faraday in the middle of the eighteenth century. The particles were attracted to each other through Van der Waals forces, which give them colloidal stability. In colloidal particles, steric stabilization is achieved by adsorbing polymers and surfactants on the surface. Nanoparticles could be further stabilized by coating them with molecules that can form chemical bonds (Fendler 2001). For food applications, nanotechnology can be applied by two different approaches, either 'bottom up' or 'top down' (as mentioned in Table 5.9). The top-down approach is achieved basically by means of a physical processing of the food materials, such as grinding and milling. For example, dry-milling technology can be used to obtain wheat flour of fine size that has a high water-binding capacity (Degant and Schwechten 2002). This technology has been used to improve antioxidant activity in green tea powder. As the powder size of green tea is reduced to 1000 nm by dry milling, the high ratio of nutrient digestion and absorption resulted in an increase in the activity of an oxygen-eliminating enzyme. By contrast, self assembly and self organization are concepts derived from biology that have inspired a bottom-up food nanotechnology. The organization of casein micelles or starch and the folding of globular proteins and protein aggregates are examples of self-assembly structures that create stable entities. Self organization on the nanometer scale can be achieved by setting a balance between the different non-covalent forces (Sozer and Kokini 2009).

Despite the lack of unifying nanotechnology guidelines, manufacturers nevertheless have to deal with existing general regulations for food products and the introduction of a new nanoingredient can be difficult and time consuming. For this reason, most expected nanoapplications in the food market will probably occur in food packaging and only few in actual food products. Already, several applications of nanotechnology are available. Up to now, most of the research on nanotechnology focused on the electronics, medicine and automation sector. The knowledge gained from these sectors could be adapted for the use of food and agriculture products, such as for applications in food safety e.g., detecting pesticides and microorganisms, in environmental protection such as water purification and in delivery of nutrients (Sozer and Kokini 2009).

Future agriculture and nanotechnology In this section, it could be summarized some potential applications of nanoscale science, engineering and nanotechnology for agriculture and food production and related issues. Despite a wide-range of industrial interest in this area, examples of available commercial products are few. Most applications are either in research and development (R & D) pipeline or at bench-top exploration stage; however, it is likely that the agriculture and food sector will see some large-scale applications of nanotechnologies in the near future.

Nanosized agricultural chemicals are mainly still at the research and developmental stage. Natural Nano, a start-up company in Rochester, N.Y., has found a way to use Halloysite, a naturally found clay nanotube, as a low cost delivery for pesticides to achieve an extended release and better contact with plants. It is estimated that using this technology could reduce the amount of pesticides applied by 70 or 80 %, a significant reduction in quantity and cost of pesticides as well as less impact on water streams (Murphy 2008).

1. ***Nanotechnologies in plant-based agricultural production and products***

Plant-based agricultural production is the basis of broad agriculture systems providing food, feed, fiber, fire (thermal energy), and fuels through advancements in materials sciences, and biomass conversion technologies. While the demand for crop yield will rapidly increase in the future, the agriculture and natural resources such as land, water and soil fertility are finite. Other production inputs including synthetic fertilizers and pesticides are predicted to be much more expensive due to the constraints of known petroleum reserve. Precision farming is hence an important area of study to minimize production inputs and maximize agricultural production outputs for meeting the increasing needs of the world sustainability. Given that nanotechnology may allow for the precise control of manufacturing at the nanometer scale, a number of novel possibilities in elevating the precision farming practices are possible (Chen and Yada 2011). It could be summarized different nanotechnogies in plant—based agricultural production within following issues:

- Nanotechnology enabled delivery of agriculture chemicals such as fertilizers, pesticides, herbicides, plant growth regulators, etc.
- Field sensing systems to monitor the environmental stresses and crop condition.
- Nanotechnology enables the study of plant disease mechanisms.

- Improving plant traits against environmental stresses and diseases.
- Lignocellulosic nanomaterials.

2. ***Nanotechnologies in animal production and animal health***

Agriculturally relevant animal production such as livestock, poultry, and aquaculture provides society with highly nutritious foods e.g. meat, fish, egg, milk and their processed products, which have been, and will continue to be, an important and integral part of human diets. There are a number of significant challenges in animal agricultural production, including production efficiency, animal health, feed nutritional efficiency, diseases including zoonoses, product quality and value, byproducts and waste, and environmental footprints. Nanotechnologies may offer effective, sometimes novel, solutions to these challenges (Kuzma 2010). Nanotechnologies in animal production and animal health include:

- Improving feeding efficiency and nutrition of agricultural animals.
- Minimizing losses from animal diseases, including Zoonoses.
- Animal reproduction and fertility.
- Animal product quality, value and safety.
- Turning animal by-products and waste and environmental concerns into value added products.

3. ***Nanotechnologies for water quality and availability***

It is reported that providing clean and abundant fresh water for human use and industry applications, including agricultural and farming uses, is one of the most daunting challenges facing the world (Vörösmarty et al. 2010). It is estimated that more than 1 billion people in the world lack access to clean water, and the situation is getting worse. Over the next two decades, the average supply of water per person will drop by a third, possibly condemning millions of people to an avoidable premature death (Savage et al. 2009). Agriculture requires considerable amount of fresh water, and in turn, often contributes substantially to pollution of groundwater through the use of pesticides, fertilizers and other agricultural chemicals. Effective technologies for remediation and purification will be needed tomanage the volume of wastewater produced by farms on a continual basis, and be cost effective for all. Technical issues in the water challenges include water quality and quantity, treatment and reuse, safety due to chemical and biological hazards, monitoring and sensors (Chen and Yada 2011).

I. ***Water quantity, quality and safety-treatment and conservation***

Accessible water resources are often contaminated with pollutants largely due to various human activities, but also natural leaching. These contaminants include, but not limited to, water-borne pathogenic microorganisms such as *Cryptosporidium*, Coliform bacteria, virus, etc., various salts and metals e.g. copper, lead, arsenic, etc., run-off agricultural chemicals, tens of thousands of compounds considered as pharmaceuticals and personal care products, and endocrine disrupting compounds, and radioactive contaminants either naturally occurring or the result of oil and gas production as well as mining activities (Chen and Yada 2011). It be included the following issues:

- Microbial disinfection
- Desalination
- Removal of heavy metals

II. ***Water conservation in agricultural crop production***

The fact that crop production requires large amounts of water has resulted in the implementation of policy and regulations in limiting agricultural production in many regions. Scientists and engineers have been working to improve water use efficiency in agricultural productions. For example, drip irrigation has been developed to conserve water. This innovation has moved precision agriculture in water usage to a much higher level of control than other irrigation technologies such as flood irrigation. New and innovative ideas will likely result in the development of more precise water delivery systems. Future technology platforms should consider the following: water storage, in situ water holding capacity, water distribution near roots, water absorption efficiency of plants, encapsulated water released on demand, interaction with field intelligence through distributed nanosensor systems, among others (Savage et al. 2009).

III. ***Detection and sensing for pollutants and impurity***

Nanotechnology based sensing and the detection of various contaminants in water has been topical over the last decade. Detection level is generally at parts per billion (ppb) for metals and organic contaminants in both laboratory and field applications. The state of science and prototyping for sensing and devices is among the most advanced in the field of nanotechnology, hence it is expected that many new technologies will be readily available in the next decade. Sensor applications for water bear many similarity to other applications, hence are not repeated here (Chen and Yada 2011).

4. ***Nanotechnology for the keeping quality of agricultural and food products***

Many agricultural products are either perishable or semiperishable. These include fresh vegetables, fruits, meats, egg, milk and dairy products, many processed foods, nutraceuticals and pharmaceuticals. The improvement of shelf-life is one of the main areas for nanotechnology research to enhance the ability to preserve the freshness; quality and safety (Chen and Yada 2011).

5. ***Nanotechnology and traceability***

A number of factors contribute to an increased demand for the traceability of foods throughout production, processing, distribution and consumption. Food safety outbreaks frequently resulted in wide spread product recalls. Advanced and improved product traceability is essential to ensure food safety by removing all the tainted products in the market and the system during the recall process. Also, product authenticity has seen an increased value in food marketing throughout the world by validating the origin, and therefore, adding to the unique inherent value proposition of the products (Chen and Yada 2011).

Traceability must meet the following five essential technical challenges:

1. Have sufficient vocabulary to distinguish all products.
2. Not compromise the products.
3. Have the same service life as the marked products.
4. Easy to read by machines such as speed, reliability, and convenience.
5. Should be very inexpensive food and agricultural products (Nightingale 2008).

6. ***Nanotechnologies and clean energy***

Access to inexpensive, safe and renewable energy is of utmost importance for worldwide sustainable development. Flexible and efficient, yet inexpensive solar cells are often highlighted as one of the most exciting areas of nanotechnology application in agriculture, as often expressed as green nanotechnology. Inexpensive systems of solar-powered electricity have long been an aspiration for tropical countries, but glass photovoltaic panels remain too expensive and delicate. Nanotechnology based photovoltaic currently is a high priority of research worldwide, including most industrialized countries. Solar energy conversion to electricity, energy storage, and other nanotechnology-enhanced solarthermal energy conversion systems are presently active areas of research and development (Chen and Yada 2011). Nanotechnology can also contribute to conversion of biomass for fuels, chemical intermediates, speciality chemicals and products. As biomass becomes an increasingly important industrial feedstock, a new generation of catalysts to reduce production cost while being economically feasible will be critical. Nanostructures have the inherent advantage as catalysts of their large surface area per unit volume, and the capability to precisely control composition, structure, functionalization, and other important properties of catalysts (Davis et al. 2009).

Therefore, it could be concluded that over the last several decades, the rapid growth in technological innovations have led to profound structural changes in the agricultural sector, including a transition from smallholder mixed farms toward large-scale specialized industrial production systems, a shift in the geographic locus of demand and supply to the developing world, and an increasing emphasis on global sourcing and marketing. The latter present challenges to the agricultural sector to provide possible improvements to its production sustainability in ways that promote food security, poverty reduction and public health improvement.

5.1.9.4 Agricultural Nanotechnology and Sustainable Development

Nanotechnologies in future agriculture Nanoscale science, engineering and technology embrace an exciting and broad scientific frontier which will have significant impacts on nearly all aspects of the global economy, industry, and people's life in the twenty-first century (Gruère et al. 2011). Nanoscale sciences reveal the properties, processes, and phenomena of matters at the nanometer (1 to approximately 100 nm) range. Nanoscale engineering renders precise capability to control and/or fabricate matters at this scale to render novel and useful properties thus leading to many new applications of nanoscale science and nanomaterials that can be used to address numerous technical and societal issues. In this section, some potential applications of nanoscale science, engineering and nanotechnology for agriculture and food production and related issues are discussed. Despite a wide-range of industrial interest in this area, examples of available commercial products are few. Most applications are either in research and development (R & D) pipeline or at bench-top exploration stage; however, it is likely that the agriculture and food sector will see some large-scale applications of nanotechnologies in the near future. Some current industrial examples are indicated in the sections below (Chen and Yada 2011).

Research advancements still provide glimpses of potential applications with clear impact on agricultural, food, and water safety that could have a significant impact on rural populations in developing countries (Waruingi and Njoroge 2008). A consultation of international experts in nanotechnology and development in 2005 identified applications to enhance agricultural productivity and food processing and storage among the 10 top areas where nanotechnology has a high potential for development (Salamanca-Buentello et al. 2005). Several developing countries already believe in the potential of nanotechnology. For instance, India has included agricultural productivity as one of its main focuses for public research in nanotechnology (Sreelata 2008); Iran launched a 35-lab research program on nanofood applications in 2005 (Joseph and Morrison 2006). At the same time, agricultural and food nanotechnologies, and especially those that could lead to reduced poverty or food insecurity, are bound to face many challenges before being commercialized and used by rural poor. As with other new technologies, all steps in the process may need to overcome constraints—from investment and research and development to regulatory approval, commercial release, distribution, access, availability, adoption, and proper use by users. But there are also specific issues with nanotechnology, such as the involvement of public research, the issue of intellectual property rights (IPRs), the management of safety and environmental risks in the presence of wide uncertainties, and the indirect effects on exports and foreign market access that could be positive or negative. Whether nanotechnologies succeed in helping the poor will largely depend on whether public research institutions, technology developers, national governments, and international donors are able to address these multiple challenges in the coming years (Gruère et al. 2011).

While we are still at the early stage of the nano revolution, current and upcoming nanotechnology applications in agriculture, food, and water already present great potential for the poor. In this section, we focus on these technologies, analyzing their potential benefits (risk mitigation, and so on) and why they have a high likelihood of increasing the benefits for the poor. Currently, in developing countries, few nanotechnology projects specifically target the needs of the poor. Nanotechnology has been identified as a scale-neutral modern technology—applicable to both large-scale commercial agriculturists and small-scale, resource-poor farmers (Lal 2007). Several applications of nanotechnology are of interest to agriculture, although even in developed countries, nanotechnology is not currently in widespread use in the agricultural sector. Applications of nanotechnology currently noted to be in the food production chain include nanosensors and nanoagricultural chemicals. Nanoparticles for soil cleaning and nanopore filters were also reported to be in the food production chain. In the food production and processing phase of the food production chain, nanoceramic devices and nanoparticles (mostly silver nanoparticles) are reported to be in use (Bouwmeester et al. 2007). Over the next 5 to 10 years, an increasing number of nanotechnology applications are expected in developed countries for food and agricultural uses. These applications include nanosensors, nano delivery systems, nanocoatings and films, nanoparticles, and quantum dots (FAO/WHO 2009). Nanosensors are capable of detecting very small amounts of chemical contaminants, viruses, and bacteria in food, water, and environmental media (Scott 2005). Nano delivery systems can precisely deliver drugs or nutrients to the site within an organism where they are needed at the time they are needed and have the potential to minimize the use of the materials they deliver. Of particular importance to developing countries are the nanotechnology applications addressing low use efficiency of inputs (such as nutrients, irrigation water, and pesticides) and stress of drought and high soil temperature. Nanoscale agrichemical formulations increase the use efficiency and decrease losses into the environment (Lal 2007). More efficient nutrient delivery can be expected to result in increased yields (Joseph and Morrison 2006). Nanoporous materials capable of storing water and slowly releasing it during times of drought can also be expected to increase yields. Applications of nanotechnology to reduce the effects of aflatoxin will increase the weight of food animals, resulting in more usable meat (Gruère et al. 2011).

The framework and databases were used to gauge the type of nanotechnology researches currently in progress and to assess them from the perspective of food security. More than 60% of records from both the databases were on R & D efforts to enhance plant/animal productivity followed by research in food processing and food packaging which address the other two components of food security systems, namely, food availability and food utilization. The type of drivers of the technological changes in various sub fields of nanoresearch (like nanodevices or nanobiotechnology), which may form base for technological trajectories that can contribute to enhancing food security, was also investigated. It was found that nanoparticles was the most widely researched area followed by nanofiltration methods/devices and nanocapsules. Formulations like capsules and particles are known to enhance target delivery, offer better control, and increase overall functional efficiency especially for inputs like fertilizers, pesticides including biopesticides, improving the management practices for enhancing productivity (Kalpana Sastry et al. 2011).

Nanotechnology by its very nature will require, and has required, a high degree of multidisciplinary and cross-sector collaboration within and between academic researchers, industry and government. Applications of nanotechnology involves many disciplines in engineering and the natural sciences, including such disciplines as physics, chemistry, biology, materials sciences, instrumentation, metrology, and others. As nanotechnology progresses from discovery to potential applications, it will require a number of tools for visualization, characterization, and fabrication, as well as methods for reproducing and controlling properties, scalability, and cost. These tools and techniques, too, are typically rooted in multiple disciplines (Chen and Yada 2011).

Therefore, it could be concluded that, research advancements still provide glimpses of potential applications with clear impact on agricultural, food, and water safety that could have a significant impact on rural populations in developing countries. A consultation of international experts in nanotechnology and development in 2005 identified applications to enhance agricultural productivity and food processing and storage among the 10 top areas where nanotechnology has a high potential for development. Applications of nanotechnology involves many disciplines in engineering and the natural sciences, including such disciplines as physics, chemistry, biology, materials sciences, instrumentation, metrology, and others.

5.1.9.5 Recent Developments, Risks and Regulation of Nanotechnology

In recent years, there has been an increased interest in the potential use of nanotechnology applications in the agriculture and food sector (Chaudhry et al. 2008). While no unified definition has been approved internationally, nanotechnology-enabled products can be broadly defined as products derived or issued from materials at scales measuring less than 100 nm in at least one dimension. At this scale, in part because of the larger surface/volume ratio, the physical, chemical and biological properties of materials can be fundamentally different from the corresponding bulk materials. In particular, nanomaterials often exhibit different thermodynamic, magnetic and optical properties than their bulk counterparts. These singular properties have opened the door to the development of new applications in all sectors. In agriculture and food, a wide range of nanotechnology applications are being developed and commercialized with different goals, ranging from improved food safety to reduced agricultural inputs, enhanced packaging and improved processing and nutrition (Yada 2009), and the potential to promote sustainable agriculture and deliver better foods globally (Gruère et al. 2011).

In agriculture, nanotechnology research and development have mostly focused on improving and better delivering input use, from water to nutrients, nano-pesticides, and nano-herbicides (Robinson and Morrison 2009). Interesting applications include the use of nanoporous zeolites to slow the release and increased efficiency of fertilizers, nanosensors to measure soil quality, smart delivery mechanisms for herbicides (Chinnamuthu and Boopathi 2009). There has also been a lot of research and development on food and water safety, with nanosilver or nanoclay products developed for improved water flitration, and nanosensors being developed to detect and help track food pathogens (Gruère et al. 2011). But much larger efforts have been undertaken in the food improvement and packaging area with nutritional supplements, additives and improved, lighter, and antibacterial food and beverage containers, among other applications (Gruère 2012).

The applications of nanotechnology in the food sector are only new emergent, but they are predicted to grow rapidly in the coming years. Many of the world's largest food companies are reported to have been actively exploring the potential of nanotechnology for use in food or food packaging. Applications in this area already span development of improved tastes, color, flavor, texture and consistency of foodstuffs, increased absorption and bioavailability of nutrients and health supplements, new food packaging materials with improved mechanical, barrier and antimicrobial properties, and nano-sensors for traceability and monitoring the condition of food during transport and storage. The rapid proliferation of nanotechnologies in a wide range of consumer products has also raised a number of safety, environmental, ethical, policy and regulatory issues (Maynard et al. 2006). The main concerns stem from the lack of knowledge with regard to the interactions of nano-sized materials at the molecular or physiological levels and their potential effects and impacts on consumer's health and the environment. The nanotechnology-derived foods are also new to consumers and it remains unclear how public perception, attitudes, choice and acceptance will impact the future of such applications in the food sector. It is, however, well known that uncertainties and lack of knowledge of potential effects and impacts of new technologies, or the lack of a clear communication of risks and benefits, can raise concerns amongst the public (Chaudhry et al. 2008).

As mentioned before, nanoscience and nanotechnology are the understanding and manipulation of materials at the atomic, molecular, and macromolecular scales. The greater surface area per mass compared with larger-sized particles of the same chemistry renders nanosized particles more active biologically (Oberdörster et al. 2005). These emerging technologies have shown great potential in nutraceuticals and functional foods for delivering bioactive compounds in functional foods to improve human health (Chen et al. 2006a). Among the nanotechnology consumer products to date, health and fitness products form the largest category, followed by electronics and computers category as well as home and garden category. The US is the overwhelming market leader, having at least three times more nanoproducts on the market than those in the East Asia and Europe (Chau et al. 2007).

Because of the limited information on the risks of handling nanomaterials, these materials of ultra-small scale have created an intense interest in their health risks. Some reports discussing the potential risks of nanomaterials have surprised the public by taking a strong precautionary tone on health and safety risks (ETC 2005a, b). The nonintervention of nanotechnology-based food products, namely nanofood, to come to the food markets in the absence of clear definition, public debate, food safety assessments, and proper food regulations may eventually jeopardize the potential benefits of nanotechnologies to food industry. The Institute of Food Science and Technology has reminded the deficiencies in current regulations concerning the impact of nanotechnology on food and packaging (IFST 2006). Owning to the lack of information about the impacts of nanotechnology on public safety, legislation, society, and food industry as well as the potential toxicity of nanomaterials, it is probably wise to take a precautionary principle to deliberate the possible regulatory control as a proactive approach until proven otherwise. More studies on the applications of nanotechnology in food processing and packaging, nanotoxicity, regulation, and risk-and-benefit analysis are necessary to fill the knowledge gaps, sustain the growth of nanofood industry, and avoid any unpredictable health hazard. From the food industry and public safety standpoints, the objective of this paper is to give a preliminary discussion on the potential applications, risks, food safety, and current regulatory situation of nanotechnology in relation to foods, thus to provide the industry, legislators, and government with some points, rather than a roadmap, that will need to be addressed as regulation for food nanotechnology moves forward (Chau et al. 2007).

In respect of the entry of nanoparticles into body and the translocation among organs due to their small size, three possible routes for nanoparticles to cause harm inside the body include dermal exposure, inhalation, or ingestion. Because of the limited information on the risks of handling nanomaterials, a stringent control on exposure should be implemented until more knowledge becomes available. For handling nanomaterials appropriately, lab safety guidelines have been provided by the Committee on Chemical Safety of the American Chemical Society (http://membership.acs.org/c/ccs/nano.htm) (Chau et al. 2007).

Many nanotechnology initiatives, commissions, or centers have been launched by governments, academia, private sectors in the United States, Europe, Japan, and some other countries around the globe to ensure rapid development and deployment of nanotechnology, promote economic growth, maintain global competitiveness, and improve the innovative capability (Chen et al. 2006a). Some of them have also participated in proposing regulation to improve the protection of human health and the environment. There are some major initiatives, centers, institutes, or government organizations are concering with nanotechnology development. These organizations or research centers mainly supported by the government sources play an essential role in performing or supporting nanotechnology researches, including the basic researches on nanotechnology, the applications of nanotechnology, safety assessment of nanomaterials, and the development of regulatory control. They also provide long-term coherence and platforms for interdisciplinary people or experts to promote nanotechnology to the public as well as to give an impetus to the nanotechnology industry. It is undoubted that the application of nanotechnology in food and packaging is growing rapidly. Currently, there is still no requirement to label food products containing nanoparticles and no regulatory standard to comply with, too. In the marketplace, different terms such as nanofood and ultrafine food have been used, whereas it has been difficult to find out which merchandise could be called "*nano*" (Chau et al. 2007).

It is well documented that available literature suggests that many uncertainties remain about nanomaterials, including the potential for bioaccumulation and potential human health risks. While proposed applications of nanotechnologies are wide and varied, developments are met with some caution, while progress may be stifled by lack of governance and potential risks. The use of nanotechnologies in the food industry may present potential risks due to the use of novel materials in novel ways, thus risk assessments must be carried out to identify and quantify these risks. All applications of this new technology must be assessed for safety of use. In the EU, the Directorate General of Health and Consumer Protection has set up the Scientific Committee on Emerging and Newly Identified Health Risks (SCENIHR). This committee provides opinions on questions concerning emerging or newly identified health and environmental risks on issues which require a comprehensive assessment of risks to consumer safety or public health (SCENIHR 2010). Nanotechnologies fit this profile. SCENIHR focuses on nanotechnologies' risk assessment with particular interest in establishing recognised terminology in the field so that research can be integrated to a certain extent. Research breakthroughs in the area of potential benefits of nanotechnologies are being published at an increasing rate so risk analysis is urgently required. Maynard (2010) put forward a view that an integrated system of research is required to fully understand the implications of nanotechnologies on human health, to pre-empt adverse health effects and to proactively minimise them (Cushen et al. 2012).

The regulation of nanotechnologies is within the scope of both so-called horizontal legislation and vertical legislation. Existing horizontal legislation is broad and happens to encompass attributes of nanotechnologies even though it does not specifically aim to do so. Vertical legislation is specifically aimed at regulating nanotechnologies and areas of industries likely to utilise nanotechnologies and so the vocabulary used makes the legislation more applicable to issues faced by users of nanotechnologies. Compared to horizontal legislation, vertical legislation for nanotechnologies is relatively recent and was non-existent until a few years ago (Cushen et al. 2012).

There are some regulation issued by the EU to regulate nanotechnologies in the food industry include: EC Cosmetics Regulation (EC No. 1223/2009), EC No. 1223/2009 (2009), EC No. 1272/2008 (2008), EC No. 1907/2006 (2006), EC No. 1935/2004 (2004), and EC No. 258/97 (1997).

Risk assessment, exposure assessment and risk management are all urgently required for existing products available on the world market. Existing uncertainties for risk assessment and exposure assessment of nanomaterials arise due to limited information on several aspects including toxicity, behaviour and bioaccumulation. These uncertainties also have implications for the effective regulation of the use of nanomaterials. Integration at a research level may serve to overcome the nomenclature and protocol issues associated with research in nanotechnology. It is argued that segregation of the various areas involved in the establishment of nanotechnology in the various industries is an inferior research model. It is undeniable that nanotechnologies present many beneficial applications to the food industry; so far some of the more developed applications include: improved supplements; novel food packaging and targeted crop pesticides. The reviewed applications may also have positive implications for people in developing countries, particularly in the area of increased agricultural productivity, improved food and water safety and nutrition. Lack of investment in these countries could mean that the benefits of these technologies may be limited to developed countries. It is clear that these new technologies, if managed and regulated correctly, can play a central role in improving product and process development to the benefit of human health and well being (Cushen et al. 2012).

Therefore, it could be summarized that, available literature suggests that many uncertainties remain about nanomaterials, including the potential for bioaccumulation and potential human health risks. While proposed applications of nanotechnologies are wide and varied, developments are met with some caution, while progress may be stifled by lack of governance and potential risks. The use of nanotechnologies in the food industry may present potential risks due to the use of novel materials in novel ways, thus risk assessments must be carried out to identify and quantify these risks. All applications of this new technology must be assessed for safety of use. The regulation of nanotechnologies is within the scope of both so-called horizontal legislation and vertical legislation.

5.1.9.6 Nano-selenium in the Environment

For years, organic forms of selenium and some salts have been used in studying its biological effects, but recently, elemental selenium (Se°) nanoparticles have gained some attention as a possible source of this beneficial component (Zhang et al. 2001). Synthesis of elemental nano-selenium employs the reduction of a selenium salt with a reducing agent, usually in the presence of a stabilizing agent to prevent the clusters of selenium atoms from growing and to obtain stabilized nanoparticles in colloidal suspension (Zhang et al. 2004). Various other methods of synthesizing selenium nanowires (Zhang et al. 2008) and nanoparticle composites (Mehdaoui et al. 2009) have been reported in literature mainly focusing on semiconductor applications (Sarin 2010).

It is known that particles of elemental Se (Se°) formed from some bacterial strains and the redox system of glutathione or ascorbate and selenite has a very low bioavailability (<5%) (Garbisu et al. 1996). It is observed that red elemental Se, formed in the redox system of selenite and glutathione or other reducing agents, was unstable, and could further aggregate into gray and black elemental Se if there were no controlling factors. It is also found that protein presented in the redox system could affect the aggregation of red elemental Se (Fig. 5.2). The size of red elemental Se formed was dependent on the amount of protein in the redox system (Zhang et al. 2001). They reported that Nano-Se at 20–60 nm had similar bioavailability to sodium selenite. Since most nano chemical compounds have size effect, i.e. with the size decreases, the physical, chemical and biological properties change (Ball and Garwin 1992). Indeed, Huang et al. (2003) reported that there was a potential size-dependent effect on scavenging various free radicals and protecting DNA *in vitro* by Nano-Se. Zhang et al. (2004) studyed whether there is a size effect of Nano-Se in the induction of Se-containing enzymes. Nano-Se was prepared by adding varying amount of bovine serum albumin (BSA). With less BSA presented, the size of Nano-Se increased, and the solubility of Nano-Se decreased. When the sizes of Nano-Se reach 200 nm and above, it forms an un-homogenous solution. Therefore, three kinds of size (5–15, 20–60, and 80–200 nm) were chosen for comparison of their biological activities. The results showed there was no size effect in the induction of Se-dependent enzymes among different sizes of Nano-Se from 5 to 200 nm in both cultured cells and animal study. This result contrasts to the *in vitro* results of Nano-Se in directly scavenging free radicals, and indicates that *in vitro* observations are not always reproducible *in vivo*, and implicates that as an novel Se form,

Fig. 5.2 Different possible colors of nano selenium suspension after using of the biological systhesis method, where MRS medium, sodium hydrogen selenite and bacteria strain were used. (Photo by Nano Food Lab, Debrecen Uni., Hungary). For LactoMicroSel we use milk as medium. The milk fat makes further processing hard, as some of the selenium containing bacteria gets attached to the fatty fraction floating at the top, while the others consolidate to the bottom. So we separate the fat from the milk using a separator seen on the top left picture, then we inoculate it with LactoBacteria. We put the inoculated milk into an incubating chamber for 2 days, on 37°. Then, we dry and mill the produced yoghurt. The resulting fine powder is the finished LactoMicroSel product, and its selenium concentration is about 3000 ppm

Nano-Se in different sizes could be considered for human supplemention since they possess equal bioavailability (Zhang et al. 2004).

It is well documented that gray and black elemental Se are biologically inert, which may be due to their insolubility. However, there is one kind of red elemental particulate selenium, observed in several kinds of bacteria, that has promising uses in the environmental protection from the pollution of the excessive selenium (Garbisu et al. 1995). These elemental selenium particles are formed in the bacteria with the detoxification of excess selenium and have only about 2 % bioavailability compared with selenite. Nano red elemental selenium (Nano-Se) with the size range of 5–100 nm, can be synthesized by reducing selenite in an environment containing BSA, which is able to adhere to Se atoms and control the size of their aggregation (Zhang et al. 2001): the sizes of the nanoparticles are dependent on BSA concentration in the preparation system. BSA at higher concentration produces smaller Nano-Se particles, hence a series of Nano-Se particles of different sizes were prepared by varying the concentration of BSA. It is well known that some materials have many new properties when particle sizes are reduced to the nanometer scale, such as large surface area, quantum effects, and high reactivity. In the work of Zhang and his colleagues, it was shown that Nano-Se has a similar bioavailability in rat, and much less acute toxicity in mice compared with selenite (Zhang et al. 2001). The results suggest that the biological activities of Nano-Se may come from the special properties of the nanoparticles (Huang et al. 2003).

Elemental Se powder in the redox state of zero is not soluble and is generally considered to be biologically inert. Nanotechnology holds promise for medication and nutrition because materials at the nanometer dimension exhibit novel properties different to those of both isolated atoms and bulk material (Albrecht et al. 2006). With bovine albumin protein as dispersing agent, nascent elemental Se atoms formed by reducing selenite with GSH can aggregate into particles in sizes of 20–60 nm (Zhang et al. 2001). Our earlier studies demonstrated that elemental Se at nano size (Nano-Se) has comparable efficacy to selenite in upregulating selenoenzymes and tissue Se levels, but is less toxic (Zhang et al. 2005). These results challenged the long-held dogma that elemental Se has no biological activities, and stimulated us to further compare Nano-Se with selenomethionine (SeMet), which has excellent bioavailability and lower toxicity among various Se forms. In comparison with SeMet, Nano-Se has lower toxicity and possesses equal efficacy in increasing the activities of selenoenzymes. Furthermore, the efficacy of GST induction by Nano-Se was higher than that of SeMet at supranutritional levels (Wang et al. 2007). These results indicate Nano-Se can serve as an antioxidant with reduced risk of Se toxicity and a potential chemopreventive agent if the induction of GST by Se is a crucial mechanism for its chemopreventive effect (Zhang et al. 2008).

Therefore, it could be concluded that, particles of elemental Se (Se°) formed from some bacterial strains and the redox system of glutathione or ascorbate and selenite has a very low bioavailability ($<5\,\%$). It is observed that red elemental Se, formed in the redox system of selenite and glutathione or other reducing agents, was unstable, and could further aggregate into gray and black elemental Se if there were no controlling factors. Protein presented in the redox system could affect the aggregation of red elemental Se. The size of red elemental Se formed was dependent on the amount of protein in the redox system.

5.1.9.7 Synthesis Methods of Nanoparticles

Different tools of nanotechnology show the capability of synthesizing nanoscale materials with specific physical, chemical, and optoelectronic properties. Various physical and chemical methods have been designed for the synthesis of nanoparticles, but the different problems related to these methods have made the researchers to search for alternative methods.

Different types of physical and chemical methods are employed for the synthesis of nanoparticles. The use of these synthesis methods requires both strong and weak chemical reducing agents and protective agents (sodium borohydride, sodium citrate and alcohols) which are mostly toxic, flammable, cannot be easily disposed off due to environmental issues and also show a low production rate (Rai et al. 2008; Sharma et al. 2009). For example, in the seeded growth method chemical reducing agents like sodium citrate and sodium borohydride (Jana et al. 2000) are used, whereas in the polyol synthesizing process methanol and ethanol are used (Kim 2007). Moreover, these are capital intensive and are inefficient in materials and energy use (Ingle et al. 2008). In addition, in many cases, synthesis is carried out at elevated temperatures, which generate a large amount of heat. For example, in thermal decomposition method, synthesizing process is carried out at very high temperature (Yang and Aoki 2005).

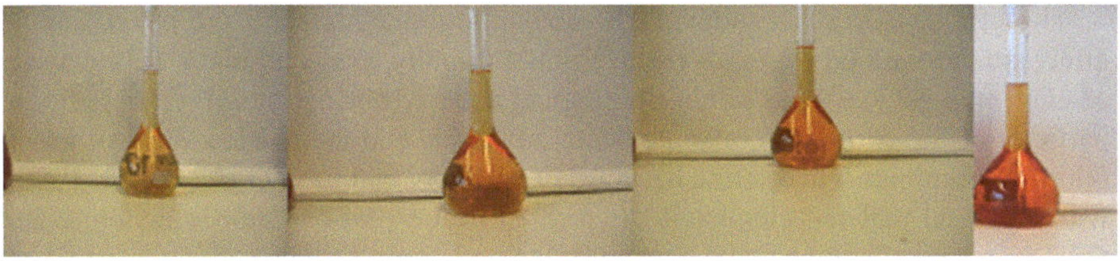

Fig. 5.3 Chemical method for nan-Se systhesis. The ascorbic acid or vitamin C was used plus sodium selenate as a source of Se. It could be distinguished formation stages of the red color of elemental nano-Se. (Photo by H. El-Ramady, Nano Food Lab, Debrecen Uni., Hungary)

It could be explained the three different methods of selenium synthesis as follows:

1. ***Chemical methods***

In general, the most often used method for the chemical synthesis of nanoparticles is the chemical reduction method, which deals with the reduction of metal particles to nanoparticles using chemical reducing agents like sodium borohydride or sodium citrate (Cao and Hu 2009). Other chemical agents utilized for the synthesis are N, N-dimethyl formamide (DMF) (Pastoriza-Santos and Liz-Marzan 2000), poly(N-vinyl pyrrolidine) (PVP), ethyl alcohol (Kim 2007), tetra-n-tetrafluoroborate (TFATFB), CTAB (Hanauer et al. 2007), etc. Seeded growth method is a colloid chemical method for the synthesis of nanoparticles which involve preparation of seeds by reducing metal ions with the suitable reducing agent (Fig. 5.3). The fine particles so formed are called seed particles which are then added to growth solutions containing metal ions and additives like L-ascorbic acid and hexadecyltrimethylammonium bromide (CTAB) (Hanauer et al. 2007). Polyol is also used in the synthesis of nanoparticles. In the polyol method a metal precursor is dissolved in a liquid polyol in the presence of capping agent. The metal sol is prepared using methanol or ethyl alcohol as a solvent and reducing agent while PVP is used as a protective and capping agent (Kim 2007). Electrochemical synthesis method induces chemical reactions in an electrolyte solution with the use of an applied voltage. A wide variety of nanomaterials could be synthesized using this method (Sau and Rogach 2010).

In the chemical way scientists don't use any living/organic material, they usually start from inorganic selenite and add some reducer agent, like ascorbic acid. Or there is a wet chemical method (Gao et al. 2003) for the preparation of α-monoclinic selenium nanowires. The method is the reduction of sodium selenite with glutathione (GSH) at room temperature in aqueous solution. Glutathione, having a thiol group, reacts with sodium selenite to form selenodiglutathione (GSSeSG), which decomposes to produce selenium molecules and diglutahtione (GSSG). Selenium molecules aggregate together to form selenium nanospheres. Positive effects of chemically synthesized selenium nanospheres were examined in several experiments succesfully.

The efficacy of Se in inducing Se-containing enzymes and the pro-oxidative effect are determined by its chemical form. Normally, gray and black bulk particle of elemental Se (Se°) has neither biological activity nor toxicity. It is known that particles of Se° formed from some bacterial strains and the redox system of glutathione or ascorbate and selenite has a very low bioavailability (Garbisu et al. 1996). It is observed that red elemental Se, formed in the redox system of selenite and GSH or other reducing agents, was unstable and could further aggregate into gray and black Se° if there were no affect the aggregation of red Se°. Se° is bright red, highly stable, soluble and of nano define size. Nano-Se is prepared by the reaction of bovine serum albumin, sodium selenite, and GSH under the Chinese Patent ZL 97107038 (Zhang et al. 2001). Transmission Electron Microscopy (TEM) showed the size of red elemental Se was between 20–60 nm (Zhang et al. 2001). The Nano-Se shows totally different biological properties contrasting to the general concepts that elemental Se is inert. In HepG2 cells, both Nano-Se and selenite have almost equal biological functions in increase of glutathione peroxidase (GPx), phospholipid hydroperoxide glutathione peroxidase (PHGPx) and thioredoxin reductase (TR), protection against free racial-mediated damage, and cell growth inhibition. Nano-Se has a 7-fold lower acute toxicity than sodium selenite in mice (LD_{50} 113 and 15 mg Se kg^{-1} body weight, respectively). In Se deficient rat, both Nano-Se and selenite were efficient and generally equal in Se uptake and GPx biosynthesis (Zhang et al. 2001).

It is reported that nano-Se taken at the dose of 180 μg Se daily, was granted as health care food by Ministry of Hygiene P. R. China in 1998. Nano-Se has comparable bioavailability of selenite, sharply lower acute toxicity, to less extent, lower subchronic toxicity at a dose therein other selenocompounds such as selenite and Se-enriched soybean could cause serious toxic changes. Supplemenat at 180 μg per day for adults is within the scope of Food and Drug Administration (2004).

2. ***Physical methods***

Physical methods used for the synthesis of nanoparticles include thermal decomposition, laser irradiation, electrolysis, condensation, diffusion, etc. The thermal decomposition method is used for the synthesis of monodisperse nanoparticles. Fatty acids are dissolved in hot NaOH solution and mixed with metal salt solution which leads to formation of metal precipitate (Yang and Aoki 2005). In diffusion method, crystals and short wires of copper are enclosed in glass ampoules and sealed at low pressure; further, the ampoules are annealed at 500 °C for 24 h. The crystals are removed from the ampoules and cooled on a metallic plate at room temperature. The so formed crystals are further characterized (Rodrıguez-Perez et al. 2006). In the UV irradiation technique, polycarbonate films are cut and placed on glass microscope slide and exposed to UV radiation which results in the formation of hydroxyl groups on polycarbonate films. Further, these polycarbonate films are silanized with 3-(aminopropyl) triethoxysilane (APS) in denatured ethanol for 2 h and rinsed with deionised water which leads to the formation of silver film on the polycarbonate film (Aslan et al. 2006). The arc-discharge method involves use of two graphite electrodes which act as cathode and anode and are immersed in metal salt solution.

The electrodes are brought in contact to strike an arc and separated immediately to sustain arc inside salt solution. The synthesis of nanoparticles is carried out at an open circuit and an optimized direct current (Ashkarran et al. 2009).

3. ***Biological methods***

Biological agents used for the synthesis of nanoparticles include mainly microbes (Ingle et al. 2008; Birla et al. 2009) and plants (Song and Kim 2009). The biological methods used for the synthesis of nanoparticles include both extracellular and intracellular methods (Rai et al. 2008; Shaligram et al. 2009). The synthesis of nanoparticles using bacteria and actinomycetes usually involves the intracellular synthesis method. In which the bacterial cell filtrate is treated with metal salt solution and kept in a shaker in dark at ambient temperature and pressure conditions (Ahmad et al. 2003a, b). For the extracellular synthesis of nanoparticles using bacteria, the bacterial culture is centrifuged at 8000 × g and the supernatant is challenged with metal salt solution (Ogi et al. 2010). In case of fungi also nanoparticles are intracellularly synthesized by treating the fungal mycelium with metal salt solution and further incubation for 24 h (Mukherjee et al. 2001). Dried mycelium of fungi is also used for synthesis of nanoparticles. In this method the fungal mycelium is harvested by centrifugation and subsequently freeze dried, and this freeze-dried mycelium is immersed in metal salt solution and kept on a shaker (Chen et al. 2003). However, in the extracellular method the filtrate of the mycelium is treated with metal salt solution and incubated for 24 h (Shaligram et al. 2009). In algal synthesis of nanoparticles washed culture of algae without the presence of any medium is treated with metal salt solution and kept in dark with controlled pH and temperature conditions (Thakkar et al. 2010). The synthesis of nanoparticles using yeast involves two steps which include firstly the synthesis of nanoparticles and next recovery of the synthesized nanoparticles (Kowshik et al. 2003). For the synthesis process, yeast culture is challenged with metal salt solution and incubated in dark for 24 h. Further, the cells are separated from the medium by centrifugation and the cell-free extract is used for recovery of nanoparticles. For recovery of nanoparticles, specifically designed apparatus (polycarbonate bottle with sampling cup) is used, which separates nanoparticles from the extract by differences in thawing temperature. The cell-free medium containing nanoparticles is filled in the bottle up to the brim and kept at −20 °C in upright position. During freezing, the nanoparticles get denser than medium and settle down. The bottle is then kept at 0 °C and allowed to thaw. The concentrated colloidal solution obtained in the sampling cup is centrifuged at 23,000 × g for 24 h, the particles are suspended in distilled water, and further the particles are dried in vacuum (Fig. 5.4; Rai et al. 2011).

In the recent years, researchers started to recognise the importance of the ability of certain microorganisms to produce nano-sized particles in the course of their metabolism. Many elements in trace concentrations are essential for the growth and reproduction of plants, animals and microorganisms, however these elements easily become toxic at concentrations higher than the physiological level.

Scientists have shown that many plants and bacteria can actively uptake and reduce metal ions from soil and solutions. Nair and Pradeep (2002) used Lactic acid

Fig. 5.4 Comparison between biological and chemical methods for nano-Se production. The right 100 measuring flask belongs the chemical method, where it used vitamin C or ascorbic acid and the other flasks are belong to the biological method using MRS media and *Lactobacillus casei* as a bacteria strain. (Photo by H. E1-Ramady, Nano Food Lab, Debrecen Uni., Hungary)

bacteria in buttermilk whey to produce gold-silver composite materials. These alloy materials, in submicron dimensions, form several well-defined crystal morphologies, and this crystal growth does not affect the viability of the bacteria. Magnetotactic bacteria, for instance, have intracellular magnetic structures, the magnetosomes, which comprise nanometer-sized, membrane-bound crystals of the magnetic iron minerals magnetite (Fe_3O_4) or greigite (Fe_3S_4). Magnetosomes (Schueler and Frankel 1999) are the results of a mineralization process with biological control over the accumulation of iron and the deposition of the mineral particle with specific size and orientation within a membrane vesicle at specific locations in the cell. Moon (Moon et al. 2007). discovered that certain thermophyl and psychrotolerant metal-reducer bacteria (*Shewanella sp.; Thermoanaerobacter sp.*) are able to produce copious amounts of extra-cellular metal (M)-substituted magnetite nano-particles using akaganeite and dopants of dissolved form (Eszenyi et al. 2011). Figure 5.5 shows some steps of the biological method of selenium synthesis.

Not only the bacteria but also the fungi are able to synthetize nano-sized products. Fungi, due to their metal accumulation and tolerant ability, were placed into the centre of attention of nano-particle production researches (Sastry et al. 2003). Their economic viability, ease in scale up in solid substrate fermentations and large-scale secretion of extracellular enzymes, makes them advantageous for nano-particle production (Gardea-Torresdey et al. 2002). As a new field of nanobiotechnology, besides bacteria and fungi, plants are also able to produce nano-particles. In a study scientists describe that they managed to fabric gold nano-particles using live plants (Gardea-Torresdey et al. 2002). In their experiment they grew alfalfa plants in an $AuCl_4$ rich environment, plants absorbed the Au metal successfully, that was confirmed by X-ray absorption studies (XAS), and transmission electron microscopy (TEM). These gold clusters surrounded by a shell of organic ligands covalently attach to proteins or other biological substances. Atomic resolution analysis confirmed that Au nanoparticles inside the plant are in a crystalline state and pure gold (Figs. 5.6 and 5.7; Eszenyi et al. 2011).

From the previous, it is clear that the biological agents in the form of microbes have emerged up as an efficient candidate for the synthesis of nanoparticles. These biogenic nanoparticles are cost efficient, simpler to synthesize, and focus toward a

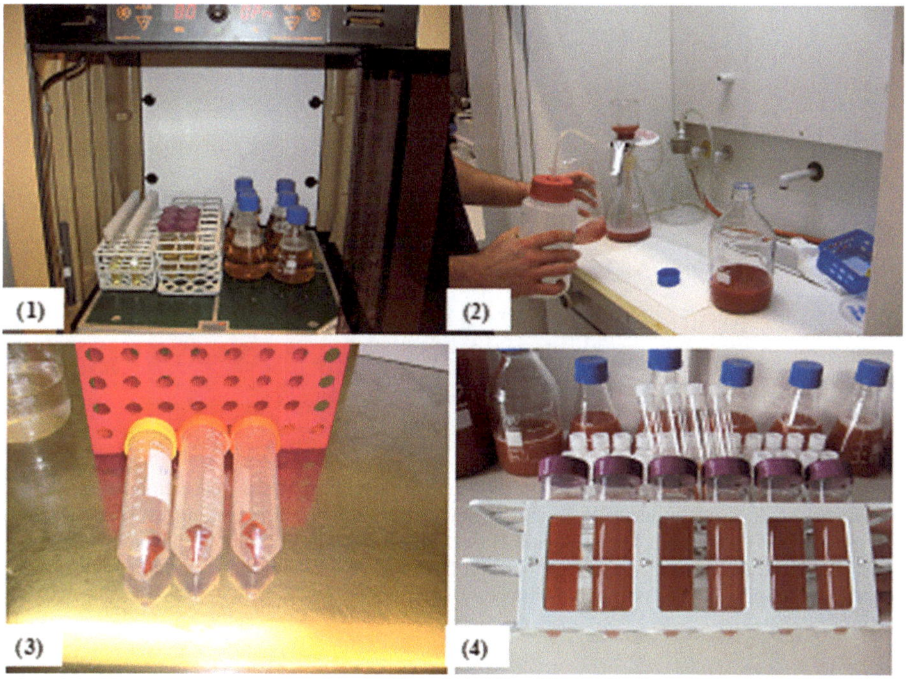

Fig. 5.5 Different steps of the biological systhesis of nano-Se using MRS broth. (*1*) The First step is the incubation at 37°C for 48 h, the second is centrifuge, (*2*) then filter to separate the filtered bacteria. (*3*) Then, use concentrated hydrochloric acid for 3–4 days, (*4*) thus purifying the selenium nanospheres, we get a solution or suspension called NanoSel, which is mainly for research. Photo (*3*) shows that the nanoselenium sphere consolidate to the bottom over time, but with a little shaking we can get back the original state. (Photo by Nano Food Lab, Debrecen Uni., Hungary)

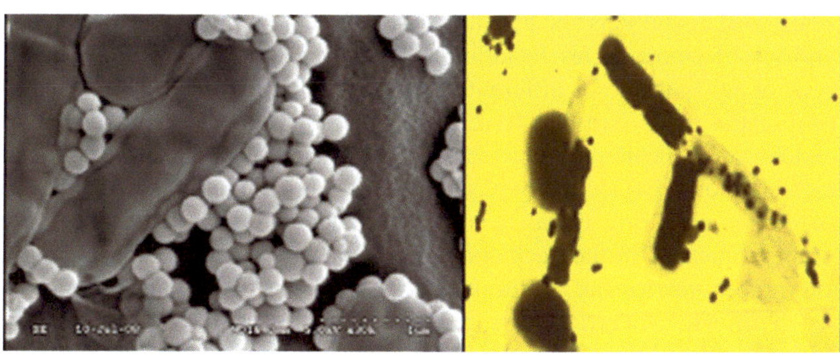

Fig. 5.6 SEM and TEM pictures of the partially digested bacteria with elemental nanosize Se. So, the LactoMicroSel is elemental selenium stored by Lactobacteria in the form of nano-sized spheres. The size of the spheres is 100–500 nm, and is dependent on the strain of bacteria used for production. The spheres of each strain are uniform in both size and form. (Photo by Nano Food Lab, Debrecen Uni., Hungary)

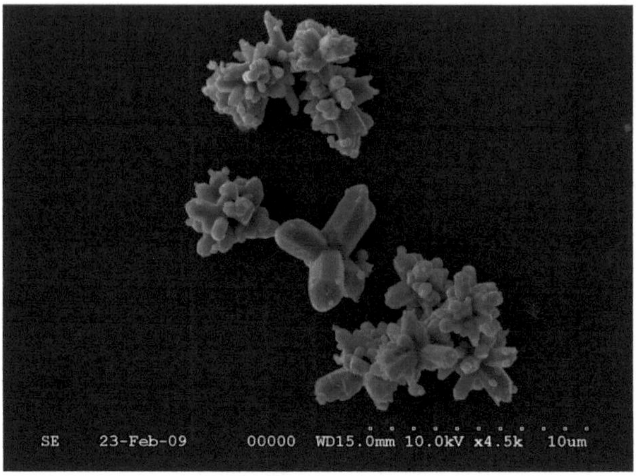

Fig. 5.7 Elemental selenium precipitation on selenium homospheres. Red elemental nano selenium spheres in water produce H_2Se and H_2SeO_3 in small amount according to $3Se + 3H_2O \leftrightarrow 2H_2Se + H_2SeO_3$ equation. The H_2Se and H_2SeO_3 are formed in the solution when the solution is dried the H_2Se and H_2SeO_3 react together and the elemental selenium precipitated, and forming crystals. (Photo by Nano Food Lab, Debrecen Uni., Hungary)

greener approach. But the exact mechanism of synthesis of biogenic nanoparticles needs to be worked out. Hence, a detailed study to elucidate the exact mechanism for the synthesis of nanoparticles needs to be carried out, as different microbial agents react differently during the synthesis of nanoparticles. So, drafting the different issues and the reducing agents related to the synthesis of biogenic nanoparticles would help in developing the biosynthetic route as the most efficient method for the synthesis of nanoparticles. The exact mechanism for the synthesis of nanoparticles using biological agents has not been devised yet as different biological agents react differently with metal ions and also there are different biomolecules responsible for the synthesis of nanoparticles. In addition, the mechanism for intra- and extracellular synthesis of nanoparticles is different in various biological agents (Rai et al. 2011).

Today, nano metal particles have drawn the attention of scientists because of their extensive application to new technologies in chemistry, electronics, medicine, and biotechnology. Beside many physical and chemical methods which have been developed for preparing metal nanoparticles, nanobiotechnology also serves as an important method in the development of clean, nontoxic, and environmentally friendly procedures for the synthesis and assembly of metal nanoparticles. To be utilized in different scientific fields, biological synthesis still requires the optimization of reaction conditions, and an understanding of the biochemical and molecular mechanisms of the reaction for obtaining better chemical composition, shape, size, and monodispersity (Shahverdi et al. 2011).

Therefore, it is concluded that the synthesis of nanoparticles using the biosynthetic route is an environmental friendly method compared with the chemical and physical methods, but there are certain key areas of research which need to be pointed out. The synthesis of biogenic nanoparticles using the biological agents involve both intracellular and extracellular methods, and the mechanism for both the methods is yet unknown. In the intracellular method of synthesis, the exact role of microbial and plant cell wall should be to be elucidated, whereas, in case of extracellular synthesis using microbes the role of other fungal enzymes in the synthesis of nanoparticles needs to be studied. In extracellular synthesis of nanoparticles a number of reducing and capping agents are found to be involved. The effect of these reducing agents on the shape and size of nanoparticles also need to be clarified. The effect of different factors on the dispersity of nanoparticles deserves further research. The size of nanoparticles also plays a key role in the determination of its activity. The yield of nanoparticles obtained using a particular system modulates the efficiency of the system for the synthesis of nanoparticles; thus, the yield is a major concern.

References

Acuña JJ, Jorquera MA, Barra PJ, Crowley DE, Mora M (2013) Selenobacteria selected from the rhizosphere as a potential tool for Se biofortification of wheat crops. Biol Fertil Soils 49:175–185. doi:10.1007/s00374-012-0705-2

Ahmad A, Mukherjee P, Senapati S, Mandal D, Khan MI, Kumar R, Sastry M (2003a) Extracellular biosynthesis of silver nanoparticles using the fungus *Fusarium oxysporum*. Coll Surf B 28:313–318

Ahmad A, Senapati S, Khan MI, Kumar R, Ramani R, Shrinivas V, Sastry M (2003b) Intracellular synthesis of gold nanoparticles by a novel alkalotolerant actinomycete, Rhodococcus species. Nanotechnology 14:824–828

Albrecht MA, Evans CW, Raston CL (2006) Green chemistry and the health implications of nanoparticles. Green Chem 8:417–432

Alva AK (1992) Micronutrient status of Florida sandy soils under citrus production. Commun Soil Sci Plant Anal 23:2493–2510

Ashkarran AA, Zad AI, Mahdavi SM, Ahadian MM, Nezhad MRH (2009) Rapid and efficient synthesis of colloidal gold nanoparticles by arc discharge method. Appl Phys A Mater Sci Process 96(2):423–428

Aslan K, Holley P, Geddes CD (2006) Metal-enhanced fluorescence from silver nanoparticledeposited polycarbonate substrates. J Mater Chem 16:2846–2852

ATSDR (2002) Draft toxicological profile for several trace elements. U.S. Dept. Health & Human Services. Agency for Toxic Substances and Disease Registry, Atlanta, GA

Ball P, Garwin L (1992) Science at the atomic scale. Nature 355:761–766

Banuelos GS, Ajwa HA, Mackey B, Wu L, Cook C, Akohoue S, Zambruzuski S (1997) Evaluation of different plant species used for phytoremediation of high soil selenium. J Environ Qual 26:639–646

Banuelos GS, Lin ZQ, Arroyo I, Terry N (2005) Selenium volatilization in vegetated agricultural drainage sediment from the San Luis Drain, Central California. Chemosphere 60:1203–1213

Baum MK, Miguez-Burbano MJ, Campa A, Shor-Posner G (2000) Selenium and interleukins in persons infected with human immunodeficiency virus type 1. J Infect Dis 182(suppl 1):S69–S73

Belzile N, Chen Y-W, Xu R (2000) Early diagenetic behavior of selenium in lake sediments. Appl Geochem 15:1439–1454

Berzelius JJ (1818) Lettre de M. Berzelius a M. Berthollet sur deux metaux nouveaux. Annales de Chimie et de Physique, Série 2, 7:199–202

Birla SS, Tiwari VV, Gade AK, Ingle AP, Yadav AP, Rai MK (2009) Fabrication of silver nanoparticles by *Phoma glomerata* and its combined effect against *Escherichia coli, Pseudomonas aeruginosa* and *Staphylococcus aureus*. Lett Appl Microbiol 48:173–179

Bouwmeester H, Dekkers S, Noordam M, Hagens W, Bulder A, De Heer C, ten Voorde S, Wijnhoven S, Marvin H, Sips A (2007) Health impact of nanotechnologies in food production. RIKILT, Wageningen. http://edepot.wur.nl/120669. Accessed 8 May 2010
Brown TA, Shrift A (1982) Selenium-toxicity and tolerance in higher-plants. Biol Rev 57:59–84
Broyer TC, Huston RP, Johnson CM (1972) Selenium and nutrition of Astragalus 1. Effects of selenite or selenate supply on growth and selenium content. Plant Soil 36:635–649
Cao J, Hu X (2009) Synthesis of gold nanoparticles using halloysites. e-J Surf Sci Nanotechnol 7:813–815
Chakraborty N, Banerjee A, Lahiri S, Panda A, Ghosh AN, Pal R (2009) Biorecovery of gold using cyanobacteria and eukaryotic alga with special reference to nanogold formation-a novel phenomenon. J Appl Phycol 21(1):145–152
Chau CF, Wu S-H, Yen G-C (2007) The development of regulations for food nanotechnology. Trends Food Sci Technol 18:269–280
Chaudhry Q, Scotter M, Blackburn J, Ross B, Boxall A, Castle L, Aitken R, Watkins R (2008) Applications and implications of nanotechnologies for the food sector. Food Addit Contam 25(3):241–258
Chen H, Roco MC (2009) Mapping nanotechnology innovations and knowledge-global and longitudinal patent and literature analysis. Integrated series in information system. Springer Science + Business Media, LLC. doi:10.1007/978-0-387-71619-0
Chen H, Yada R (2011) Nanotechnologies in agriculture: new tools for sustainable development. Trends Food Sci Technol 22:585–594
Chen JC, Lin ZH, Ma XX (2003) Evidence of the production of silver nanoparticles via pretreatment of *Phoma sp.* 32883 with silver nitrate. Lett Appl Microbiol 37:105–108
Chen H, Weiss J, Shahidi F (2006a) Nanotechnology in nutraceuticals and functional foods. Food Technol 3:30–36
Chen L, Remondetto GE, Subirade M (2006b) Food protein based materials as nutraceuticals delivery systems. Trends Food Sci Technol 17:272–283
Chen Y-W, Li L, D'Ulivo A, Belzile N (2006c) Extraction and determination of elemental selenium in sediments—A comparative study. Analytica Chimica Acta 577:126–133. doi:10.1016/j.aca.2006.06.020
Chinnamuthu CR, Boopathi PM (2009) Nanotechnology and agroecosystem. Madras Agric J 96(1–6):17–31
Combs GF Jr (2005) Global importance of selenium and its relation to human health. In: Welch RM, Ãakmak I (eds) Impacts of agriculture on human health and nutrition. In: Encyclopedia of Life Support Systems (EOLSS), Developed under the Auspices of the UNESCO, EOLSS Publishers, Oxford, UK. http://www.eolss.net. Accessed 25 Aug 2012
Combs GF, Gray WP (1998) Chemopreventive agents: selenium. Pharmacol Ther 79:179–192
Crombie AC (1959) Medieval and early modern science, vol 1. Doubleday, New York, p 137, 230
Crystal RG (1973) Elemental selenium: structure and properties. In: Klayman DL, Gunther WHH (eds) Organic selenium compounds: their chemistry and biology. Wiley, New York, pp 107–167
Cushen M, Kerry J, Morris M, Cruz-Romero M, Cummins E (2012) Nanotechnologies in the food industry-recent developments, risks and regulation. Trends Food Sci Technol 24:30–46
Cutter GA, Bruland KW (1984) The marine biogeochemistry of selenium: a reevaluation. Limnol Oceanogr 29:1179–1192
Davda J, Labhasetwar V (2002) Characterization of nanoparticle uptake by endothelial cells. Int J Pharm 233:51–59
Davis R, Guliants VV, Huber G, Lobo RF, Miller JT, Neurock M et al (2009) An international assessment of research in catalysis by nanostructure and materials. WTEC, Baltimore. http://www.wtec.org/catalysis/WTEC-CatalysisReport-6Feb2009-color-hi-res.pdf. Accessed 21 March 2013
Deepak V, Kalishwaralal K, Pandian SRK, Gurunathan S (2011) An insight into the bacterial biogenesis of silver nanoparticles, industrial production and scale-up. In: Rai M, Duran N (eds) Metal nanoparticles in microbiology. Springer-Verlag, Berlin, pp 17–35. doi:10.1007/978-3-642-18312-6-1

Degant O, Schwechten D (2002) Wheat flour with increased water binding capacity and process and equipment for its manufacture. German Patent DE10107885A1

Denina B, Simmons D, Wallascheeger D (2005) A critical review of the biogeochemistry and ecotoxicology of selenium in lotic and lentic environments. Environ Toxicol Chem 24:1331–1343

Dhillon KS, Dhillon SK (1999) Adsorption-desorption reactions of selenium in some soils of India. Geoderma 93:19–31

Di Gregorio S (2008) Selenium: a versatile trace element in life and environment. In: Prasad AS (ed) Trace elements in human health and disease, vol 2. Academic, New York, pp 593–622

Di Gregorio S, Lampis S, Vallini G (2005) Selenite precipitation by a rhizospheric strain of *Stenotrophomonas* sp. isolated from the root system of *Astragalus bisulcatus*: a biotechnological perspective. Environ Int 31:233–241

Domokos-Szabolcsy E, Marton L, Sztrik A, Babka B, Prokisch J, Fari M (2012) Accumulation of red elemental selenium nanoparticles and their biological effects in *Nicotinia tabacum*. Plant Growth Regul 68:525–531. doi:10.1007/s10725-012-9735-x

Duffus JH (2002) "Heavy metals"—a meaningless term? Pure Appl Chem 74:793–807

EC No. 258/97 (1997). Regulation (EC) No 258/97 of the European Parliament and of the Council of 27 January 1997 concerning novel foods and novel food ingredient. Official Journal of the European Union

EC No. 1935/2004 (2004). Regulation (EC) No 1935/2004 of the European Parliament and of the Council of 27 October 2004 on materials and articles intended to come into contact with food and repealing Directives 80/590/EEC and 89/109/EEC. Official Journal of the European Union

EC No. 1907/2006 (2006). Regulation (EC) No 1907/2006 of the European Parliament and of the Council of 18 December 2006 concerning the Registration, Evaluation, Authorisation and Restriction of chemicals (REACH), establishing a European Chemicals Agency, amending Directive 1999/45/EC and repealing Council Regulation (EEC) No 793/93 and Commission Regulation (EC) No 1488/94 as well as Council Directive 76/769/EEC and Commission Directives 91/155/EEC, 93/67/EEC, 93/105/EC and 2000/21/EC. Official Journal of the European Union

EC No. 1272/2008 (2008). Regulation (EC) No 1272/2008 of the European Parliament and of the Council of 16 December 2008 on classification, labelling and packaging of substances and mixtures, amending and repealing Directives 67/548/EEC and 1999/45/EC, and amending Regulation (EC) No 1907/2006. Official Journal of the European Union

EC No. 1223/2009 (2009) Regulation (EC) No 1223/2009 of the European Parliament and of the Council of 30 November 2009 on cosmetic products. Official J Eur Union

Ellis DR, Salt DE (2003) Plants, selenium and human health. Curr Opin Plant Biol 6:237–279

Elrashidi MA, Adriano DC, Lindsay WL (1989) Solubility, speciation and transformation of selenium in soils. In: Jacobs LW (Ed.) Selenium in agriculture and the environment. American Society of Agronomy and Soil Science Society of America, Madison, WI, pp. 51-69

Enghag P (2004) Encyclopedia of the elements-technical data, history, processing and applications. WILEY-VCH Verlag GmbH & Co. KGaA, Weinheim

Eszenyi P, Sztrik A, Babka B, Prokisch J (2011) Elemental, nanosized (100–500 nm) selenium production by probiotic lactic acid bacteria. Inter J Biosci Biochem Bioinform 1:148–152

ETC, Action Group on Erosion, Technology and Concentration (2005a) ETC group report: NanoGeoPolitics. ETC Group Surveys the Political Landscape. http://www.etcgroup.org/upload/publication/51/01/com89specialnanopoliticsjul05eng.pdf. Accessed 21 March 2013

ETC, Action Group on Erosion, Technology and Concentration (2005b) The potential impacts of nano-scale technologies on commodity markets: the implications for commodity dependent developing countries. http://www.etcgroup.org/upload/publication/45/01/southcentre.commodities.pdf. Accessed 21 March 2013

Fairweather-Tait SJ, Collings R, Hurst R (2010) Selenium bioavailability: current knowledge and future research requirements. Am J Clin Nutr 91:1484S–1491S

Fairweather-Tait SJ, Bao Y, Broadley MR, Collings R, Ford D, Hesketh JE, Hurst R (2011) Selenium in human health and disease. Antioxid Redox Sign 14(7):1337–1383. doi:10.1089/ars.2010.3275 (Mary Ann Liebert, Inc)

FAO (1996) Report of the World Food Summit. FAO, Rome

FAO/WHO (1998) Preparation and use of food-based dietary guidelines. Report of a joint FAO/WHO consultation. Geneva, World Health Organization (WHO Technical Report Series, No. 880)

FAO/WHO (2004) Vitamin and mineral requirements in human nutrition, 2nd edn. Report of a joint FAO/WHO expert consultation, Bangkok, Thailand, 21–30 September 1998. Geneva, World Health Organization. http://whqlibdoc.who.int/publications/2004/9241546123.pdf. Accessed 21 March 2013

FAO/WHO (2009) FAO/WHO joint expert meeting on the application of nanotechnologies in the food and agriculture sectors: potential food safety applications. Meeting Report, Rome

Fendler JH (2001) Colloid chemical approach to nanotechnology. Korean J Chem Eng 18:1–13

Fernández-Martánez A, Charlet ĆL (2009) Selenium environmental cycling and bioavailability: a structural chemist point of view. Rev Environ Sci Biotechnol 8:81–110. doi:10.1007/s11157-009-9145-3

Fesharaki PJ, Nazari P, Shakibaie M, Rezaie S, Banoee M, Abdollahi M, Shahverdi AR (2010) Biosynthesis of selenium nanoparticles using *Klebsiella pneumoniae* and their recovery by a simple sterilization process. Braz J Microbiol 41:461–466

Feynman R (1959) The classic talk by Richard Feynman entitled '*There's plenty of room at the bottom*' delivered at the annual meeting of the American Physical Society at the California Institute of Technology in 1959 is possibly the first serious exposition on the problem of manipulating nanoscale objects (the talk is available on the web at http://www.zyvex.com/nanotech/feynman.html). Accessed 21 March 2013

Finley JW, Ip C, Lisk DJ, Davis CD, Hintze KJ, Whanger PD (2001) Cancer-protective properties of high-selenium broccoli. J Agric Food Chem 49:2679–2683

Fisinin VI, Papazyan TT, Surai PF (2009) Producing selenium-enriched eggs and meat to improve the selenium status of the general population. Crit Rev Biotechnol 29:18–28

Fordyce FM (2005) Selenium deficiency and toxicity in the environment. In: Selinus O, Alloway B, Centeno JA, Finkelman RB, Fuge R, Lindh U, Smedley P (eds) Essentials of medical geology. Elsevier, London, pp 373–415

Fordyce F (2007) Selenium geochemistry and health. Ambio 36:94–97

Frankenberger WT, Engberg RA (1998) Environmental chemistry of selenium. Marcel Dekker, New York, p 736

Frankenberger WT, Karlson U (1994) Microbial volatilization of selenium from soils and sediments. In: Frankerberger WT, Benson S (eds) Selenium in the environment. Marcel Dekker, New York, pp 369–387

Freeman JL, Lindblom SD, Quinn CF, Fakra S, Marcus MA, Pilon-Smits EAH (2007) Selenium accumulation protects plants from herbivory by Orthoptera via toxicity and deterrence. New Phytol 175:490–500

Frost DV (1972) Two faces of selenium-can selenophobia be cured? In: Hemphill D (ed) CRC critical reviews in toxicology. CRC, Boca Raton, pp 467–514

Gade AK, Bonde P, Ingle AP, Marcato PD, Duran N, Rai MK (2008) Exploitation of Aspergillus niger for synthesis of silver nanoparticles. J Biobased Mater Bioenergy 2:243–247

Gaillardet J, Viers J, Dupré B (2003) Trace elements in river waters. In: Drever JI (ed) Surface and ground water, weathering and soils In: Holland HD, Turekian KK (eds) Treatise on geochemistry, vol 5. Elsevier, Oxford, pp 225–227

Gajbhiye M, Kesharwani J, Ingle A, Gade A, Rai M (2009) Fungus-mediated synthesis of silver nanoparticles and their activity against pathogenic fungi in combination with fluconazole. Nanomed Nanotechnol Biol Med 5:382–386

Gao XY, Hiroshi M (2005) Peptide-based nanotubes and their applications in bionanotechnology. Adv Mater 17:2037–2050

Gao XY, Zhang JS, Zhang LD (2002) Hollow sphere selenium nanoparticles: their in vitro anti hydroxyl radical effect. Adv Mater 14:290–293

Gao X, Gao T, Zhang L, Mater J (2003) Solution-solid growth of α-monoclinic selenium nanowires at room temperature. J Mater Chem 13:6–8

Garbisu C, Gonzalez S, Yang WH (1995) Physiological mechanisms regulating the conversion of selenite to elemental selenium by *Bacillus subtilis*. Biofactors 5:29–37
Garbisu C, Ishii T, Leighton T, Buchanan BB (1996) Bacterial reduction of selenite to elemental selenium. Chem Geol 132:199–204
Gardea-Torresdey JG, Parsons GE, Peralta-Videa J, Troiani HE, Santiago P, Yacaman MJ (2002) Formation and growth of Au nanoparticles inside live alfalfa plants. Nanoletters 2:397–401. doi:10.1021/nl015673
Gates B, Mayers B, Cattle B, Xia Y (2002a) Synthesis and characterization of uniform nanowires of trigonal selenium. Adv Funct Mater 12:219–227
Gates B, Mayers B, Grossman A, Xia Y (2002b) A sonochemical approach to the synthesis of crystalline selenium nanowires in solutions and on solid supports. Adv Mater 14:1749–1752
Govender Y, Riddin T, Gericke M, Whiteley CG (2009) Bioreduction of platinum salts into nanoparticles: a mechanistic perspective. Biotechnol Lett 31:95–100
Gregory PJ, Ingram JSI, Brklacich M (2005) Climate change and food security. Philos Trans Roy Soc B 360:2139–2148
Gruère GP (2012) Implications of nanotechnology growth in food and agriculture in OECD countries. Food Policy 37:191–198. doi:10.1016/j.foodpol.2012.01.001
Gruère G, Narrod C, Abbott L (2011) Agriculture, food, and water nanotechnologies for the poor: opportunities and constraints. International. http://www.ifpri.org/sites/default/files/publications/ifpridp01064.pdf/27.1.2013. Food Policy Research Institute (IFPRI), NW, Washington, USA
Hanauer M, Lotz A, Pierrat S, Sonnichsen C, Zins I (2007) Separation of nanoparticles by gel electrophoresis according to size and shape. Nano Lett 7(9):2881–2885
Hartikainen H (2005) Biogeochemistry of selenium and its impact on food chain quality and human health. J Trace Elem Med Biol 18:309–318
Haudin CC, Fardeau ML, Amenc L, Renault P, Ollivier B, Leclerc-Cessac E, Staunton S (2007) Responses of anaerobic bacteria to soil amendment with selenite. Soil Biol Biochem 39:2408–2413
Haygarth PM (1994) Global importance and cycling of selenium. In: Frankenberger WT, Benson S (eds) Selenium in the environment. Marcel-Dekker, New York, pp 1–28
He ZL, Yang XE, Stoffella PJ (2005) Trace elements in agroecosystems and impacts on the environment. J Trace Elem Med Biol 19:125–140. doi:10.1016/j.jtemb.2005.02.010
Hoefer F (1842) Histoire de la Chimie, vol 1, p 389 (Cited in: Mellor JW (1934) A comprehensive treatise on inorganic and theoretical chemistry, vol x. Longmans Green, London, pp 693–696)
Holben DH, Smith AM (1999) The diverse role of selenium within selenoproteins: a review. J Am Diet Assoc 99:836–843
Hu CH, Li YL, Xiong L, Zhang HM, Song J, Xia MS (2012) Comparative effects of nano elemental selenium and sodium selenite on selenium retention in broiler chickens. Anim Feed Sci Tech 177:204–210
Huang B, Zhang J, Hou J, Chen C (2003) Free radical scavening effiency of nano-Se in vitro. Free Radical Bio Med 35(7):805–813. doi:10.1016/S0891-5849(03)00428-3
Huang J, Chen C, He N, Hong J, Lu Y, Qingbiao L, Shao W, Sun D, Wang XH, Wang Y, Yiang X (2007) Biosynthesis of silver and gold nanoparticles by novel sundried *Cinnamomum camphora* leaf. Nanotechnology 18:105–106
Hunter WJ, Manter DK (2008) Bio-reduction of selenite to elemental red selenium by *Tetrathiobacter kashmirensis*. Curr Microbiol 57:83–88. doi:10.1007/s00284-008-9160-6
Hurd DL, Kipling JJ (1964) Origins and growth of physical science, Part 2. Penguin Books, Harmondsworth
Hurst R, Armah CN, Dainty JR, Hart DJ, Teucher B, Goldson AJ, Broadley MR, Motley AK, Fairweather-Tait SJ (2010) Establishing optimal selenium status: results of a randomized, double-blind, placebo-controlled trial. Am J Clin Nutr 91:923–931
IFST, Institute of Food Science and Technology (2006) Nanotechnology. http://www.ifst.org/uploadedfiles/cms/store/ATTACHMENTS/Nanotechnology.pdf. Accessed 21 March 2013

Ingle A, Gade A, Pierrat S, Sonnichsen C, Rai MK (2008) Mycosynthesis of silver nanoparticles using the fungus *Fusarium acuminatum* and its activity against some human pathogenic bacteria. Curr Nanosci 4:141–144

Jana NR, Pal T, Sau TK, Wang ZL (2000) Seed- mediated growth method to prepare cubic copper nanoparticles. Curr Sci 79(9):1367–1370

Jiang W (1994) Delineating the distribution of selenium in bacterial cultures that reduce and methylate oxyanions of this toxic metalloid. Master of Science, The Faculty of the Department of Chemistry, Sam Houston State University, Huntsville, Texas

Johnson CC, Fordyce FM, Rayman MP (2010) Symposium on "Geographical and geological influences on nutrition": factors controlling the distribution of selenium in the environment and their impact on health and nutrition. Proc Nutr Soc 69:119–132

Joseph T, Morrison M (2006) Nanotechnology in agriculture and food. Woodrow Wilson International Center for Scholars. Project on Emerging Nanotechnologies. www.nanoforum.org. Accessed 20 May 2010

Juniper DT, Phipps RH, Ramos-Morales E, Bertin G (2008) Effect of dietary supplementation with selenium-enriched yeast or sodium selenite on selenium tissue distribution and meat quality in beef cattle. J Anim Sci 86:3100–3109

Kabata-Pendias E (2011) Trace elements in soils and plants, 4th edn. LLC, CRC Press/Taylor & Francis Group, Boca Raton

Kabata-Pendias A, Mukherjee AB (2007) Trace elements from soil to human. Springer Science + Business Media, Berlin

Kabata-Pendias E, Pendias H (2001) Trace elements in soils and plants, 3rd edn. LLC, CRC Press/ Taylor & Francis Group, Boca Raton

Kalpana Sastry R, Rao NH, Cahoon R, Tucker T (2007) Can nanotechnology provide the innovations for a second green revolution in Indian agriculture? In: Paper presented in NSF nanoscale science and engineering grantees conference, 3–6 December 2007

Kalpana Sastry R, Rashmi HB, Rao NH, Ilyas SM (2010) Integrating nanotechnology into agri-food systems research in India: a conceptual framework. Technol Forecast Soc Change 77:639–648

Kalpana Sastry R, Rashmi HB, Rao NH (2011) Nanotechnology for enhancing food security in India. Food Policy 36:391–400. doi:10.1016/j.foodpol.2010.10.012

Kim JS (2007) Antibacterial activity of Ag^{+}ion-containing silver nanoparticles prepared using the alcohol reduction method. J Ind Eng Chem 13(4):718–722

Klayman DL, Gunther WHH (1973) Organic selenium compounds: their chemistry and biology. Wiley, New York

Kowshik M, Ashataputre S, Kharrazi S, Kulkarni SK, Paknikari KM, Vogel W, Urban J (2003) Extracellular synthesis of silver nanoparticles by a silver-tolerant yeast strain MKY3. Nanotechnology 14:95–100

Kumar V, Yadav SK (2009) Plant-mediated synthesis of silver and gold nanoparticles and their applications. J Chem Technol Biotechnol 84:151–157

Kumar AS, Ansary AA, Ahmad A, Khan MI (2007a) Extracellular biosynthesis of CdSe quantum dots by the fungus, *Fusarium Oxysporum*. J Biomed Nanotechnol 3:190–194

Kumar SA, AbyanehMK, GosaviSW, Kulkarni SK, Pasricha R, Ahmad A, Khan MI (2007b) Nitrate reductase-mediated synthesis of silver nanoparticles from $AgNO_3$. Biotechnol Lett 29:439–445

Kuzma J (2010) Nanotechnology in animal production-Upstream assessment of applications. Livest Sci 130:14–24

Kyriakopoulos A, Behne D (2002) Selenium-containing proteins in mammals and other forms of life. Rev Physiol Biochem Pharmacol 145:1–46

Lal R (2007) Ushering soil science into the 21 century. President's message, Soil Science Society of America. Nov 7, 2007, Madison Wisconsin. http://www.soils.org/about-society/presidents-message/archive/16. Accessed 22 May 2010

Lam JCW, Tanabe S, Wong BSF, Lam PKS (2004) Trace element residues in eggs of little egret (*Egretta garzetta*) and black-crowned night heron (*Nycticorax nycticorax*) from Hong Kong, China. Marine Pollut Bull 48:378–402

Lawson S, Macy JM (1995) Bioremediation of Se (IV) in oil refinery wastewater. Appl Microbiol Biotechnol 43:762–765

Lemly AD (1997) Ecosystem recovery following selenium contamination in a freshwater reservoir. Ecotoxicol Environ Safety 36:275–281

Lenz M, Gmerek A, Lens PNL (2006) Selenium speciation in anaerobic granular sludge. Inter J Environ Anal Chem 86:615–627

Levander OA (1976) Selective aspects of the comparative metabolism and biochemistry of selenium and sulfur. In: Prasad AS (ed) Trace elements in human health and disease, vol 2. Academic, New York, pp 135–163

Li Y-H (2000) A compendium of geochemistry: from solar nebula to the human brain. Princeton University Press, Princeton

Liao CD, Hung WL, Jan KC, Yeh AI, Ho CT, Hwang LS (2010) Nano/sub-microsized lignan glycosides from sesame meal exhibit higher transport and absorption efficiency in Caco-2 cell monolayer. Food Chem 119:896–902

Lin Z, Zayed A, Terry N (1999) Role of selenium volatilization in the management of selenium-laden agricultural drainage water. 5th International Conference on the Biogeochemistry Trace Elements, Vienna, pp 878–879

Lipiec E, Siara G, Bierla K, Ouerdane L, Szpunar J (2010) Determination of selenomethionine, selenocysteine, and inorganic selenium in eggs by HPLC-inductively coupled plasma mass spectrometry. Anal Bioanal Chem 397:731–741

Marmiroli N, Maestri E (2008) Health implications of trace elements in the environment and the food chain. In: Prasad MNV (ed) Trace elements as contaminants and nutrients: consequences in ecosystems and human health. Wiley, Hoboken, pp 23–53

Masscheleyn PH, Delaune R, Patrik WH Jr (1990) Transformations of selenium as affected by sediment oxidation-reduction potential and pH. Environ Sci Technol 1:91–96

Mayland HF, James LJ, Panter KE, Sonderegger JL (1989). Selenium in seleniferous environments. In: Jacobs LW (ed). Selenium in Agriculture and the Environment. SSSA Special Publication 23. Madison, WI: ASA and SSSA, pp. 15–50

Maynard A (2010) Presentation: nanotechnology and human health impact. A framework for strategic research. http://www.nanotechproject.org/process/files/2741/18_nanotechnologyhuman-healthimpactframeworkstrategicresearch.pdf. Accessed June 2010

Maynard AD, Aitken RJ, Butz T, Colvin V, Donaldson K, Oberdorster G, Philbert MA, Ryan J, Seaton A, Stone V et al (2006) Safe handling of nanotechnology. Nature 444:267–269

McNeal JM, Balistrieri LS (1989) Geochemistry and occurrence of selenium: an overview. In: Jacob LW (ed) Selenium in agriculture and the environment. Soil Science Society of America Special Publication no. 23. American Society of Agronomy and Soil Science Society of America, Madison, pp 1–14

Mehdaoui S, Benslim N, Aissaoui O, Benabdeslem M, Bechiri L, Otmani A, Portier X, Nouet G (2009) Study of the properties of $CuInSe_2$ materials prepared from nanoparticle powder. Mater Charact 60:451–455

Mildau G, Huber B (2010) The new EC cosmetics regulation 1223/2009- contents and first explanations. SOFW J Engl Edition 136:39–59

Moon JW, Roh Y, Lauf RJ, Vali H, Yeary LW, Phelps TJ (2007) Microbial preparation of metal-substituted magnetite nanoparticles. J Microbiol Meth 70(1):150–158

Moore Jr PA, Daniel TC, Gilmour JT, Shreve BR, Edwards DR, Wood BH (1998) Decreasing metal runoff from poultry litter with aluminum sulfate. J Environ Qual 27:92–99

Mukherjee P, Ahmad A, Mandal D, Senapati S, Sainkar SR, Khan MI, Ramani R, Parischa R, Ajayakumar PV, Alam M, Sastry M, Kumar R (2001) Bioreduction of AuC_4^- ions by the fungus *Verticillium* sp and surface trapping of the gold nanoparticles formed. Angew Chem Int Ed 40(19):3585–3588

Muniz-Naveiro O, Dominguez-Gonzalez R, Bermejo-Barrera A, Bermejo-Barrera P, Cocho JA, Fraga JM (2007) Selenium speciation in cow milk obtained after supplementation with different selenium forms to the cow feed using liquid chromatography coupled with hydride generationatomic fluorescence spectrometry. Talanta 71:1587–1593

Murphy K (2008) Nanotechnology: agriculture's next industrial revolution, pp 3–5. Financial Partner, Spring. 8–9

Mustacich D, Powis G (2000) Thioredoxin reductase. Biochem J 346 (Pt 1):1–8

Myneni SCB, Tokunaga TK, Brown GE Jr (1997) Abiotic selenium redox transformations in the presence of Fe(I, III) oxides. Science 278(5340):1106–1109. doi:10.1126/science.278.5340.1106

Nair B, Pradeep T (2002) Coalescence of nanoclusters and the formation of sub-micron crystallites assisted by *Lactobacillus* strains. Cryst Growth Des 2:293

Narayanan KB, Sakthivel N (2010) Biological synthesis of metal nanoparticles by microbes. Adv Colloid Interface Sci 156:1–13

Neal RH, Sposito G (1989) Selenate adsorption on alluvial soils. Soil Sci Soc Am J 53:70–74

Nightingale SD (2008) How and where might nanotechnology improve food safety? IFT international food nanoscience conference, June 27–28, New Orleans

Nowak J, Kaklewski K, Klódka D (2002) Influence of various concentrations of selenic acid (IV) on the activity of soil enzymes. Sci Total Environ 291:105–110

Nozaki Y (2005) A fresh look at element distribution in the North Pacific, AGU. http://www.agu.org/eos.elec/97025e.html. Accessed 21 March 2013

Nriagu JO (1989) Global cycling of selenium. In: Milan I (ed) Occurrence and distribution of selenium. CRC, Boca Raton, pp 327–340

Oberdörster G, Oberdörster E, Oberdörster J (2005) Nanotoxicology: an emerging discipline evolving from studies of ultrafine particles. Environ Health Perspect 113:823–839

Ogi T, Saitoh N, Nomura T, Konishi Y (2010) Room-temperature synthesis of gold nanoparticles and nanoplates using *Shewanella algae* cell extract. J Nanopart Res. doi:101007/s11051-009-9822-8

Oldfield JM (1990) Selenium: its uses in agriculture, nutrition and health and the environment. Selenium-Tellurium development association, Grimbergen, Belgium

Olin A, Nolang G, Osadchii E, Ohman LO, Rosen E (2005) Chemical thermodynamics of selenium. North Holland Elsevier Science Publishers B. V, The Netherlands

Opara L (2004) Emerging technological innovation triad for smart agriculture in the 21st century. Part I. Prospects and impacts of nanotechnology in agriculture, Agricultural Engineering International: the CIGR Journal of Scientific Research and Development, 2004, invited overview paper. vol VI. July

Oremland RS, Herbal MJ, Blum JS, Langely S, Beveridge TJ, Ajayan PM, Sutto T, Ellis AV (2004) Structural and spectral features of selenium nanospheres produced by Se-respiring bacteria. Appl Environ Microbiol 70:52–60

Parikh RY, Singh S, Prasad BLV, Patole MS, Sastry M, Shouche YS (2008) Extracellular synthesis of crystalline silver nanoparticles and molecular evidence of silver resistance from *Morganella* sp: towards understanding biochemical synthesis mechanism. Chem Bio Chem 9:1415–1422

Parker DR, Page AL (1994) Vegetation management strategies for remediation of selenium contaminated soils. In: Frankenberger WT Jr, Benson S (eds) Selenium in the environment. Marcel Dekker, New York, pp 327–341

Pastoriza-Santos I, Liz-Marzan LM (2000) Reduction of silver nanoparticles in DMF formation of monolayers and stable colloids. Pure Appl Chem 72(1–2):83–90

Pauling L (1932) The nature of the chemical bond. IV. The energy of single bonds and the relative electronegativity of atoms. J Am Chem Soc 54(9):3570–3582. doi:10.1021/ja01348a011

Pearce CI, Coker VS, Charnock JM, Pattrick RAD, Mosselmans JFW, Law N, Beveridge TJ, Lloyd JR (2008) Microbial manufacture of calcogenide-based nanoparticles via reduction of selenite using *Veillonella atypica*: an in situ EXAFS study. Nanotechnology. doi:10.1088/0957-4484/19/15/155603

Pilon-Smits EAH, Quinn CF (2010) Selenium metabolism in plants. In: Hell R, Mendel R-R (eds) Cell biology of metals and nutrients. Plant cell monographs 17. Springer, Berlin, pp 225–241 (doi:10.1007/978-3-642-10613-2_10)
Plant JA, Kinniburgh DG, Smedley PL et al (2003) Arsenic and selenium. In: Lollar BS, Holland HD, Turekian KK (eds) Treatise on geochemistry, vol 9, environmental geochemistry. Elsevier, London, pp 17–66
Plant JA, Kinniburgh DG, Smedley PL, Fordyce FM, Klinck BA(2004). Arsenic and selenium. In: Dollar BS (ed.) Environmental geochemistry 9:17–66
Prakash NT, Sharma N, Prakash R, Raina K, Fellowes J, Pearce C, Lloyd J, Pattrick R (2010) Aerobic microbial manufacture of nanoscale selenium: exploiting nature's bio-nanomineralization potential. Biotechnol Lett 31:1857–1862
Prokisch J, Zommara M (2008) Process for producing elemental selenium nanospheres PCT/IB2008/052838 date of receipt: 15 July 2008 receiving office: International Bureau of the World intellectual property organization your reference: P104315
Prokisch J, Széles E, Kovács B, Daróczy L, Zommara M (2008) Formation of metal selenium nanospheres in bacteria: is it a possible detoxification mechanism? Cer Res Commun 36:947–995
Rai M, Yadav A, Gade A (2008) Current trends in phytosynthesis of metal nanoparticles. Crit Rev Biotechnol 28(4):277–284
Rai M, Yadav A, Bridge P, Gade A (2009) Myconanotechnology: a new and emerging science. In: Rai MK, Bridge PD (eds) Applied mycology, vol 14. CAB International, New York, pp 258–267
Rai M, Gade A, Yadav A (2011) Biogenic nanoparticles: an introduction to what they are, how they are synthesized and their applications. In: Rai M, Duran N (eds) Metal nanoparticles in microbiology. Springer-Verlag, Berlin, pp 1–14 (doi:10.1007/978-3-642-18312-6-1)
Rayman MP (2000) The importance of selenium to human health. Lancet 356:233–241
Rayman MP, Infante HG, Sargent M (2008) Food-chain selenium and human health: spotlight on speciation. Br J Nutr 100:238–253
Reilly C (2006) Selenium in food and health, 2nd edn. Springer Science + Business Media, LLC. New York, USA
Reimann C, de Caritat P (1998) Chemical elements in the environment. Springer, Berlin
Reyes LH, Mar JL, Rahman GM, Seybert B, Fahrenholz T, Kingston HM (2009) Simultaneous determination of arsenic and selenium species in fish tissues using microwaveassisted enzymatic extraction and ion chromatography inductively coupled plasma mass spectrometry. Talanta 78:983–990
Robinson DKR, Morrison MJ (2009) Nanotechnology developments for the agrifood sector-report of the observatory NANO
Rodrıguez-Perez A, Flores-Acosta M, Peirez-Salas R, Rodrıguez-Mijangos R (2006) Cu halide nanoparticle formation by diffusion of copper in alkali halide crystals. Rev Mexi de Fisica 52(2):151–154
Rosenfeld I, Beath OA (1964) Selenium geobotany, biochemistry, toxicity and nutrition. Academic, New York, p 7
Roy M, Kiremidjian-Schumacher L, Wishe HI, Cohen MW, Stotzky G (1994) Supplementation with selenium and human immune cell functions. I. Effect on lymphocyte proliferation and interleukin 2 receptor expression. Biol Trace Elem Res 41:103–114
Salamanca-Buentello F, Persad DL, Court EB, Martin DK, Daar AS, Singer PA (2005) Nanotechnology and the developing world. PLoS Med 2(5):383–386, 97
Salerno M, Landoni P, Verganti R (2008) Designing foresight studies for nanoscience and nanotechnology (NST) future developments. Technol Forecast Soc Change 75:1202–1223
Sanguansri P, Augustin MA (2006) Nanoscale materials development—a food industry perspective. Trends Food Sci Technol 17:547–556
Sarin L (2010) Nano-selenium: novel formulations for biological and environmental applications. Ph. D. Thesis. Division of Engineering at Brown University, Providence, Rhode Island, USA

Sastry M, Ahmad A, Islam NI, Kumar R (2003) Biosynthesis of metal nanoparticles using fungi and actinomycete. Curr Sci 85:162–170
Sau TK, Rogach AL (2010) Nonspherical noble metal nanoparticles: colloid-chemical synthesis and morphology control. Adv Mater 22(16):1781–1804
Savage N, Diallo M, Duncan J, Street A, Sustich R (2009) Nanotechnology applications for clean water. William Andrew. Micro and nano technologies Series, Norwich
SCENIHR, Scientific Committee on Emerging and Newly-Identified Health Risks (2010) http://ec.europa.eu/health/scientific_committees/emerging/index_en.htm#. Accessed Jan 2010
Scharz K, Foltz CM (1957) Selenium as an integral part of Factor 3 against dietary necrotic liver degeneration. J Am Chem Soc 79:3292–3293
Schlekat CE, Dowdle PR, Lee BG, Luoma SN, Oremland RS (2000) Bioavailability of particleassociated Se to the bivalve *Potamocorbula amurensis*. Environ Sci Technol 34:4504–4510
Schrauzer GN (2004) Selenium. In: Merian E, Anke M, Ihnat M, Stoeppler M (eds) Elements and their compounds in the environment, 2nd edn. Wiley-VCH, Weinheim, pp 1365–1406
Schueler D, Frankel RB (1999) Bacterial magnetosomes: microbiology, biomineralization and biotechnological applications. Appl Microbiol Biotechnol 52(4):464–473
Scott NR (2005) Nanotechnology and animal health. Rev Sci Tech Off int Epiz 24(1):425–432
Seby F, Potin-Gautier M, Giffaut E, Borge G, Donard OFX (2001) A critical review of thermodynamic data for selenium species at 25 °C. Chem Geol 171:173–194
Shahi SK, Patra M (2003) Microbially synthesized bioactive nanoparticles and their formulation active against human pathogenic fungi. Rev Adv Mater Sci 5:501–509
Senesi GS, Baldassarre G, Senesi N, Radina B (1999) Trace element inputs into soils by anthropogenic activities and implications for human health. Chemosphere 39:343–377
Shahverdi AR, Minaeian S, Shahverdi HR, Jamalifar H, Nohi AA (2007) Rapid synthesis of silver nanoparticles using culture supernatants of Enterobacteria: a novel biological approach. Process Biochem 42(5):919–923
Shahverdi N, Wong CW, Nur Yasumira AA (2009) Rapid biosynthesis of silver nanoparticles using culture supernatant of bacteria with microwave irradiation. Eur J Chem 6(1):61–70
Shahverdi A-R, Shakibaie M, Nazari P (2011) Basic and practical procedures for microbial synthesis of nanoparticles. In: Rai M, Duran N (eds) Metal nanoparticles in microbiology. Springer-Verlag, Berlin, pp 177–196 (doi:10.1007/978-3-642-18312-6-1)
Shaligram NS, Bule M, Bhambure RM, Singhal RS, Singh SK, Szakacs G, Pandey A (2009) Biosynthesis of silver nanoparticles using aqueous extract from the compactin producing fungal strain. Process Biochem 44:939–948
Shao S, Zheng B (2008) The biogeochemistry of selenium in Sunan grassland, Gansu, Northwest China, casts doubt on the belief that Marco Polo reported selenosis for the first time in history. Environ Geochem Health 30:307–314
Sharma VK, Yngard RA, Lin Y (2009) Silver nanoparticles: green synthesis and their antimicrobial activities. Adv Colloid Interface Sci 145:83–96
Shehata SA, El-Ramady H (2012) Micronutrients: uptake and its roles in plants. Print house Fares Abd El-Hameed for Printing and Publishing, ISBN-No. 978-977-403-505-0 (in Arabic)
Shrift A (1969) Aspects of selenium metabolism in higher plants. Annu Rev Plant Physiol 20:475–495
Shriver DF, Atkins PW (1999) Inorganic chemistry, 3rd edn. Oxford University, Oxford
Singaravelu G, Arockiamary JS, Ganesh Kumar V, Govindraju K (2007) A novel extracellular synthesis of monodisperse gold nanoparticles using marine alga, Sargassum wightii Greville. Colloid Surf B Biointerface 57:97–101
Smith MI, Franke KW, Westfall BB (1936) The selenium problem in relation to public health. Public Health Rep 51:1496–1505
Song JY, Kim BS (2009) Rapid biological synthesis of silver nanoparticles using plant leaf extracts. Bioprocess Biosyst Eng 44:1133–1138. doi:10.1007/s00449-008-0224-6
Sozer N, Kokini JL (2009) Nanotechnology and its applications in the food sector. Trends Biotechnol 27(2):82–89. doi:10.1016/j.tibtech.2008.10.010

Spallholz JE (1994) On the nature of selenium toxicity and carcinostatic activity. Free Radical Biol Med 17:45–64

Spallholz JE, Hoffman DJ (2002) Selenium toxicity: cause and effects in aquatic bird. Aquat Toxicol 57:27–37

Sreelata M (2008) India looks to nanotechnology to boost agriculture. Sci-Dev.net. http://www.scidev.net/en/news/india-looks-to-nanotechnology-to-boost-agriculture.html. Accessed 18 May 2010

Steinnes E (2003) Biogeochemiacl cycling of iodine and selenium and potential geomedicval relevance. In: Skinner HC, Berger AR (eds) Geology and health: closing gap. Oxford University Press, Oxford, pp 57–60

Stroud JL, Broadley MR, Foot I, Fairweather-Tait SJ, Hart DJ, Hurst R, Knott P, Mowat H, Norman K, Scott P, Tucker M, White PJ, McGrath SP, Zhao F-J (2010) Soil factors affecting selenium concentration in wheat grain and the fate and speciation of Se fertilisers applied to soil. Plant Soil 332:19–30

Taniguchi N (1974) Proceedings of International Conference on Precision Engineering (ICPE), Tokyo, Japan

Terry N, Zayed AM, de Souza MP, Tarun AS (2000) Selenium in higher plants. Annu Rev Plant Physiol Plant Mol Biol 51:401–432

Thakkar KN, Mhatre SS, Parikh RY (2010) Biological synthesis of metallic nanoparticles. Nanomedicine 6(2):257–262

Thompson-Eagle ET, Frankenberger WT Jr, Karlson U (1989) Volatilization of selenium by *Alternaria alternata*. Appl Environ Microbiol 55:1406–1413

Thornton I (1981) Geochemical aspects of the distribution and forms of heavy metals in soils. In: Lepp NW (ed) Effect of heavy hetal pollution on plants: metals in the environment, vol II. Applied Science Publishers, London, pp 1–34

Uchida H, Shimoishi Y, Toei K (1980) Gas chromatographic determination of selenium (-II, 0), -(IV), and (VI) in natural waters. Environ Sci Technol 14: 541

USGS, U.S. Geological Survey (2013) Mineral commodity summaries 2013. http://minerals.usgs.gov/minerals/pubs/mcs/2013/mcs2013.pdf/. Accessed 26 Jan 2013

van der Krogt P (2004) Elementology and elements multidictionary. http://vanderkrogt.net/elements/elem/se.html. Accessed 21 March 2013

Velinsky DJ, Cutter GA (1990) Determination of elemental and pyrite selenium in sediments. Anal Chim Acta 235:419–425

Vinceti M, Wei ET, Malagoli C, Bergomi M, Vivoli G (2001) Adverse health effects of selenium in humans. Rev Environ Health 16:233–251

Vörösmarty CJ, McIntyre PB, Gessner MO, Dudgeon D, Prusevich A, Green P et al (2010) Global threats to human water security and river biodiversity. Nature 467:555–561

Wang D, Alfthan G, Aro A (1994) The impact of selenium fertilization on the distribution of selenium in rivers in Finland. Agric Ecosyt Environ 50:133–149

Wang H, Zhang J, Yu H (2007) Elemental selenium at nano size possesses lower toxicity without compromising the fundamental effect on selenoenzymes: comparison with selenomethionine in mice. Free Radic Biol Med 42:1524–1533

Waruingi M, Njoroge J (2008) Nanotech in Kenya. Business Daily Africa. Accessed 21 May 2010

Watanabe J, Iwamoto S, Ichikawa S (2005) Entrapment of some compounds into biocompatible nano-sized particles and their releasing properties. Coll Surf B Biointerfaces 42:141–146

Wei S, Zhou Q (2008) Trace elements in agro-ecosystems. In: Prasad MNV (ed) Trace elements as contaminants and nutrients: consequences in ecosystems and human health. Wiley, Hoboken, pp 55–79

Whanger PD (2002) Selenocompounds in plants and animals and their biological significance. J Am Coll Nutr 21:223–232

White PJ, Bowen HC, Parmaguru P, Fritz M, Spracklen WP, Spiby RE, Meacham MC, Mead A, Harriman M, Trueman LJ, Smith BM, Thomas B, Broadley MR (2004) Interactions between selenium and sulphur nutrition in *Arabidopsis thaliana*. J Exp Bot 55:1927–1937. doi:10.1093/jxb/erh192

WHO, World Health Organization (2011) Selenium in drinking-water-Background document for development of WHO Guidelines for Drinking-water Quality. WHO/HSE/WSH/10.01/14
WHO (1987) Environmental Health Criteria 58—Selenium. Geneva: WHO
Wolffram S, Anliker E, Scharrer E (1986) Uptake of selenate and selenite by isolated intestinal brush-border membrae vesicles from pig, sheep, and rat. Biol Trace Elem Res 10:293–306
Wright MT, Parker C, Amrhein DR (2003) Critical evaluation of the ability of sequential extraction procedures to quantify discrete forms of selenium in sediments and soils. Environ Sci Technol 37:4709–4716
Yada R (2009) Nanotechnology: a new frontier in foods, food packaging, and nutrient delivery. In: Pray L, Yaktine A (eds) Nanotechnology in food products. National Academies, Washington DC
Yamada H, Kase Y, Usuki M, Kajiyama S (1999) Selective determination and formation of elemental selenium in soils. Soil Sci Plant Nutr 45(2):403–408
Yang N, Aoki K (2005) Voltammetry of the silver alkylcarboxylate nanoparticles in suspension. Electrochim Acta 50:4868–4872
Yokota A, Shigeoka S, Onishi T, Kitaoka S (1988) Selenium as inducer of glutathione peroxidase in low-CO_2-grown *Chlamydomonas reinhardtii*. Plant Physiol 86:649–651
Zachara BA, Pileexi A (2000) Selenium concentration in the milk of breast-feeding mothers and its geographic distribution. Environ Health Perspect 108:1043–1046
Zayed A, Pilon-Smits E, de Souza M, Lin ZQ, Terry N (2000) Remediation of selenium polluted soils and waters by phytovolatilization. In: Terry NN, Banuelos G (eds) Phytoremediation of contaminated soil and water. Lewis, Boca Raton, pp 61–83
Zduôska H, Bloch-BogusÝawska E, Przygoôska J, Ţliwka K, Zachara BA (1994) Selenium concentration and glutathione peroxidase activity in the organs of people died a violent death. In: Kabata-Pendias A, Szteke B (eds) Arsenic and selenium in the environment. Ecologic analyt problem (Polish Ac Sci, pp 190–193) (in Polish and cited from Kapata-Pendias and Mukherjee 2007)
Zha LY, Xu ZR, Wang MQ, Gu LY (2008) Chromium nanoparticle exhibits higher absorption efficiency than chromium picolinate and chromium chloride in Caco-2 cell monolayers. J Anim Physiol Anim Nutr 92:131–140
Zhang YQ, Frankenberger WT Jr (2003) Determination of selenium fractionation and speciation in wetland sediments by parallel extraction. Int J Environ Anal Chem 83:315–326
Zhang YQ, Frankenberger WT Jr (2005) Removal of selenium from river water by a microbial community enhanced with *Enterobacter taylorae* in organic carbon coated S and columns. Sci Total Environ 346:280–285
Zhang JS, Gao XY, Zhang LD, Bao YP (2001) Biological effects of a nano red elemental selenium. Biofactors 15:27–38
Zhang J, Wang H, Bao Y, Zhang L (2004) Nano red elemental selenium has no size effect in the induction of seleno-enzymes in both cultured cells and mice. Life Sci 75:237–244. doi:10.1016/j.lfs.2004.02.004
Zhang J, Wang H, Yan X, Zhang L (2005) Comparison of short-term toxicity between Nano-Se and selenite in mice. Life Sci 76:1099–1109
Zhang J, Wang X, Xu T (2008) Elemental selenium at nano size (nano-Se) as a potential chemopreventive agent with reduced risk of selenium toxicity: comparison with Se-Methylselenocysteine in mice. Toxicol Sci 101(1):22–31. doi:10.1093/toxsci/kfm221
Zingaro RA, Cooper R (1974) Selenium. Van Nostrand Reinhold, New York, pp 788–807 (cited from Reilly 2006)

Chapter 6
Biofuels: Bioethanol, Biodiesel, Biogas, Biohydrogen from Plants and Microalgae

Mahdi Gharabaghi, Hossein Delavai Amrei, Alireza Moosavi Zenooz, Jafar Shahrivar Guzullo and Farzin Zokaee Ashtiani

Contents

M. Gharabaghi (✉) · J. Shahrivar Guzullo
School of Mining Engineering, College of Engineering,
University of Tehran, Tehran, Iran
e-mail: gharabaghi@ut.ac.ir, m.gharabaghi@gmail.com

H. Delavai Amrei · A. Moosavi Zenooz · F. Zokaee Ashtiani
Department of Chemical Engineering, Amirkabir University of Technology,
Tehran, Iran

© Springer International Publishing Switzerland 2015

E. Lichtfouse et al. (eds.), *CO_2 Sequestration, Biofuels and Depollution*,
Environmental Chemistry for a Sustainable World 5, DOI 10.1007/978-3-319-11906-9_6

Abstract The recent issues of global warming has recently prompted an intense research for sustainable fuels as alternatives to fossil fuels. This chapter presents an overview of current biofuels including bioethanol, biodiesel, biogas and biohydrogen. Biofuels are classified into three generations, the first from agriculture, the second from lignocellulosic materials, the third from microalgae. The definition of a renewable resource is presented. We describe the various feedstocks and processes to produce biofuels. We discuss advantages and disadvantages of biofuels.

Keywords Biofuel · Sustainable energy · Biodiesel · Bioethanol · Biogas · Lignocellulosic compounds · Pyrolysis · Feremnataion · Bio-oil · Microalgae

6.1 Introduction

Since the start of modern petroleum history in nineteenth century, the world has changed its way of life due to reliance to the oil and its derivatives. This too much reliance was due to the unique properties of the oil-based products, which have been led to dramatic changes in the life-style of every human being all around the world. Before the discovery of oil and after the start of industrial revolution and the

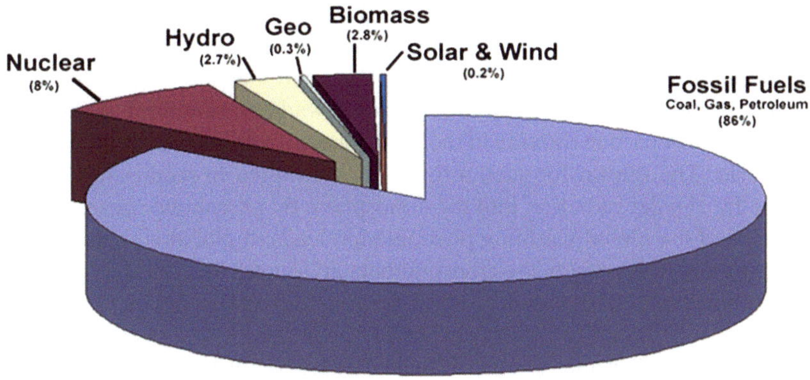

Fig. 6.1 An overall look at world energy sources. (World Energy Look 2008)

invention of steam engine, the coal has enjoyed a great deal of attention for being the highest rank of energy source of those days for moving the industry of those days (Maugeri 2006; Vassiliou 2009).

Figure 6.1 shows the different available energy sources and share of each source to fulfill the energy demand (World Energy Look 2008).

According to Fig. 6.1, the supplier sources of more than 80% of overall energy demand are oil, natural gas and coal, which are being called the fossil fuels (World Energy Look 2008). This deal of energy demand has worked suitably for at least a century since 1900's but things must be changed about the energy and our perspective on it. Our planet is now facing an increased population every day, which leads to greater energy demand. However, there are serious problems and questions, which lead us to change our point of view toward the following energy consumption regime.

Until now and by increasing the total energy demand, the countries which have resources for fossil fuels like oil-producing countries have invested more and more and have increased their production along with the energy demand but this trend cannot be continued for very long. There are many indications that show the possibility of declining in the overall resources of fossil fuels in the world. Such a story is happening more dramatically for oil, this unique source of energy. It is now estimated that more than 95% of available oil wells have been discovered and many of them has reached to the point that cannot produce high-quality, cheap and economically feasible oil for industrial usage. According to an estimate, the reserves of fossil fuels will last 218 years for coal, 41 years for oil and 63 years for natural gas with current scenario of energy demand and consumption (Sheehan and N.R.E. Laboratory 1998; Reijnders 2006; Pandey 2008; Arumugam et al. 2007).

Along with the described problems of availability for cheap fossil fuels, there are now more important issues, which lead us to consider other resources of energy more seriously. These major issues are the serious environmental impacts that are caused by burning fossil fuels. Keep burning the fossil fuels is the main cause of accumulation of carbon dioxide in the atmosphere (Mabee et al. 2005). The result of having more and more amounts of CO_2 is the greenhouse effect and the global warming phenomenon. Carbon dioxide is one of the main components of greenhouse gases. The carbon dioxide content in the world has increased from 280 ppm in the start of industrial revolution to 397 ppm nowadays (Houghton 2008). Increasing the amount of carbon dioxide in the atmosphere leads to increase in temperature of the world. The reason for such a thing is contributed to reemission of infrared wavelengths of solar radiation into the atmosphere by greenhouse gases molecules. Occurrence of the global warming phenomenon leads to global climate change and such a thing severely affects the environments all around the world (Howarth et al. 2011; Pironon et al. 2010; Hondo 2005; Intergovernmental panel on climate change 2007).

All of these reasons and severity of them by passing the time leads us to consider other sources of energy.

6.1.1 Renewable Energy Sources

To avoid repetition of the current problems with fossil fuels in relation to future energy sources, there are some points that should be taken into account. A new energy source should be reliable, available, be environmental-friendly and economically feasible. By employing such a definition to the future energy sources, the concept of renewable energy sources could emerge. A renewable energy is the one resulted from sustainable sources that are not depleted with consumption. Also, consuming these new energies would not result in problematic issues like greenhouse gases with fossil fuels or dangerous wastes like nuclear energy. Some of the renewable energy sources are water, solar, wind and biofuels. Employing these energy sources would result in a sustainable energy output, which does have lesser-unwanted and harmful effects on the environment along with having the concept of being sustainable and renewable (Hanjalic et al. 2008; International Energy Agency 2006; Boyle 2004; Puppan 2002).

Water-derived renewable energy is in the form of hydroelectricity and is resulted from both rotating turbines in the dams with water flowing power and tidal waves in the offshore facilities (Pimentel 2008; Deane et al. 2010).

Wind power converts wind energy to a variety of useful energy forms like electricity from wind turbines, mechanical power from windmills and water suction or drainage with wind pumps (Pimentel 2008; Dincer 2000).

Solar power is the usage of solar energy or energy coming from Sun to different forms of applications like utilization of solar energy for water heating in solar thermal collectors and producing electricity with solar photovoltaic cells (Pimentel 2008; Nozik 1978; Green 1982).

None of the renewable energy sources except biofuels is transportable. Rather than fossil fuels economic feasibility, it is their very unique specification of transportability, which makes them so popular until now (Arumugam et al. 2007; Murugesan et al. 2009; Rickeard and Thompson 1993).

Transportation industry is one of the most energy demanding industries with consumption of more than 33 % of total energy in the EU in 2009. Also, this energy consumer is the biggest emitting source of greenhouse gases specially carbon dioxide to the atmosphere. In the concept of employing energy from renewable sources, it should be taken into account that the renewable energy source should be able to replace fossil fuels in its greatest consumption area—transportation. Though there were efforts to introduce solar cars or incorporate electrical energy for use in vehicles, until now only biofuels have the same specification of transportability as petroleum fuels to be used the same in the available technologies of transportation systems (Abdullah et al. 2009).

6.1.2 Definition of Biofuels

The biofuels are referred to those kinds of fuels, which are incorporated biomass derived from plant-based products being processed to form a sustainable energy source (Pandey 2008). Different kinds of biofuels generate a wide variety of fuel types (in forms of liquid, solid and gaseous) for producing heat, electricity, chemicals and other materials (Jaccard 2006).

Biofuels seems as a promising new source of energy with some similar properties to petroleum fuels of which the transportability is the most important. This property would make biofuels superiority in comparison to other renewable energy sources, which is the potential to incorporate biofuels as driving force for the transportation. Regarding the environmental concerns and greenhouse effect, due to plant-based origin of biofuels, burning this kind of fuels would not lead to increase in the net amount of carbon dioxide in the atmosphere because the stored carbon materials inside biofuels had been added through photosynthesis from adsorption of carbon dioxide from the atmosphere (Stevens and Verhé 2004).

In general and due to origin of biomass which has been used for production of biofuels, the different kinds of biofuels are being classified in three generation. In following sections, the differences of each generation of biofuels along with the origin and advantages and disadvantages would be discussed.

6.2 First Generation Biofuels

First generation biofuels are those biofuels, which are originated from food crops like sugar, starch and oilseeds or food wastes originated from crops. The most important first generation biofuels are bioethanol, biodiesel and biogas (Boyle 2004; Dincer 2000).

Bioethanol is the product of fermentation of sugars, starch and sugar biomass. The used biomass in fermentation is cereal crops like corn, maize and sugarcane. Bioethanol could be utilized in pure form in some specialized vehicles or blended with gasoline up to a ratio that fuel specifications are met according to regulations (Minteer 2006).

Biodiesel is the result of transesterification reaction of vegetable oils like rapeseed, soya and palm fruits with methanol and alkali or acid agents. Like the bioethanol, biodiesel can also be used in both forms of pure or blended with automotive diesel (Mittelbach and Remschmidt 2004).

Biogas is obtained by anaerobic digestion of manure and humid biomass materials like food waste. The resulting biogas which is generally in form of methane, could be utilized in every natural gas derived system and also, in natural gas vehicles (Deublein and Steinhauser 2011).

6.2.1 Biodiesel from Vegetable Oils

The first generation biodiesel is resulted from the transesterification of fatty acid methyl esters, which are originated from vegetable oils and animal fats. Different kinds of catalysts are employed for the reaction completion such as homogenous, heterogeneous and biocatalysts. The commercialized reaction for biodiesel production includes transesterification reaction of triglycerides of the fatty acid with methanol with using of basic catalysts like sodium hydroxide; which yields methyl ester of fatty acid and that is called biodiesel.

In addition, other types of biofuels could be produced from vegetable oils and fats:

For instance, employing straight vegetable oils (SVO) as fuels, this is not so promising due to different properties of vegetable oils in comparison to diesel fuels.

Another example includes biodiesel obtained from vegetable oils via hydrocracking reaction. Such kind of biodiesel mainly contains alkanes and is similar to petroleum diesel but is lesser developed in comparison to fatty acid methyl ester-based biodiesel and could be considered as a competitor to it in the future.

Feedstocks for biodiesel include vegetable oil obtained from oil crops, recycled oil, animal fats and algae. For different feedstocks, different kinds of pretreatment and treatment are required to perform and the depending on these processes, the quality of the produced biodiesel and its cost could be varied.

In this section, we will take a brief look at different kinds of biodiesel and different technologies adapted for different feedstocks.

6.2.2 Biodiesel as Fatty Acid Esters

For production of methyl esters from oils, there are three general ways:

- Transesterification of triglycerides (of oils) with methanol under basic conditions (base as catalyst)
- Esterification of free fatty acids with methanol by employing acid-based catalysts
- Conversion of oil to free fatty acids after their esterification as mentioned above

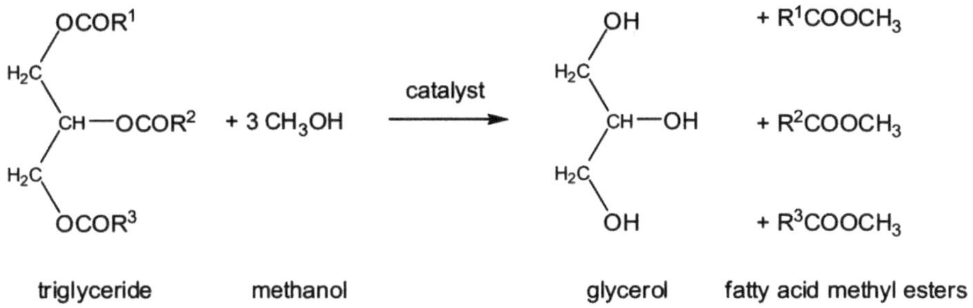

Fig. 6.2 Transesterification process

The most applicable and economic way for production of methyl esters through basic catalyzed reaction due to:

- Low temperature and pressure of reaction medium;
- High yields and short reaction time;
- Direct conversion process;
- Simplicity in operation and control;
- Being environmental friendly.

Transesterification is the reaction between a triglyceride (oil) with an alcohol (e.g. methanol or ethanol) in the presence of a catalyst, such as sodium hydroxide or potassium hydroxide. During this reaction, molecule of oil is chemically broken into methyl or ethyl esters. Glycerol, also known as glycerin, is the by-product of this reaction. The process is somehow similar to hydrolysis, except the employing alcohol rather than water. In Fig. 6.2, the general scheme of this reaction is depicted:

Transesterification consists of a number of consecutive, reversible reactions. During the process, diglycerides and monoglycerides are produced as the intermediates. The triglyceride is converted stepwise to diglyceride, monoglyceride and finally glycerol. The reactions are all reversible, although by using a little excess of alcohol, the equilibrium progresses to the production of fatty acid esters and glycerol. With addition of excess amount of alcohol, forward reaction exhibits pseudo-first order behavior and reverse reaction is found to be second order. It was also observed that transesterification is faster when catalyzed by alkali (Meher et al. 2006).

The mechanism of alkali-catalyzed transesterification is shown in Fig. 6.3 (Arumugam et al. 2007; Ma and Hanna 1999). It is a three-step reaction; First step involves the attack of the alkoxide ion to the carbonyl carbon of the triglyceride molecule, which results in the formation of a tetrahedral intermediate compound. In second step, the formed intermediate compound of step one is reacted with an alcohol and produces the alkoxide ion. Third step involves the rearrangement of the tetrahedral intermediate, which is resulted in formation of an ester and a diglyceride (Arumugam et al. 2007; Schwab et al. 1988).

Among other catalysts for transesterification, are Bronsted acids, preferably sulfonic and sulfuric acids. Employing these catalysts prompt very high reaction

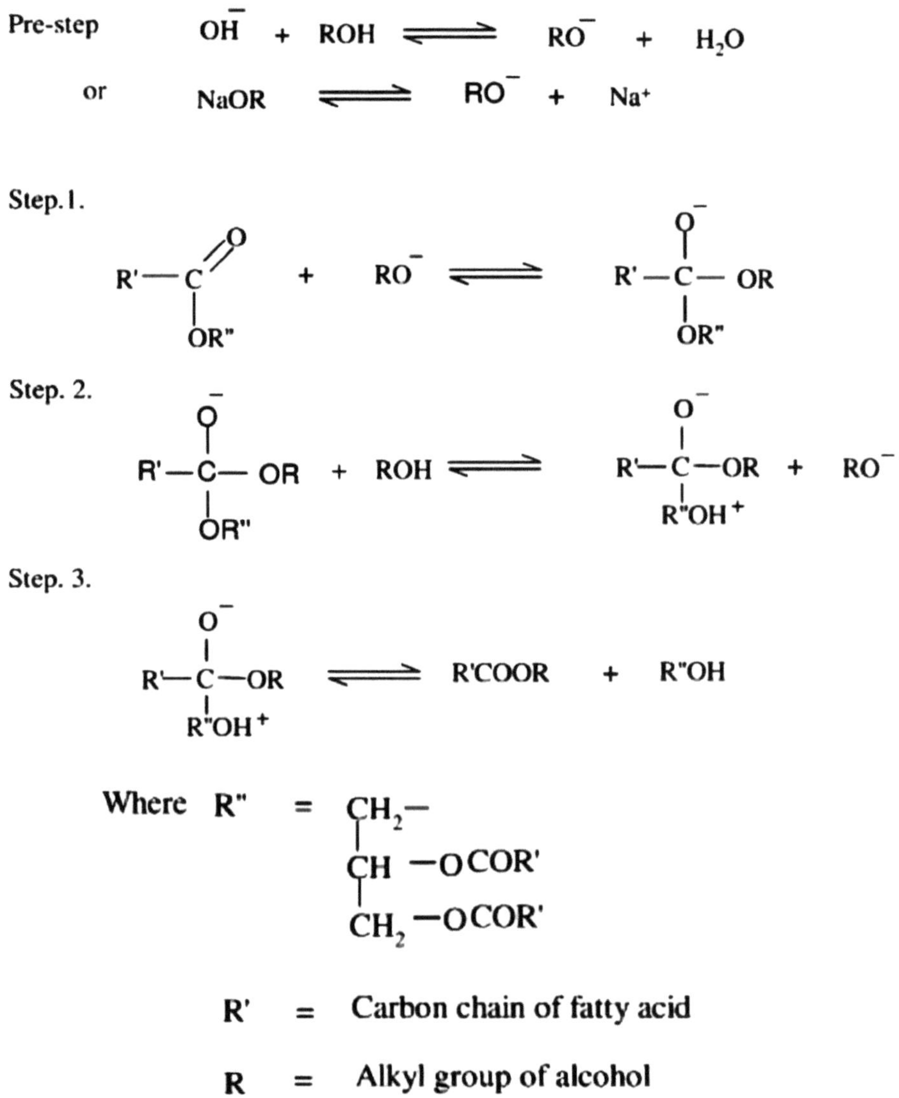

Fig. 6.3 Mechanism of base catalyzed transesterification. (Ma and Hanna 1999)

yields in alkyl esters, though the reactions are generally slow with necessity of temperatures above 100 °C and more than 3 h time to complete the conversion (Arumugam et al. 2007; Schuchardta et al. 1998).

The mechanism of acid catalyzed transesterification of vegetable oil (for a monoglyceride) is shown in Fig. 6.4. The protonation of carbonyl group of the ester leads to the carbocation, which after a nucleophilic attack of the alcohol produces a tetrahedral intermediate.

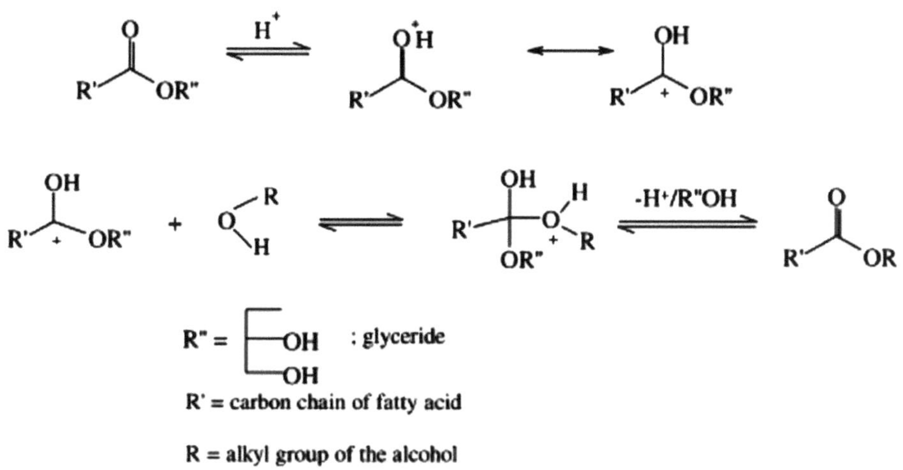

Fig. 6.4 Mechanism of acid catalyzed transesterification. (Schuchardta et al. 1998)

This intermediate compound changes glycerol to form a new ester and to regenerate the catalyst. This mechanism can be extended to di- and tri-glycerides (Kulkarni et al. 2006).

The process of transesterification is affected by various factors depending upon the reaction conditions. The effects of these factors are described below. The most relevant variables are: the reaction temperature, the ratio of alcohol to vegetable oil, the amount and the type of catalyst, the mixing intensity and the raw oils used (Bulack 1985).

6.2.3 Transesterification Process Technological Arrangement

Figure 6.5 presents the basic transesterification process flow diagram. Critical quality parameters of the process are:

- Complete reaction;
- Removal of glycerol;
- Removal of catalyst;
- Removal of alcohol;
- Absence of free fatty acids;
- Low sulfur content.

By employing pre-defined standards, it would be ensure that these factors in the fuel production process are satisfied. Basic industrial tests to determine whether the products conform to the standards typically include gas chromatography, a test that verifies only the more important of the variables above. More tests that are complete are more expensive. Fuel meeting the quality standards is very non-toxic, with a toxicity rating of greater than 50 ml kg^{-1}.

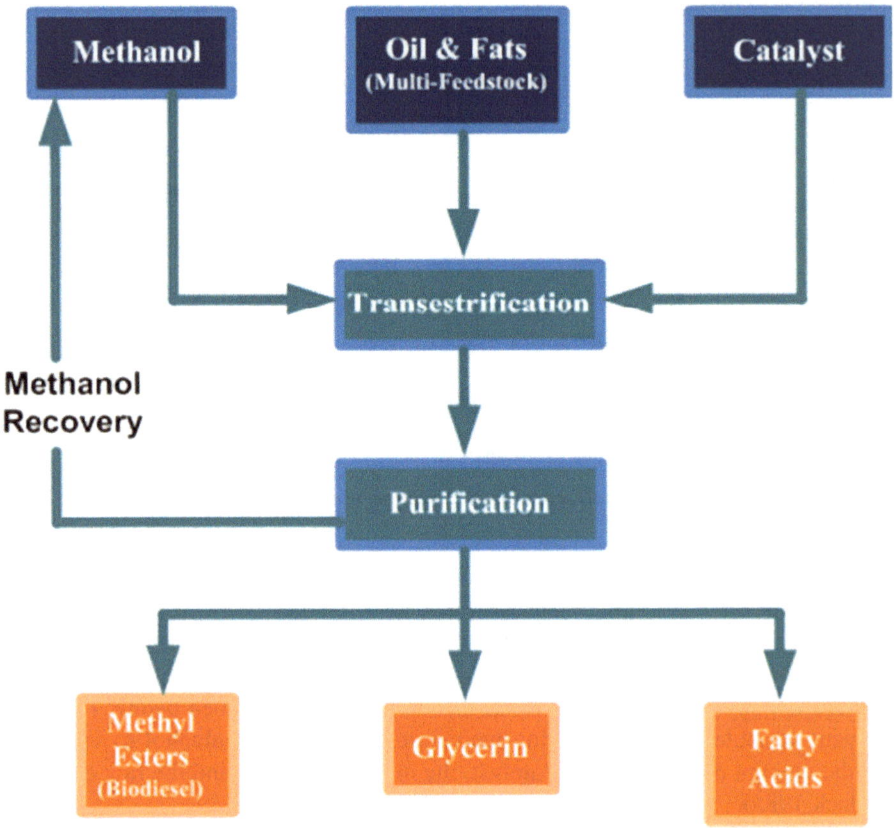

Fig. 6.5 Flowchart of the esterification process

Figure 6.6 illustrates that the first step for biodiesel production is the supply chain of vegetable oils or provision of other feedstocks. Then the extracted and purified oils and fats are undergoing the conversion process to biodiesel in the production plant. At the end and by purification of result, the product is ready to be distributed to the end users.

6.2.4 Type of Feedstock

The feedstock of oil, and its quality and cost is the most important factor for technical design of a transesterification plant. It affects the corresponding material and energy flows in the plant, which shows the technical and economic efficiency of biodiesel production plant. In Table 6.1, the most significant parameters of feedstock and their relevance for produced biodiesel are presented.

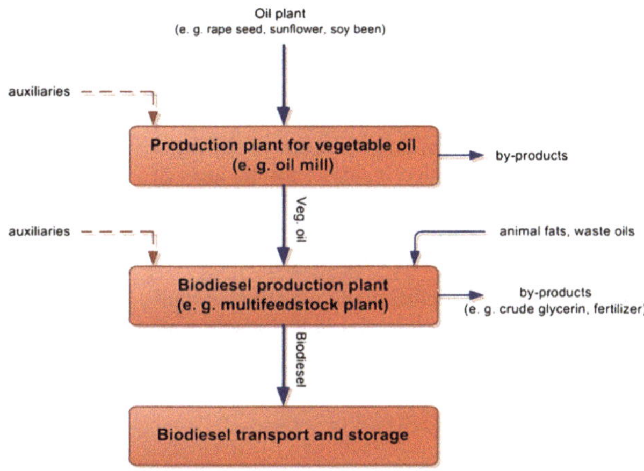

Fig. 6.6 Feedstock supply and biodiesel production and distribution

6.2.5 By-products Issues

The principle by-product of biodiesel production is glycerol. Vegetable oils contain an approximate amount of 10 % glycerol. Due to impurities in crude glycerol, it has low value for usage; however, by growing the amount of produced biodiesel, its utilization becomes important. Crude glycerol refining depends on production scale and availability of purification facility.

The method for glycerol purification consists of employing filtration, addition of chemical and fractional vacuum distillation to yield various commercial grade products. Small to moderate biodiesel producers cannot afford the high cost of crude glycerol purification and the disposal way to be a problem.

Larger scale biodiesel producers refine their crude glycerol and move it markets in the food, pharmaceutical and cosmetic industries. Also, glycerol could be burnt as fuel. Another way is to etherify glycerol with alcohols (e.g. methanol or ethanol) or alkenes (e.g. isobutene) and produce branched oxygen-containing components that have usage as fuel or solvents. Employing produced glycerol as fuel has the advantage of being a bio-component and thus could be included in the renewables energy category (Arumugam et al. 2007; Kemp 2006).

6.2.6 Effect of Free Fatty Acid and Moisture Content

The amount of free fatty acid and moisture are important factors on the viability of the vegetable oil transesterification process. Completion of transesterification reaction in basic catalyzed environment requires amounts of free fatty acid values lower than 3 % is needed. The higher amount of free fatty acids and higher the acidity of the oil; smaller is the conversion efficiency in the reaction. It should be noted that both of excess and insufficient amounts of catalyst in medium may cause

Table 6.1 Feedstock parameters and their relevance for biodiesel production and use. (Arumugam et al. 2007)

Parameters	Characterization	Relevance for biodiesel production and use
Free fatty acids (FFA)	Significant measure of feedstock quality, indicator for the level of hydrolysis; FFA of native unrefined oils and fat can be above 20, refined oils/fats have FFA up to 1.0	Influence degree of required processing (e.g. catalyst demand) and biodiesel quality (primarily cold and properties)
Total contamination	Proportion of unresolved impurities (particles in the oils/fats; mainly affected by the oil production procedure	Relevant to glycerin quality and unwanted glycerin caking within the process; high total contaminations lead to clogging the fuel filters and increase the danger of damage to the injection pump and to injection nozzle as well as of deposits in the combustion chamber
Water content	Mainly affected by the moisture of the seeds and refined oils; with all oils/fats the water content can rise through storage and transport	At high temperature water can hydrolyze the triglycerides to diglycerides and form free fatty acids; relevance for disturbing the transesterification by catalyst loss and unwanted soap production
Cinematic viscosity	Physical-mechanical characteristic, depending of specific melting point	Influenced by the temperature, fatty acid profile and oil-aging degrees, whereas the kind of the oil production procedure does not affected viscosity
Cold flow properties	Strongly affected by temperature; saturated fatty compounds with a significantly higher melting points than unsaturated fatty acid compounds	Cloud point (CP) and the cold filter plugging point (CFPP) or pour point (PP) for fuels not suitable to describe cold flow properties of oils/fats, since the transition of the liquid to the solid phase cannot be definite
Iodine number	Indicator for double bonding in molecular structure of oils/fats. The higher the iodine value, the more unsaturation acids are present in the oils/fats	High iodine number in oils/fats for less air resisting than oils/fats with high degree of saturation; informs about the tendency of deposit in the combustion chamber and at injection nozzle
Phosphorous content	Present in vegetable oils in the form of phospholipids; depends on refining grad of oils/fats (influenced by oil production process)	Decreasing oxidation stability with rising portion of phospholipids; high amount of phospholipids leading to disturbance with technical process (e.g. blockage of filters and injection nozzles); avoidance of phosphorous compounds in waste water
Oxidation stability	Value, which describes the aging condition and the shelf-life of oils/fats	Oxidation and polymerization procedures during fuel storage; which can lead to formation of insoluble compounds and thus cause e.g. filter blockage

soap formation (Dorado et al. 2002). Also, the triglycerides should be anhydrous. Higher acidity content is compensated by addition of more sodium hydroxide catalyst but it results in soap formation and causes an increase in viscosity or formation of gels that interferes in the reaction as well as separation of glycerol (Freedman et al.1984). When the reaction conditions do not meet the above requirements, ester yields are significantly reduced.

Biodiesel is mostly produced from edible oils by using methanol and alkaline catalyst. However, there are large amounts of low cost oils and fats that could be converted to biodiesel. The problems with processing these low cost oils and fats are that they often contain large amounts of free fatty acids that cannot be converted to biodiesel using alkaline catalyst. Therefore, two-step esterification process is required for these feedstocks. Initially free fatty acids of these can be converted to fatty acid methyl esters by an acid catalyzed pretreatment and in the second step transesterification is completed by using alkaline catalyst to complete the reaction (Arumugam et al. 2007).

The acid level of the high free fatty acids feed stocks could be reduced to less than 1 % with a two-step pretreatment reaction. The reaction mixture was allowed to settle between steps so that the water containing phase could be removed. After reducing the acid levels to less than 1 %, the transesterification reaction was completed with an alkaline catalyst to produce fuel grade biodiesel. If higher amounts of free fatty acids would be present in reaction medium and transesterification reaction was completed by employing basic catalysts, substantial amount of free fatty acid reacted with basic catalyst, one part of free fatty acids led to soap formation and other part led to catalyst neutralization.

The oils with high content of free fatty acids should be processed with an immiscible basic glycerol phase to neutralize the free fatty acids and cause them to pass over into the glycerol phase by means of monovalent alcohols.

6.2.7 Catalyst Type and Concentration

As described before, different types of catalyst are employed for the transesterification of triglycerides; alkali, acid, enzyme of heterogeneous origin and among which alkali catalysts like sodium hydroxide, sodium methoxide, potassium hydroxide, potassium methoxide are more effective (Ma and Hanna 1999). If the oil has high free fatty acid content and more water, acid catalyzed transesterification is suitable. Among acidic catalysts is sulfuric acid, phosphoric acid, hydrochloric acid or organic sulfonic acid.

Chemical transesterification using an alkaline or acidic catalysis has advantage of high conversion levels of triglycerides to their corresponding methyl esters in short reaction times. However, the reaction has several drawbacks: it is energy intensive with difficult recovery of glycerol is difficult, and the acidic or alkaline catalyst has to be removed from the product; alkaline wastewater requires treatment, and free fatty acid and water interfering with the reaction.

Table 6.2 Comparison of the different technologies to produce biodiesel. (Arumugam et al. 2007)

Variable	Alkali catalysis	Lipase catalysis	Supercritical alcohol	Acid catalysis
Reaction T (°C)	60–70	30–40	239–385	55–80
FFA in raw material	Saponified products	Methyl esters	Esters	Esters
H_2O in raw materials	Interference with reaction	No influence		Interference with reaction
Yield of methyl esters	Normal	Higher	Good	Normal
Recovery of glycerol	Difficult	Easy		Difficult
Purification of methyl esters	Repeated washing	None		Repeat washing
Production cost of catalyst	Cheap	Relatively expensive	Medium	Cheap

To overcome the mentioned problems above, enzymatic catalysts like lipases are able to effectively be employed as catalyst for the transesterification of triglycerides in either aqueous or non-aqueous systems (Fukuda et al. 2001). Also, about the by-products, glycerol can be easily removed without any complex process, and free fatty acids contained in waste oils and fats can be completely converted to alkyl esters. On the other hand, in general the production cost of a lipase catalyst is significantly greater than that of an alkaline one. Table 6.2 summarizes the differences between the various technologies used to produce biodiesel (Arumugam et al. 2007; Kemp 2006).

6.2.8 Molar Ratio of Alcohol to Oil and Type of Alcohol

The molar ratio of alcohol to triglyceride is one of the most important factors that affecting the yield of ester. The stoichiometric ratio for transesterification requires three moles of alcohol and one mole of triglyceride to yield three moles of fatty acid alkyl esters and one mole of glycerol. However, transesterification is an equilibrium reaction and a large excess of alcohol is mandatory for drive the reaction toward development. It is observed that employing a molar ratio of 6:1 would maximize conversion to ester. In addition, it is noticed that molar ratio has no effect on acid, peroxide, saponification and iodine value of produced methyl esters (Tomasevic and Siler-Marinkovic 2003). However, due to increase in solubility, the separation of glycerin in higher molar ratios of alcohol to vegetable oil would be more difficult. Remaining the glycerin in the reaction medium would drive the equilibrium reaction back wise which lowers the yield of esters.

Under using basic catalysts, formation of ethyl ester is more difficult compared to the formation of methyl esters. Specifically the formation of stable emulsion during ethanolysis is a problem. Methanol and ethanol are both immiscible with triglycerides at ambient temperature, and the reaction mixtures have to stir mechanically to enhance mass transfer. During the reaction, the formation of emulsions is usually occurred. Usage of methanol as alcohol in the reaction lead to formation of emulsions, which are quickly and easily breakable to form a lower glycerol rich layer and upper methyl ester rich layer. On the hand, with ethanol in the reaction medium, those emulsions are more stable and severely complicate the separation and purification of esters (Zhou et al. 2003). The occurrence of emulsions could be partly attributed to formation of the intermediate compounds such as monoglycerides and diglycerides. They have both polar hydroxyl groups and non-polar hydrocarbon chains and this makes them strong surface-active agents. During the alcoholysis stage of reaction, the catalyst, either sodium hydroxide or potassium hydroxide, is dissolved in polar alcohol phase, in which triglycerides must transfer in order to react. At early stages, the reaction is mass-transfer controlled and does not develop according to expected homogeneous kinetics pattern. By increasing the concentrations of these intermediates to a critical level, emulsions form. The presence of larger non-polar group in ethanol in comparison to methanol is assumed the critical factor in stabilizing the emulsions. However, the concentrations of mono- and di-glycerides are extremely low, and then the emulsions become unstable. This shows the necessity for the reaction to be as complete as possible, thereby reducing the concentrations of mono- and di-glycerides (Arumugam et al. 2007).

6.2.9 Effect of Reaction Time and Temperature

The rate of conversion rate increases with reaction time and reaches to a maximum after some time. It was observed that during early stages of reaction (1 min), the most of reaction would be completed (up to yield of 80 %) and passing more time just leads to increase the yield (95 % after an hour) (Freedman et al.1984).

The same story goes for the effect of temperature on the final yield and is observed that during early stages, temperature has a meaningful effect on reaction yield but passing time would lead the reaction to almost same yield in different temperatures (Arumugam et al. 2007; Ma and Hanna 1999).

6.2.10 Mixing Intensity

Mixing is a very important factor in the transesterification reaction, since oils or fats are immiscible with sodium hydroxide–methanol solution. Once the two phases are mixed and the reaction is started, stirring is no longer needed.

6.2.11 Product Properties and Quality

The presence of contaminants from production process or other sources would influence the suitability of any material as fuel, including biodiesel. Some of the

Table 6.3 Properties of biodiesel from different oils. (Barnwal and Sharma 2005)

Vegetable oil Methyl esters (biodiesel)	Kinematic viscosity (mm^2/s)	Cetane no.	Lower heating value (MJ/kg)	Cloud point (°C)	Pour point (°C)	Flash point (°C)	Density (kg/l)
Peanut	4.9	54	33.6	5	–	176	0.883
Soya bean	4.5	45	33.5	1	−7	178	0.885
Babassu	3.6	63	31.8	4	–	127	0.875
Palm	5.7	62	33.5	1	–	183	0.880
Sunflower	4.6	49	33.5	1	–	183	0.860
Tallow	–	–	–	12	9	96	–
Diesel	3.06	50	43.8	–	−16	76	0.855
20% Biodiesel blend	3.2	51	43.2	–	−16	128	

properties included as specifications in standards can be traced to the structure of the fatty esters comprising biodiesel. However, due to biodiesel being consist of fatty acid esters, not only the structure of the fatty acids but also structure of the ester moiety derived from the alcohol can influence the fuel properties of biodiesel. Since the transesterification reaction of oil or fat leads to a biodiesel fuel corresponding in its fatty acid profiles with that of the parent oil or fat, biodiesel is a mixture of fatty esters with each ester component contributing to the properties of the fuel. The properties of a biodiesel fuel that are determined by the structure of its component fatty esters include ignition quality, heat of combustion, cold flow, oxidative stability, viscosity and lubricity. In Table 6.3, the properties of biodiesel from different origins are presented and their similarity to diesel properties makes them a good alternative for diesel fuels (Arumugam et al. 2007).

6.2.12 Global Warming and Pollution Reduction

By displacing biodiesel instead of petroleum fuels, it would lead to reduction in levels of greenhouse gases such as carbon dioxide, which in turn would cause in lowering the global warming phenomenon. The reason for this is the origin of biodiesel, which has been resulted from plants.

When plants like soybeans grow, they take carbon dioxide from atmosphere and by performing photosynthesis; they make use of this carbon source to the stems, roots, leaves and seeds. After extraction of oil from the plant, it is refined into biodiesel and, when burned, produces carbon dioxide and other emissions, which are eventually returned to the atmosphere. However due to origin of this produced carbon dioxide which had been from the atmosphere, there would be no addition of excess carbon dioxide in the atmosphere and this cycle would repeat itself by other plants reusing the carbon dioxide again and make fatty acids in their organs. Another important environmental issue lies on the fact that biodiesel usage causes in decrease in amount of other contaminating materials like HC and carbon monoxide. Such a thing happens due to presence of 11 % by weight of oxygen in the biodiesel. The occurrence of oxygen allows the fuel to burn more smoothly and more completely, resulting in fewer emissions of unburned hydrocarbons and carbon monoxide. This would cause to reduction in air toxicity, which is related to amount of free hydrocarbons and carbon monoxide. In addition, it is observed this reduction of HC and CO is independent of type of vegetable oil used for making the biodiesel.

6.2.13 Safety

Biodiesel is moreover than being a non-toxic compound, has another advantage of being almost not flammable. It has to heat up to over 150 before it starts to combust. The only reason it works in the diesel engine is because diesels compress the vaporized fuel and air mixture so much that it combusts (without the need for spark plugs).

6.2.14 Concerns/Barriers

Biodiesel exhibits many advantages like lowering emissions, raising the fuel's cetane number and flashpoint, excellent lubricity, and has the attractive advantage of its direct usage in diesel engine without further modifications. However, Biodiesel has some drawbacks as well. Some issues should be bear in mind prior its usage, which includes its cold flow properties, stability, energy content, elevated NO_x levels and material compatibility.

6.2.15 Mixing Biodiesel with Conventional Diesel Fuels

Biodiesel is generally used in blended form with petroleum-based diesel. The most common blend is a mixture consisting of 20 % biodiesel and 80 % petroleum diesel, called B20. The motive for blending the fuels is to gain some of the advantages of biodiesel while avoiding higher costs and lower the disadvantages. Biodiesel is currently higher in price than conventional diesel fuel, so by blending it with conventional diesel, the environmental effects of biodiesel would be in the mixed fuel with lower price. The following is a look at different blends of biodiesel:

- B100 (100 % biodiesel) offers the most overall environmental benefits. Use of B100 may require some engine and fuel system components to be modified and can cause operating problems, especially in cold weather.
- B20 (20 % biodiesel) offers about one fifth of the environmental benefits of B100, but can be more broadly applied to existing engines with little or no modification.

- B2 (2% biodiesel) offers little environmental or petroleum dependence benefit and could be potentially used an environmental marketing tool.

6.2.16 Bioethanol from Sugar Crops

Fermentation process of sugar byproducts leads to production of ethanol, which is called bioethanol due to its origin. Ethanol can also be produced chemically through reaction of ethylene and steam. It is a colorless liquid compound with low toxicity and exhibits biodegradability. Employing of bioethanol as transport fuel could lead in lesser environmental pollution in comparison to other fuels. It has a content of 35% oxygen, which reduces the emissions of NOX and particulate matters from combustion. It is commonly blended on basis of 10% bioethanol and 90% petrol and is called E10. Current engines have no problem running with E10 and vehicle warranties remain unaffected, as well. Flexible vehicles can run on up to 85% ethanol and 15% petrol blends (E85) (Arumugam et al. 2007).

6.2.17 Bioethanol from Sugar Feedstocks

Easiest way to produce ethanol is from C6 sugars using fermentation process. The most common sugar is glucose (C6) or biomasses containing higher levels of glucose or precursors to glucose are the easiest to convert to ethanol. Many microorganisms like fungi, bacteria, and yeast can be used for fermentation of sugars but *Saccharomyces cerevisiae* also known as Baker's yeast is frequently used to ferment sugar to ethanol. Sugarcane is a typical example for sugar feedstock.

Brazil as one of the biggest producers of sugar in the world is one of the highest producers of bioethanol for fuel and uses sugarcane as feedstock. Other sugar rich biomass feedstocks are sweet sorghum, sugar beet, and different types of fruits. However, all these materials are in the human food chain, and are usually too expensive to use for ethanol production but their waste residues can be used for alternative feedstock for the production of bioethanol (Arumugam et al. 2007; Huang et al. 2008).

6.2.18 Bioethanol from Starch

Another feedstock for bioethanol production is starch. It is made up of long chains of glucose molecules, which can be fragmented into simple sugars before fermentation to produce ethanol. Starch biomass feedstocks are including tubers like sweet potato, potato, cassava and cereal grains. These feedstocks are undergone hydrolysis to breakdown the starch into fermentable sugars i.e., saccharification. The hydrolysis process is consisted of mixing water with feedstocks to form slurry, the slurry then is heated to break cell walls, at this stage, different enzymes added to break chemical bonds present in the starch materials and complete the hydrolysis process (Erbaum 2009).

6.2.19 *Milling Processes*

For processing the cereals into ethanol, they should undergo milling either wet or dry.

Wet milling process includes the steeping the cereal like corn kernel in warm water, which helps to break down the proteins and release the starch present in the corn and helps to soften the kernel for the milling process. Then the resultant corn is milled to produce germ, fiber and starch products. The germ is extracted to result in corn oil; the starch fraction undergoes centrifugation and saccharification to produce gluten wet cake. Then ethanol is separated by distillation. The method of wet milling is used in factories producing several 100 million gallons of ethanol every year.

The dry milling process involves cleaning and breaking down the cereals kernel into fine particles using a hammer mill process. This creates a powder with a flour type consistency. The powder contains the cereals germ, starch and fiber. To gain a sugar solution, the mixture undergoes hydrolysis or breaks down into sucrose sugars using enzymes or dilute acid. After cooling the mixture, yeast would be added for its fermentation to ethanol. The dry milling process is used in facilities producing less than 50 million gallons of ethanol every year.

6.2.20 *Sugar Fermentation Process*

By employing hydrolysis process, the cellulosic part of biomass or corn would convert to into sugar solutions, and then the solution would be fermented into ethanol.

By addition of yeast to solution, it is heated. The yeast contains an enzyme called invertase, which acts as a catalyst and converts the sucrose sugars into glucose and fructose, which are both C6 sugars (both with chemical formula of $C_6H_{12}O_6$).

The chemical reactions are as following:

$$\underset{\text{Sucrose}}{C_{12}H_{22}O_{11}} + H_2O \xrightarrow{\text{Invertase}} \underset{\text{Fructose}}{C_6H_{12}O_6} + \underset{\text{Glucose}}{C_6H_{12}O_6}$$

After this stage, fructose and glucose react with an enzyme called zymase. This enzyme is contained in the yeast, as well and is responsible for completion of fermentation. The products of this reaction would be ethanol and carbon dioxide.

$$\underset{\text{Frucose/Glucose}}{C_6H_{12}O_6} \xrightarrow{\text{Zymase}} \underset{\text{Ethanol}}{C_2H_5OH} + 2CO_2$$

Table 6.4 Feedstocks and enzymes used for production of bioethanol. (Arumugam et al. 2007)

Carbohydrate feedstock	Main carbohydrate to be converted	Process utilizing added enzymes	Required enzymes
Sugar cane or molasses	Sucrose	Dextran hydrolysis	Dextranase
Grain and tubers	Starch (α-1,4 linked glucose molecules)	Starch, beta glucan and pentosan hydrolysis	Amylase, glucoamylase, betagulcanase, xylanase

At temperature between 250 and 300 °C, fermentation would take three days to complete. Table 6.4 lists different types of enzymes, which are used for various types of feedstocks for bioethanol production.

6.2.21 Fractional Distillation Process

The produced ethanol from fermentation process contains a significant amount of water that must be removed. Due to difference between boiling points of water (100 °C) and ethanol (78.3 °C), the separation could be achieved by fractional distillation.

6.2.22 Advantages of Bioethanol as Fuel

Employing bioethanol as fuel has some advantages in comparison to conventional fuels. It is originated from renewable resources and its usage could reduce greenhouse gases emission. Encouraging bioethanol use could boost the rural economy growing the necessary crops. It is also biodegradable and far less toxic that fossil fuels. In addition, using bioethanol in older engines can reduce the amount of carbon monoxide produced by the vehicle thus improving air quality. Another advantage of bioethanol is its ease of usage due to its integration to current transport fuel system. The blend of 5 % bioethanol with 95 % conventional fuel can be utilized in cars without need of engine modification. Also, bioethanol is produced using familiar methods, such as fermentation, and it can be distributed using the same petrol forecourts as before (Arumugam et al. 2007; Demirbas 2008).

6.2.23 Disadvantages of Bioethanol as Fuel

Though ethanol based fuels has many advantages over fossil fuels, it has some compability problems with current fuel system components. It can cause corrosion in ferrous-made components, deposit of jelly-like compounds on fuel strainer screens and salt deposits. Water content in ethanol should be less than 1.0% otherwise, phase separation occurs during blending with gasoline. In addition, presence of lesser amounts of water could lead ethanol to mix with either water or gasoline but not in both. Ethanol blended gasoline can also affect the electric fuel pumps by causing internal weariness and undesirable spark generation (Arumugam et al. 2007; Demirbas 2008).

6.2.24 Biogas

Biogas is the result and product of anaerobic digestion of disposals and animal wastes, which can be used a renewable source of energy. Farmers prefer to use the digesters more due to environmental issues rather than electrical and thermal energy generation potential. However, the newly built digesters can produce biomethane because of anaerobic digestion process. In developing rural areas, the lack of need to complicated infrastructures for anaerobic digestion makes this technology highly attractive.

Furthermore, employing anaerobic digestion can effectively reduce fecal coliform bacteria in manure by more than 99%, and can cause in eliminating a major source of water pollution. In addition, the potential of this method to produce and capture methane from the manure reduces the amount of methane that otherwise would enter the atmosphere. Methane is among greenhouse gases and its emission can accelerate global warming (Arumugam et al. 2007).

6.2.25 The Process of Anaerobic Digestion

Anaerobic digestion occurs in a sequence of stages and each stage involves distinct types of bacteria. At first, hydrolytic and fermentative bacteria break down the carbohydrates, proteins and fats present in biomass feedstock and convert them into fatty acids, alcohol, carbon dioxide, hydrogen, ammonia and sulphides. This stage is called hydrolysis.

Next, acidogenic (acid-forming) bacteria digest the products of hydrolysis stage more and change them into acetic acid, hydrogen and carbon dioxide.

At last, methanogen (methane-forming) bacteria convert these products into biogas.

After filtration and drying, digester gas can be used as fuel in an internal combustion engine. The engine combined with a generator, can produce electricity. Future applications of digester gas may include electric power production from gas turbines or fuel cells. Digester gas can substitute for natural gas or propane in space heaters, refrigeration equipment, cooking stoves or other equipment. Compressed digester gas can be used as an alternative transportation fuel.

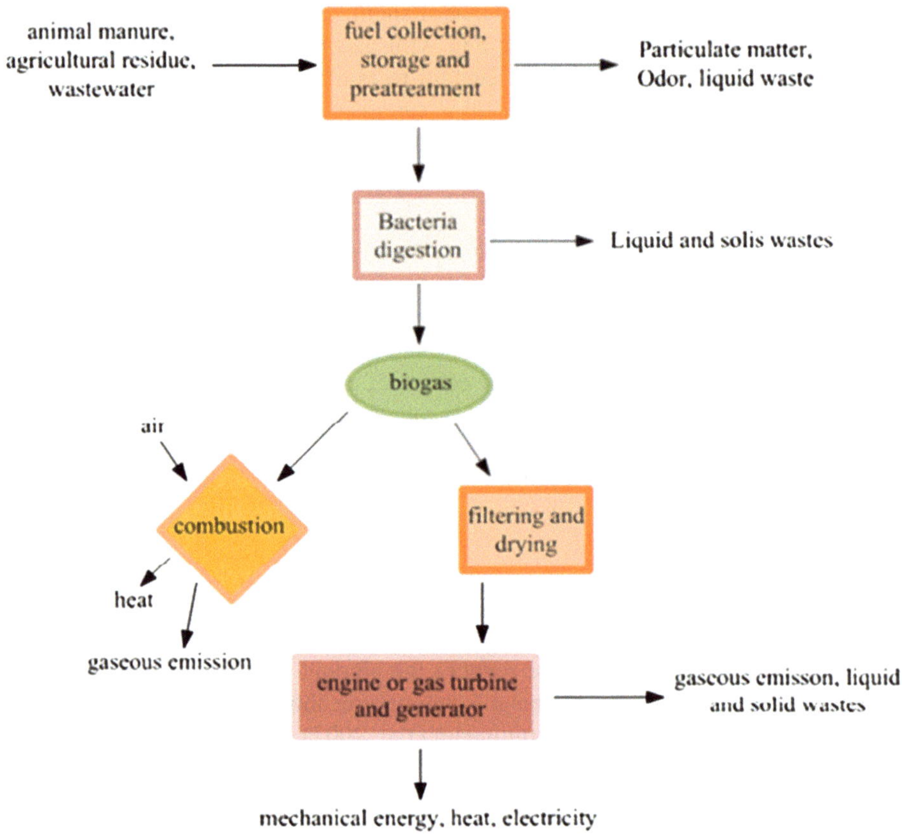

Fig. 6.7 Anaerobic digestion

6.2.26 Digester Technology

Biomass with high content of moisture content like animal manure and food-processing wastes is a suitable feedstock for biogas production using anaerobic digester technology. During anaerobic digestion, bacteria digest biomass in an oxygen-free environment. As described earlier, symbiotic groups of bacteria perform different functions at different stages of the digestion process. Figure 6.7 shows a glimpse of this process (Mudhoo 2012).

There are four basic types of microorganisms for anaerobic digestion. Hydrolytic bacteria break down complex organic wastes into sugars and amino acids. Fermentative bacteria convert those products into organic acids. Acidogenic microorganisms convert the acids into hydrogen, carbon dioxide and acetate. Finally, the methanogenic bacteria produce biogas from acetic acid, hydrogen and carbon dioxide.

To perform anaerobic digestion under controlled conditions, an airtight chamber is required, which is called a digester. Maintaining a temperature of at least 18 °C in digester is necessary for promoting bacterial activity. Higher temperatures, up to 65 °C, shorten processing time and reduce the required volume of the tank by 25–40 %. However, increasing temperature would lead to include a limited range of bacterial species (thermophilic bacteria) in the digestion. Most of anaerobic bacteria species prefer the temperature range of a standard design (mesophilic bacteria).

The produced biogas in a digester (also known as "digester gas") is actually a mixture of gases, with methane and carbon dioxide making up more than 90 % of the total. Small amounts of other gases such as hydrogen sulphide, nitrogen, hydrogen, methylmercaptan and oxygen can be present in biogas.

For individual farms, small-scale plug-flow or covered lagoon digesters of simple design can produce biogas for on-site electricity and heat generation. For example, a plug-flow digester could process 8000 gallons of manure every day, the amount produced by a herd of 500 dairy cows. By using digester gas to fuel an engine-generator, a digester of this size would produce more electricity and hot water than the dairy consumes.

Larger scale digesters are suitable for manure volumes of 25,000–100,000 gallons per day. In Denmark and in several other European countries, central digester facilities use manure and other organic wastes collected from individual farms and transported to the facility (Abbasi 2011).

6.2.27 Types of Anaerobic Digesters

For anaerobic digestion, there are three basic digester designs. All of them are common in the specification of trapping methane and reducing fecal coli form bacteria, but they are different in cost, climate suitability and the concentration of manure solids they can digest. They are covered lagoons, complete mix and plug-flow digesters.

In a covered lagoon digester, there is a storage lagoon, which would be filled with manure and then, the lagoon would be covered. The cover acts as a trap and keeps gas produced during decomposition of the manure. This kind of digester is the cheapest type of the three. The covering lagoon technology is suitable for liquid manure with less than 3 % solids. The used cover in this kind of digesters is an impermeable floating cover of industrial fabric and it covers all parts of the lagoon. The airtight sealed cover is kept in place by a concrete footing along the edge of the lagoon. The methane gas that produced in the lagoon would be exhausted to a suction pipe. Though covered lagoon is a rather cheap design, it needs large volume of lagoon and a warm climate. It is not suitable for cool climates or locations where a high water table exists.

Complete mix digester is another type of anaerobic digester. In a heated tank above or underground, suspended solids of organic wastes would convert to biogas. The tank must be kept mixing by mechanical or as mixers. This kind of digester is the most expensive type with running costs greater than plug-flow digesters. Complete mix digesters can handle large volumes of manure with solid content of 3–10 %. The design of reactor is in shape of circular container made of steel or poured concrete. By mixing the manure slurry during the digestion, the solids are kept in suspension. Biogas eventually accumulates at the top of the reactor, which can be used for running an engine generator for electricity production or a boiler for steam production. The heat from burning biogas can be employed for digestion chamber heating up, which in turn would lead to reduction in slurry retention time to less than 20 days.

Plug-flow digester is the last type of anaerobic digesters, which are appropriate for digestion the manure with a solid concentration of 11–13 %. The plug-flow system is generally composed of a manure collection system, a mixing pit and the digester itself. By water addition, the solid concentration inside the digester would be adjusted to optimal amounts. The shape of digester is a long, rectangular container with an expandable airtight cover. By addition new raw materials to a tank at one end of the plug-flow reactor, old manure inside the digester would push to the opposite end. By completion the digestion process, the remnants of manure form a vicious material inside the plug and this causes some limitations in solid operation inside the digester. So the materials flow through the plug during a period of 20–30 days. The resulted biogas from digestion would release by material flow through the plug. Biogas would be collected at the ends of digester and pipes beneath the covers would carry it to an engine generator. The important point about the plug flow digesters is the requirement for minimal maintenance. The heat of engine generator would be carried inside the plug through heating pipes with running hot water inside to keep the slurry at 25– 40 °C, which is the suitable temperature for anaerobic digestion (Arumugam et al. 2007).

6.2.28 Biogas from Wastes

Many municipal sewage and wastewater treatment plants use anaerobic digestion as a method for reduction in volume of organic biomass solids. Employing this process leads to stablizaton of sewage sludge and destruction of pathogen organisms. The result of such activity is a biogas containing 60–70 % methane with an energy content of about 600 Btu per cubic foot.

In wastewater treatment plants, which are using anaerobic digestion, the produced biogas would be burnt to for maintaining the digester temperature. The excess amount of biogas would be burned off as waste but could be produces electricity in an engine generator. Prior to use, biogas should be cleaned to remove impurities like hydrogen sulphide, halogens (fluorine, chlorine and bromine), moisture, bacteria and solids (Arumugam et al. 2007).

6.2.29 Landfill Gas

Decomposition of cellulose solid waste could result in production of biogas. This kind of digestion often occurs in landfills and that is an uncontrolled decay process. Waste composition, moisture content and temperature maintained in the landfill are the parameters that controlled the efficiency of this process. The released biogas from landfills is called "landfill gas". It is mainly composed of 50 % methane, 45 % carbon dioxide and 5 % other gases. The landfill gas could contain energy of 14.90–20.50 MJ m^{-3}.

To use this source of energy, it is necessary to capture and collect the landfill gas prior to its escape to atmosphere. However, the landfill should have some specifications that gas collection could be feasible for power production. A suitable landfill for such an activity should be at least 40 ft deep and contain at least 1 million tons of waste. The landfill gas capturing would not only produce energy but also can lead in reduction of greenhouse gas emissions (Arumugam et al. 2007).

6.2.30 Biogas Upgrading to Biomethane

In order to feed the produced biogas to gas grid, it is necessary to undergone the raw biogas through two major processes, cleaning and upgrading, to obtain natural gas quality. The cleaning stage includes processes for removal and purification the biogas from trace components such as hydrogen sulphide, ammonia and water. The upgrading stage consists of separation processes for removal of carbon dioxide from biogas and increasing the amount of methane in it. The heating value, Wobbe index (WI) and other parameters are highly depending on the methane content. By employing these processes, methane content would be adjusted to pipeline specifications by removal the carbon dioxide and cleaning trace impurities. At the end, it is necessary for biogas to be odorized prior to injection in natural gas grid due to safety reasons.

6.3 Second Generation Biofuels

Second generation biofuels are seem to be superior to many of first generation biofuels in many terms like energy balance, greenhouse gases emission reduction, land use requirement, and competition with food market demands in terms of water, land and food crops. However, this kind of biofuels with all these known advantages in comparison with first generation biofuels, have a very little portion of total produced biofuels. The main problem is that the necessary conversion technologies are not developed completely and thus, second generation biofuels have not seen commercial usage like first generation ones. To employ them commercially, it is required to find and maintain more reliable technologies to produce biofuels from second generation feedstocks with lesser end-price to be comparable with other biofuels. In addition, it is a need to conduct further researches on terms of required land use, co-products, water and energy needed for processing the biofuels.

It is known that second generation biofuels currently are responsible for less than 0.1 % of total bioethanol production in the world. For second generation biofuels, there are only just some pilot plants working and some demonstration plants under development or planned.

6.3.1 Feedstocks

Second generation biofuel incorporates renewable non-edible sources which are resulted from crops, forest and wood processing facilities especially the lignocellulosic feedstocks. Lignocellulose is the term is being used in botanical science for defining the biomass from woody and fibrous plant materials. It is composed of lignin, cellulose and hemicellulose with variation in composition of each component according to different type of plant. Though this variation with different plant types, a total of 50–75 % of total dry mass is cellulose and hemicellulose and the rest is composed of lignin.

Cellulose is a polysaccharide consisting of D-glucose units connected through β (1 → 4) glycoside bondage with formula of $C_6H_{10}O_5$.

Hemicellulose is a heterogeneous material composed of cellulose and other compounds like arabinoxylan (a copolymer of xylose and arabinose). It is almost present in all plant cell walls. In contrast to cellulose which only contains anhydrous glucose units, hemicellulose includes different polysaccharides include xylan, glucuronoxylan, arabinoxylan, glucomannan, and xyloglucan. Thus hemicellulose has other monomers like xylose, mannose, galactose, rhamnose, and arabinose along with glucose. Although cellulose is resistant to hydrolysis, hemicellulose due to presence of other monomers is easily hydrolyzed by dilute acid or base.

Lignin is a complicated chemical compound which is derived from wood and acts as an integral component of secondary cell wall of plant cells. It contains some phenolic compounds which would act as a barrier and inhibitor to hydrolysis or fermentation process.

Lignocellulosic biomass is among renewable feedstocks, which considers as an abundant material without any specific usage. Its annual worldwide production stands at 10–50 billion dry tones, but a very small portion of could be utilized. This enormous feedstock includes cereal straw, wheat chaff, rice husk, corn cob, corn stover, sugarcane bagasse, nut shell, forest harvest residue and wood process residue.

There are many challenges for incorporation of lignocellulosic feedstocks. Among them is the variety in the composition of feedstock due to variation in the origin plant. This variation along with the differentiation in behavior or three major components—cellulose, hemicellulose and lignin—towards bacterial breakdown and enzymatic activation makes whole thing more complicated (Arumugam et al. 2007).

6.3.2 Synthesis Methods

The known conversion pathways for employing lignocellulosic biomass for biofuels productions are categorized in two main ways:

1. Hydrolysis of lignocellulosic biomass to form sugars, which can be used for fermentation to alcohol. This method is known as 'cellulosic bioethanol'. The hydrolysis could be undergone chemically or biochemically.

2. Thermochemical process of lignocellulose to a raw gas or oil. The thermochemical pathway includes employing of high temperature to pyrolysis or gasifies biomass. The resulted raw gas is then treated and converted to syngas, which is comprised of carbon monoxide and hydrogen. By different fuel synthesis routes, this gas can be processed into different types of liquid and gaseous fuels. The resulting fuels of these pathways are called 'synthetic biofuels'. Among liquid synthetic fuels are BtL (biomass-to-liquid), biomethanol and Fischer-Tropsch fuels. Examples of gaseous synthetic biofuels are dimethyl ether (DME) and Bio-SNG (a form of biomethane which is similar to biogas and can be used as natural gas). Also, the product gas can be converted to hydrogen. Bio-oil which would be obtained from pyrolysis can be treated via deoxygenation method to high quality liquid oil (like HTU diesel).

6.3.3 *Pyrolysis and Gasification of Biomass*

Employing pyrolysis on biomass can lead is a method has being interested for the past two decades. Pyrolysis has the advantage of variety of different potential resulted products over gasification. They can be a range from transportation liquid fuel to chemical feedstocks.

Pyrolysis is a reaction occurs in total or partial absence of oxygen under temperatures from 400–650 °C. By adjusting the operational parameters like temperature and pressure of the reactor, residence time of different phases in the reactor, time and rate of heating the biomass, reactor internal shape and initial conditions of biomass, different portions of gases, liquids and solids would be prepared in the reactor.

One of the most recently developed technologies for biomass conversion to maximum bio-oil is known as fast pyrolysis. It is based on the short heating time of materials in the reactor along with high heating rates and mass transfer coefficients; and moderate temperature from heat source. Residence for vapor phase should be kept less than 1 min. The occurrence of decomposition reactions included isomerization, cracking (split), and recombination by condensation, polymerization, depolymerisation and fragmentation of biomass, would lead in the production of bio-oil. It has high content of water in its composition.

Another technology similar to pyrolysis is gasification. It results in production of high quality bio-syngas with higher yield. It should be carried out with high temperature and in presence of low oxygen or air and steam, at last. During gasification, the biomass is heated and it undergoes a combination of drying, pyrolysis, oxidation, and reduction processes (see Fig. 6.8).

6.3.4 *Cellulosic Bioethanol*

Employing the lignocellulosic feedstocks for ethanol production seems promising since these feedstocks do not interfere with human food chain and also, they are considered as abundant waste without specific usage.

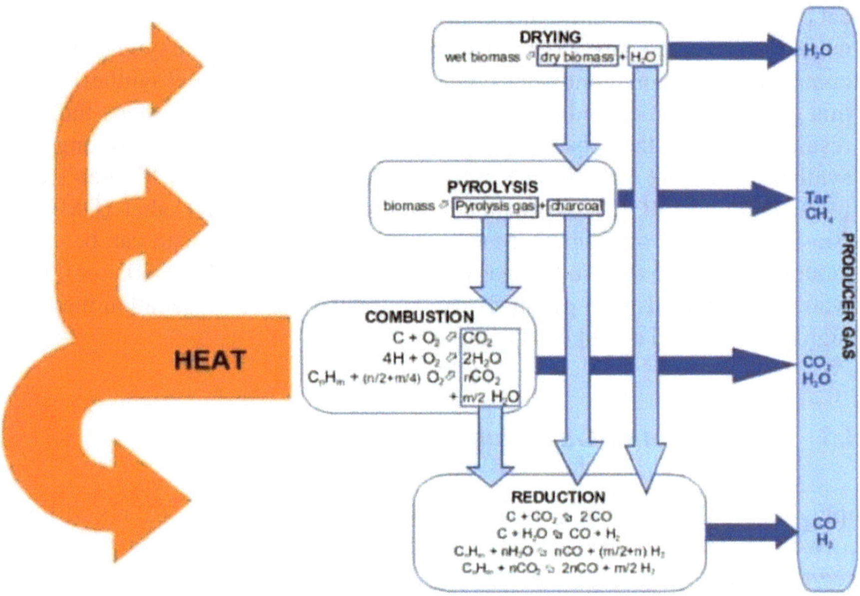

Fig. 6.8 Schematic representation of biomass gasification. (Arumugam et al. 2007)

Cellulose like starch comprises long chains of glucose with different structural configuration. In the lignocellulose, this structural different compound has been encapsulated lignin, so cellulosic would be more difficult than hydrolyze than starch.

Along with cellulose and lignin, the presence of hemicellulose makes the matter worst, because the exact composition of hemicellulose itself and also, the composition of all components of lignocellulosic feedstocks differ deeply from one plant to another.

6.3.5 *Process Description*

For manufacturing ethanol form cellulosic biomass feedstocks, the feedstock should be first treated with acidic or enzymatic hydrolysis and then by thermo chemical treatment. Usually the acid hydrolysis is accomplished with sulfuric acid, due to it is being less expensive. In Fig. 6.9 different steps of biomass conversion to bioethanol has been illustrated.

6.3.6 *Bioethanol Production from Hydrolyzed Cellulosic Biomass*

After hydrolysis of cellulosic biomass, the fermentation process could be performed with different enzymes according to the feedstock of sugar moieties. In Table 3.2,

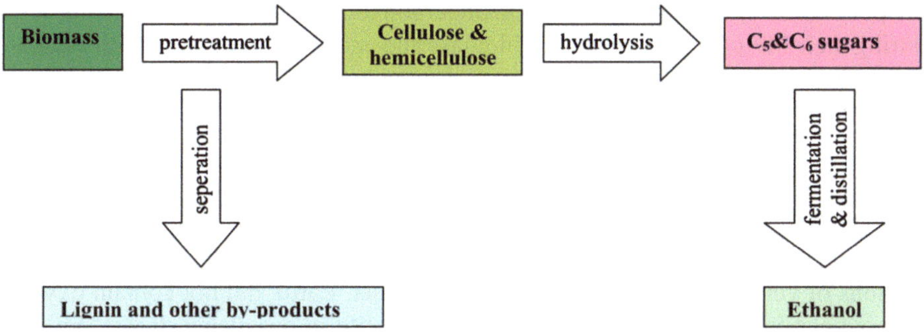

Fig. 6.9 Schematic representation of production of ethanol from cellulosic biomass. (Arumugam et al. 2007)

a list of these enzymes is presented. Conversion of cellulose and hemicellulose could be performed by cellulases and hemicellulases enzymes, respectively.

6.4 Third Generation biofuels

Although first and second generation biofuels show some promising aspects of being renewable energy sources which have potential to be employed instead of petroleum fuels, there are some problems and concerns regarding their widespread usage. The most important concern lies on their ability to act as a sustainable feedstock for a reliable energy source. For the first generation biofuels, the problem is their production capability and their interference with human food chain. As production capacities for first generation biofuels increases to meet the growing demand of energy, so does their competition with agriculture industry for arable land, which used to farm food crops. Such a thing would lead to a renewed and more pressure on the agriculture industry and arable lands, which currently have some difficulties to meet the increasing requirement for supplying food demands of human beings. The insisting for such a thing may lead to severe food shortages in market and increase the price of food crops. In addition, the intensive usage of land with high fertilizer and pesticide applications can cause significant environmental problems (Schenk et al. 2008). About second-generation biofuels, conversion of lignocellulosic biomass into fermentable sugars or synthetic fuels requires costly technologies that have not developed to produce economically feasible fuels on a large scale (Brennan and Owende 2010).

With such unanswered questions and so many concerns, third generation biofuels derived from microalgae, have been considered to be a viable alternative energy source. Microalgae are photosynthetic microorganisms are able to produce lipids, proteins and carbohydrates inside their organs and cultivate in large scales over short periods. These useful compounds can be processed into biofuels and other desired chemical products (Brennan and Owende 2010; Demirbas 2011). Also, due to the potential of microalgae for fixation of carbon dioxide by photosynthesis process during their life-cycle, the produced biomass and energy derived from their biomass would have no negative effect in terms of greenhouse gases emission (de Morais and Costa 2007; Wang et al. 2008).

6.4.1 *Microalgae*

Billions of years ago, the atmosphere of our planet, Earth, did not contain oxygen. The oxygen is an essential piece for developing life in our planet. It was believed that some Cyanobacteria, prokaryote cells with photosynthesis ability, has triggered for the start of life with conversion of carbon dioxide to oxygen. By their continuous activity with photosynthesis in Earth's atmosphere, the levels of carbon dioxide started to decrease and at the same time, oxygen levels raised and the new balance of atmosphere was a great step to start the life that we are known (Demirbas 2011). Microalgae is being defined as any living microorganisms with ability to photosynthesis and without other organs of a plant like leaf or root.

The microalgae are in two distinct categories:

- Prokaryote cells without nucleus; in this category only blue-green algae or cyanobacteria are present.
- Eukaryote cells with nucleus in their structure; in this category, other groups of microalgae are being categorized like green, brown, yellow, red and brown algae.

Due to their small size and their being float in water sources, these organisms can grow in high rate. It is estimated that the biomass productivity of microalgae could be 50 times more than that of switch grass, which is the fastest growing terrestrial plant. Their photosynthetic mechanism is similar to land-based plants, but being afloat in watery environments like oceans, along with their small size and having sufficient access to carbon dioxide and other nutrients make them to grow at very high rates with high efficiency in photosynthesis process for converting solar energy to biomass (Chisti 2007; Gouveia 2011).

From the cultivation aspect, microalgae can grow under either autotrophic or heterotrophic circumstances. Under autotrophic growth condition, microalgae acquire needed inorganic compounds as carbon source. Autotrophic condition can be either photoautotrophic, using light as a source of energy, or chemoautotrophic, oxidizing inorganic compounds for energy. Heterotrophic growth condition incorporates microalgae would use organic compounds for growth. Heterotrophs can be photoheterotrophs, using light as a source of energy, or chemoheterotrophs, oxidizing organic compounds for energy. Also, photosynthetic microalgae can be grown under mixotrophic condition, combining heterotrophy and autotrophy by photosynthesis (Mata et al. 2010).

After microalgae cells being cultivated, they grow with doing photosynthesis and some compounds are being accumulated inside their cells. In Table 6.5, there is a glimpse in overall composition of some of known microalgae species. In terms of

Table 6.5 Biomass composition of microalgae expressed on a dry matter basis (%). (Pimentel 2008)

Algal species	Lipid	Protein	Carbohydrate	Nucleic acid
Scenedesmus obliquus	12–14	50–56	10–17	3–6
Chlamydomonas rheinhardii	21	48	17	–
Chlorella vulgaris	14–22	51–58	12–17	4–5
Spirulina platensis	4–9	46–63	8–14	2–15
Euglena gracilis	14–20	36–61	14–18	–
Dunaliella salina	6	57	32	–
Prymnesium parvum	22–38	28–45	25–33	1–2
Synechoccus sp.	11	63	15	5

Table 6.6 Oil yield, land use an biodiesel productivity of microalgae and seed plants. (Chisti 2007; Mata et al. 2010; Abou Kheira and Atta 2009)

Source	Oil content (% oil by wt. in biomass)	Oil yield (l oil/ha year)	Land use (m^2year/kg biodiesel)	Biodiesel productivity (kg biodiesel/ha year)
Jatropha	28	741	15	656
Sunflower	40	1070	11	946
Corn/Maize	44	17	66	152
Palm oil	36	5366	2	4747
Microalgae (low oil content)	30	58,700	0.2	51,927
Microalgae (high oil content)	70	136,900	0.1	121,104

use microalgae as biofuel source, the important point is the ability of oil accumulation in microalgae cells.

The oil productivity of some microalgae cells is greater than conventional oilseeds. For instance, some microalgae species can accumulate oil in amounts of 250 times greater than soybean per acre. With such high productivity, here lies the more important feature of microalgae which makes them somewhat superior to their rivals in agricultural plants; no need for arable land and freshwater for microalgae cultivation. So there is some species with ability to grow in non-arable lands without need to freshwater, both are bottle-necks and essential requirements for growing agricultural plants, along with a whole new definition of growth with very rapid growth cycle and ability to growth with less nutrient in comparison to plants. These are microalgae, the key for solving the energy demand of human being (Chisti 2007; Mata et al. 2010). Also, for the production of 1 l of biofuel from fuel crops, approximately 10,000 l of water are needed. Microalgae need much less water. For photosynthesis alone, ~0.75 l of water is needed per kg of biomass produced. Per liter of biofuel, assuming a lipid content of 50%, 1.5 l of water are required (Wijffels and Barbosa 2010).

Another important point regarding microalgae is their oil content. The oil content in microalgae cells is dependent to some factors: microalgae strain, concentration, cell age and growth conditions. In Table 6.6, there is a comparison between general oil content of microalgae cells and some conventional oilseeds. Although,

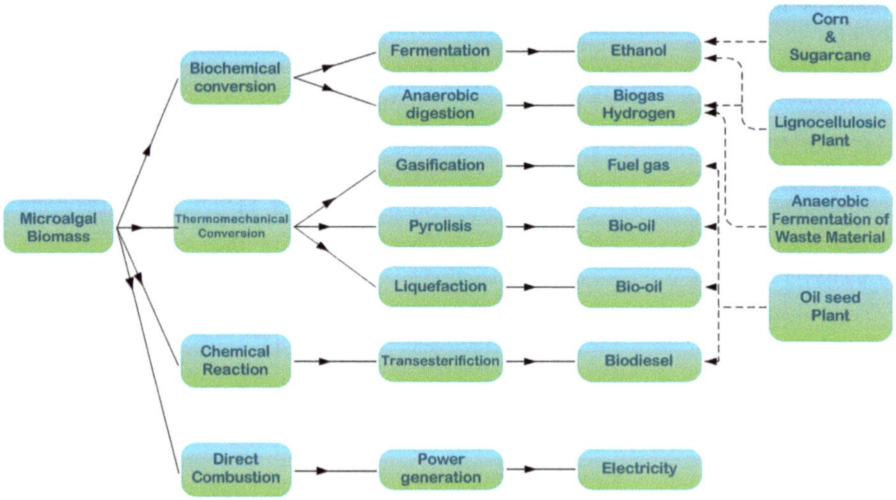

Fig. 6.10 Different renewable biofuels which can be derived from microalgae and other generation biofuel sources. (Wang et al. 2008; Jones and Mayfield 2011)

the oil contents are similar between oilseeds and microalgae, due to the difference in microalgae productivity and required land use, there is a huge difference in biodiesel productivity of microalgae and conventional plants. Here it can be concluded that microalgae is much better choice as a sustainable feedstock for biofuel production in comparison to conventional plants.

The oil content in microalgae is not only related to the kind of specie but also to the growth conditions. As described earlier, microalgae are photosynthetic organisms with requirement of being under light (artificial or sun light). For maximizing the oil content, one conventional method is to maximize biomass production for specific microalgae by providing its optimum culturing conditions like temperature, culture medium composition, aeration rate and an appropriate design to have uniform light distribution in the medium (Chisti 2007; Mata et al. 2010).

6.4.2 *Biofuel from Microalgae*

Biodiesel, bioethanol, biogas and biohydrogen are common energy sources that could be produced from microalgae biomass. Photosynthetic process in microalgae by synthesis of electron and proton produces biohydrogen. Production of sugar and starch in photosynthesis process are the main source for bioethanol production. Also, Produced oil is main source for biodiesel production and residue biomass can be used for biogas production through anaerobic digestion. Biochemical, thermochemical, chemical and direct combustion processes are usual method for production of energy from microalgae (Wang et al. 2008; Jones and Mayfield 2011). Figure 6.10 shows a glimpse in microalgae-derived different biofuels and

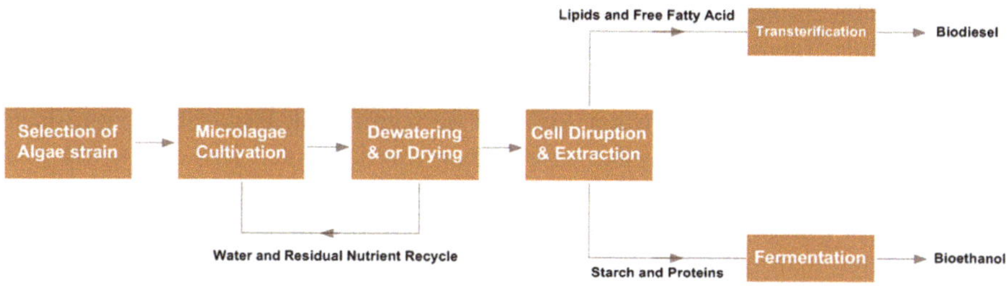

Fig. 6.11 Main process for biofuel production from microalgae. (Mata et al. 2010; Lam and Lee 2012)

the method for their production along with the similar renewable sources for those biofuels from other generation feedstocks.

6.4.3 Microalgae Culturing Steps

The main steps for biofuel production consist of microalgae cultivation, cell separation from culture medium, lipid extraction and biodiesel production through trans-esterification reaction. Figure 6.11 represents the whole of this process.

6.4.4 Microalgae Cultivation Systems

Microalgae cultivation in large scale is a major step towards large-scale biofuel. For this, the bioreactor for microalgae cultivation, which is called photobioreactor, should be optimized to be able to produce vast amounts of biomass. To design a commercial plant for microalgae cultivation, the following points should be considered (Gouveia 2011):

- Growing rate, normally measured by total amount of biomass accumulated per unit time and unit volume;
- Nutrients availability, in particular carbon dioxide source, when the goal of carbon sequestration is also deemed relevant;
- Robustness and resistance to environmental conditions changes, such as nutrients, light, temperature and contamination from other microorganisms;
- Biomass harvesting and downstream processing technique.

6.4.5 Microalgae Large-scale Cultivation

The methods for large-scale cultivation of microalgae are open pond and closed photobioreactor. Open ponds for mass culture of microalgae have been used since the 1950s. The classical open cultivation systems comprise lakes and natural ponds, circular ponds, raceway ponds and inclined systems. In open system, any cooling is achieved only by evaporation. Temperature fluctuates within a diurnal cycle and seasonally. Evaporative water loss can be significant. Because of significant losses to atmosphere, open system use carbon dioxide much less efficiently than photobioreactors. Productivity is affected by contamination with unwanted other microalgae or microorganisms that feed on algae. Open ponds are perceived to be less expensive than photobioreactors, because they cost less to build and operate. Although open pond are low-cost, they have a low biomass productivity compared to photobioreactors (Chisti 2007).

Unlike open systems, photobioreactors permit essentially single-species culture of microalgae for prolonged durations. Many different designs have been developed, but there are mainly categorized in three different designs: (1) tubular (e.g. helical, manifold, serpentine, and shaped); (2) flat plates and (3) column (e.g. bubble columns and airlift). Much research has been carried out in order to optimize different PBR systems for microalgae cultivation (Chisti 2007; Chaumont 1993; Janssen et al. 2003; Carvalho et al. 2006; Tredici 2007; Kunjapur and Eldridge 2010). Comparison of different techniques for microalgae mass cultivation is presented in Table 6.7.

It has been reported biomass productivity rate in both open ponds and closed photobioreactors are in the range of 20–30 gm^{-2} Day^{-1}, but it has observed amounts of peak productivities of 50 g m^{-2} day^{-1} in both open ponds and also, in natural algal blooms (Field 1998; Lee 2001). In a closed photobioreactor and due to higher surface-to-volume ratio, higher biomass productivity rate can be achieved in a closed photobioreactor than in comparison with an open pond. However in large scale production, both open pond and closed photobioreactors have limitations because of limited amount of incident sunlight on the Earth's surface and cannot exceed a maximum biomass productivity rate of 100 gm^{-2} day^{-1} (Schenk et al. 2008).

6.4.6 *Harvesting*

The harvesting process is energy dependent and a major bottleneck towards the industrial scale processing of microalgae for biofuel production. The cost of biomass recovery from the broth can make up to 20–30 % of the total cost of producing the biomass.

Most common harvesting methods include gravity sedimentation, centrifugation, filtration and microscreening, ultrafiltration, flotation, sometimes with an additional flocculation step or with a combination of flocculation–flotation, and electrophoresis techniques. The selection of harvesting technique is dependent on

Table 6.7 Comparison of different techniques for microalgae mass cultivation. (Chisti 2007; Chaumont 1993; Janssen et al. 2003; Carvalho et al. 2006; Tredici 2007; Kunjapur and Eldridge 2010)

Culturing system	Advantageous	Limitation
Open system	Easy to clean up	Little control of culture conditions
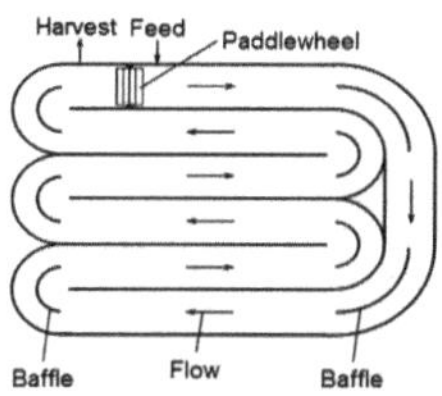	Low maintenance cost	Poor mixing, light and CO_2 utilization
	Utilization of nonagricultural land	Poor productivity
	Low energy consumption	Limited to few strains
	Low fixed cost	Cultures are easily contaminated
Tubular PBR	Large illumination surface area	Oxygen accumulation
	Suitable for outdoor cultures	Fouling
	Relatively cheep	Some degree of wall growth
	High volumetric biomass density	Most land use
	Rapid cycling of fluid between light and dark zone	Photoinhibition
Flat PBR	Relatively cheap	Low photosynthetic efficiency
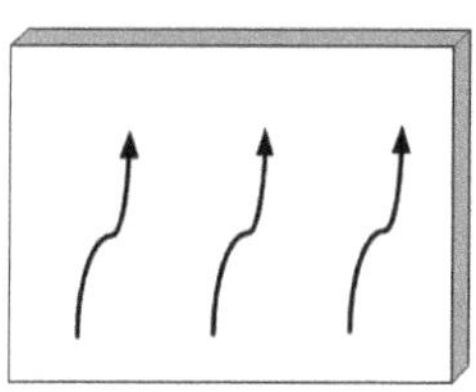	Easy to clean up	Shear damage from aeration
	Large illumination surface area	Difficult scale up
	Good light penetration	Difficult temperature control
	Suitable for outdoor cultures	Some degree of wall growth
	Low power consumption	
	High volumetric biomass density	
	Low oxygen build up	
	Shortest oxygen path	
Column PBR	Low energy consumption	Small illumination surface area
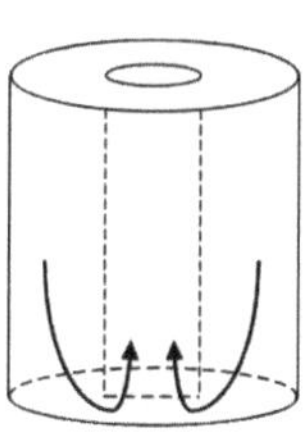	Good mixing	Shear stress to algal cultures
	Best exposure to light–dark cycles	Decrease of illumination surface area
	Low shear stress	Expensive compared to open ponds
	Reduced photoinhibition	Support costs
	Reduced photooxidation	Modest scalability
	High photosynthetic efficiency	

Table 6.8 Comparison between some harvesting methods. (Chisti 2007; Molina Grima 2003; Harun et al. 2010a)

Harvesting method	Advantageous	Limitation
Gravity sedimentation	Simple process	Very slow
	Good result when combined with flocculation	Unusable for low-density microalgal particle
	Suitable for low value product	
Centrifugation	Easily sterilized	Time consuming
	Very useful secondary harvesting method[a]	Energy intensive and costly
Flocculation	Small dosages of cationic flocculants lead to neutralization of microalgae surface charge and lead them to form flocs for ease of separation from culture medium	Employing chemical compounds for flocculation may lead to addition of toxic chemical compounds in harvested microalgae.
		The mount of chemical flocculant may not be economical especially for marine microalgae
Filtration	Cost-effective for filamentous or large colonial of microalgae	Membrane clogging
	Simplicity in function and construction	Formation of compressible filter cake
	Ease of operation	High maintenance cost

[a] Concentrate an initial slurry (10–20 g/l) to an algal paste (100–200 g/l)

the properties of microalgae, such as density, size and value of the desired products. Table 6.8 represents different methods for harvesting process.

6.4.7 Cell Disruption and Extraction

Cell disruption is necessary for extraction intracellular products from microalgae, such as oil and starch for biodiesel and bioethanol production. It is performed along with chemical solvent extraction for the recovery the lipid effectively.

Cell disruption methods that have been used include bead milling, freezing, alkali and organic solvents, osmotic shocks, sonication, microwave and high-pressure homogenization. Some of the cell disruptive methods such as ultra-sonication and microwave posed several safety and health hazards and need to be addressed before up-scaling to commercial stage. Between these methods, bead milling has been used for years to disrupt microorganisms. When compared with high-pressure methods of cell disruption wet bead milling is low in shearing forces. There are more bead mills types, such as shaking and rotor types, however, with less application on the microalgae cell disruption (Brennan and Owende 2010; Lam and Lee

Table 6.9 Energy efficiency in extracting lipid from *Botryococcus* sp. through various cells disruptive methods. (Lam and Lee 2012)

Cell disruptive technique	Lipid extracted (% dry microalgae biomass)	Total energy in microalgae biodiesel (kJ)	Additional energy consumed due to cell disruptive process (kJ)	Net energy (kJ)
Non-disruptive	8	1.5	0.00	1.51
Autoclave	11	1.89	4.28	−2.39
Bead-beater	28	5.29	1.58	3.71
Microwave	28.5	5.39	2.09	3.30
Ultrasonication	8.5	1.61	5.63	−4.02

2012; Mendes-Pinto et al. 2001). Energy efficiency in extracting lipid from a type of microalgae through various cells disruptive methods is presented in Table 6.9.

6.4.8 *Microalgae Lipid Extraction*

Numerous methods for extraction of lipids after microalgae cell disruption have been applied like expeller/oil press, liquid–liquid extraction (solvent extraction), supercritical fluid extraction (SFE) and ultrasound technique. In solvent extraction, organic solvents, such as benzene, hexane, cyclohexane, acetone, chloroform are added to algae paste, which is resulted of previous harvesting techniques like centrifuge or filtration.

After extraction of oil from microalgae paste, solvent extract can then be subjected to distillation process to separate oil from solvent (Harun et al. 2010a). In supercritical extraction method high temperature and pressure used to rapture the cell. This method has proved to be extremely time-efficient and is commonly employed. Also, ultrasound method is already in extensive use at laboratory scale but there is not sufficient information on its feasibility and cost for a commercial-scale operation. Though, this approach seems to have a high potential, but more research is needed about it (Harun et al. 2010a). Generally, processes built upon dry biomass are unlikely to be economical due to the energy inputs involved, and so methods that work with algal slurries or wet paste are preferred (Hejazi and Wijffels 2004).

6.4.9 *Biodiesel and Bioethanol Production*

Many species of microalgae produce large amounts of lipids as much as 50–60 % of their dry weight. The lipid contain fatty acid and triglyceride (TAG) compounds, which like their terrestrial seed-oil counterparts, oil can be converted into biodiesel through transesterification process. The more on transesterification process was discussed earlier in first generation biofuels of the current chapter.

Microalgae based biodiesel has not economically feasible in comparison with petroleum diesel with former costs $ 1.25 per pound and latter costs only $ 0.43 (Jones and Mayfield 2012).

Other than oil, microalgae provide carbohydrates (in the form of glucose, starch and other polysaccharides) and proteins that can be used as carbon sources for fermentation by bacteria, yeast or fungi (Harun et al. 2010b). For instance, *Chlorella vulgaris* has been considered as a potential raw material for bioethanol production because it can accumulate high levels of starch. *Chlorococum sp.* was also used as a substrate for bioethanol production under different fermentation conditions. Results showed a maximum bioethanol concentration of 3.83 g/l obtained from 10 g/l of lipid-extracted microalgae debris (Harun et al. 2010b).

Production of ethanol by using microalgae can be performed according to the following procedure. First, the starch content of microalgae is released from the cells with the aid of mechanical equipment or an enzyme. When the cells start to degrade, *Saccharomyces cerevisiae* yeast is added to the biomass for beginning of fermentation. The product of fermentation is ethanol. The ethanol is drained from the tank and pumped to a holding tank to be fed to a distillation unit. Fermentation process requires less consumption of energy and simplified process compared to biodiesel production system. Besides, carbon dioxide produced as by-product from fermentation process can be recycled as carbon sources to microalgae in cultivation process thus reduce the greenhouse gases emissions. However, the production of bioethanol from microalgae is still under investigation and this technology has not yet been fully commercialized (Harun et al. 2010b; John et al. 2011).

6.4.10 Further Research Perspective

To have an economically feasible process for production of microalgae based biofuel, there are some issues that have to be solved completely, especially in the field of microalgae biomass recovery.

Currently, there are significant problems about microalgae growth stage. There is a need for a constant amount of microalgae with almost same properties like concentration, age and cell wall properties after every growth cycle so a production line for biodiesel production can work successfully. Another problem which is dependent to the growth stage is to obtain a suitable method for dewatering and harvesting microalgae for biomass recovery. There are many reports which have suggested flocculation process as an appropriate harvesting method but even this process widely needs optimization in the term of needed amount of flocculants and amount of produced biomass. Other major bottle-neck is choosing the best means of lipid extraction from microalgae harvested cells. After all these steps being fully researched and understood, there is a further need to design a biomass recovery line to let all these units work together in a feasible method to regain the most appropriate efficiency (Jones and Mayfield 2012; Pienkos and Darzins 2009).

At last, it must always be in mind that industrial production of biofuel from microalgae would be largely depend on the interdisciplinary research and collaboration between engineers, biologists and chemists so all sections of growth, harvesting, and processing of these microorganisms could be efficiently done and with such a cooperation, a real sustainable platform for bioenergy could be gained.

6.5 Conclusion

As the mankind goes every day to his life, the amount of fossil fuels decreases and volume of the products of their burning increases in the atmosphere. The greenhouse effect and warming of Earth has made this inevitable that we must look for a better and cleaner source of energy.

Biofuels have the same unique property of fossil fuels, the transportability and their development could be a way to get out of this situation. We have discussed regarding three generations of biofuels, according to the kind of feedstocks of biofuel.

In short term we can call all of them photosynthetic: first generation with using agricultural products as source, second generation with lignocellulosic materials as feedstock and third generation microalgae biofuel.

The major blow to the first and second generation is the need of them to agricultural lands and water sources. Both of them will compete with our food sources and although nowadays, they are being used in some parts of the world, their further development is restricted due to increase in world population and global food demand. In recent years, most research on oil extraction is focused on microalgae to produce economically biodiesel from algal-oil. Algal-oil processes into biodiesel as easily as oil derived from land-based crops. It seems that algae biomass can play an important role in solving the problem between the production of food and that of biofuels in the near future.

References

Abbasi T (2011) Biogas energy. Springer, Berlin

Abdullah AZ et al (2009) Current status and policies on biodiesel industry in Malaysia as the world's leading producer of palm oil. Energy Policy 37(12):5440–5448

Abou Kheira AA, Atta NMM (2009) Response of *Jatropha curcas L.* to water deficits: yield, water use efficiency and oilseed characteristics. Biomass Bioenergy 33(10):1343–1350

Arumugam S et al (2007) Bio-fuels, technology status and future trends, technology assessment and decision support tools. ICS-UNIDO, p. 142

Barnwal BK, Sharma MP (2005) Prospects of biodiesel production from vegetable oils in India. Renew Sustain Energy Rev 9(4):363–378

Boyle G (2004) Renewable energy. Oxford University Press, USA

Brennan L, Owende P (2010) Biofuels from microalgae—a review of technologies for production, processing, and extractions of biofuels and co-products. Renew Sustain Energy Rev 14(2):557–577

Bulack SC (1985) Phosphate diestertransesterification: general base catalysis and solvent effects on specific ion catalysis. Cornell University, Ithaca
Carvalho AP, Meireles LA, Malcata FX (2006) Microalgal reactors: a review of enclosed system designs and performances. Biotechnol Progr 22(6):1490–1506
Chaumont D (1993) Biotechnology of algal biomass production: a review of systems for outdoor mass culture. J Appl Phycol 5(6):593–604
Chisti Y (2007) Biodiesel from microalgae. Biotechnol Adv 25(3):294–306
de Morais MG, Costa JAV (2007) Biofixation of carbon dioxide by Spirulina sp. and *Scenedesmus obliquus* cultivated in a three-stage serial tubular photobioreactor. J Biotechnol 129(3):439–445
Deane JP, Gallachóir BPÓ, McKeogh EJ (2010) Techno-economic review of existing and new pumped hydro energy storage plant. Renew Sustain Energy Rev 14(4):1293–1302
Demirbas A (2008) Biofuels: securing the planet's future energy needs. Springer, New York
Demirbas MF (2011) Biofuels from algae for sustainable development. Appl Energy 88(10):3473–3480
Deublein D, Steinhauser A (2011) Biogas from waste and renewable resources: an introduction. Wiley-VCH, Weinheim
Dincer I (2000) Renewable energy and sustainable development: a crucial review. Renew Sustain Energy Rev 4(2):157–175
Dorado MP et al (2002) An alkali-catalyzed transesterification process for high free fatty acid waste oils. Trans Am Soc Agric Eng 5(3):525–529
Erbaum JB (2009) Bioethanol: production, benefits and economics. Nova Science Publ., New York
Field CB et al (1998) Primary production of the biosphere: integrating terrestrial and oceanic components. Science 281(5374):237–240
Freedman B, Pryde E, Mounts T (1984) Variables affecting the yields of fatty esters from transesterified vegetable oils. J Am Oil Chem Soc 61(10):1638–1643
Fukuda H, Kondo A, Noda H (2001) Biodiesel fuel production by transesterification of oils. J Biosci Bioeng 92(5):405–416
Gouveia L (2011) Microalgae as a feedstock for biofuels. Springer, Berlin, pp 1–69
Green MA (1982) Solar cells: operating principles, technology, and system applications. Prentice-Hall, Inc., Englewood Cliffs, p 288
Hanjalic K et al (2008) Sustainable energy technologies options and prospects. Available from: http://public.eblib.com/EBLPublic/PublicView.do?ptiID=337939. Accessed 25 September 2012.
Harun R et al. (2010a) Bioprocess engineering of microalgae to produce a variety of consumer products. Renew Sustain Energy Rev 14(3):1037–1047
Harun R, Danquah MK, Forde GM (2010b) Microalgal biomass as a fermentation feedstock for bioethanol production. J Chem Technol Biotechnol 85(2):199–203
Hejazi MA, Wijffels RH (2004) Milking of microalgae. Trends Biotechnol 22(4):189–194
Hondo H (2005) Life cycle GHG emission analysis of power generation systems: Japanese case. International Symposium on CO2 Fixation and Efficient Utilization of Energy (CandE 2002) and the International World Energy System Conference (WESC-2002). 30(11–12), pp 2042–2056
Houghton R (2008) Carbon flux to the atmosphere from land-use changes: 1850–2005, in trends: a compendium of data on global change. Carbon Dioxide Information Analysis Center, Oak Ridge
Howarth R, Santoro R, Ingraffea A (2011) Methane and the greenhouse-gas footprint of natural gas from shale formations. Clim Change 106(4):679–690
Huang H-J et al (2008) A review of separation technologies in current and future biorefineries. Sep Purif Technol 62(1):1–21
Intergovernmental panel on climate change (2007) Climate change: the physical science basis. In: Solomon S, et al (eds). Cambridge University Press, Cambridge
International Energy Agency (2006) World energy outlook 2006: organisation for economic co-operation and development OECD

Jaccard M, (2006) Sustainable Fossil Fuels: The unusual suspect in the quest for clean and enduring energy. Cambridge University Press, Cambridge
Janssen M et al (2003) Enclosed outdoor photobioreactors: light regime, photosynthetic efficiency, scale-up, and future prospects. Biotechnol Bioeng 81(2):193–210
John RP et al (2011) Micro and macroalgal biomass: a renewable source for bioethanol. Bioresour Technol 102(1):186–193
Jones CS, Mayfield SP (2011) Algae biofuels: versatility for the future of bioenergy. Curr Opin Biotechnol 23(3):346–351
Jones CS, Mayfield SP (2012) Algae biofuels: versatility for the future of bioenergy. Curr Opin Biotechnol 23(3):346–351
Kemp WH (2006) Biodiesel basics and beyond: a comprehensive guide to production and use for the home and farm. Aztext Press, Ontario
Kulkarni MG et al (2006) Solid acid catalyzed biodiesel production by simultaneous esterification and transesterification. Green Chem 8(12):1056–1062
Kunjapur AM, Eldridge RB (2010) Photobioreactor design for commercial biofuel production from microalgae. Ind Eng Chem Res 49(8):3516–3526
Lam MK, Lee KT (2012) Microalgae biofuels: a critical review of issues, problems and the way forward. Biotechnol Adv 30(3):673–690
Lee Y-K (2001) Microalgal mass culture systems and methods: their limitation and potential. J Appl Phycol 13(4):307–315
Ma F, Hanna MA (1999) Biodiesel production: a review. Bioresour Technol 70(1):1–15
Mabee W, Gregg D, Saddler J (2005) Assessing the emerging biorefinery sector in Canada. Appl Biochem Biotechnol 123(1):765–778
Mata TM, Martins ANA, Caetano NS (2010) Microalgae for biodiesel production and other applications: a review. Renew Sustain Energy Rev 14(1):217–232
Maugeri L (2006) The age of oil: the mythology, history, and future of the world's most controversial resource. Praeger, Westport
Meher LC, Vidya Sagar D, Naik SN (2006) Technical aspects of biodiesel production by transesterification—a review. Renew Sustain Energy Rev 10(3):248–268
Mendes-Pinto MM et al (2001) Evaluation of different cell disruption processes on encysted cells of & lt;b & gt; & lt;i & gt;Haematococcus pluvialis & lt;/i & gt; & lt;/b & gt;: effects on astaxanthin recovery and implications for bio-availability. J Appl Phycol 13(1):19–24
Minteer S (2006) Alcoholic Fuels. Taylor & Francis, Boca Raton
Mittelbach M, Remschmidt C (2004) Biodiesel: the comprehensive handbook. Martin Mittelbach, Graz
Molina Grima E et al (2003) Recovery of microalgal biomass and metabolites: process options and economics. Biotechnol Adv 20(7–8):491–515
Mudhoo A (2012) Biogas production: pretreatment methods in anaerobic digestion. Wiley, New York
Murugesan A et al (2009) Bio-diesel as an alternative fuel for diesel engines—a review. Renew Sustain Energy Rev 13(3):653–662
Nozik AJ (1978) Photoelectrochemistry applications to solar energy conversion. Annu Rev Phys Chem 29(1):189–222
Pandey A (2008) Handbook of plant-based biofuels. Taylor & Francis, Boca Raton
Pienkos PT, Darzins A (2009) The promise and challenges of microalgal-derived biofuels. Biofuels, Bioprod Biorefining 3(4):431–440
Pimentel D (2008) Biofuels, solar and wind as renewable energy systems: benefits and risks. Springer, Dordrecht
Pimentel D et al (2008) Biofuel impacts on world food supply: use of fossil fuel, land and water resources. Energies 1(2):41–78
Pironon J et al (2010) On-line greenhouse gas detection from soils and rock formations. Int J Greenh Gas Control 4(2):217–224
Puppan D (2002) Environmental evaluation of biofuels. Period Polytech Soc Manage Sci 10:95–116

Reijnders L (2006) Conditions for the sustainability of biomass based fuel use. Energy Policy 34(7):863–876
Rickeard DJ, Thompson ND (1993) A review of the potential for bio-fuels as transportation fuels. SAE International, Warrendale
Schenk PM et al (2008) Second generation biofuels: high-efficiency microalgae for biodiesel production. Bioenergy Res 1:20–43
Schuchardta U, Serchelia R, Vargas RM (1998) Transesterification of vegetable oils: a review. J Braz Chem Soc 9(3):199–210
Schwab A et al (1988) Diesel fuel from thermal decomposition of soybean oil. J Amer Oil Chem Soc 65(11):1781–1786
Sheehan J, Laboratory NRE (1998) An overview of biodiesel and petroleum diesel life cycles1998. National Renewable Energy Laboratory, Golden
Stevens CV, Verhé R (2004) Renewable bioresources: scope and modification for non-food applications. Wiley, New York
Tomasevic AV, Siler-Marinkovic SS (2003) Methanolysis of used frying oil. Fuel Process Technol 81(1):1–6
Tredici MR (2007) Mass production of microalgae: photobioreactors, in handbook of microalgal culture. Blackwell Publishing Ltd, Oxford, pp 178–214
Vassiliou MS (2009) Historical dictionary of the petroleum industry. Scarecrow Press, Lanham
Wang B et al (2008) CO2; bio-mitigation using microalgae. Appl Microbiol Biotechnol 79(5):707–718
Wijffels RH, Barbosa MJ (2010) An outlook on microalgal biofuels. Science 329(5993):796–799
World Energy Look (2008) Organisation for Economic Co-operation Development (International Energy Agency), Paris
Zhou W, Konar S, Boocock D (2003) Ethyl esters from the single-phase base-catalyzed ethanolysis of vegetable oils. J Am Oil Chem Soc 80(4):367–371

Chapter 7
Digestion Tests to Measure Heavy Metal Bioavailability in Soils

Yi Li, Walelign Demisie and Ming-kui Zhang

Contents

Y. Li (✉) · W. Demisie · M.-k. Zhang
Department of Resource Science, College of Environmental and Natural Resource Sciences, Zhejiang University, 310058 Hangzhou, China
e-mail: liyi1343116@163.com

School of Geography and Tourism, Guizhou Normal College, 550018, Guiyang, China

W. Demisie
Department of Dry Land Crop Science, Jigiga University, Jigiga, Ethiopoa

© Springer International Publishing Switzerland 2015

E. Lichtfouse et al. (eds.), *CO_2 Sequestration, Biofuels and Depollution*,
Environmental Chemistry for a Sustainable World 5, DOI 10.1007/978-3-319-11906-9_7

Abstract In-vitro digestion tests have been recently used to assess the health risk of oral ingestion of heavy metals from soils. These tests measure the solubility of soil heavy metals in simulated human digestive juice. Here we first introduce the origin, development and merits of the in-vitro digestion test, and related definitions and terms. We present the various procedures such as mouth, stomach and intestine; and effects of test parameters: soil particle size, temperature, mixing, pH, solid-to-liquid ratio, retention time, digestive compositions, food addition, analysis and calculation. We describe the six commonly used in-vitro digestion tests are (1) the physiologically-based extraction test (PBET), (2) the simple bioaccessibility extraction test (SBET), (3) the unified barge methods (UBM) from the bioaccessibility research group of Europe (BARGE), (4) the United States pharmacopeia methodology (USPM), (5) the in vitro gastrointestinal method (IVG) and (6) the diluted HCl solution test. We also review the effects of soil properties and ageing on bioaccessibility. We present the applications of in-vitro digestion test in soil metal pollution research with focus on remediation and bioaccessibility. The challenges of comparing results from in-vitro digestion test research are due to the differing extraction abilities of individual in-vitro digestion test; and the lack of regulation allowing operators to modify the composition and procedure of in-vitro digestion test.

Keywords Soil·metal · In-vitro digestion test · Bioaccessibility · Relative bioavailability

7.1 Introduction

Heavy metals can enter human body along with orally ingested soils, accumulate in the body and eventually harm health. There are three groups of people susceptible to harm from heavy metals in orally ingested soils. The first group are sufferers of chthonophagia: soil-eating, which may be a result of nutrition and minerals deficiencies (Hamilton et al. 2001). This condition is globally prevalent, especially amongst women from developing countries, such as Africa and Bengal (Kawai et al. 2009; Al-Rmalli et al. 2010). It was reported that the amounts of soil ingested by sufferers of chthonophagia differed from person to person, and was ranging from 2.5 to 219.0 g/day, averaging 41.5 g/day (Geissler et al. 1999). The second affected group are young children as they often play on the ground and may use their mouths as tool to explore the world. As a result, hand-to-mouth or object-to-mouth behaviors are a major pathway for exposure to soil (Duggan and Inskip 1985; Watt et al. 1993; Hubal et al. 2000). Xue et al. (2007) found average indoor hand-to-mouth behavior of children ranged from 6.7 to 28.0 contacts/hour and average outdoor hand-to-mouth frequency of children ranged from 2.9 to 14.5 contacts/hour. Some children were observed to ingest up to 25–60 g soil during a single day (Calabrese et al. 1997). The third group of people likely to be affected are those adults subjected to occupational exposure. The amount of soil ingested by adults is thought to be less than that ingested by children, since their personal hygiene habits are better, thereby the study on ingested soil by adults is limited (Davis and Mirick 2006). However, the amount of soil ingested by adults would increase when his/her occupation is soil related. Adult mean soil ingestion rates measured by Davis and Mirick (2006) varied between 23–625 mg/day with the highest value based on using titanium as a tracer. The negative effects of soil heavy metals on human health would be overestimated if the total content of soil heavy metals is used to assess the health risk of soil heavy metals, since the total content of soil heavy metals does not represent the actual amount available to the body (Freeman et al. 1994, 1996). As a result of this, it is recommendable to take into account the bioavailability of heavy metals present in the soils. Bioabsorption of heavy metals into the body from soils occurs in two distinct phases: a physicochemically driven desorption process and a physiologically driven uptake process (Peijnenburg et al. 1997). In theory, therefore, the determination of the bioavailability of soil heavy metals should also consists of two parts (Fig. 7.1). The first stage is the determination of the relative bioavailability, which can be obtained by determining the solubility of soil heavy metals in digestive juice. The second stage is the determination of the absolute bioavailability, which refers to the fraction of heavy metals mobilized from soil to digestive juice, absorbed by the body and involved in the circulation system (Ruby et al. 1999). The term soil heavy metal bioavailability is usually used to denote the absolute bioavailabilty, and the relative bioavailability of soil heavy metals is generally referred to as bioaccessibility (Ruby et al. 1999). Since mobilization of heavy metals from soil is the first step for soil heavy metals to be absorbed by the body, the soil heavy metals bioaccessibility is always less than (or equal to) the total content of heavy metals in the soil and higher than the absolute bioavailability of heavy metals in the soil.

The absolute bioavailability of heavy metals in soil can only be determined by in-vivo test, i.e. animal (or human) experimentation. But many experimentalists in this area believe that inappropriate animal choice can lead to misrepresentations of the relevance of results due to differences in behavior, anatomy, physiology and pharmacokinetics between the animal model and humans (Ruby et al. 1999). Besides, In-vivo testing is acknowledged to be expensive, time-consuming and cruel to the tested animals (Finch et al. 1978; Cabanero et al. 2004; Intawongse and Dean 2008). With the rise of the animal protection movement, animal welfare began to receive significant attention and unnecessary animal experimentation was subjected to challenge under these circumstances. The 3Rs program (Replacement, Reduction and Refinement) was proposed (Flecknell 2002; Vicky 2005), and in-vitro test for metal bioaccessibility, as an alternative to animal experimentation, has been developed and applied. The in-vitro digestion test mimicking the human digestion procedure was first used to estimate the bioaccessibility of iron in food for the purpose of determining nutritional value (Rao and Prabhavathi 1978; Miller et al. 1981). The procedure was then introduced into environmental science to assess the bioaccessibility of pollutants for the purpose of protection and precaution. Compared to the animal experimentation required to determine absolute bioavailability an in-vitro digestion test for bioaccessibility is relatively easy, cheap and reproducible.

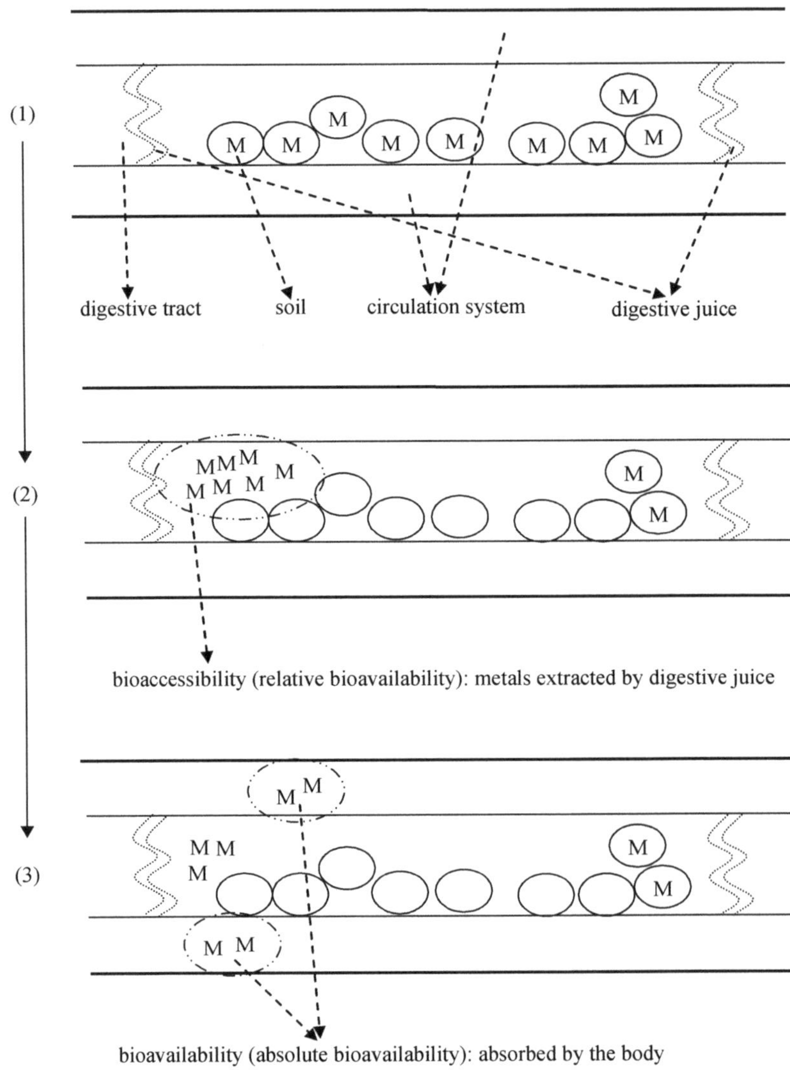

Fig. 7.1 The descriptions of relative bioavailability and absolute bioavailability *M* metal

7.2 Compositions of In-Vitro Digestion Test

The design of in-vitro digestion test follows the process of human digestion, which is composed of mouth, stomach and intestine in a particular sequence.

Mouth is the first step of digestion, in which ingested food is fragmented by mastication and mixed with saliva, and initial digestion of starches and lipids occurs (Pedersen et al. 2002). The pH value of saliva is nearly neutral and the retention time is short. Hence, considering the solution pH is main factor governing the

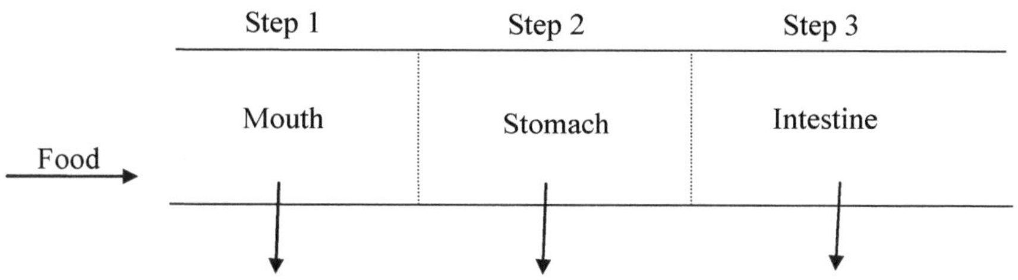

pH	neutral (5.5-7.8)	acid (1-5)	neutral (6-8)
Retention time	<2 min	3.5-5 h	15 min-5 h
Digestive enzyme	α-amylase	pepsin	bile and pancreatin
Role	fragment food initial digestion	major part of digestion	absorption
Influences on the solubility of soil metals	little effects due to the neutral environment and short retention time	the mobility of soil metals is promoted as a result of the low pH	after enter the intestine, the dissolved soil metals in stomach may precipitate due to the change of environmental pH

Fig. 7.2 The descriptions of compositions of in-vitro digestion test (*1*) gastric compartment; (*2*) small intestine; (*3*) pH electrodes; (*4*) secretion of lipases and pepsin; (*5*) secretion of pancreatic juice and bile; and (*6*) hollow fiber membranes simulating the absorption of digested products (Kong and Singh 2008

dissolution of soil heavy metals, many in-vitro digestion tests do not include mouth digestion. The second step of digestion is stomach where food is mainly digested. The hydrochloric acid (HCl) secreted by stomach keeps the pH of gastric juice low and facilitates the mobility of soil heavy metals, thereby the gastric digestion is the most important part of in-vitro digestion test. The last step of digestion is intestine. The elements soluble in gastric juice can be really absorbed by the body only after passing through the intestine cells. Because elements are mostly absorbed in small intestine, the intestinal digestion of in-vitro digestion test usually refers to the digestion of small intestine. The small intestine comprises duodenum, jejunum and ileum in succession from up. The pH of intestinal juice is nearly neutral, so dissolved soil heavy metals by gastric juices may be deposited when enter the intestine due to the change of environmental pH.

The descriptions of compositions of in-vitro digestion test were seen in Fig. 7.2.

7.3 In-Vitro Digestion Test Parameters

To simulate the real physiological environment of human body, the incubation condition of the test should be as close to human digestive tract as possible, moreover, several parameters should be fixed.

7.3.1 Soil Particle Size

There are different views on the particle size that adheres to fingers and is likely to be ingested. Duggan and Inskip (1985) found that most of the particles on the children's hands were less than 10 μm. However, the in-vitro digestion tests are usually performed on <250 μm soil samples (Yang et al. 2001; Schroder et al. 2004; Tang et al. 2004). This is because predominately smaller particles than 250 μm (i.e., 100 μm) are considered to be likely to adhere to children's hands, promoting ingestion, and are subject to wind transport and inhalation (Ruby et al. 1992; Fendorf et al. 2004). However using <250 μm particle size fraction has the potential to underestimate metal exposure because smaller particles have higher metal concentration and preferential adhesion to hands (Juhasz et al. 2011). Therefore, soil particles less than 125 μm (Hamel et al. 1999), 180 μm (Bosso and Enzweiler 2008; Bamett and Turner 2001) or 200 μm (Madrid et al. 2008a) were chosen for some in-vitro digestion tests.

7.3.2 Temperature and Mixing

The incubation temperature is usually 37 °C, as it is the normal temperature of human body. But the soil Hg bioaccessibility at 37 °C was not significantly higher compared to the results at 23 °C (Bamett and Turner 2001).

There are many ways to simulate the mixing of human digestive tract, such as shaking, mechanical stirring or Ar gas dispersion. The purposes of mixing are: (1) to mimic the movement of digestive tract, and (2) to make samples and digestive juice fully contact. There are few studies discussing the influences of different mixing methods on soil metals bioaccessibility.

7.3.3 pH

The pH is the major factor controlling the mobilization of soil heavy metals.

Saliva: Its pH ranged from 5.5 to 7.8 (Dodds and Johnson 1993; Kou and Takahama 1995)

Gastric juice: Its pH is usually between 1 and 5 (Samloff and O'Dell 1985; Zhang and Ohta 1991). The selection of gastric juice pH value determines to a large extent the extraction ability of in-vitro digestion test. High bioaccessiblity values are typically observed for a simple in-vitro digestion test only including gastric digestion and applying a gastric pH of 1.5 and the lowest bioaccessibility values are observed for an in-vitro digestion test consisting of gastrointestinal digestion and employing a high gastric pH of 4.0 (Oomen et al. 2002). The Pb dissolution in the gastric juice averagely decreased 57 % when gastric pH was raised from 1.3 to 2.5 and a further 66 % decrease in stomach-solubilized Pb was found when the gastric pH was increased from 2.5 to 4.0 (Ruby et al. 1996). Turner and Radford (2010) noticed the reduction effect of increasing the stomach pH from 1 to 4.5 on bioaccessibility was propagated into the intestinal phase in most cases. The increasing gastric pH from 1.2 to 1.7 resulted in: (1) no significant variation of Cd bioaccessibility in the gastric phase but a decrease in the gastrointestinal phase; (2) a decrease of Pb bioaccessibility in the gastric phase and a significant variation of Pb bioaccessiblity in the gastrointestinal phase (Pelfrene et al. 2011a).

Intestinal juice: Its pH is around 6 in duodenum and slowly increases to 8 in lower ileum (Borgstrom et al. 1957).

7.3.4 Solid-to-Liquid Ratio and Retention Time

The volume of digestive tract is flexible, especially for stomach, whose volume can expand to accommodate food up to a volume of about 4 L (Kong and Singh 2008). Solid-liquid ratios between 1:5 and 1:25 (g/ml) affected the mobilization of metals in synthetic digestive juice (Ruby et al. 1996). General trends of increased bioaccessiblity with increasing solid-liquid ratios ranging from 1:25 to 1:1000 were observed (Smith et al. 2010). However the results of Hamel et al (1998) indicated that solid-liquid ratios in the range of 1:100 to 1:5000 (g/ml) only had slight effects on the bioaccessibility. The types of in-vitro digestion tests used by Smith et al. (2010) and Hamel et al. (1998) were SBET (Simple Bioaccessibility Extraction Test, described in 5.2) and USPM (United States Pharmacopeia Methodology, described in 5.4), respectively. The different types of methods used by these researchers could be the reason for the variations in their results.

The retention times in different digestion parts distinctly vary. The retention time in mouth is different from person to person, mainly dependent on personal eating habit and usually no more than 2 min. The gastric emptying time was observed ranging from 3.5 to 5 h in health people (Bolondi et al. 1985). The transit time of small intestine was found between 15 min to 5 h, less than 2 h for 83 % of cases studied (315 adult cases) and 84 min on average (Kim 1968).

7.3.5 Digestive Compositions

Saliva: The digestive function of saliva is basically dependent on the presence of α-amylase (Schenkels et al. 1995; De Almeida et al. 2008).

Gastric juice: Pepsin is the only digestive enzyme present in normal stomach (Correa 1988).

Intestinal juice: Mixture of digestive enzymes (pancreatin) produced by pancreas is delivered to the small intestine for the hydrolysis of complex nutrients (Whitcomb and Lowe 2007). In addition, bile is also important for the in-vitro digestion test and often added to synthetic intestinal juice along with pancreatin, although technically bile is not regarded as digestive enzyme. Bile has great effects on the dissolution of soil heavy metals by forming complexes with heavy metals and decreasing interface tension (Miller 1995; Oomen et al. 2003b). Oomen et al. (2003a) observed no decreasing or increasing trend of bioaccessibility when the concentration of bile varied. Oomen et al. (2004) studied the influences of different animal origin bile on bioaccessibility and considered the result differences because of bile type to be irrelevant for risk assessment purpose.

The addition of digestive enzymes to synthetic digestive juice is important because the existence and content of digestive enzymes in digestive juice are relatively stable. It dose not mean digestive enzymes are the only component present in digestive juice.

7.3.6 Food Addition

It is believed fasting state causes greater harm when soil is ingested (Ruby et al. 1996; Oomen et al. 2003b), because more metals could be dissolved in digestive juice and/or absorbed by the body. Hence, most in-vitro digestion tests are designed to simulate the fasting state. However, there are still researches on the impacts of additional food on soil heavy metal bioaccessibility. Schroder et al. (2004) reported the existence of dough decreased the bioaccessible soil Pb because of phytic acid associated with dough addition. Marshcner et al. (2006) found the bioaccessibility of soil Pb increased when powdered milk was added and this was because soluble milk constituents competed with soil organic ligands for Pb.

7.3.7 Analysis and Calculation

Centrifugation and filtration are usually used to separate the digestive juice from soil, and the metal concentrations in digestive juice (stomach or intestine) are determined by AAS (atomic absorption spectrometry), ICP-MS (inductively coupled plasma-mass spectrometry) or ICP-AES (inductively coupled plasma-atomic emission spectrometry). The metals extracted by saliva are usually not analyzed due to the short digestion time in mouth. The soil metal bioaccessibility is often calculated (or defined) by the following equation:

Bioaccessibility (%) = (metal concentration in digestive juice (stomach or intestine) after digestion/ the total metal concentration in soil before digestion) $\times$ 100 %

7.4 In-Vitro Digestion Test Classification

7.4.1 Dynamic and Static In-Vitro Digestion Tests

As a whole, in-vitro digestion test falls into two categories: dynamic in-vitro digestion test and static in-vitro digestion test. The dynamic in-vitro digestion test simulates not just the digestive juice but also the peristalsis, by which chyme is

Fig. 7.3 The dynamic in-vitro digestion test: TIM

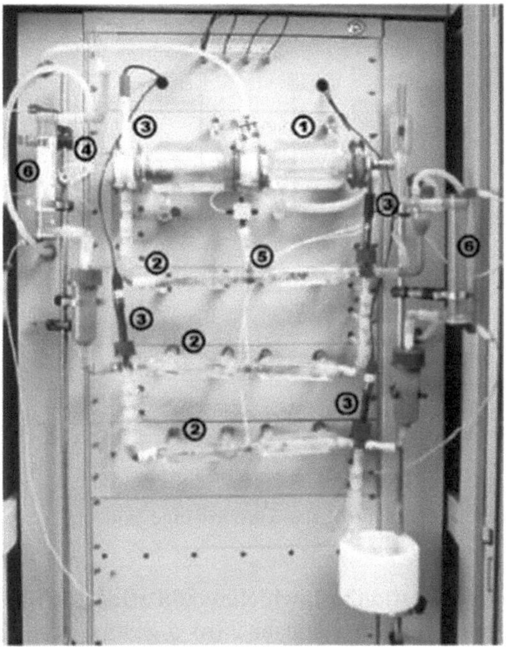

(1): gastric compartment; (2): small intestine; (3): pH electrodes; (4): secretion of lipases and pepsin; (5): secretion of pancreatic juice and bile; and (6): hollow fiber membranes simulating the absorption of digested products (Kong and Singh 2008)

propelled through the digestive tract. For example, the dynamic in-vitro digestion test: TIM (TNO Intestinal Model, TNO Nutrition, The Netherlands) (Fig. 7.3), is designed to mimic the whole movement of chyme passing through the stomach, duodenum, jejunum and ileum in sequence, and the corresponding physiological conditions, such as pH change, digestive enzymes secretion and absorption (Minekus et al. 1995; Krul et al. 2000; Krul et al. 2001; Oomen et al. 2002; Kong and Singh 2008). The static in-vitro digestion test focuses on the chemical digestion process and omits the chyme transit process. Oomen et al. (2002) compared the results of static in-vitro digestion test with dynamic in-vitro digestion test. They observed the bioaccessibility of Cd was similar in both tests, however, the bioaccessibility of Pb in dynamic in-vitro digestion test was relatively lower that that of static in-vitro digestion test. Two reasons were given to explain this observation. First, in the dynamic in-vitro digestion test, the soil is gradually emptied from the stomach when the pH is decreasing, thus soil cannot be fully exposed to the final low pH. Second, after digestion, ultra-filtration membrane is used in dynamic in-vitro digestion test to separate the solution from digested soil, whereas centrifugation and/or micro-filtrtion are chosen in static in-vitro digestion test, and thus smaller and fewer particles are included in the bioaccessible fraction of dynamic

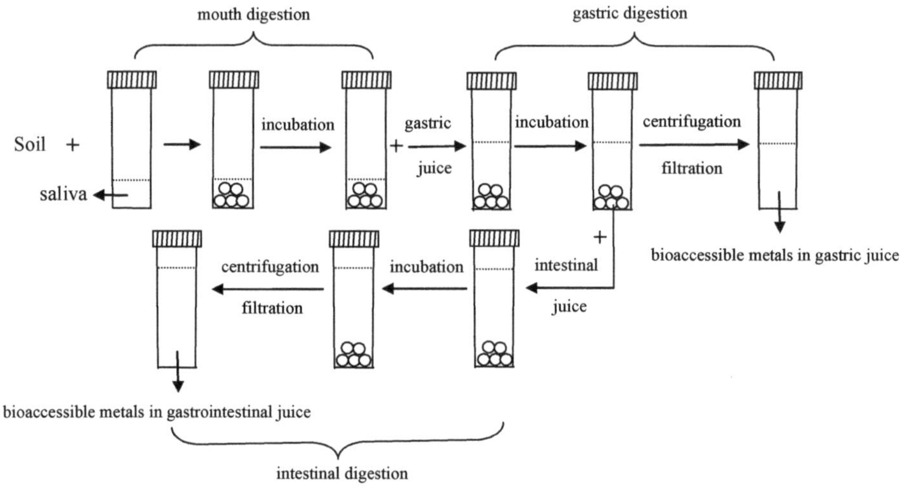

Fig. 7.4 The scheme of static in-vitro digestion test

in-vitro digestion test, which would affect the bioaccessibility of metals associated with particles. In static in-vitro digestion test, the incubation condition of each digestion phase (mouth, stomach and intestine) stays stable, the soil metals and synthetic digestive juice can fully contact, and the operation is easy and suitable for batch analysis. Therefore, the static in-vitro digestion test has a wide application in the assessment of soil metals bioaccessibility as compared to the dynamic in-vitro digestion test. In this paper, the in-vitro digestion test discussed mainly refers to the static in-vitro digestion test. The scheme of static in-vitro digestion test was present in Fig. 7.4.

7.4.2 One-Step, Two-Step and Three-Step In-Vitro Digestion Tests

In-vitro digestion test is divided into three types according to the procedure: one-step in-vitro digestion test, two-step in-vitro digestion test and three-step in-vitro digestion test. The gastric digestion is the most important and indispensable part for the mobilization of soil heavy metals, whereas the digestion of mouth and intestine is optional based on the intention of researcher: risk assessment or physiologically simulation. Then the in-vitro digestion test only including gastric digestion is one-step in-vitro digestion test; the one comprising the sequential digestion of mouth and stomach, or stomach and intestine, is two-step in-vitro digestion test; the one consisting of sequential digestion of mouth, stomach and intestine is three-step in-vitro digestion.

7.4.3 In-Vitro Digestion Test with and Without Digestive Enzymes

Although the composition of human digestive juice is different from person to person because of diet, gender and/or age differences, the existence of digestive enzymes is common to all human digestive juice. The role of digestive enzymes in in-vitro digestion tests was discussed. Digestive enzymes are helpful in keeping heavy metal soluble by forming complexes with metal ions (Ruby et al. 1993; Oomen et al. 2003b; Wragg et al. 2011), however, our previous studies showed that after digestive enzyme addition, no clear trend was observed in changes of bioaccessibility of soil Cu, Zn ad Pb (Li et al. 2013). It is understandable the addition of digestive enzymes to synthetic digestive juice is accordant with human digestive juice and reproduces a situation closer to human digestive tract. But as a screening-level method, sometimes emphasis of in-vitro digestion test is to measure the bioaccessibility easily, quickly and reproducibly for the purpose of precaution and protection. That is why digestive enzymes are not included in some synthetic digestive juice. For example, Mercier et al. (2002) decided to omit the addition of pepsin considering the complexity of method and variability of results along with pepsin addition. Hence, in-vitro digestion test is grouped into two kinds: in-vitro digestion test with digestive enzymes and in-vitro digestion test without digestive enzymes.

7.5 In-Vitro Digestion Test Used Commonly

There have been many in-vitro digestion tests developed and applied in the past two decades. In this paper, detailed descriptions will be given about the six in-vitro digestion tests: PBET (Physiologically Based Extraction Test), SBET (Simple Bioaccessibility Extraction Test), UBM (Unified BARGE Method), USPM (United States Pharmacopeia Methodology), IVG (In-Vitro Gastrointestinal method) and diluted HCl solution. Here, two points should be noted. Firstly, the six methods are chosen because they were widely used and there are other in-vitro digestion test developed and applied, such as the GJST (Gastric Juice Simulation Test) by Mercier et al. (2002) and EHS (Extraction of Heavy metals in Stomach and small intestine) by Lim et al. (2008). Secondly, the synthetic digestive juice composition and procedure of different in-vitro digestion tests are distinct. Because the gastric digestion is the most important part of in-vitro digestion test, the composition of synthetic gastric juice is the key to identify if two in-vitro digestion tests are the same method or from the same method.

7.5.1 PBET (Physiologically Based Extraction Test)

PBET, which was designed around pediatric gastrointestinal tract parameters for a child 2–3 years old, was proposed and validated using in-vivo test (rabbit and rat) by Ruby et al. (1993, 1996). One liter synthetic gastric juice of PBET was comprised of 1.25 g pepsin, 0.50 g citrate, 0.50 g malate, 0.42 ml lactic acid and 0.50 ml acetic acid. When the gastric digestion of 1 h was completed, the intestinal juice was prepared by adding 70 mg bile salts and 20 mg pancreatin to the gastric juice after the pH was adjusted to 7.0 with dialysis bag containing $NaHCO_3$. The digestion time of intestinal phase was 3 h. The solid-to-liquid of PBET was 1:100 (0.4 g sample/ 40 ml stomach solution) and the argon gas (1 L/min) was chosen to provide mixing in a water bath at 37 °C. To study the effects of gastric pH on Pb bioaccessibility, the bioaccessibility was determined using PBET at gastric pH values of 1.3, 2.5 and 4.0 (adjusting with 12 N HCl), respectively, and it was found the results of PBET at stomach pH of 2.5 had a strong linear correlation ($r^2 = 0.93$) with the results of in-vivo test. The addition of organic acids was because malate and citrate were the most concentrated carboxylic acids in the rabbit stomachs, and the latter two organic acids (lactate and acetate) were most concentrated in the small intestine (Ruby et al. 1993). There were some differences between operation specifics of PBET described in Ruby et al. (1993) and (1996). The samples tested in Ruby et al. (1993) were mine-wastes, whereas two out of seven samples tested in Ruby et al. (1996) were composite residential soil samples. Therefore, the procedure of PBET described above was from Ruby et al. (1996).

PBET was modified in some researches. NaCl was added to the gastric juice besides pepsin and the four organic acids (Tang et al. 2004, 2006, 2008; Man et al. 2010; Cui et al. 2010, 2011; Cui and Chen 2011). The synthetic gastric juice prepared by Sialelli et al. (2010, 2011) consisted of 1.25 g pepsin, 0.5 g tri-sodium citrate, 0.5 g malic acid and 0.42 ml lactic acid. After the gastric digestion was completed, the pH was adjusted to 7 and then only 500 mg/L pancreatin was added. The acetic acid in gastric juice and bile salts in intestinal juice were omitted. Moreover, the concentrations of the constituents in PBET and solid-to-liquid ratio were changed. For instance, the amounts of bile salts and pancreatin added in intestinal phase were 175 and 50 mg per 100 ml by Abrahams et al. (2006), Carrizales et al. (2006), Turner et al. (2009), Gbefa et al. (2011) and Karadas and Kara (2011, 2012); 20 and 6 mg per 100 ml by Tang et al. (2004, 2006, 2008), Cui et al. (2010, 2011), Man et al. (2010) and Cui and Chen (2011). The solid-to-liquid ratio used by Man et al. (2010) was 1:30 (g/ml).

7.5.2 SBET (Simple Bioaccessibility Extraction Test)

SBET, which was adaped from PBET, was used by the British Geological Survey (BGS, United Kingdom) (Oomen et al. 2002). It is a one-step in-vitro digestion test without pepsin. The typical synthetic gastric juice of SBET is 0.4 M glycine solution of pH 1.5 (adjusting with concentrated HCl). The digestion is carried out by shaking the mixture of sample and synthetic gastric juice (1 g/100 ml) for 1 h at 37 °C. SBET was validated using in-vivo test (juvenile swine) by Drexler and Brattin (2007).

The SBET, in which the 1 h gastric digestion was followed by a 4 h neutral (pH 6.5) intestinal digestion with digestive enzymes (175 mg/100 ml bile and 50 mg/100 ml pancreatin), is called as the Solubility Bioavailability Research Consortium (SBRC) in vitro assay (Juhasz et al. 2011; Smith et al. 2011). Juhasz et al. (2009) chose the SBRC to predict the in vivo Pb bioavailability (swine) in contaminated soils and found the gastric results of SBRC provided the best estimate of in vivo Pb bioavailability for soils. Poggio et al. (2009) also applied a modified SBET, which had an additional 3 h neutral (pH 7.0) intestinal digestion without intestinal digestive enzymes. Compared with the results of single gastric digestion, the bioaccessibility of Cu and Ni in the gastrointestinal digestion was higher, whereas the bioaccessibility of Pb and Zn was lower (Poggio et al. 2009).

7.5.3 UBM (Unified BARGE Method)

The UBM proposed by the Bioaccessibility Research Group of Europe (BARGE) with the aim of producing a validated and standardized procedure was derived from the RIVM (In-Vitro Digestion Model, National Institute of Public Health and the Environment) physiologically based in vitro extraction test, which was based on the in-vitro digestion test by Rotard et al. (1995) (Oomen et al. 2002; Cave et al. 2006). This method is composed of mouth, stomach and intestine digestion. The composition of UBM digestive juice (saliva, gastric juice, intestinal juice and bile) is listed in Table 7.1. In the UBM procedure, 0.6 g sample is mixed with 9 ml saliva (pH 6.5) for 5 min (manually shaking). Then 13.5 ml gastric juice (pH 0.9–1.0) is added to obtain a final pH ranging from 1.2 to 1.7, and the mixture is shaken for 1 h at 37 °C. After the gastric digestion, 27 ml intestinal juice and 9 ml bile are added giving a final pH ranging from 5.8 to 6.8, and the mixture is shaken for 4 h at 37 °C to finish the intestinal digestion. Wragg et al. (2011) carried out an inter-laboratory trial (seven laboratories, five European and two North American) of UBM for soil As, Cd and Pb, and pointed out the pH tolerance for UBM is too wide, which is probably a main source of between-labaoratory variability. Denys et al. (2012) validated the UBM to assess soil As, Sb, Cd and Pb bioaccessibility using in vivo test (juvenile swine) and found UBM was not suitable for soils to assess the Sb bioaccessiblity.

7.5.4 USPM (United States Pharmacopeia Methodology)

The USPM is a one-step in-vitro digestion test with pepsin. The 1 L gastric juice of USPM includes 2.0 g NaCl, 3.2 g pepsin and 7 ml concentrated HCl. Hamel et al. (1998) chose the USPM to study the influence of solid-to-liquid ratio on soil metal bioaccessibility. The procedure was shaking the mixture of soil and USPM gastric juice for 2 h at 37 °C. Hamel et al. (1999) modified the USPM by adding a

Table 7.1 Composition of the UBM digestive juices. (Cave et al. 2006; Pelfrene et al. 2011b)

Saliva	Gastric juice	Intestinal juice	Bile
Inorganic solution (500 ml)			
KCl 896 mg	NaCl 2752 mg	NaCl 7012 mg	NaCl 5.59 mg
NaH_2PO_4 888 mg	NaH_2PO_4 266 mg	$NaHCO_3$ 5607 mg	$NaHCO_3$ 5785 mg
KSCN 200 mg	KCl 824 mg	$KH_2PO_4$80 mg	KCl 376 mg
Na_2SO_4 570 mg	$CaCl_2$ 400 mg	KCl 564 mg	
NaCl 298 mg	NH_4Cl 306 mg	$MgCl_2$ 50 mg	180 µl HCl 37%
1.8 ml NaOH 1.0 M	8.3 ml HCl 37%	180 µl HCl 37%	
Organic solution (500 ml)			
Urea 200 mg	Glucose 650 mg Glucuronic acid 20 mg Urea 85 mg Glucosamine hydro-chloride 330 mg	Urea 100 mg	Urea 250 mg
Added compounds			
Amylase 145 mg	Bovine albumin 1 g	$CaCl_2$ 200 mg	$CaCl_2$ 222 mg
Mucin 50 mg	Mucin 3 g	Bovine albumin 1 g	Bovine albumin 1.8 g
Uric acid 15 mg	Pepsin 1 g	Pancreatin 3 g Lipase 500 mg	Porcine bile 6 g

5 s saliva digestion (pH 5.5) and a 2 h intestinal digestion. The 1 L saliva was made up of 4.0 g mucin, 1.0 g urea, 0.6 g Na_2HPO_4, 0.99 g $CaCl_2 \cdot 4H_2O$, 0.4 g KCl and 0.4 g NaCl. The intestinal juice was 0.2 M $NaHCO_3$ solution. Ellickson et al. (2001) used this modified three-step USPM to evaluate soil Pb bioaccessibility, compared the results with the soil Pb bioavailability (rat) and found the bioaccessibility was greater than the bioavailability. Bosso and Enzweiler (2008) simplified the three-step USPM by omitting the saliva digestion.

7.5.5 IVG (In-Vitro Gastrointestinal Method)

The IVG was developed by Rodriguez et al. (1999) to assess As bioaccessibility in soils and solid media. The IVG is a two-step (stomach and intestine) in-vitro digestion test with digestive enzymes and food. The gastric juice of IVG includes 0.15 M NaCl and 1% pepsin. Soil (4 g) and dough (200 g, 5 g dough per 100 mg soil) was added to 600 ml gastric juice and the pH of this mixture was adjusted to 1.8 using HCl. After gastric digestion (mechanical stirring for 1 h at 37 °C), the pH of gastric juice was adjusted to 5.5 with NaHCO3 followed by the addition of bile extract (2.10 g) and pancreatin (0.21 g) to start the intestinal digestion (mechanical stirring for 1 h at 37 °C). Schroder et al. (2003, 2004) validated the IVG using in-vivo test (swine).

7.5.6 Diluted HCl Solution

Davis et al. (1994) used a modified PBET method to study the effects of constituents added to the gastric juice on soil Pb bioaccessibility and noticed the absence of either organic acids or enzyme, or both components together, resulted in only a slight decrease in maximum Pb dissolution in the stomach, which substantiated the theory that HCl concentration is the primary factor controlling Pb dissolution in the stomach. Therefore, diluted HCl solution is employed to mimic the human gastric juice. For example, 0.07 M HCl solution at pH 1.5 (digestion at 37 °C for 2 h at a solid-to-liquid ratio of 1:50) is used by the European Standard Toy Safety Protocol (EN-71 1995) to assess the risk of metal in toys (Ruby et al. 1999; Rasmussen et al. 2008). The CDM (Camp Dresser and McKee Inc. Method) also uses a diluted HCl solution (pH 2.5) in the gastric phase (digestion at room temperature for 4 h at a solid-to-liquid ratio of 0.37 g:0.5 L), but an intestinal digestion (4 h) is carried out after the gastric digestion is completed and the pH of digestive juice is adjusted to 6.5 with NaOH (Bamett and Turner 2001; Welfringer and Zagury 2009). Similarily, a two-step diluted HCl solution was employed by Laird et al. (2011), and the most important difference is the intestinal enzymes (oxgall and pancreatin) were added to the intestinal juice.

7.5.7 Comparison and Discussion of These Six Methods

The Comparison of specifics of these six methods was listed in Table 7.2. Because the design of each method is not always the same, the comparison was made based on the original and/or commonly used design of each method. The modifications of in-vitro digestion test were primarily shown in the addition or omission of digestion phase (mouth and intestine) and digestive enzymes. According to the observation of Li and Zhang (2012), the results of gastrointestinal digestion, as compared to the gastric digestion, showed more differences which resulted from element and soil types. Li et al. (2013) also found digestive enzymes in PBET can help soil Cu stay soluble in intestinal phase and the results of method with digestive enzymes reflected more variations resulting from element and soil types. The impacts of digestive enzymes on heavy metal dissolution are mostly seen in the intestinal phase, therefore the addition of digestive enzyme to the gastrointestinal digestion methods is indispensable. However, addition of pepsin is not important for the methods only comprised of gastric digestion. It cannot be easily confirmed which method is better. Each method has its own theory foundation and application scope. For instance, the emphasis of PBET, UBM, USPM and IVG is on the reproduction of incubation condition of human digestive tract and the acquisition of the bioaccessibility closest to the amounts of absorbed metals, while SBET and diluted HCl solution pay more attention to the risk assessment and obtaining the maximum amounts of metals

Table 7.2 Comparison of six commonly used in-vitro digestion tests

		PBET	SBET	UBM	USPM	IVG	Diluted HCl solution
Composition		Stomach and intestine	stomach	Mouth, stomach and intestine	Stomach	Stomach and intestine	Stomach
Digestive enzymes addition		Pepsin Bile salts Pancreatin	No	Amylase Pepsin Bile Pancreatin Lipase	Pepsin	Pepsin Bile extract Pancreatin	No
Food addition		No	No	No	No	Dough	No
Saliva		No	No	Table 7.1, pH 6.5	No	No	No
Gastric juice		1.25 g pepsin, 0.50 g citrate, 0.50 g malate, 0.42 ml lactic acid, and 0.50 ml acetic acid in 1 L, pH 2.5	0.4 M glycine solution, pH 1.5	Table 7.1, pH 1.2–1.7	2.0 g NaCl, 3.2 g pepsin and 7 ml concentrated HCl in 1 L	0.15 M NaCl, 1 % pepsin and 5 g dough per 100 mg soil, pH 1.8	0.07 M HCl solution, pH 1.5
Intestinal juice		70 mg bile salts and 20 mg pancreatin, pH 7.0	No	Table 7.1, pH 5.8–6.8	No	2.10 g bile extract and 0.21 g pancreatin, pH 5.5	No
Retention time	Mouth	no	No	5 min	No	No	No
	Stomach	1 h	1 h	1 h	2 h	1 h	2 h
	Intestine	3 h	no	4 h	No	1 h	No
Reference		Ruby et al. (1996)	Oomen et al. (2002)	Pelfrene et al. (2012)	Hamel et al. (1998)	Rodriguez et al. (1999)	Ruby et al. (1999)

soluble in digestive tract; PBET, SBET, UBM, USPM and diluted HCl solution simulate the fast state of human digestive tract, but IVG considers the effects of food and dough is contained in its gastric juice. Therefore, the choice to the type and design of in-vitro digestion test mainly depends on the purpose of research.

7.6 Effects of Soil on Metal Bioaccessibility

7.6.1 Soil Properties

The soil metal bioaccessibility is greatly affected by the total content of corresponding metal in soil. A significant correlation was observed between the total concentrations and bioaccessible concentrations of soil metals (Mercier et al. 2002; Carrizales et al. 2006; Poggio et al. 2009; Roussel et al. 2010; Sialelli et al. 2010, 2011). The variations in concentrations of bioaccessible metals are mostly explained by the total metal contents in soils (Luo et al. 2012a, 2012b). This phenomenon demonstrates the strong extraction ability of in-vitro digestion test. Madrid et al. (2008b) found bioaccessible forms of soil Cu, Pb and Zn are distributed among the three sequential fractions of BCR, and even the fraction considered as residual is also bioaccessible to a significant extent. Adsorption is an important mechanism by which metals can be retained in soils. These soil properties influencing the adsorption ability of soil for metals also have effects on the soil metal bioaccessibility. The bioaccessibility was affected by various physicochemical parameters, such as sand, carbonated, organic matter, assimilated P, free Al oxides, and total Fe contents (Pelfrene et al. 2011b). Total carbonate, organic matter, sand, P_2O_5, free Fe-Mn oxide, and total Al and trace element contents appeared as the main variables governing metal bioaccessibility (Pelfrene et al. 2012). Correlation studies show a weak negative relationship between soil clay content and bioaccessible Pb, while there was no correlation between soil OM and bioaccessible Pb (Kim et al. 2009). Bioaccessibility of Pb and Zn in bulk soils correlated significantly with metal concentrations in fine silt and/or very fine sand fractions (Luo et al. 2011). The bioaccessibility of Cr in soil was related to the clay and total inorganic carbon (TIC) content of the soil. Bioaccessibility decreased as the soil TIC content increase and as the clay content decreased (Stewart et al. 2003). But, Hansen et al. (2007) did not found apparent correlation between bioaccesibility and the soil parameters.

7.6.2 Soil Ageing

The temporal variations of soil metal bioaccessibility were studied. In general, the soil metal bioaccessibility significantly decreases with time at the beginning of incubation (Fendorf et al. 2004; Zapusek and Lestan 2009), and then reaches a steady level. The bioaccessibility of Cr (III) varied widely as a function of soil type with most soils limiting bioaccessibility to <45 and 30% after 1 and 100 days soil-Cr ageing, respectively (Stewart et al. 2003). The bioaccessibility of Cd in strong acidic (pH 4.5) soils reached nearly steady levels (76.5–76.9% and 52.0–52.6% in the gastric and intestinal phases, respectively) after a sharp decline in the first week of ageing; in contrast, the bioaccessibility of Cd in higher pH (>6.0) soils was found to be much lower (53.3–72.7% and 29.9–43.4% in gastric and intestinal phases, respectively) and took 2 weeks of ageing to reach steady levels (Tang et al. 2006). During the soil aging process, Pb bioaccessibility decreased exponentially to nearly steady levels in mildly acidic or alkali (pH 6.09–7.43) soils, for both gastric (69.91–71.75%) and intestinal (7.53–9.63%) phases within the first 2–4 weeks and 1–2 months of incubation, respectively; however, it took only 1–2 weeks for strongly acidic (pH 4.5) soils to reach nearly steady levels of Pb bioaccessibility (73.01–74.46% and 10.30–10.98% in the gastric and intestinal phases, respectively) (Tang et al. 2008).

7.7 Applications of In-Vitro Digestion Tests

7.7.1 Remediation Assessment of Metal Contaminated Soil

The addition of P is considered to be an effective way to immobilize soil Pb. It was found the addition of H_3PO_4 (rototilling, surface application, pressure injection) markedly reduced Pb bioaccessibility in the soil and adding 10,000 mg of P kg^{-1} reduced bioaccessible Pb by 60% (Yang et al. 2001, 2002; Yang and Mosby 2006). The effectiveness of P amendments at the intestinal phase was higher than at the gastric phase, and the effectiveness of various P treatments in the intestinal phase generally followed this order at the equivalent P addition level: hydroxyapatite > phosphate rock >hydroxyapatite + single super-phosphate > single super-phosphate (Tang et al. 2004). Similarly, Cao et al. (2009) observed all P amendments (phosphoric acid and/or phosphate rock) significantly reduced Pb bioaccessibility by 28–92% compared to the control, however the bioaccessibility of Cu and Zn were elevated by up to 48 and 40%, respectively, in the H_3PO_4 treatments (phosphoric acid, phosphoric acid + phosphate rock). Besides the effect of P amendments on Pb bioaccessibility, the impacts of other remediation ways on soil metal bioaccessibility were studied. Contin et al. (2008) tested a chemical treatment consisting in repeated cycles of soil saturation with 0.1 M $FeSO_4$, air drying and pH neutralization with $Ca(OH)_2$, and found the bioaccessibility of toxic elements (Cd, Cu, Ni, Pb and Zn) showed a marked decrease for all metals ranging between 61% for Ni to 80% for Cu after the 8th cycle. After 3 months incubation, the 1, 2 and 5% bone char addition (by weight) significantly decreased the concentration of bioaccessible

Pb, however, only the treatment with 5 % bone char addition significantly decreased the concentration of bioaccessible Zn (Xiao-Wei et al. 2010). Mustard leaf addition caused the bioaccessibility of Pb to decrease in the gastric phase, whereas the values increased in the intestinal phase; the Cd bioaccessibility increased with mustard leaf addition in both the gastric and intestinal phases (Cui et al. 2011). The red mud addition markedly reduced the concentration of bioaccessible Zn by 53.1–56.7 % after 3 months incubation compared with the control, while having little effect on the concentration of bioaccessible Pb in soil (Huang and Hao 2012).

7.7.2 Evaluation of Soil Metal Bioaccessibility

The soil metal bioaccessibility reported in previous researches was summarized and listed in Table 7.3. Usually, soil Cd, Cu, Mn, Pb and Zn are more bioaccessible compared to Cr, Ni and Fe (Le Bot et al. 2010; Sialelli et al. 2011). Furthermore, the bioaccessibility of soil Cd, Pb and Zn are obviously pH-dependent because the bioaccessible Cd, Pb and Zn in the gastrointestinal phase are clearly lower than those in the gastric phase due to the environmental change from acid gastric phase to neutral intestinal phase, while the solubility of soil Cu and Cr in the gastric phase does not necessarily decrease after enter the intestinal phase.

7.8 Conclusions and Prospects

In-vitro digestion test has been developed rapidly and used widely in soil metal pollution researches, and efforts were made to establish a unified in-vitro digestion test that is validated, accepted and physiologically-based, but several issues need to be pointed out: (1) the comparability of results from different in-vitro digestion tests is questionable even if all the in-vitro digestion tests are validated using in-vivo test, because the extraction abilities of different in-vitro digestion tests are dissimilar as a result of the varied design; (2) the composition and procedure of in-vitro digestion test are not strictly regulated, and in-vitro digestion tests are modified at will. Consequently, results from the same kind of in-vitro digestion test may be incomparable; (3) the development of in-vitro digestion test is somewhat incompatible with the application of in-vitro digestion test. The purpose of in-vitro digestion test development is to simulate the human physiological situation and obtain the results best representing the real amounts available for absorption. But researchers prefer the simple in-vitro digestion test, such as SBET, for fast screening in the application. Therefore, the guidance for choose and use of in-vitro digestion test should be proposed and a unified in-vitro digestion test should be quickly established in the future research.

Table 7.3 Soil metal bioaccessibility reported in previous researches

Element	Method	Sample	Particle size (μm)	Gastric phase		Gastrointestinal phase		Reference
				Range	Average	Range	Average	
Pb	PBET	Soils near a copper-smelter (Mexico)	–			13–64%	32%	Carrizales et al. 2006
		Soils from abandoned mine (Spain)	<250	0.5–85.7%	25.3%	0.0–80.2%	11.5%	Navarro et al. 2006
		Soils purchased from ethnic shops (UK)	–			3.0–3.4%	3.2%	Abrahams et al. 2006
		Soils from abandoned mining district (Brazil)	<180		64%			Bosso and Enzweiler 2008
		Contaminated soils (Turkey)	<200	10.3–66.4%	40.7%	3.4–73.1%	31.8%	Karadas and Kara 2011
		Soils from south China	<250	24.6–82%	55.1%	2.3–57.5%	12.4%	Cui and Chen 2011
	SBET	Soils from a farm (USA)	–	72–98%				Liu and Zhao 2007
		Soils from abandoned mining district (Brazil)	<80		67%			Bosso and Enzweiler 2008
		Top soils from green area of Torino (Italy)	<200	46–60%	53%			Madrid et al. 2008a
		Top soils from green area of Sevilla (Spain)	<200	30–51%	41%			Madrid et al. 2008a
		Top soils of Torino (Italy)	<200		25%		5%	Poggio et al. 2009
		Lead-contaminated soils	<250	35.7–64.1%	57%	1.2–2.7%	2%	Juhasz et al. 2009
		Surface soils from the region surrounding smelter	<250	32–100%	70%			Kim et al. 2009
		Lead-contaminated soils (Australia and New Zealand)	<250	26.8–105.2%	66%	0.6–11.1%	4.4%	Smith et al. 2011
		Soils from mining area (Chingola)	<250	11–81%	41%			Ettler et al. 2012

Table 7.3 (continued)

Element	Method	Sample	Particle size (μm)	Gastric phase		Gastrointestinal phase		Reference
				Range	Average	Range	Average	
		Soils from smelting area (Kitwe)	<250	25–100 %	73 %			Ettler et al. 2012
		Urban park soils of Xiamen (China)	–	23–68 %	49 %			Luo et al. 2012a
		Urban soils of Hong Kong (China)	–		59 %			Luo et al. 2012b
	UBM	Urban soils contaminated by smelters (France)	<250	33–76 %	62 %	14–63 %	32 %	Roussel et al. 2010
		Contaminated agricultural soils (northern France)	<250		55 %		20 %	Pelfrene et al. 2011b
		Smelter-contaminated soils (northern France)	<250		58±10 %		21±9 %	Pelfrene et al. 2012
	USPM	Contaminated soils	<125			39–70 %	60 %	Hamel et al. 1999
		NIST standard reference soil material	<74		76.1 %		10.7 %	Ellickson et al. 2001
		Soils from a superfund site in Puerto Rico	<250	51–80 %				Kientz et al. 2003
		Soils from abandoned mining district (Brazil)	<180		70 %			Bosso and Enzweiler 2008
	IVG	Contaminated soils	<250	1.4–64.4 %	32.2 %	0.03–3.23 %	1.06 %	Schroder et al. 2004
		Urban soils of China	<250	16.4–64.9 %	39.1	0.25–7.02 %	1.87 %	Lu et al. 2011
Cd	PBET	Soils from abandoned mine (Spain)	<250	9–99 %	47 %	6.6–52.4 %	27.8 %	Navarro et al. 2006
		Contaminated soils of Turkey	<200	0–98.5 %	67.38 %	0–94.2 %	48.7 %	Karadas and Kara 2011

Table 7.3 (continued)

Element	Method	Sample	Particle size (μm)	Gastric phase		Gastrointestinal phase		Reference
				Range	Average	Range	Average	
	SBET	Urban park soils of Xiamen (China)	–	48–100 %	92 %			Luo et al. 2012a
	UBM	Urban soils contaminated by smelters (France)	<250	58–81 %	68 %	16–59 %	31 %	Roussel et al. 2010
		Contaminated agricultural soils (northern France)	<250		82 %		45 %	Pelfrene et al. 2011b
		Smelter-contaminated soils (northern France)	<250		78±14 %		46±19 %	Pelfrene et al. 2012
Zn	PBET	Soils purchased from ethnic shops (UK)	–			9.8–33.3 %	21.6 %	Abrahams et al. 2006
		Contaminated soils of Turkey	<200	22.4–66.5 %	42.4 %	0.07–39.9 %	22 %	Karadas and Kara 2011
	SBET	Top soils from green area (Torino)	<200	34–43 %	40 %			Madrid et al. 2008a
		Top soils from green area (Sevilla)	<200	32–83 %	55 %			Madrid et al. 2008a
		Top soils of Torino (Italy)	<200		16 %		11 %	Poggio et al. 2009
		Urban park soils of Xiamen (China)	–	14–66 %	39 %			Luo et al. 2012a
		Urban soils of Hong Kong (China)	–		38 %			Luo et al. 2012b
		Soils from mining area (Chingola)	250	23–83 %	45 %			Ettler et al. 2012
		Soils from smelting area (Kitwe)	250	16–79 %	49 %			Ettler et al. 2012

Table 7.3 (continued)

Element	Method	Sample	Particle size (μm)	Gastric phase		Gastrointestinal phase		Reference
				Range	Average	Range	Average	
	UBM	Urban soils contaminated by smelters (France)	<250	17–85 %	47 %	8–47 %	23 %	Roussel et al. 2010
		Contaminated agricultural soils (northern France)	<250		33 %		10 %	Pelfrene et al. 2011b
		Soil around smelters (France)	<250		32±10 %		9±4 %	Pelfrene et al. 2012
	USPM	Soils from a superfund site in Puerto Rico	<250	30–61 %				Kientz et al. 2003
Cu	PBET	Soils purchased from ethnic shops (UK)	–			19.7–54.4 %	37.1 %	Abrahams et al. 2006
		Contaminated soils of Turkey	< 200	5.2–38.6 %	20.67 %	23.6–46.1 %	34.5 %	Karadas and Kara 2011
	SBET	Top soils from green area (Torino)	<200	38–57 %	44 %			Madrid et al. 2008a
		Top soils from green area (Sevilla)	<200	13–24 %	20 %			Madrid et al. 2008a
		Top soils of Torino (Italy)	<200		20 %		35 %	Poggio et al. 2009
		Urban park soils of Xiamen (China)	–	31–84 %	54 %			Luo et al. 2012a
		Urban soils of Hong Kong (China)	–		58 %			Luo et al. 2012b
		Soils from mining area (Chingola)	250	38–83 %	57 %			Ettler et al. 2012
		Soils from smelting area (Kitwe)	250	45–80%	60 %			Ettler et al. 2012

Table 7.3 (continued)

Element	Method	Sample	Particle size (μm)	Gastric phase		Gastrointestinal phase		Reference
				Range	Average	Range	Average	
Cr	PBET	Contaminated soils of Turkey	<200	0.9–6.1 %	2.7 %	1.8–19.4 %	6.4 %	Karadas and Kara 2011
	SBET	Top soils from green area (Torino)	<200	1–6 %	3 %			Madrid et al. 2008a
		Top soils from green area (Sevilla)	<200	4–16 %	9 %			Madrid et al. 2008a
		Urban park soils of Xiamen (China)	–	2.3–35 %	10 %			Luo et al. 2012a
Co	PBET	Soils purchased from ethnic shops (UK)	–			6.8–8.3 %	7.6 %	Abrahams et al. 2006
	SBET	Urban park soils of Xiamen (China)	–	8.3–78 %	27 %			Luo et al. 2012b
		Soils from mining area (Chingola)	250	12–58 %	34 %			Ettler et al. 2012
		Soil from smelting area (Kitwe)	250	7–65 %	38 %			Ettler et al. 2012
Mn	PBET	Soils purchased from ethnic shops (UK)	–			12.0–57.8 %	34.9 %	Abrahams et al. 2006
		Contaminated soils of Turkey	<200	33.6–75.5 %	47.07 %	3.3–65.9 %	37.9 %	Karadas and Kara 2011
	SBET	Urban park soils of Xiamen (China)	–	15–80 %	45 %			Luo et al. 2012a
Mg	PBET	Soils purchased from ethnic shops (UK)	–			5.5–51.7 %	28.6 %	Abrahams et al. 2006
	SBET	Urban park soils of Xiamen (China)	–	4.9–33 %	16 %			Luo et al. 2012a

Table 7.3 (continued)

Element	Method	Sample	Particle size (μm)	Gastric phase		Gastrointestinal phase		Reference
				Range	Average	Range	Average	
Ni	PBET	Soils purchased from ethnic shops (UK)	–			5.2–12.9 %	9.1 %	Abrahams et al. 2006
		Contaminated soils of Turkey	<200	11.8–27.4 %	19.0 %	3.9–22.7 %	14.09 %	Karadas and Kara 2011
	SBET	Top soils from green area (Torino)	<200	8–14 %	12 %			Madrid et al. 2008a
		Top soils from green area (Sevilla)	<200	34–86 %	60 %			Madrid et al. 2008a
		Top soils of Torino (Italy)	<200		7 %		12 %	Poggio et al. 2009
		Urban park soils of Xiamen (China)	–	6.3–64 %	26 %			Luo et al. 2012a
Fe	PBET	Soils purchased from ethnic shops (UK)	–			0.2–4.1 %	2.2 %	Abrahams et al. 2006
	SBET	Urban park soils of Xiamen (China)	–	0.8–12 %	2.9 %			Luo et al. 2012a
Ca	PBET	Soils purchased from ethnic shops (UK)	–				40.1 %	Abrahams et al. 2006
	SBET	Urban park soils of Xiamen (China)	–	51–155 %	88 %			Luo et al. 2012a
Ba	PBET	Contaminated soils (Turkey)	<200	7.8–43.4 %	26.9 %	5.1–28.2 %	17.2 %	Karadas and Kara 2011
K	PBET	Soils purchased from ethnic shops (UK)	–			0.2–5.8 %	3.0 %	Abrahams et al. 2006
Hg	HCl	Surface soils	<180	0.1–29 %	3.5 %	0.2–17 %	1.8 %	Bamett and Turner 2001

– indicates the particles size of soil sample used in in-vitro digestion test was not specifically demonstrated

References

Abrahams PW, Follansbee MH, Hunt A, Smith B, Wragg J (2006) Iron nutrition and possible lead toxicity: an appraisal of geophagy undertaken by pregnant women of UK Asian communities. Appl Geochem 21:98–108

Al-Rmalli SW, Jenkins RO, Watts MJ, Haris PI (2010) Risk of human exposure to arsenic and other toxic elements from geophagy: trace element analysis of baked clay using inductively coupled plasma mass spectrometry. Environ Health 9:79

Bamett MO, Turner RR (2001) Bioaccessibility of mercury in soils. Soil Sediment Contam 10(3):301–316

Bolondi L, Bortolotti M, Santi V, Calletti T, Gaiani S, Labo G (1985) Measurement of gastric emptying time by real-time ultrasonography. Gastroenterology 89(4):752–759

Borgstrom B, Dahlqvist A, Lundh G, Sjovall J (1957) Studies of intestinal digestion and absorption in the human. J Clin Invest 36(10):1521–1536

Bosso ST, Enzweiler J (2008) Bioaccessible lead in soils, slag, and mine wastes from an abandoned mining district in Brazil. Environ Geochem Health 30:219–229

Cabanero AI, Madrid Y, Camara C (2004) Selenium and mercury bioaccessibility in fish samples: an in vitro digestion method. Anal Chim Acta 526:51–61

Calabrese EJ, Stanek EJ, James RC, Roberts SM (1997) Soil ingestion: a concern for acute toxicity in children. Environ Health Perspect 105:1354–1358

Cao X, Wahbi A, Ma L, Li B, Yang Y (2009) Immobilization of Zn, Cu, and Pb in contaminated soils using phosphate rock and phosphoric acid. J Hazard Mater 164:555–564

Carrizales L, Razo I, Tellez-Hernandez JI, Torres-Nerio R, Torres A, Batres LE, Cubillas AC, Diaz-Barriga F (2006) Exposure to arsenic and lead of children living near a copper-smelter in San Luis Potosi, Mexico: importance of soil contamination for exposure of children. Environ Res 101:1–10

Cave M, Wragg J, Klinck B, Gron C, Oomen A, Van de Wiele T, Ollson C, Koch I, Reimer K, Basta N, Tack K (2006) Preliminary assessment of a unified bioaccessibility method for potentially harmful elements in soils. Epidemiology 17:39

Contin M, Mondini C, Leita L, Zaccheo P, Crippa L, De Nobili M (2008) Immobilisation of soil toxic metals by repeated additions of Fe (II) sulphate solution. Geoderma 147:133–140

Correa P (1988) A human model of gastric carcinogenesis. Cancer Res 48:3554–3560

Cui Y, Chen X (2011) Lead (Pb) and arsenic (As) bioaccessibility in various soils from south China. Environ Monit Assess 177:481–492

Cui Y, Du X, Weng L, Van Riemsdijk WH (2010) Assessment of in situ immobilization of lead (Pb) and Arsenic (As) in contaminated soils with phosphate and iron: solubility and bioaccessibility. Water Air Soil Pollut 213:95–104

Cui Y, Fu J, Chen X (2011) Speciation and bioaccessibility of lead and cadmium in soil treated with metal-enriched Indian mustard leaves. J Environ Sci 23(4):624–632

Davis S, Mirick DK (2006) Soil ingestion in children and adults in the same family. J Expo Sci Environ Epidemiol 16:63–75

Davis A, Ruby MV, Bergstrom PD (1994) Factors controlling lead bioavailability in the Butte mining district, Montana, USA. Environ Geochem Health 16(3–4):147–157

De Almeida PDV, Gregio AMT, Machado MAN, De Lima AAS, Azevedo LR (2008) Saliva composition and functions: a comprehensive review. J Contemp Dent Pract 9(3):72–80

Denys S, Caboche J, Tack K, Rychen G, Wragg J, Cave M, Jondreville C, Feidt C (2012) In vivo validation of the unified BARGE method to assess the bioaccessibility of arsenic, antimony, cadmium, and lead in soils. Environ Sci Technol 46:6252–6260

Dodds MWJ, Johnson DA (1993) Influence of mastication on saliva, plaque pH and masseter muscle activity in man. Arch Oral Biol 38(7):623–626

Drexler JW, Brattin WJ (2007) An in vitro procedure for estimation of lead relative bioavailability: with validation. Hum Ecol Risk Assess 13:383–401

Duggan MJ, Inskip MJ (1985) Childhood exposure to lead in surface dust and soil: a community health problem. Public Health Rev 13 (1–2):1–54

Ellickson KM, Meeker RJ, Gallo MA, Buckley BT, Lioy PJ (2001) Oral bioavailability of lead and arsenic from a NIST standard reference soil material. Arch Environ Contam Toxicol 40:128–135

Ettler V, Kribek B, Majer V, Knesl I, Mihaljevic M (2012) Differences in the bioaccessibility of metals/metalloids in soils from mining and smelting areas (Copperbelt, Zambia). J Geochem Explor 113:68–75

Fendorf S, La Force MJ, Li Guangchao (2004) Temporal changes in soil partitioning and bioaccessibility of arsenic, chromium, and lead. J Environ Qual 33:2049–2055

Finch CA, Ragan HA, Dyer IA, Cook JD (1978) Body iron loss in animals. Proc Soc Exp Biol Med 159:335–338

Flecknell P (2002) Replacement, reduction and refinement. Altex 19(2):73–78

Freeman GB, Johnson JD, Liao SC, Davis AO, Ruby MV, Schoof RA, Chaney RL, Bergstrom PD (1994) Absolute bioavailability of lead acetate and mining waste lead in rats. Toxicology 91:151–163

Freeman GB, Dill JA, Johnson JD, Kurtz PJ, Parham F, Matthews HB (1996) Comparative absorption of lead from contaminated soil and lead salts by weanling Fischer 344 rats. Fundam Appl Toxicol 33:109–119

Gbefa BK, Entwistle JA, Dean JR (2011) Oral bioaccessibility of metals in an urban catchment, Newcastle upon Tyne. Environ Geochem Health 33:167–181

Geissler PW, Prince RJ, Levene M, Poda C, Beckerleg SE, Mutemi W, Shulman CE (1999) Perception of soil-eating and anaemia among pregnant women on the Kenyan coast. Soc Sci Med 48:1069–1079

Hamel SC, Buckley B, Lioy PJ (1998) Bioaccessibility of metals in soils for different liquid to solid ratios in synthetic gastric fluid. Environ Sci Technol 32:358–362

Hamel SC, Ellickson KM, Lioy PJ (1999) The estimation of the bioaccessibility of heavy metals in soils using artificial biofluids by two novel methods: mass-balance and soil recapture. Sci Total Environ 243/244:273–283

Hamilton S, Rothenberg SJ, Khan FA, Manalo M, Norris KC (2001) Neonatal lead poisoning from maternal pica behavior during pregnancy. J Natl Med Assoc 93:317–319

Hansen JB, Oomen AG, Edelgaard I, Gron C (2007) Oral bioaccessibility and leaching: tests for soil risk assessment. Eng Life Sci 7(2):170–176

Huang YZ, Hao XW (2012) Effect of red mud addition on the fractionation and bio-accessibility of Pb, Zn and As in combined contaminated soil. Chem Ecol 28(1):37–48

Hubal EAC, Sheldon LS, Burke JM, McCurdy TR, Berry MR, Rigas ML, Zartarian VG, Freeman NC (2000) Children's exposure assessment: a review of factors influencing children's exposure, and the data available to characterize and assess that exposure. Environ Health Perspect 108:475–486

Intawongse M, Dean JR (2008) Use of the physiologically-based extraction test to assess the oral bioaccessibility of metals in vegetable plants grown in contaminated soil. Environ Pollut 152:60–72

Juhasz AL, Weber J, Smith E, Naidu R, Marschner B, Rees M, Rofe A, Kuchel T, Sansom L (2009) Evaluation of SBRC-gastric and SBRC-intestinal methods for the prediction of in vivo relative lead bioavailability in contaminated soils. Environ Sci Technol 43:4503–4509

Juhasz AL, Weber J, Smith E (2011) Impact of soil particle size and bioaccessibility on children and adult lead exposure in peri-urban contaminated soils. J Hazard Mater 186:1870–1879

Karadas C, Kara D (2011) In vitro gastro-intestinal method for the assessment of heavy metal bioavailability in contaminated soils. Environ Sci Pollut Res Int 18:620–628

Karadas C, Kara D (2012) Chemometric evaluation for the relation of BCR sequential extraction method and in vitro gastro-intestinal method for the assessment of metal bioavailability in contaminated soils in Turkey. Environ Sci Pollut Res 19:1280–1295

Kawai K, Saathoff E, Antelman G, Msamanga G, Fawzi WW (2009) Geophagy (soil-eating) in relation to anemia and helminth infection among HIV-infected pregnant women in Tanzania. Am J Trop Med Hyg 80:36–43

Kientz K, Jimenez BD, Perez L, Rodriguez-Sierra CJ (2003) In vitro bioaccessibility of metals in soils from a superfund site in Puerto Rico. Bull Environ Contam Toxicol 70:927–934

Kim SKY (1968) Small intestine transit time in the normal small bowel study. Am J Roentgenol Radium Ther Nucl Med 104(3):522–524

Kim KR, Owens G, Naidu R (2009) Heavy metal distribution, bioaccessibility, and phytoavailability in long-term contaminated soils from Lake Macquarie, Australia. Aust J Soil Res 47:166–176

Kong F, Singh RP (2008) Disintegration of solid foods in human stomach. J Food Sci 73(5):R67–R80

Kou F, Takahama U (1995) Hydrogen peroxide-induced luminescence and evolution of molecular oxygen in human saliva. Arch Oral Biol 40(1):15–21

Krul C, Luiten-Schuite A, Baan R, Verhagen H, Mohn G, Feron V, Havenaar R (2000) Application of a dynamic in vitro gastrointestinal tract model to study the availability of food mutagens, using heterocyclic aromatic amines as model compounds. Food Chem Toxicol 38:783–792

Krul C, Luiten-Schuite A, Tenfelde A, Van Ommen B, Verhagen H, Havenaar R (2001) Antimutagenic activity of green tea and black tea extracts studied in a dynamic in vitro gastrointestinal model. Mutat Res 474:71–85

Laird BD, Peak D, Siciliano SD (2011) Bioaccessibility of metal cations in soil is linearly related to its water exchange rate constant. Environ Sci Technol 45:4139–4144

Le Bot B, Gilles E, Durand S, Glorennec P (2010) Bioaccessible and quasi-total metals in soil and indoor dust. Eur J Mineral 22:651–657

Li Y, Zhang M (2012) A comparison of physiologically based extraction test (PBET) and single-extraction methods for release of Cu, Zn, and Pb from mildly acidic and alkali soils. Environ Sci Pollut Res. doi:10.1007/s11356-012-1234-0

Li Y, Demisie W, Zhang M (2013) The function of digestive enzymes on Cu, Zn, and Pb release from soil in in vitro digestion tests. Environ Sci Pollut Res. doi:10.1007/s11356-013-1472–9

Lim HS, Lee JS, Chon HT, Sager M (2008) Heavy metal contamination and health risk assessment in the vicinity of the abandoned Songcheon Au-Ag mine in Korea. J Geochem Explor 96:223–230

Liu R, Zhao D (2007) Reducing leachability and bioaccessibility of lead in soils using a new class of stabilized iron phosphate nanoparticles. Water Res 41:2491–2502

Lu Y, Yin W, Huang L, Zhang G, Zhao Y (2011) Assessment of bioaccessibility and exposure risk of arsenic and lead in urban soils of Guangzhou City, China. Environ Geochem Health 33:93–102

Luo X, Yu S, Li X (2011) Distribution, availability, and sources of trace metals in different particle size fractions of urban soils in Hong Kong: implications for assessing the risk to human health. Environ Pollut 159:1317–1326

Luo XS, Ding J, Xu B, Wang YJ, Li HB, Yu S (2012a) Incorporating bioaccessibility into human health risk assessment of heavy metals in urban park soils. Sci Total Environ 424:88–96

Luo X, Yu S, Li X (2012b) The mobility, bioavailability, and human bioaccessibility of trace metals in urban soils of Hong Kong. Appl Geochem 27:995–1004

Madrid F, Biasioli M, Ajmone-Marsan F (2008a) Availability and bioaccessibility of metals in fine particles of some urban soils. Arch Environ Contam Toxicol 55:21–32

Madrid F, Diaz-Barrientos E, Madrid L (2008b) Availability and bio-accessibility of metals in the clay fraction of urban soils of Sevilla. Environ Pollut 156:605–610

Man YB, Sun XL, Zhao YG, Lopez BN, Chung SS, Wu SC, Cheung KC, Wong MH (2010) Health risk assessment of abandoned agricultural soils based on heavy metal contents in Hong Kong, the world's most populated city. Environ Int 36:570–576

Marschner B, Welge P, Hack A, Wittsiepe J, Wilhelm M (2006) Comparison of soil Pb in vitro bioaccessibility and in vivo bioavailability with Pb pools from a sequential soil extraction. Environ Sci Technol 40:2812–2818

Mercier G, Duchesne J, Carles-Gibergues A (2002) A simple and fast screening test to detect soils polluted by lead. Environ Pollut 118:285–296
Miller RM (1995) Biosurfactant-facilitated remediation of metal-contaminated soils. Environ Health Perspect 103(Suppl 1):59–62
Miller DD, Schricker BR, Rasmussen RR, Campen DV (1981) An in vitro method for estimation of iron availability from meals. Am J Clin Nutr 34:2248–2256
Minekus M, Marteau P, Havenaar R, Huis in't Veld JHJ (1995) A multicompartmental dynamic computer-controlled model simulating the stomach and small intestine. ATLA 23(2):197–209
Navarro MC, Perez-Sirvent C, Martinez-Sanchez MJ, Vidal J, Marimon J (2006) Lead, cadmium and arsenic bioavailability in the abandoned mine site of Cabezo Rajao (Murcia, SE Spain). Chemosphere 63:484–489
Oomen AG, Hack A, Minekus M, Zeijdner E, Cornelis G, Verstraete W, De Wiele TV, Wragg J, Rompelberg CJM, Sips AJAM, Van Wijnen JH (2002) Comparison of five in vitro digestion models to study the bioaccessibility of soil contaminants. Environ Sci Technol 36:3326–3334
Oomen AG, Tolls J, Sips AJAM, Groten JP (2003a) In vitro intestinal lead uptake and transport in relation to speciation. Arch Environ Contam Toxicol 44:116–124
Oomen AG, Rompelberg CJM, Bruil MA, Dobbe CJG, Pereboom DPKH, Sips AJAM (2003b) Development of an in vitro digestion model for estimating the bioaccessibility of soil contaminants. Arch Environ Contam Toxicol 44:281–287
Oomen AG, Rompelberg CJM, Kamp EV, Pereboom DPKH, Zwart LLD, Sips AJAM (2004) Effect of bile type on the bioaccessibility of soil contaminants in an in vitro digestion model. Arch Environ Contam Toxicol 46:183–188
Pedersen AM, Bardow A, Jensen SB, Nauntofte B (2002) Saliva and gastrointestinal functions of taste, mastication, swallowing and digestion. Oral Dis 8:117–129
Peijnenburg WJGM, Posthuma L, Eijsackers HJP, Allen HE (1997) A conceptual framework for implementation of bioavailability of metals for environmental management purposes. Ecotoxicol Environ Saf 37:163–172. doi:10.1006/eesa.1997.1539
Pelfrene A, Waterlot C, Douay F (2011a) In vitro digestion and DGT techniques for estimating cadmium and lead bioavailability in contaminated soils: influence of gastric juice pH. Sci Total Environ 409:5076–5085
Pelfrene A, Waterlot C, Mazzuca M, Nisse C, Bidar G, Douay F (2011b) Assessing Cd, Pb, Zn human bioaccessibility in smelter-contaminated agricultural topsoils (northern France). Environ Geochem Health 33:477–493
Pelfrene A, Waterlot C, Mazzuca M, Nisse C, Cuny D, Richard A, Denys S, Heyman C, Roussel H, Bidar G, Douay F (2012) Bioaccessibility of trace elements as affected by soil parameters in smelter-contaminated agricultural soils: a statistical modeling approach. Environ Pollut 160:130–138
Poggio L, Vrscaj B, Schulin R, Hepperle E, Marsan FA (2009) Metals pollution and human bioaccessibility of topsoils in Grugliasco (Italy). Environ Pollut 157(2):680–689
Rao BSN, Prabhavathi T (1978) An in vitro method for predicting the bioavailability of iron from foods. Am J Clin Nutr 31:169–175
Rasmussen PE, Beauchemin S, Nugent M, Dugandzic R, Lanouette M, Chenier M (2008) Influence of matrix composition on the bioaccessibility of copper, zinc, and nickel in urban residential dust and soil. Hum Ecol Risk Assess 14:351–371
Rodriguez RR, Basta N (1999) An in vitro gastrointestinal method to estimate bioavailable arsenic in contaminated soils and solid media. Environ Sci Technol 33:642–649
Rotard W, Christmann W, Knoth W, Mailahn W (1995) Bestimmung der resorptionsverfugbaren PCDD/PCDF aus Kieselrot. UWSF-Z Umweltchem Okotox 7:3–9
Roussel H, Waterlot C, Pelfrene A, Pruvot C, Mazzuca M, Douay F (2010) Cd, Pb and Zn oral bioaccessibility of urban soils contaminated in the past by atmospheric emissions from two lead and zinc smelters. Arch Environ Contam Toxicol 58:945–954
Ruby MV, Davis A, Kempton JH, Drexler JW, Bergstrom PD (1992) Lead bioavailability: dissolution kinetics under simulated gastric conditions. Environ Sci Technol 26:1242–1248

Ruby MV, Davis A, Link TE, Schoof R, Chaney RL, Freeman GB, Bergstrom P (1993) Development of an in vitro screening test to evaluate the in vivo bioaccessibility of ingested mine-waste lead. Environ Sci Technol 27:2870–2877
Ruby MV, Davis A, Schoof R, Eberle S, Sellstone CM (1996) Estimation of lead and arsenic bioavailability using a physiologically based extraction test. Environ Sci Technol 30:422–430
Ruby MV, Schoof R, Brattin W, Goldade M, Post G, Harnois M, Mosby DE, Casteel SW, Berti W, Carpenter M, Edwards D, Cragin D, Chappell W (1999) Advances in evaluating the oral bioavailability of inorganics in soil for use in human health risk assessment. Environ Sci Technol 33(21):3697–3705
Samloff IM, O'Dell C (1985) Inhibition of peptic activity by sucralfate. Am J Med 79(2):15–18
Schenkels LCPM, Veerman ECI, Amerongen AVN (1995) Biochemical composition of human saliva in relation to other mucosal fluids. Crit Rev Oral Biol Med 6(2):161–175
Schroder JL, Basta NT, Si J (2003) In vitro gastrointestinal method to estimate relative bioavailable cadmium in contaminated soil. Environ Sci Technol 37:1365–1370
Schroder JL, Basta NT, Casteel SW, Evans TJ, Payton ME, Si J (2004) Validation of the in vitro gastrointestinal (IVG) method to estimate relative bioavailable lead in contaminated soils. J Environ Qual 33:513–521
Sialelli J, Urquhart GJ, Davidson CM, Hursthouse AS (2010) Use of a physiologically based extraction test to estimate the human bioaccessibility of potentially toxic elements in urban soils from the city of Glasgow, UK. Environ Geochem Health 32:517–527
Sialelli J, Davidson CM, Hursthouse AS, Ajmone-Marsan F (2011) Human bioaccessibility of Cr, Cu, Ni, Pb and Zn in urban soils from the city of Torino, Italy. Environ Chem Lett 9:197–202
Smith BA, Kirk JL, Stephenson GL (2010) The influence of liquid to soil ratios on arsenic and lead bioaccessibility in reference and field soil. Hum Ecol Risk Assess 16:149–162
Smith E, Weber J, Naidu R, McLaren RG, Juhasz AL (2011) Assessment of lead bioaccessibility in peri-urban contaminated soils. J Hazard Mater 186:300–305
Stewart MA, Jardine PM, Barnett MO, Mehlhorn TL, Hyder LK, McKay LD (2003) Influence of soil geochemical and physical properties on the sorption and bioaccessibility of chromium (Đ¨). J Environ Qual 32:129–137
Tang XY, Zhu YG, Chen SB, Tang LL, Chen XP (2004) Assessment of the effectiveness of different phosphorus fertilizers to remediate Pb-contaminated soil using in vitro test. Environ Int 30:531–537
Tang XY, Zhu YG, Cui YS, Duan J, Tang L (2006) The effect of ageing on the bioaccessibility and fractionation of cadmium in some typical soils of China. Environ Int 32:682–689
Tang XY, Cui YS, Duan J, Tang L (2008) Pilot study of temporal variations in lead bioaccessibility and chemical fractionation in some Chinese soils. J Hazard Mater 160(1):29–36
Turner A, Radford A (2010) Bioaccessibility of trace metals in boat paint particles. Ecotoxicol Environ Saf 73:817–824
Turner A, Singh N, Richards JP (2009) Bioaccessibility of metals in soils and dusts contaminated by marine antifouling paint particles. Environ Pollut 157:1526–1532
Vicky R (2005) Finding alternatives: an overview of the 3Rs and the use of animals in research. School Sci Rev 87(319):111–114
Watt J, Thornton I, Cotter-Howells J (1993) Physical evidence suggesting the transfer of soil Pb into young children via hand-to-mouth activity. Appl Geochem 8:269–272
Welfringer B, Zagury G (2009) Evaluation of two in vitro protocols for determination of mercury bioaccessibility: influence of mercury fractionation and soil properties. J Environ Qual 38:2237–2244
Whitcomb DC, Lowe ME (2007) Human pancreatic digestive enzymes. Dig Dis Sci 52:1–17
Wragg J, Cave M, Basta N, Brandon E, Casteel S, Denys S, Gron C, Oomen A, Reimer K, Tack K, Wiele TV (2011) An inter-laboratory trial of the unified BARGE bioaccessibility method for arsenic, cadmium and lead in soil. Sci Total Environ 409:4016–4030.
Xiao-Wei H, Yi-Zong H, Yan-Shan C (2010) Effect of bone char addition on the fractionation and bio-accessibility of Pb and Zn in combined contaminated soil. Acta Ecologica Sinica 30:118–122

Xue J, Zartarian V, Moya J, Freeman N, Beamer P, Black K, Tulve N, Shalat S (2007) A meta-analysis of children¢s hand-to-mouth frequency data for estimating nondietary ingestion exposure. Risk Anal 27(2):411–420

Yang J, Mosby D (2006) Field assessment of treatment efficacy by three methods of phosphoric acid application in lead-contaminated urban soil. Sci Total Environ 366:136–142

Yang J, Mosby DE, Casteel SW, Blanchar RW (2001) Lead immobilization using phosphoric acid in a smelter-contaminated urban soil. Environ Sci Technol 35:3553–3559

Yang J, Mosby ED, Casteel SW, Blanchar RW (2002) In vitro lead bioaccessibility and phosphate leaching as affected by surface application of phosphoric acid in lead-contaminated soil. Arch Environ Contam Toxicol 43:399–405

Zapusek U, Lestan D (2009) Fractionation, mobility and bio-accessibility of Cu, Zn, Cd, Pb and Ni in aged artificial soil mixtures. Geoderma 154:164–169

Zhang XB, Ohta Y (1991) In vitro binding of mutagenic pyrolyzates to lactic acid bacterial cells in human gastric juice. J Dairy Sci 74(3):752–757

Chapter 8
Biosafety Risk of Genetically Modified Crops Containing *Cry* Genes

Mehboob-ur-Rahman, Tayyaba Shaheen, Samra Irem and Yusuf Zafar

Contents

Abstract Genomic technologies started in the early 1980s to improve the genomes of cultivated crop species. For example the term "Bt" comes from the soil bacterium *Bacillus thuringiensis* containing genes, e.g. *Cry1Ac*, *Cry2Ab*, *Cry1F*, *Cry3Bb1*, that provides protection against lepidopteran insect pests. Those genes have been

Mehboob-ur-Rahman (✉) · T. Shaheen · S. Irem · Y. Zafar
Plant Genomics & Molecular Breeding Lab, National Institute for Biotechnology & Genetic Engineering (NIBGE), P.O. Box 577, Jhang Road, Faisalabad, Pakistan

Department of Bioinformatics and Biotechnology, Government College University, Faisalabad, Pakistan
e-mail: mehboob_pbd@yahoo.com

© Springer International Publishing Switzerland 2015

E. Lichtfouse et al. (eds.), *CO_2 Sequestration, Biofuels and Depollution*,
Environmental Chemistry for a Sustainable World 5, DOI 10.1007/978-3-319-11906-9_8

inserted in crops such as corn, cotton, soybean, rice, potato and canola released for cultivation in mid 1990s in USA, and later in many other countries like China and India. About 29 countries commercialized genetically-modified (GM) or 'transgenic' crops while 30 countries granted regulatory approvals for planting GM-crops; together making 75 % of the world population. Potential harmful effects of the Bt-crops on non-targets were quantified before releasing such non-conventional crops into the environment. The cultivation of Bt-crops were most commonly found safe, based on various studies including the insertional impact of transgene and its regulatory elements on plant phenotype and agronomic performance, effect on non-target organisms (NTOs) and nutritional impacts on multiple experimental models. Albeit the studies were conducted for limited durations. However, the skeptics always claim for conducting extensive clinical as well as field trials, and also doubt on methods and procedures of calculating the ecological risks. This debate is still on-going, especially after reports on substantial reduction of monarch butterfly caterpillars exposed to Bt-maize pollen, though later nullified; and detection of traces of transgene in various tissues of experimental animals. Procedures, methods and protocols for evaluating potential risks of GM-crops and foods should be standardized as the first step to build trust of researchers and end-users. Many efforts should be exerted in deploying genes of interest, marker genes and regulatory sequences invoking no or little issues of potential risks to the ecosystem.

Keywords GM-crops • Bt-crops • Cry genes • Risk assessment • Safety evaluation • Genotoxicity • Blood biochemistry • Allergic response • Non-target organisms • Mammals • Birds • Human

Abbreviations

GM	Genetically modified
GE	Genetically engineered
Bt	Bacillus thuringiensis
NTOs	Non-target organisms
PIPs	Plant-incorporated protectants
CaMV35 S	Cauliflower mosaic virus 35 S promoter
GFP gene	Green fluorescent protein gene
nptII gene	Neomycin phosphotransferase gene
IgE	Immunoglobulin E
IgG	Immunoglobulin G
ELISA	Enzyme-linked immunosorbent assay
PCR	Polymerase chain reaction
MBC	Biomass carbon
MBN	Biomass nitrogen

8.1 What are Transgenic Crops?

Genomic techniques, such as genetic engineering—emerged rapidly over the last two decades, has made possible the introduction of gene(s) isolated from alien species or even synthetic gene(s), into a plant species, called as genetically engineered or genetically modified (GM) plant, and their product or byproducts are used as food are referred as GM-food. Plant derived GM-foods comprising staples such as soybean, maize, canola, rice, and potatoes, have been commercialized (Table 8.1). Expression of unique and favorable traits which are beneficial for the consumers is attributed to this technology (Magana-Gomez and Barca 2009).

First GM-crop was commercialized in 1996, and since that many other crops like GM- soybean, maize, cotton, potato, and canola have been made public. We have witnessed the rapid expansion in global area of GM-crops with a sustained growth 8 % (160 million hectares, James 2011). GM-crops have been classified on the basis of the introduced trait. First generation GM-crops are derived for enhanced production; however, the crops are not considerably different from their conventional equivalents except these crops have genes for combating plant disease, insect pests, viruses and herbicides, exhibiting that these are similar in taste, appearance and nutritional value for the consumers. While, second generation of GM plants comprises of crops containing new traits of direct value to the consumers. It offer benefits to the processor, end-user, and consumer.

Researchers are going to introduce a third generation of GM plants by manipulating their genomes. These plants will have a greater ability to combat abiotic stresses such as drought, high temperatures, and salinity. Moreover, some modified crops are able to provide food with supplemental health benefits or renewable raw materials. This generation also includes "pharmaplants", which are used as biological production systems for producing high-grade active pharmaceutical elements (Magana-Gomez and Barca 2009).

8.2 Spectrum of Bt Genes Diversity

8.2.1 Discovery of Bt Genes

The bacterium *Bacillus thuringiensis* (*Bt*) was first discovered by Japanese biologist, Shigetane Ishiwatari, trying to interpret the cause of the sotto disease (sudden-collapse disease) that was challenging the populations of silkworms in 1901. Later in 1911 Ernst Berliner found a bacterium that killed a Mediterranean flour moth, named as *Bacillus thuringiensis*, after the name of German town Thuringia where the moth was found. Presence of crystal was discovered in Bt in 1915 (Berliner, 1915), but its activity was described much later. In the US, Bt was registered as a pesticide in 1961. In the 1980's, deployment of Bt sprays was substantially increased when insect pests became increasingly resistant to the synthetic insecticides. Also

Table 8.1 Countries growing more than 0.10 million hectares of biotech crops. (Source: Clive James 2011)

Sr No	CN	Total Area (million hectares)	Bt-Crops	Insect toxin genes	Year of adoption of Bt-crops
1	Argentina	22.9	Cotton	*Cry1Ac*	1998
			Corn	*Cry1Ab*	1996
			Corn	*Cry1F*	2005
2	Australia	0.7	Cotton	*Cry1Ac*	1996
			Cotton	*Cry1Ac+ Cry2Ab*	2002
3	Brazil	25.4	Cotton	*Cry1Ac*	2005
			Cotton	*Cry1Ac+ Cry2Ab*	2009
			Cotton	*Cry1Ac+ Cry1F*	2009
			Corn	*Cry1Ab*	2008
			Corn	*Cry1F*	2009
			Corn	*VIP3Aa20*	2009
			Corn	*Cry1A.105 + Cry2Ab*	2009
4	Burkina Faso	0.3	Cotton	*Cry1Ac+ Cry2Ab*	2008
5	Canada	8.8	Corn	*Cry1Ab*	1997
			Corn	*Cry1F*	2002
			Corn	*Cry3Bb1*	2003
			Corn	*Cry34Ab1 + Cry35Ab1*	2005
			Corn	*mCry3A*	2007
			Corn	*Cry1A.105 + Cry2Ab2*	2008
6	China	3.5	Cotton	*Cry1Ac*	1997
			Cotton	*Cry1A*^{b} *+ CPTI*	1999
			Rice	*Cry1Ab/cry1Ac*	2009
			Poplar	*Cry1Aa*	2008
7	India	9.4	Cotton	*Cry1Ac*	2002
			Cotton	*Cry1Ac + Cry2Ab*	2006
			Cotton	*Cry1Ac + Cry1Ab*	2006
8	Mexico	0.1	Cotton	*Cry1Ac*	1997
9	Myanmar	0.3	Cotton	NA	2006
10	Pakistan	2.4	Cotton	*Cry1Ac*	2010
11	Philippines	0.5	Corn	*Cry1Ab*	2002
12	South Africa	2.2	Corn	*Cry1Ab*	2001
			Cotton	*Cry1Ac*	1997
			Cotton	*Cry1Ac + Cry2Ab*	2005
13	Spain	0.1	Corn	*Cry1Ab*	2003
14	USA	66.8	Cotton	*Cry1Ac*	1995
			Cotton	*Cry1Ac + Cry2Ab*	2002
			Cotton	*Cry1Ac + Cry1F*	2004
			Cotton	*Cry1F*	2004

Table 8.1 (continued)

Sr No	CN	Total Area (million hectares)	Bt-Crops	Insect toxin genes	Year of adoption of Bt-crops
			Corn	*Cry1Ab*	1995
			Corn	*Cry3Bb1*	1996
			Corn	*Cry1F*	2001
			Corn	*Cry34Ab1* + *Cry35Ab1*	2005
			Corn	*mCry3A*	2007
			Corn	*Cry1A.105* + *Cry2Ab2*	2008
			Soybean	*Cry1Ac*	2011
			Potato	*Cry3A*	1996
15	Uruguay	1.1	Corn	*Cry1Ab*	2003
			Corn	*Cry1F*	2006

NA Not Available

the chemicals were found to be hazardous to the environment. (www.bt.ucsd.edu/bt_history.html).

Bacillus thuringiensis (*Bt*) produces insecticidal crystal proteins, solubilized in the larval midguts, are activated by the midgut proteases. Numerous kinds of Cry proteins have been isolated which were found to be toxic for different orders of the insect family. To achieve high level of expression in plants, extensive modifications (with respect to codon-bias) of the *Cry* genes are required (Sudha et al. 1999).

A number of Bt genes have been introduced in crops such as cotton (*Cry1Ac, Cry2Ab2, Cry1Fa2*), maize (*Cry1Ab, Cry1Ac, Cry1Fa2, Cry3Bb1, Cry9C*), and potato (*Cry3Aa)* (Federici 2002; Shelton et al. 2002; Hellmich and Hellmich 2012, Fig. 8.1). Engineered chimeric Bt toxins in PIPs (e.g., a *Cry1Ac/Cry1Fa* hybrid protein; Perlak et al. 2001), binary Bt toxins (Baum et al. 2004; Ellis et al. 2002), as well as hybrid Bt toxins targeting multiple insect orders were introduced. Moreover, crops like apple, broccoli, cabbage, tobacco, tomato, soybean, and rice have also been engineered to express Bt genes (Huesing and English 2004).

8.2.2 Bt-Crops

GM-crops are the most popular commodities in agriculture, and at the same time are the most controversial from biosafety point of view (Tabashnik 2010). In the early 1980s, GM plants were developed by four groups independently at Washington University in St. Louis, Missouri, the Rijksuniversiteit in Ghent, Belgium, Monsanto Company in St. Louis, Missouri, and the University of Wisconsin. The first three groups claimed at a conference in Miami, Florida on the same day in January 1983 that they had introduced bacterial genes into the plants. On the other hand, in 1983, the fourth group claimed in a conference held in Los Angeles, California, for the

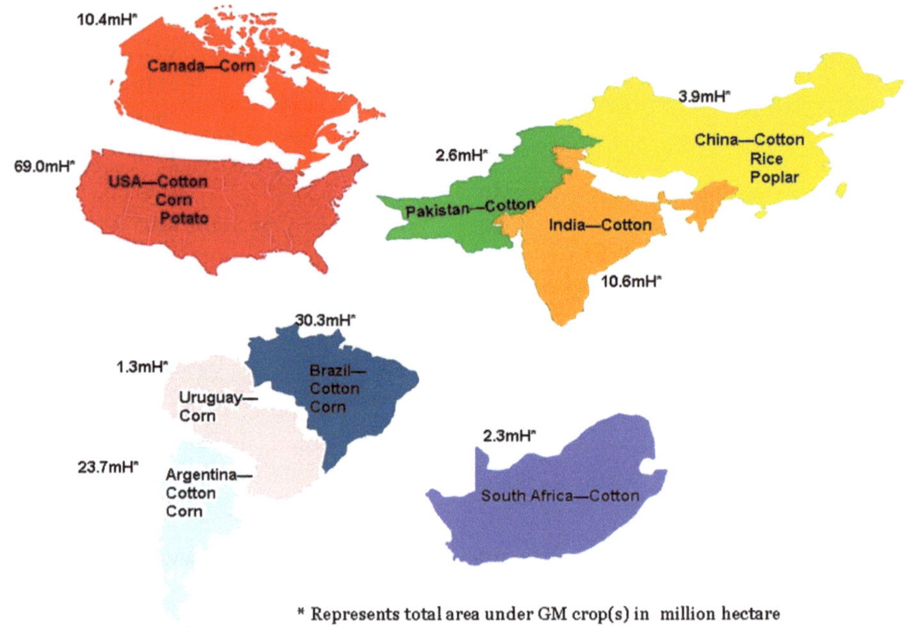

Fig. 8.1 Genetically modified (GM) Bt crops grow on more that 1 million hectares worldwide

insertion of a gene from one species to another. Similarly, a GM tobacco containing kanamycin resistant gene was produced by a group headed by Marry Dell Chilton (Framond et al. 1983). Jeff Schell and Marc Van Montagu, working in Belgium, produced tobacco plants showing resistance to kanamycin and to methotrexate-a drug used to treat cancer and rheumatoid arthritis (Schell et al. 1983).

In the 1990s, a first commercially grown GM-tomato was produced by California based Company called the FlavrSavr, for improving the shelf life (takes longer to decompose after being picked). A variety of the tomato was used to make tomato puree that was sold in Europe in the mid-1990s, but later many safety concerns were raised over GM-crops. Since 1995, many GM-crops including soybean, barley, potato, cotton, corn etc were commercialized (Rahman et al. 2012) (www.gmcrops.ewebsite.com/articles/history). Cotton and corn have been genetically modified to mitigate utility of insecticide sprays. Before the Bt-corn was introduced, insect pests have caused losses of $ 1 billion per year in the United States (Tabashnik 2010).

8.2.3 *Targeted Insect Pests Species*

Bacillus thuringiensis produces a diverse group of Cry and Cyt proteins which act as toxin to most of the insect pest species belonging to orders Lepidoptera, Coleoptera and Diptera. These toxins cause lysis of midgut epithelial cells that leads to pore formation by implanting into the target membrane of the insect pests. Similar to other pore-forming toxins, Cry toxins also attach precisely with receptors present on the host cell surface. These receptors are stimulated by the host proteases after receptor binding which result into a pre-pore oligomeric structure that is insertion competent (Bravo et al. 2007). These toxins are very specific to a small range of insect pests, and this specificity is attributed to specific pH levels, enzymes, and furthermore specific midgut receptors (Federici 2002). This specificity can be explained by a "lock and key" theory. Insect death will only occur if the lock and key matches. For example, midgut receptor can be considered as "lock" and the Cry protein can be considered as "key" (Hellmich and Hellmich 2012).

8.3 Benefits of the Adoption of Bt-Crops

GM-crops are being cultivated on 160 million hectares in about 29 countries worldwide (James 2011). An important trait used in GM plants is resistance to insect pests; which is attributed to several Bt proteins. Development of GM–crops containing Bt genes is a step towards making agricultural system profitable for the producers including small farming community through increased earnings by reduction in chemical pesticides as well as farm labor required to protect crops from the insect pests infestation (Ismael et al. 2002; Pray et al. 2002; Huang et al. 2002; Morse et al. 2004; Qaim and de Janvry 2005; Gandhi and Namboodiri 2006; Crost et al. 2007; Dev and Rao 2007; Pray and Naseem 2007). Profitability earned by cultivating GM-crops may be diverted to improve the quality of life (Pray et al. 2001; James 2002; Mal et al. 2011; Hellmich and Hellmich 2012). Also, the GM technology helps in saving time of the women and children working as a farm labor in most developing countries, sparing them to engage in household and educational activities—may have high social significance for a society (Purcell and Perlak 2004; FAO 2004; Zilberman et al. 2007; Huang et al. 2008; Krishna and Qaim 2007).

Both the macroeconomic level outcomes (de Janvry and Sadoulet 2002; Elbehri and Macdonald 2004; Huang et al. 2004; Anderson et al. 2008), and micro-economic level effects (Subramanian and Qaim 2009) of cultivating Bt-crops investigated in different countries; showed that whole farming community in India including small and big farmers can reap benefits by cultivating Bt-cotton. Later, a significant impact of Bt-cotton cultivation to mitigate poverty was observed in India (Subramanian and Qaim 2010). Such commonalities of increased yield per-hectare were observed in China (Huang et al. 2010; Hu et al. 2009) and Pakistan.Fluctuations in yield are largely due to weather conditions and pest pressure. Similarly, cultivation of other Bt-crops like egg plant and rice, compared to their non-Bt counterpart, will also add in the farm income by cutting down the cost of pesticides and farm labor. In India, the Bt-egg plant will be commercialized once it gets approval from biosafety regulators (Pray and Nagarajan 2011).

Other indirect benefit of Bt-crops is the substantial reduction in lepidopteron insect pest populations—not requiring chemical pesticides to apply on non-Bt-crops. For example, lepidopteron populations in cotton have been substantially declined in China (Huang et al. 2010) as well as in corn in the US (Hutchison et al. 2010). In contrary to this, a few insect pests species appeared to be minor in the past, are emerging as major pests because of less use of insecticides on Bt-crops. For instance, Mirids in China (Wang et al. 2008; Lu et al. 2010) and mealy bug in Pakistan (Arif et al. 2009) have been reported as emergent potential pests.

8.4 Assessment of Potential Risks of Bt-Crops to the Ecosystem

8.4.1 Procedures and Methods

Since the development of first transgenic plant, debates and discussions on the safe release and their uses have been initiated which resulted in formulating guidelines for assessing the safety of foods derived from GM-crops by a group of international experts on food safety evaluation. Some non-GM activists still have divergent views, e.g. demand for long term safety assessment by adopting high stringent conditions which are even more rigorous than for any other foods. Methods for testing the safety of GM-crops and their byproducts have strengths as well as weaknesses. Guidelines designed to regulate the introduction of GM microbes and plants into the environment, found to have some critical gaps in the scientific knowledge concerning the compositional effects of genetic transformation and also in the safety testing procedures (Malik 2011). Similarly, the concept of substantial equivalence was introduced in 1993 for comparing the properties of GM-food with its conventional counterpart, and the GM-food will be regarded as safe as its conventional counterpart after establishing the substantial equivalence and no further safety consideration is needed—concluded by the a Joint Food and Agriculture Organization (FAO)/World Health Organization (WHO), Expert Consultation on Biotechnology and Food Safety. It was found imperative for growing GM and its parental varieties under similar conditions for making comparisons of key compounds as well as the genotypic and phenotypic differences. It is much likely that some low-content compounds of plants with biological activity may remain unknown. Secondly, unintended consequences of GM-crops have been reported in many GM-crops. For example, higher lignin contents in Bt-maize than its non-Bt-maize, depleted plant flavonoids in herbicide tolerant soybean etc. have been reported (for details please consult the review of Kuiper et al. 2001). Hence, the concept of substantial equivalence is not an acceptable method for GM evaluation because of its inability to detect unintended effects. Theoretically, the unintended changes may arise due to the insertion of genetic construct, gene regulation, gene-gene interactions, and also possible interferences in metabolic pathways. For predicting such changes, DNA-based technologies such as DNA analysis, DNA/mRNA microarray hybridization, and proteomics and chemical fingerprinting (metabolomics) are required for quantifying the differences in GM-crops and their non-GM counterpart. In early days of GM-crop development, these methods were not available or were at their infancy. Now it is accepted that for safety assessment of any GM-crop, substantial equivalence together with other parameters such as molecular characterization, phenotypic characteristics, key nutrients, toxicants, and allergens should be considered—based on the guide lines prepared by the International Life Sciences Institute Europe and FAO/WHO in 1996. Despite the official standards for food safety evaluation published by the *Codex Alimentarius* Commission of FAO/WHO in 2003, risk assessment guidelines of GM-foods have not adopted as described. In the present review, we summarize the results of multiple published studies of safety assessment of GM-crops containing different Bt-genes (Table 1). It is agreed that the safety evaluation of GM-crops will be conducted on a case-by-case basis (Pakistan national biosafety rules 2005). Also, it was emphasized that standardized methods (Malik 2011), designs and statistical analysis for conducting animal feeding trials should be followed. Now, the deployment of high-tech tools, (transcriptomics, proteomics, metabolomics etc.) for assessing the risk of GM-crops are getting much more attention for comparing the profiles of GM with its non GM counterpart.

8.4.2 DNA-Based Genotoxity Test

The comet assay, described by Singh et al. in 1988, is used to detect the extent of DNA damage at individual cell level—one of the indicators for evaluating genotoxicity of GM-crops. In this test, amount of sheared genomic DNA (genomic DNA degrades after exposing to various mutagens) which forms a tail like structure is calculated by fluorescence. A numerical value is assigned to each of the migrating genomic DNA for quantifying the extent of genotoxicity (Tice et al. 2000). In this particular test, tail length and the percentage of DNA damaged cells are important parameters for estimating the impact of genotoxicity. A study was conducted using the organ samples of rabbits fed on Bt-cotton as well as conventional cotton type. No significant difference for damaged cell (2–3 %) within and between the normal and Bt-cotton exposed groups were found, highlighting that the transgenic cotton containing *Cry1Ac* gene is quite safe for other than target organisms (Rahman and Co-workers, unpublished results).

8.4.3 Potential Harmful Effects of Bt-Foods to Mammals

Globally, numerous studies for assessing the environmental risk of Bt technology have been conducted in different countries on multiple Bt-crop species such as maize, potato, soybean, brinjal, popular, rice and cotton (Fares and El-Sayed 1998; Marvier 2001; Shelton et al. 2002; Conner et al. 2003; Hilbeck et al. 2006; Jayaraman 2009; James 2011). However, studies elucidating the harmful effect of Bt gene on the feeding safety of non-target organisms have not been conducted in systematic manner (Magaña-Gómez et al. 2008), and the information generated is limited

(Dona and Arvanitoyannis 2009). Earlier, rodents (rats) were exposed for 90 days to the semi-synthetic diet containing 10% (w/w) of lyophilized powder of Bt-tomato and non-Bt-tomato (Noteborn and Kuiper 1994), and the average consumption was ~200 g tomato day^{-1} per rodent. Based on multiple clinical, toxicological or histopathological studies, it was found that the group of rats fed on diet containing Bt-tomato is safe. In another study, no toxic impact was found by feeding rats for 90 days with diet containing transgenic tomato (*Cry1Ab*) through studying various morphological features such as body weight and different organs weight, feed consumption rate, and blood chemistry and histopathology (Noteborn et al. 1995). Similarly, sheep were exposed to GM-corn (containing *Cry 1A*) and conventional corn, which was found safe as the values for change in body weight gain and feeding were nonsignificantly different (Barriere et al. 2001). Such commonalities were found in many other studies conducted on different animals, such as chicken fed on GM-corn containing *Cry9c* gene (Yonemochi et al. 2002) and dairy cattle exposed to GM-corn containing *Cry1Ab* gene (Donkin et al. 2003). Similarly, risk assessment studies on GM-rice containing *Cry1Ac* gene was conducted by exposing rat to the GM-rice as well as to the non-GM rice. No substantial differences among the two groups of rats were found in the clinical trials including animal behavior, weight gain, hematological and biochemical parameters, macroscopic and histopathological tests of organs (Schroder et al. 2007). Also, in another set of experiments, the non-toxic impacts of Bt pesticidal protein to aquatic animals like fish, and mammals and invertebrates were reported (Xu-Chongren and Chang 2001).

In multiple investigations, various experimental animals such as mice, zebra fish and eelworms were exposed to Bt-cotton plants/seeds/leaves. Based on acute and chronic toxicity trials, and also the genotoxicity experiments, each of the animals responded normally when fed on Bt-transgenic cotton plants/seeds/parts. Such conclusions have also been drawn by feeding catfish and Northern Bobwhite Quail to Bt-cotton seed meal (Li and Robinson 2000; Campbell 1985; Gallagher et al. 2000).

Recently, a study was conducted for evaluating the safety assessment of Bt-cotton (containing *Cry 1Ac*, Mon531) in Pakistan, the fourth largest producer of cotton in the world. In this study, various clinical trials such as sign of allergenicity, weekly weight gain, hematological parameters, histopathological studies etc. were conducted on two groups of rabbits (one group fed on Bt-cotton leaves/seeds and the other was exposed to non-GM seeds and leaves of cotton), and it was declared that Bt-cotton cultivation has no toxic impact on the health of rabbit (Rahman and Co-workers published data).

8.4.4 Potential Impact on Non-targeted Organisms

Harmful impact of Bt-crops, particularly on non-target organisms, was a major apprehension (Romeis et al. 2008). A unique quality of *Cry* genes is their specificity for certain orders of insects. In multiple studies, not a single negative report revealed harmful impact of Bt sprays on the population dynamics of predators or parasitoids and on non-target herbivores was reported (Schoenly et al. 2003). After the development of Bt-crops, relative impact of transgenics and their control was estimated on the population of NTOs (Fitt et al. 1994; Sims 1995; Extension Toxicology Network 1996; Orr and Landis 1997; Bashir et al. 2004a, b), and generally notoxic impact of the transgenic crops were observed. For example, the harmful effect of Bt-rice on the populations of NTOs was not found (Chen et al. 2006; Rahman et al. 2007). Similarly, no harmful impact on arthropod community was found while comparing the data (family composition, diversity index etc.) collected from the transgenic and non-transgenic rice (Chen et al. 2003; Liu et al. 2002; Liu et al. 2003; Chen et al. 2006). Such commonalties were also found while comparing the influence of Bt-corn versus its conventional counterpart on communities of NTOs such as predators and parasitoids.

A total of five predator populations were monitored in various Bt and non-Bt-corn plots of Iowa State; a significant depression (29–60 %) of the *M. cingulum* population was found in Bt-corn field. Since the *M. cignulum* is a predator species of the European corn borer, thus it is much likely that the predator species would prefer to visit non-Bt-corn field because of finding relatively more prey population than Bt-corn (Pilcher et al. 2005).

Only major contrary report showed substantial reduction of the population of monarch butterfly caterpillars, *Danaus plexippus*, due to feeding on milkweed leaves treated with Bt-maize pollen (Losey et al. 1999; Jesse and Obrycki 2000), however, later insignificant influence on the population of monarch butterfly was reported (Hellmich et al. 2001; Sears et al. 2001; Stanley-Horn et al. 2001; Dively et al. 2004; Hellmich and Hellmich 2012). However, the positive impact of Bt-corn cultivation was found on biodiversity in comparison with the corn treated with chemical insecticides (O'Callaghan et al. 2005; Romeis et al. 2008). In subsequent years, a comprehensive study conducted jointly by the scientists of US and Canada, showed no acute toxic effects at different pollen densities in lab as well as in the field due to low level of Bt protein expression in pollens of the Bt-hybrids (Bt Insect Resistant Technology 2011) http://www.isaaa.org/resources/publications/pocketk/6/default.asp.

8.4.5 Potential Threats to Human Health

Bt proteins are target specific and their specificity lies in their receptor mediated responses. Thus Bt protein can harm the organisms having receptor sites in their gut, making the protein receptor-mediated. By chance, most of the beneficial insects and human lack these receptors.

Prior to commercialization, Bt-crops must go through stringent regulatory tests for evaluating toxicity and allergic responses. Bt proteins have been assessed at high dosage for evaluating toxicology by the U.S. Environmental Protection Agency (US-EPA). Also, the Extension Toxicology Network (Extoxnet), mutli-universities project in the US dealing with the pesticide information, did not report any complaint of toxicity/poison when a group of 18 humans were exposed to one gram of commercial Bt preparation for five but on alternate days or for three consecutive days. Moreover, in vitro studies revealed a rapid degradation of Bt proteins in human gastric fluid (Extension Toxicology Network 1996). http://www.isaaa.org/resources/publications/pocketk/6/default.asp.

8.4.6 Potential Risk of the Introduced Gene Cassette

The level of anti-nutrients, a few of which are difficult to degenerate by heat (Bakke-McKellep et al. 2007), can be escalated due to insertion of new genes which may cause infertility in sheep and cattle (Liener 1994). These anti-nutrients (heat stable) also cause allergenic reactions and bind to phosphorus and zinc—making inaccessible to the animals (Adams 1995). In most of the GM-crops, cauliflower mosaic virus 35S promoter (CaMV35S) has been used extensively, that can be transferred horizontally which may cause disease, carcinogenesis and mutagenesis. In few cases, the promoter sequences may lead to reactivate the dormant viruses as well as to generate new viruses (Hodgson 2000). In contrary to this hypothesis, CaMV, present in the normal food, cannot cause infections and thus mammals cannot absorb it (Ho et al. 2000). No disease or recombination with human viruses has ever been reported, irrespective of the fact that humans have been ingesting high levels of CaMV and its 35S promoter (Paparini and Romano-Spica 2004). In mammalian cells, transient expression of the transgenes transcribed by the CaMV35S promoter may possibly let the genes controlled by 35S promoter to express (Tepfer et al. 2004). Contrary to this hypothesis, recent studies conducted using mice as an experimental animal, were unable to detect DNA transfer as well as transcriptional activity of the CaMV35S quantified through real time PCR (Paparini and Romano-Spica 2006). This area of research needs further explorations (Dona and Arvanitoyannis 2009).

In most Bt-crops, antibiotic resistance genes have been used as selectable markers which may potentially transfer to microflora, comprising of 500–1000 distinct bacterial species, of human gastrointestinal tract, thus reducing the efficacy of antimicrobial treatment (Dona and Arvanitoyannis 2009). In nature, transfer of DNA from plants to microbes can occur (Bertolla and Simonet 1999), called as horizontal gene transfer, as bacteria integrate with free DNA in their surrounding (Nielsen et al. 1998). Chances of such type of gene transfer from transgenic plant to microbial community are extremely low (Halford and Shewry 2000). It requires release of the introduced gene(s) from the plant tissues, its stability, presence of competent bacterial species, uptake of trangene, recombination with the host genome etc. (Bertolla and Simonet 1999). In multiple studies, attempts were made to transfer gene from plants to bacterial species but could not be demonstrated successfully (Schlüter et al. 1995; Coghlan 2000). However, under very stringent laboratory conditions, very low frequency of plant DNA transfer to bacterial species has been demonstrated between the homologous sequences(De Vries and Wackernagel 1998; Gebhard and Smalla 1998; Mercer et al. 1999; De Vries et al. 2001). In other study, it was shown that without introducing homologous sequences in the recipient strain, uptake of the transgene is not possible. These phenomena were demonstrated on multiple crop species like sugarbeet, tomato, potato and oilseed rape containing the *nptII* gene (Nielsen et al. 2000; De Vries et al. 2001). Also, a jellyfish green fluorescent protein (GFP) gene, another marker gene, was utilized but did not find any risk of toxicity and allergenicity (Richards et al. 2003).

There are strong concerns about the aforementioned potential impacts of genes that can cause gene silencing, changes in expression level or, can turn on the existing silent genes (Conner and Jacobs 1999). Alternatively, expression of the Bt proteins may potentially alter the metabolism and biochemical pathways of the plants. For example, interaction of two genetically produced foods, tryptophan and g-linolenic acid, has been created new toxic compounds (Hill et al. 1993; Sayanova et al. 1997). Also, the epigenetic changes may happen in GM-organism that may raise concerns like unpredictability of genetic modifications, non-reproducible results and instability of the products, thus together suggest that in animals, toxicity assessment of whole food should be evaluated instead of the single novel protein. Though it is very well conceived that it is difficult to generate a dose-response relationship (Kuiper et al. 2004).

8.4.7 Horizontal Gene Transfer to the Consumer Species

Another important potential hazard of the GM-crops or GM-foods is associated with their capability to transfer the transgene to animals, including humans through their guts. GM-soybean containing glyphosate resistant gene was fed to pigs, however, DNA fragments were not detected in tissues of the pigs (Jennings et al. 2003a; b). Whereas, short DNA fragments were detected in the gastrointestinal tract of pigs when exposed to Bt-corn but were absent in the blood stream (Chowdhury et al. 2003). Also, the M13 phage DNA (administered orally) was detected in the blood stream of mice (Schubbert et al. 1994), and short DNA fragments of the transgene in the white blood cells and in milk of cows were also identified. Such type of DNA fragments were also reported in various tissues of mice and chicken when fed on GM- soybean and corn, respectively (Beever and Kemp 2000; Einspanier et al. 2001; Hohlweg and Doerfler 2001; Phipps and Beever 2001). All these studies suggest that a minute proportion of the trangene cassette is not degraded in the digestive tract, and this small quantity is difficult to amplify with PCR from the genomic DNA isolated from the blood because of their low level, but can easily be amplified in animal tissues (Pusztai 2001). Thus PCR assays may affect the interpretations (Murray et al. 2007), thus needs to be optimized. In spite of the fact that DNA fragments were detected but it is much unlikely that the DNA taken up by the cells of gastrointestinal tract will be integrated into the host genome, usually degraded in the cell (Flachowsky et al. 2005). Possibilities of horizontal gene transfer from Bt-crops to soil microflora were also explored because of the evidence of such transfer reported after conducting several planned experiments to facilitate the transfer. However, such conditions are not possible to occur in open environment. Furthermore, gene incorporated in Bt-crops is already present in most of soil bacteria. Therefore, it was concluded that horizontal gene transfer is a rare event in Bt-crops (Mendelsohn et al. 2004).

8.4.8 Potential Allergic Response

GM-foods derived from GM-crops including GM-soybean expressing methionine from Brazil nut (Nordlee et al. 1996) and GM-corn expression Bt protein (Bernstein et al. 2003) may cause allergic hypersensitivity (Taylor and Hefle 2002; Conner et al. 2003). It has also been conceived that the transgene expressing non-allergenic protein such as GM field pea, expressing alpha-amylase inhibitor-1, may have potential to produce product with allergenicity (Prescott et al. 2005). Thus each of the GM case should be treated separately (case-to-case basis).

In mice, low IgE response was observed because of the expression of choline oxidase gene (transgene) in *Brassica juncea.* Whereas, in another study, expression of the gene did not cause any allergic hypersensitivity (Singh et al. 2006), highlighting the need to undertake safety evaluation test on multiple experimental models for establishing a valid correlation between IgE response and toxicity. Farmers may have allergic sensitivity when exposed to GM-crops containing various Bt genes as skin sensitization and IgG antibodies were detected in farm workers exposed to Bt pesticide (Bernstein et al. 2003).

8.4.9 Allergenicity Assessment

According to a decision tree approach, formulated in 1996 (Metcalfe et al. 1996) and later it was revised (FAO/WHO 2001; Metcalfe 2003), if the conventional counter part of the GM-plant species is known for causing allergy and or toxicity, then the whole GM-plant of that particular species should be evaluated for quantifying the chances, if any, of increasing in toxicity. A 90-day long study for toxicity testing is required in rodents through comparing concentration of allergens in non-GM-crop versus GM-crop. Possibilities for the differential accumulations of toxic compounds or allergens in GM-crops containing single transgene with GM-crops containing stacked events should be considered as in each case interaction with the host genomes may vary. For example, interaction of the transgene conferring regulatory proteins if transferred into an entirely different background may change (De Schrijver et al. 2007). At this point of time, sequence analysis of amino acids is difficult to predict, thus limiting their utility for comparing their sequences with the known allergens (Alinorm 2003; Prescott and Hogan 2005). Also quantifying their degradation in *vitro* system has been the major challenge in establishing valid correlations with allergens (Bannon et al. 2003) which set a stage for conducting such experiments in *vivo* systems (Pusztai et al. 2003). It has also been shown that no single animal model can help in testing allergencity responses of various GM-foods as different animal species respond differentially to the allergens, indicating that animal models should be validated (Tryphonas et al. 2003). A comprehensive study addressing the allergenicity in human in response to GM-food has been discussed by Germolec et al. (2003, please consult for detail study).

8.5 Potential Impact to the Environment

8.5.1 Pollen Flow

Gene flow or transmission of genetic material between GM-crops and their wild types is a potential threat (Messeguer 2003). Maize is an open pollinated crop, and pollen can travel miles of distances through air currents. Thus cultivation of GM-maize should be separated from the related species that have tendency for hybridizing with maize. Farmers who are growing organic crops raise objection on cultivation of GM-plants for avoiding contamination from GE-pollen or seed (Hellmich and Hellmich 2012). In another study, chances of pollen-mediated gene flow from transgenic lines of rice to their untransformed counterparts through natural cross-pollination were found to be very low (0.14 %) (Rahman et al. 2007).

Bt-cotton, another important crop commercialized in 1996, is predominantly a self-pollinated crop (usually 2–5 % cross pollination was reported mainly through insects) in most of the cotton growing countries. However, in a few countries like Panama, its cross-pollination rate may even increase to 80 %. Its pollens are sticky, eliminating the chances of traveling through wind (Poehlman 1994). Hence, crossing is only possible when honeybees collect pollens (Oosterhuis and Jernstedt 1999). Also, propensity of shifting pollen from one flower to another could substantially be minimized by increasing the distance between the two cotton genotypes. We studied that pollen transfer rate reached to < 1 % if the distance between the two genotypes is more than 100 ft (Rahman and Co-workers, unpublished result). In another study, chances of gene flow between transgenic lines and their untransformed counterparts through cross-pollination are found to be low (0.14 %; Rahman et al. 2007). Secondly, a chance of transfer of pollen to other species even with in the same genus is extremely low as phyletic barriers exist among different species. Lastly, Bt-crops have no sound potential of transferring transgene to near by cultivated wild relatives because of difference in chromosome number, phenology and territory. In a few states of America, Bt-cotton cultivation was restricted to Hawaii, Florida, Puerto Rico and the US Virgin Islands due to chances of transfer of the Bt gene from the cultivated Bt-cotton to their wild relatives (Mendelsohn et al. 2003).

8.5.2 Grain Yield

Grain yield is one of the most important parameters for studying the agronomic performance of the crop species. In multiple studies, Bt-crop varieties out yielded their non-Bt counterparts. For example, Bt corn hybrids exhibited 11 % more grain yield than non-Bt corn hybrids (Dillehay et al. 2004; Subedi and Ma 2007) while few reports have shown no differences in any of the parameters including grain yield and chemical composition from non-Bt corn hybrids (Yanni et al. 2011).

Like other Bt-crops, cultivation of Bt-cotton has also got popularity among the farming community because of the increased protection against lepidopteron insect pests, ultimately resulted in high yield especially in developing countries like China (Qaim and Zilberman 2003; Hu et al. 2009; Huang et al. 2010), India (Karihaloo and Kumar 2009) and Pakistan (Zaman and Co-workers). The Bt gene(s) have been bred into the elite cotton cultivars from the Bt-coker genotype through attempting several backcrosses with the adopted cotton cultivars. It has been shown that the Bt cotton cultivar is much like their parental varieties by comparing traits like germination rate, establishment, rate of vegetative growth, flowering duration, fruiting potential, fibre yield and fibre quality. In another study, Bt-cotton variety IR-NIBGE-901 was grown along with its conventional variety FH-901 for a period of four years in Pakistan, it was found that Bt-cotton and the parental variety were similar in all morphological and quality characteristics (Zaman and co-workers, unpublished results).

While differences in morphological parameters such as plant height, flowering duration and lodging resistance have been reported in Bt-rice developed in China (Jiang et al. 2000) and Pakistan (Bashir et al. 2005). Such fluctuations in morphological traits are generally attributed to the insertion of transgene in the host genome (Van Lijsebettens et al. 1991) which may cause gene silencing (Matzke et al. 2000). However, somaclonal variations (Larkin and Scowcroft 1981) are the much likely cause of creating variations in transgenic lines (Kaeppler et al. 2000). Also, chemicals like hygromycin may also induce variations in rice (Wu et al. 2000). In contrary to this, characters like panicle length, aroma and flag leaf area were found to be similar in Bt-rice/non-Bt-rice, where fluctuations in average number of tillers, plant height and maturity were reported (Rahman et al. 2007). Small differences in physiochemical properties among the transgenic and non-transgenic lines were observed due to fluctuation in the prevailing environmental conditions—late maturing lines find slightly different environment than the early maturing lines (Rahman et al. 2007).

8.5.3 Weediness

Weediness indicates that if a cultivated crop species establishes as a weed—survival beyond economic life, in the succeeding crop or neighbouring crop. Potential indicators of weediness can be numerous. For example, seed related characters (prolonged and high seed production with discontinuous germination under different environmental conditions), and physiological and morphological traits, together lead to evolve or enhance the capability of producing allelochemicals, special seed dispersal mechanisms, unusual high growth rate etc. Chances of producing such traits are very much unlikely as most domesticated crops species have lost, if not all, many traits which add in weediness traits (Becker et al. 1992). Also, the traits make the plant species to be domesticated render them unsuitable for sustaining in a wide range of environments. Detailed experimental studies revealed that Bt trait did not add in the fitness of the Bt-plant, except that Bt gene confers resistance to the lepidopteron insect pest species. For example, in Pakistan, a study was conducted for a period of four years, investigated the potential weediness trait of Bt-cotton, showed nonsignificant differences in agronomic characteristics between Bt-cotton and its parental variety. Bt-cotton meets all morphological, yield, and quality characteristics of non-Bt-*cotton* varieties produced in Pakistan. Based on such mechanistic arguments and field experiences, insertion of the *Cry1Ac* gene into the cotton genome would not add any effect toward the weediness trait of the cotton (Zaman and colleagues, unpublished data).

8.5.4 Persistence of Cry Proteins in Soil

Persistence of Bt protein in the soil is moderate, thus considered immobile due to its less mobility and leaching with groundwater. However, it is not persistent in acidic soils as it degrades rapidly upon exposure of UV radiations in the sunlight (http://www.isaaa.org/resources/publications/pocketk/6/default.asp).

In another experiment, presence of *Cry1Ac* protein, assayed by ELISA and bioassay, was not detected in soil samples collected from Bt-cotton fields (Head et al. 2002). A substantial rapid degradation of Bt proteins in soil cultivated with Bt-cotton (*Cry1Ac*), Bt-potatoes (*Cry3Aa*) and Bt-corn (*Cry1Ab*), is a major cause for not reaching the concentration of biologically significant levels (Palm et al. 1994; Sims and Holden 1996; Head et al. 2002). In few countries like Australia, where cotton is cultivated on soils with pH ranging from 7.5 to 8.5 (Tapp and Stotzky 1998) that helps in rapid degradation of Bt endotoxins by soil microorganisms. Pakistan, another important cotton growing country, where pH of the soil is in the range of 8.5–9.5, is likely to degrade Bt proteins relatively faster. It may be concluded that there are meager chances of accumulation of *Cry1Ac* proteins in soils as a result of repeated rounds of Bt-cotton cultivation.

It is much likely that soil microorganisms can be exposed to Bt proteins because of the occurrence of root exudations or during the decomposition of Bt-plant in the soil as this phenomenon has been reported in Bt-corn containing *Cry1Ab* gene (Saxena et al. 1999; Stotzky 2000). Some studies also confirmed the release of Bt protein in soil cultivated with Bt-cotton (Gupta et al. 2002).

Multiple studies conducted to evaluate the impact of Bt-crops on soil organisms, showed that Bt proteins have no harmful impact on the soil microbs even at far higher concentration of the Bt proteins. In another study, variations were not found in the soil microbiota of the fields with Bt plant material versus the fields with conventional plant material (Donegan, et al. 1995, 1996). http://www.isaaa.org/resources/publications/pocketk/6/default.asp. Also, no substantial changes in the counts of soil micorbs were found from the fields cultivated with Bt-cotton and non-Bt-cotton in Pakistan (unpublished reports).

However, recently a report appeared, showing a significant reduction in actinobacteria (17 %), bacterial (14 %) count as well as acid phosphatase (27 %), phytase (18 %), nitrogenase (23 %) and dehydrogenase (12 %) activities in the Bt- cotton fields versus non-Bt cotton fields of India. Fungal and nitrifier counts, and esterase and alkaline phosphatase activities were not affected by the introduction of Bt-cotton in fields. Nonetheless, substantial decline between 8 and 9 % in biomass carbon (MBC) and biomass nitrogen (MBN) was observed (Jagadish et al. 2012).

8.5.5 *Allelopathic Impact*

To explore the allelopathic effects of Bt-crops is important especially in developing countries because most farmers observe various rotations worldwide. For example, in subcontinent, cotton wheat or cotton rice rotations are very popular for harvesting maximum profitability for the farming community. Multiple planned experiments were conducted for testing the allelopathic effect of Bt-crops including rice, cotton etc., and it was shown that the cultivation of Bt rice has no harmful effect on germination of wheat (Rahman et al. 2007). Similarly, field experiments were conducted for three years to assess the impact of plant residue containing Bt protein on weed population of the Bt-cotton field at various intervals. Weeds were allowed to grow in one big plot of Bt-cotton field and non-Bt cotton field in various locations of Pakistan. Nonsignificant differences were observed between the weed populations of Bt and non-Bt cotton fields (Zaman & Co-workers, unpublished results).

8.6 Conclusions

Cultivation of GM-crops has been gaining popularity worldwide every year among the farmer community that resulted in growing GM-crops on 160 million hectares in 2011. Introduction of a series of *Cry* genes in different crops confers resistance to the lepidopteron insect pests, which is instrumental in achieving sustainability in agricultural system and also paves the way for protecting environment by reducing the number of chemical applications. Beneficial impact of cultivating Bt-crops has been found relatively high in developing countries than in the industrialized countries. The indirect benefit of cultivating Bt-crops is a substantial suppression in insect pest populations which may help in controlling pests on their non-Bt counterparts with fewer inputs. However, cultivation of Bt-crops may help minor pests to emerge as major pests because of reduced insecticides application on Bt-crops. Thus, this phenomenon may arise much faster in developing countries where farmers are not much educated about IPM programs. For harvesting maximum benefits of the Bt-crops, public sector organizations should make deliberate efforts to educate farmers for controlling insect pests by supplementing with some other control measures.

So far, numerous crop varieties modified with genes expressing Cry toxins have been developed, and no detrimental impacts of the Bt-crops on NTOs populations were found in experiments conducted at lab scale as well as in the field. Furthermore, populations of beneficial insects are increasing on Bt-crops—further strengthening the defense umbrella of crops.

In future, new genes derived from different wild species, preferably belonging to the same genus, should be kept on incorporating in major domesticated crops. It will help in releasing crop varieties with little potential risks of developing resistance in the target insect pests species. Secondly, the efficacy of the *Cry* genes in various genetic backgrounds should be tested as the expression of the gene(s) fluctuates in different backgrounds; it will facilitate in designing strategies regarding "when to introduce new genes or stacked genes with different mode of action". It will help in cultivating crops containing diverse genes which will set a stage for designing IPM strategies for combating resistance concerns in insect pests. Thirdly, even after two decade of Bt toxins deployment, their mode of action is not fully explored. Many Bt toxins are active against insects of more than one order. Thus it is vital to characterize thoroughly any new Bt gene before introducing into a crop variety.

Antibiotic resistant gene has been used extensively as a marker gene in most Bt-crops, which needs to be replaced with other reporter genes like green fluorescent protein gene, herbicide tolerant gene etc. that will set a stage for building confidence of most of the skeptics regarding the safe use of GM-foods. Often, these marker genes have been tested for toxicity and allergenic responses for short duration, which is not sufficient to deduce valid conclusions. It is suggested that potential risk of every marker genes should be tested for a longer period of time. Also, in most studies, limited number of animals were exposed to GM-foods which should be increased to draw trustworthy conclusions.

For studying harmful impact of Bt-crops on soil microbial communities, it is imperative to carry out experiments in different ecological zones as we know that microbial communities fluctuate in various ecological zones. Such studies will help to identify any negative impact of the Bt proteins on microbial populations. Also, for each of the new gene of the same family, such studies should be carried out individually.

For assessing the safety of Bt-food, evaluation of the allergic responses should be treated case-to-case basis. Also, the individuals having some allergic issues should orally be given GM-foods expressing known allergens. While studying the allergic response of GM-foods in human, both allergy history as well as immunodeficiency problems of individuals should be considered to avoid erroneous conclusions.

Genotoxicity studies should be undertaken on each of the animal species without considering the specific toxic properties. Some antagonistic effects of two genes of the same family have been reported. Thus it is imperative to study the interactions of the genes not only in the GM-plant but also in the GM-food. Similarly, synergistic effects of Bt toxins with chemicals such as pesticides were reported. Thus, the new Bt genes should be thoroughly studied before releasing into the environment.

In a few studies, small traces of ingested DNA were found. It is likely that the ingested DNA may get into the blood stream or be excreted. For addressing such issues, intensive scientific inputs as well as the influx of funds are required for predicting and exploring the possible consequences on NTOs including humans and animals. Also, the post release monitoring of the GM-crops should be undertaken stringently for studying allergic issues especially in infants and individuals.

In the present scenario, guidelines addressing the safety evaluation of GM-foods are too general, lacking any detailed methodology for testing. For this, lot of investments should be made to comprehensively draw guidelines for proving the safety of the GM-foods. For example, DNA-based toxicity tests like Comet assay should be used for finding any harmful impact of the transgene on the integrity of the genomic DNA.

Acknowledgements We are extremely grateful to the funding agency Pakistan Science Foundation for providing funds through a project "Exploration of Cotton Germplasm Potential Against Drought Stress Using Genomic Approaches"—Project No. PSF/NSLP/P-NIBGE (19)

References

Adams NR (1995) Detection of the effects of phytoestrogens on sheep and cattle. J Anim Sci 73:1509–1515

Alinorm (2003) Joint FAO/WHO Food Standard Programme CAC: Appendix III: Guideline for the conduct of food safety assessment of foods derived from recombinant-DNA plants; Appendix IV: Annex on the assessment of possible allergenicity

Anderson K, Valenzuela E, Jackson LA (2008) Recent and prospective adoption of genetically modified cotton: a global computable general equilibrium analysis of economic impacts. Econ Dev Cult Change 56:265–296

Arif MI, Rafiq M, Ghaffar A (2009) Host plants of cotton mealybug (Phenacoccus solenopsis): a new menace to cotton agroecosystem of Punjab. Int J Agric Biol 11:163–167

Bakke-McKellep AM, Koppang EO, Gunnes G, Senden M, Hemre GI, Landsverk T, Krogdahl A (2007) Histological, digestive, metabolic, hormonal and some immune factor responses in Atlantic salmon, Salmo salar L, fed genetically modified soybeans. J Fish Dis 30:65–79

Bannon G, Fu TJ, Kimber I, Hinton DM (2003) Protein digestibility and relevance to allergenicity. Environ Health Persp 111:1122–1124

Barriere Y, Verite R, Brunschwig P, Surault F, Emile JC (2001) Feeding value of corn silage estimated with sheep and dairy cows is not altered by genetic incorporation of Bt1376 resistance to Ostrinia nubilalis. J Dairy Sci 84:1863–1871

Bashir K, Husnain T, Fatima T, Latif Z, Mehdi SA, Riazuddin S (2004a) Field evaluation and risk assessment of transgenic indica basmati rice. Mol Breeding 13(4):301–312

Bashir K, Husnain T, Riazuddin S (2004b) Response of transgenic rice expressing two Bt genes to nontarget insects. Int Rice Res Notes 29(2):15–16

Bashir K, Husnain T, Fatima T, Riaz N, Makhdoom R, Riazuddin S (2005) Novel indica basmati line (B-370) expressing two unrelated genes of Bacillus Thuringiensis is highly resistant to two lepidopteron insects in the field. Crop Prot 24:870–879

Baum KA, Haynes KJ, Dillemuth FP, Cronin JT (2004) The matrix enhances the effectiveness of corridors and stepping stones. Ecology 85:2672–2675

Becker HC, Karle R, Han SS (1992) Environmental variation for outcrossing rates in rapeseed (Brassica napus). Theor Appl Genet 84:303–306

Beever DE, Kemp F (2000) Safety issues associated with the DNA in animal feed derived from genetically modified crops. A review of scientific and regulatory procedures. Nutr Revs 70:197–204

Berliner E (1915) Uber die Schalffsuchi der Mehlmottenraupo und thren Erreger, Bacillus thuringiensis n. sp. Zeitschfit fur Angewandte Entomologie 2:29–56
Bernstein JA, Bernstein IL, Bucchini L, Goldman LR, Hamilton RG, Lehrer S, Rubin C (2003) Clinical and laboratory investigation of allergy to genetically modified foods. Environ Health Persp 111:1114–1121
Bertolla F, Simonet P (1999) Horizontal gene transfers in the environment: natural transformation as a putative process for gene transfers between transgenic plants and microorganisms. Res Microbiol 150:375–384
Bravo A, Gill SS, Soberón M (2007) Mode of action of Bacillus thuringiensis cry and cyt toxins and their potential for insect control. Toxicon 49(4):423–435
Bt Insect Resistant Technology (2011) http://www.isaaa.org/resources/publications/pocketk/6/default.asp
Campbell DR (1985) Pollen and gene dispersal: the influences of competition for pollination. Evolution 39:418–431
Chen M, Ye GY, Yao HW, Hu C, Shu QY (2003) Impact evaluation of insect-resistant transgenic rice on the feeding and oviposition behavior of its non-target insect, the brown planthopper, Nilaparvata lugens (Homptera: Delphacidae). Agric Sci China 2:1000–1006
Chen M, Ye GY, Liu ZC, Yao HW, Chen XX, Shen ZC, Hu C, Datta SK (2006) Field assessment of the effects of transgenic rice expressing a fused gene of cry1Ab and cry1Ac from Bacillus thuringiensis Berliner on Nontarget Planthopper and Leafhopper Populations. Environ Entomol 35(1):127–134
Chowdhury EH, Kuribara H, Hino A, Sultana P, Mikami O, Shimada N, Guruge KS, Saito M, Nakajima Y (2003) Detection of corn intrinsic and recombinant DNA fragments Cry1Ab protein in the gastrointestinal contents of pigs fed genetically modified corn Bt11. J Anim Sci 81:2546–2551
Coghlan A (2000) So far so good: for the moment, the gene genie is staying in its bottle. New
Conner AJ, Jacobs JME (1999) Genetic engineering of crops as potential source of genetic hazard in the human diet. Mutat Res 443:223–234
Conner AJ, Glare TR, Nap JP (2003) The release of genetically modified crops into the environment: part II overview of ecological risk assessment. Plant J 33:19–46
Crost B, Shankar B, Bennett R, Morse S (2007) Bias from farmer self-selection in genetically modified crop productivity estimates: evidence from Indian data. J Agric Econ 58:24–36
de Janvry A, Sadoulet E (2002) World poverty and the role of agricultural technology: direct and indirect effects. J Dev Studies 38:1–26
de Vries J, Wackernagel W (1998) Detection of nptII (kanamycin resistance) genes in genomes of transgenic plants by marker-rescue transformation. Mol Gen Genet 257:606–613
de Vries J, Meier P, Wackernagel W (2001) The natural transformation of the soil bacteria pseudomonas stutzeri and acinetobacter sp by transgenic plant DNA strictly depends on homologous sequences in the recipient cells. FEMS Microbiol Lett 195:211–215
De Schrijver A, Devos Y, van den Bulcke M, Cadot P, De Loose M, Reheul D, Sneyers M (2007) Risk assessment of GM stacked events obtained from crosses between GM events. Trends Food Sci Tech 18:101–109
Dev SM, Rao NC (2007) Socioeconomic impact of Bt cotton. CESS Monographs 3, Centre for Economic and Social Studies. Hyderabad
Dillehay BL, Roth GW, Calvin DD, Kratochvil RJ, Kuldau GA, Hyde JA (2004) Performance of Bt corn hybrids, their near isolines, and leading corn hybrids in Pennsylvania and Maryland. Agron J 96:818–824
Dively GP, Rose R, Sears MK, Hellmich RL, Stanley-Horn DE, Calvin D, Russo JM, Anderson PL (2004) Effects on monarch butterfly larvae (Lepidoptera Danaidae) after continuous exposure to Cry1Ab-expressing corn during anthesis. Environ Entomol 33:1116–1125
Dona A, Arvanitoyannis IS (2009) Health risks of genetically modified foods. Crit Rev Food Sci Nutr 49:164–175

Donegan KK, Palm CJ, Fieland VJ, Porteous LA, Ganio LM, Schaller DL, Bucao LQ, Seidler RJ (1995) Changes in levels, species and DNA fingerprints of soil microorganisms associated with cotton expressing the *Bacillus thuringiensis* var kurstaki endotoxin. Appl Soil Ecol 2:111–124

Donegan KK, Schaller DL, Stone JK, Ganio LM, Reed G, Hamm PB, Seidler RJ (1996) Microbial populations, fungal species diversity and plant pathogen levels in field plots of potato plants expressing the Bacillus thuringiensis var tenebrionis endotoxin. Transgenic Res 5:25–35

Donkin SS, Velez JC, Totten AK, Stanisiewski EP, Hartnell GF (2003) Effects of feeding silage and grain from glyphosate-tolerant or insect-protected corn hybrids on feed intake, ruminal digestion, and milk production in dairy cattle. J Dairy Sci 86:1780–1788

Einspanier R, Klotz A, Kraft J, Aulrich K, Poser R, Schwaegle F, Jahreis G, Flachowsky G (2001) The fate of forage plant DNA in farm animals, a collaborative case-study investigating cattle and chicken fed recombinant plant material. Eur Food Res Technol 212:129–134

Elbehri A, Macdonald S (2004) Estimating the impact of transgenic Bt cotton on west and central Africa: a general equilibrium approach. World Dev 32:2049–2064

Ellis C, Karafyllidis L, Turner JG (2002) Constitutive activation of jasmonate signaling in an Arabidopsis mutant correlates with enhanced resistance to Erysiphe cichoracearum, Pseudomonas syringae, and Myzus persicae. Mol Plant Microbe Interact 15:1025–1030

Extension Toxicology Network (1996) Pesticide information profile, *Bacillus thuringiensis.* Accessed 23 August 2012.

FAO (2004) The state of food and agriculture 2003–2004; agricultural biotechnology: meeting the needs of the poor? FAO Agriculture Series No. 35. Rome

FAO/WHO (2001) Evaluation of allergenicity of genetically modified foods report of a joint FAO/WHO expert consultation on foods derived from biotechnology, 22–15 January 2001, Rome Italy available: food and agriculture organisation of the United Nations, Rome. http://www.fao.org/es/esn/allergygm.pdf

Fares NH, El-Sayed AK (1998) Fine structural changes in the ileum of mice fed on delta-endotoxin-treated potatoes and transgenic potatoes. Nat Toxins 6:219–233

Federici B (2002) Case study: Bt crops. In: Atherton K (Ed) Genetically modified crops: assessing safety (chapter 8). Taylor & Francis, New York

Fitt GP, Mares CL, Llewellyn DJ (1994) Field evaluation and potential ecological impact of transgenic cotton (Gossypium hirsutum) in Australia. Biocontrol Sci Tech 4:535–548

Flachowsky G, Chesson A, Aulrich K (2005) Animal nutrition with feeds from genetically modified plants. Arch Anim Nutr 59:1–40

Framond AJ, Bevan MW, Barton KA, Flavell F, Chilton MD (1983) Mini-Ti plasmid and a chimeric gene construct: new approaches to plant gene vector construction. Advances in gene technology: molecular genetics of plants and animals. Miami Winter Symposia 20:159–170

Gallagher S, Grimes J, Beavers JB (2000) Insect protection protein 2 in cottonseed meal, a dietary toxicity study with the northern Bobwhite. Report No MSL-1678, Monsanto Company, St Louis

Gandhi VP, Namboodiri NV (2006) The adoption and economics of Bt cotton in India: preliminary results from a study Work Pap 2006–09-04, Indian Inst Manag

Gebhard F, Smalla K (1998) Transformation of Acinetobacter sp strain BD413 by transgenic sugar beet DNA. Appl Environ Microbiol 64:1550–1554

Germolec DR, Kimber I, Goldman L, Selgrade M (2003) Key issues for the assessment of the allergenic potential of genetically modified foods: breakout group reports. Environ Health Persp 11:1131–1139

Gupta VVSR, Roberts GN, Neate SM, McClure SG, Crisp P, Watson SK (2002) Impact of Bt-cotton on biological processes in Australian soils. In: Akhurst RJ, Beard CE, Hughes PA (Eds) Proceedings of the fourth pacific rim conference on the biotechnology of bacillus thuringiensis and its environmental impacts. CSIRO, Australia, pp. 191–194

Halford NG, Shewry PR (2000) Genetically modified crops: methodology, benefits, regulation and public concerns. Brit Med Bullet 56:71

Head G, Surber JB, Watson JA, Martin JW, Duan JJ (2002) No detection of Cry1Ac protein in soil after multiple years of transgenic Bt cotton (Bollgard) use. Environ Entomol 31:30–36
Hellmich RL, Hellmich KA (2012) Use and impact of Bt maize. Nat Edu Knowledge 3(5):4
Hellmich RL, Siegfried BD, Sears MK, Stanley-Horn DE, Mattila HR, Spencer T, Bidne KG, Daniels MJ, Lewis LC (2001) Monarch larvae sensitivity to Bacillus thuringiensis purified proteins and pollen. Proc Natl Acad Sci USA 98:11925–11930
Hilbeck A, Andow DA, Fontes EMG (2006) Environmental risk assessment of genetically modified organisms Volume 2: Methodologies for Assessing Bt Cotton in Brazil. CABI Publishing, Wallingford, ISBN-10: 1-84593-000-2, ISBN-13: 978-1-84593-000-4
Hill RH, Caudill SP, Philen RM, Bailey SL, Flanders WD, Driskell WJ (1993) Contaminants in L-tryptophan associated with eosinophilia myalgia syndrome. Arch Environ Contam Toxicol 25:134–142
History of Bt http://www.bt.ucsd.edu/bt_history.html
Ho MW, Ryan A, Cummins J (2000) Hazards of transgenic plants containing the cauliflower mosaic virus promoter. Microb Ecol Health Dis 12(3):189–198
Hodgson J (2000) Scientists avert new GMO crisis. Nat Biotechnol 18:13
Hohlweg U, Doerfler W (2001) On the fate of plant or other foreign genes upon the uptake in food or after intramuscular injection in mice. Mol Genet Genomics 265:225–233
http://www.gmcrops.ewebsite.com/articles/history.html
Hu R, Pray CE, Huang J, Rozelle S, Fan C, Zhang C (2009) Reforming intellectual property rights and the Bt cotton seed industry in china: who benefits from policy reform? Res Policy 38:793–801
Huang J, Rozelle S, Pray C, Wang Q (2002) Plant biotechnology in China. Science 295:674–677
Huang J, Hu R, van Meijl H, van Tongeren F (2004) Biotechnology boosts to crop productivity in China: trade and welfare implications. J Dev Econ 75:27–54
Huang J, Hu R, Rozelle S, Pray CE (2008) Genetically modified rice, yields, and pesticides: assessing farm-level productivity effects in China. Econ Dev Cult Change 56:241–263
Huang J, Mi JW, Lin H, Wang Z, Chen R, Hu R, Rozelle S, Pray CE (2010) A decade of Bt cotton in chinese fields: assessing the direct effects and indirect externalities of Bt cotton adoption in China. Sci China Life Sci 53:981–991
Huesing J, English L (2004) The Impact of Bt crops on the developing world. AgBioForum 7(1&2):84–95
Hutchison WD, Burkness EC, Mitchell PD, Moon RD, Leslie TW, Fleischer SJ, Abrahamson M, Hamilton KL, Steffey KL, Gray ME, Hellmich RL, Kaster LV, Hunt TE, Wright RJ, Pecinovsky K, Rabaey TL, Flood BR, Raun ES (2010) Area wide suppression of European corn borer with Bt maize reaps savings to non-Bt maize growers. Sci 330:222–225
Ismael Y, Bennett R, Morse S (2002) Bt cotton, pesticides, labour and health: a case study of smallholder farmers in the Makhatini Flats, Republic of South Africa Paper presented at the 6th International ICABR Conference, Ravello, Italy
Jagadish CT, Indira R, Vandana S (2012) Effect of Bt-transgenic cotton on soil biological health. Appl Biol Res 14(1):5–23
James C (2002) Global review of commercialized transgenic crops: 2001 feature: Bt cotton. ISAAA briefs No. 26. ISAAA, Ithaca
James C (2011) Global status of commercialized biotech/GM crops. ISAAA, Ithaca
Jayaraman KS (2009) Transgenic aubergines put on ice. Nature 461:1041
Jennings JC, Kolwyck DC, Kays SB, Whetsell AJ, Suber JB, Cromwell GL, Lirette RP, Glenn KC (2003a) Determining whether transgenic and endogenous plant DNA and transgenic protein are detectable in muscle from swine fed roundup ready soybean meal. J Anim Sci 81:1447–1455
Jennings JC, Albee LD, Kolwyck DC, Surber JB, Taylor ML, Hartnell GF, Lirette RP, Glenn Monsanto KC (2003b) Attempts to detect transgenic and endogenous plant DNA and transgenic protein in muscle from broilers fed YieldGard Corn Borer Corn. Poult Sci 82:371–380
Jesse LCH, Obrycki JJ (2000) Field deposition of Bt transgenic corn pollen: lethal effects on the monarch butterfly. Oecologia 125:241–248

Jiang J, Linscombe SD, Wang J, Oard JH (2000) Field evaluation of transgenic rice (Oryza sativa L.) produced by agrobacterium and particle bombardment methods. In: Plant and Animal Genome VIII Conference (January 9–12, 2000, Town and County Hotel, San Diego, CA) (Abstr)
Kaeppler SM, Kaeppler HF, Rhee Y (2000) Yong Epigenetic aspects of somaclonal variation in plants. Plant Mol Biol 43:179–188
Karihaloo JL, Kumar PA (2009) Bt cotton in India—a status report (second edition) Asia-Pacific Consortium on agricultural biotechnology. (APCoAB), New Delhi
Krishna VV, Qaim M (2007) Estimating the adoption of Bt eggplant in India: who benefits from public-private partnership? Food Policy 32:523–543
Kuiper HA, Kleter GA, Noteborn HPMJ, Kok EJ (2001) Assessment of the food safety issues related to genetically modified foods. Plant J 27:503–528
Kuiper HA, Konig A, Kleter GA, Hammes WP, Knudsen I (2004) Concluding remarks. Food Chem Toxicol 42:1195–1202
Larkin PJ, Scowcroft WR (1981) Somaclonal variation—a novel source of variability from cell cultures for plant improvement. Theor Appl Genet 60:197–214
Li MH, Robinson EH (2000) Evaluation of cottonseed meal derived from insect protected cotton lines 15813 and 15985 as a feed ingredient for catfish. Report No MSL-16179, Trad Cochran National Warm water Aquaculture Centre (Testing facility) Mississippi State University Stoneville MS 38776–40197
Liener IE (1994) Implications of anti-nutritional components in soybean food. Crit Rev Food Sci Nutr 34:31–67
Liu ZC, Ye GY, Hu C, Datta SK (2002) Effects of Bt transgenic rice on population dynamics of main nontarget insect pests and dominant spider species in rice paddies. Acta Phytophylacica Sin 29:138–144
Liu ZC, Ye GY, Fu Q, Zhang ZT, Hu C (2003) Indirect impact assessment of transgenic rice with cry1Ab gene in predations by the wolf spider, Pirata subpiraticus. Chinese J Rice Sci 17:175–178
Losey JE, Rayor LS, Carter ME (1999) Transgenic pollen harms monarch larvae. Nature 399–214
Lu Y, Wu K, Jiang Y, Xia B, Li P, Feng H, Wyckhuys KAG, Guo Y (2010) Mirid bug outbreaks in multiple crops correlated with wide-scale adoption of Bt cotton in China. Sci 328:1151–1153
Magaña-Gómez JA, de la Barca AM (2009) Risk assessment of genetically modified crops for nutrition and health. Nutri Rev 67:1–16
Magaña-Gómez JA, Cervantes GL, Yepiz-Plascencia G, Barca AMCdl (2008) Pancreatic response of rats fed genetically modified soybean. J App Toxico 28:217–226
Mal P, Manjunatha AV, Bauer S, Ahmed MN (2011) Technical efficiency and environmental impact of Bt cotton and non-Bt cotton in North India. AgBioForum 14(3):164–170
Malik TH (2011) Standard operating protocols (SOPs) for testing, evaluation, monitoring, approval, release and registration of genetically modified (GM) crop varieties in Pakistan with emphasis on Bt Cotton. The Pak Cotton 55:45–60
Marvier M (2001) Ecology of transgenic crops. Am Scientist 89:160–167
Matzke MA, Mette MF, Matzke AJM (2000) Transgene silencing by the host genome defense: implications for the evolution of epigenetic control mechanisms in plants and vertebrates. Plant Mol Biol 43:401–415
Mendelsohn M, Kough J, Vaituzis Z, Matthewset K (2003) Are Bt crops safe? Nat. Biotechnol 21:1003–1009
Mendelsohn M, Kough J, Vaituzis Z, Matthews K (2004) Are Bt crops safe? AgBioForum 7(1&2):30
Mercer DK, Scott KP, Bruce-Johnson WA, Glover LA, Flint HJ (1999) Fate of free DNA and transformation of the oral bacterium streptococcus gordonii DL1 by plasmid DNA in human saliva. Appl Environ Microb 65:6–10
Messeguer J (2003) Gene flow assessment in transgenic plants. Plant Cell Tiss Org 73:201–212
Metcalfe DD (2003) Introduction: what are the issues in addressing the allergenic potential of genetically modified foods? Environ Health Persp 11:1110–1113

Metcalfe DD, Astwood JD, Townsend R, Sampson HA, Taylor SL, Fuchs RL (1996) Assessment of the allergenic potential of foods derived from genetically engineered crop plants. Crit Rev Food Sci Nutr 36:s165–s186
Morse S, Bennett R, Ismael Y (2004) Why Bt cotton pays for small-scale producers in South Africa. Nat Biotechnol 22:379–380
Murray SR, Butler RC, Hardacre AK, Timmerman-Vaughan GM (2007) Use of quantitative real-time PCR to estimate maize endogenous DNA degradation after cooking and extrusion or in food products. J Agric Food Chem 55:2231–2239
Nielsen KM, Bones AM, Smalla K, Van Elsas JD (1998) Horizontal gene transfer from transgenic plants to terrestrial bacteria—a rare event? FEMS Microbiol Rev 22:79–103
Nielsen K, Van Elsas J, Smalla K (2000) Transformation of Acinetobacter sp strain BD413 (pFG4deltanptII) with transgenic plant DNA in soil microcosms and effects of kanamycin on selection of transformants. Appl Environ Microbiol 66:1237–1242
Nordlee JA, Taylor SL, Townsend JA, Thomas LA, Bush RK (1996) Identification of a Brazil nut allergen in transgenic soybeans. N Engl J Med 334:688–692
Noteborn HPJM, Kuiper HA (1994) Safety assessment strategies for genetically modified plant products case study: Bacillus thuringiensis-toxin tomato. In: Jones DD (ed) Proc 3rd Intl Symp on the biosafety results of field tests of genetically modified plants and microorganisms Oakland. CA University of California, Division of Agriculture and Natural Resources, California, 199–207
Noteborn HPJM, Bienenmann-Ploum ME, Van Den Berg JHJ, Alink GM, Zolla L, Reynaerts A, Pensa M, Kuiper HA (1995) Safety assessment of the Bacillus thuringiensis insecticidal crystal protein CryIAb expressed in transgenic tomatoes. In: Engel KH, Takeoka GR, Teranishi R (eds) Genetically Modified Foods Safety Aspects Washington, DC American Chemical Society. 134–147
O'Callaghan M, Glare TR, Burgess EP, Malone LA (2005) Effects of plants genetically modified for insect resistance on nontarget organisms. Ann Rev Entomol 50:271–292
Oosterhuis DM, Jernstedt J (1999) Morphology and anatomy of cotton plant. In: Smith CW, Cothren JT (eds) Cotton: origin, history, technology and production john willie and sons. Darvers, MA, 170–206
Orr DB, Landis DA (1997) Oviposition of Eauropean corn borer (Lepidoptera: pyralidae) and impact of natural enemy populations in transgenic versus isogenic corn. J Econ Entomol 90(4):905–909
Pakistan national biosafety rules (2005)
Palm CJ, Donegan KK, Harris D, Seidler RJ (1994) Quantification in soil of Bacillus thuringiensis var kurstaki d-endotoxin from transgenic plants. Mol Ecol 3:145–151
Paparini A, Romano-Spica V (2004) Public health issues related with the consumption of food obtained from genetically modified organisms. Biotechnol Ann Rev 10:85–122
Paparini A, Romano-Spica V (2006) Gene transfer and cauliflower mosaic virus promoter 35 S activity in mammalian cells. J Environ Sci Health 41:437–449
Perlak FJ, Oppenhuizen M, Gustafson K, Voth R, Sivasupramaniam S, Heering D, Carey B, Ihrig RA, Roberts JK (2001) Development and commercial use of Bollgard® cotton in the USA: early promises versus today's reality. The Plant J 27(6):489–501
Phipps RH, Beever DE (2001) Detection of transgenic DNA in bovine milk: preliminary results for cows receiving a TMR containing Yieldguard TM MON810. Proc Int Anim Agr Food Sci Conf Indianapolis pp Abst. 476
Pilcher CD, Rice ME, Obrycki JJ (2005) Impact of transgenic *Bacillus thuringiensis* corn and crop phenology on five nontarget arthropods. Environ Entomol 34:1302–1316
Poehlman JM (1994) Breeding field crops—third edition. Iowa State University Press, Ames
Pray C, Naseem, A (2007) Supplying crop biotechnology to the poor: opportunities and constraints. J Dev Stud 43(1):192–217
Pray CE, Nagarajan L (2011) Price controls and biotechnology innovation: are state government policies reducing research and innovation by the Ag biotech industry in India? Forthcoming in Agbioforum

Pray C, Ma D, Huang J, Qiao F (2001) Impact of Bt cotton in China. World Dev 29:1–34
Pray CE, Huang J, Hu R, Rozelle S (2002) Five years of Bt cotton in China—the benefits continue. Plant J 31:423–430
Prescott VE, Hogan SP (2005) Genetically modified plants and food hypersensitivity diseases: usage and implications of experimental models for risk assessment. Pharmacol Therap 111:374–383
Prescott VE, Campbell PM, Mattes J, Rothenberg ME, Foster PS (2005) Transgenic expression of bean alpha-amylase inhibitor in peas results in altered structure immunogenicity. J Agric Food Chem 53:9023–9030
Purcell JP, Perlak FJ (2004) Global impact of insect-resistant (Bt) cotton. AgBioForum 7(1&2) 27–30
Pusztai A (2001) Genetically modified foods: are they a risk to human/animal health action bioscience
Pusztai A, Bardocz S, Ewen SWB (2003) Genetically modified foods: potential human health effects. In: D'Mello JPF (ed) Food safety: contaminants and toxins. CAB International, Wallingford Oxon, pp 347–372
Qaim M, de Janvry A (2005) Bt cotton and pesticide use in Argentina: economic and environmental effects. Environ Dev Econ 10:179–200
Qaim M, Zilberman D (2003) Yield effects of genetically modified crops in developing countries. Science 299:900–902
Rahman M, Rashid H, Shahid AA, Bashir K, Husnain T, Riazuddin S (2007) Insect resistance and risk assessment studies of advanced generations of basmati rice expressing two genes of *Bacillus thuringiensis*. Electron J Biotechnol 10(2):241–251
Rahman M, Shaheen T, Ashraf M, Zafar Y (2012) Bridging genomic and classical breeding approaches for improving crop productivity. Crop Production for Agricultural Improvement, vol 1. Springer Netherland, pp 19–41
Richards HA, Chung-Ting Han RG, Hopkins ML, Failla WW, Ward CN, Stewart Jr (2003) Safety assessment of recombinant green fluorescent protein orally administered to weaned rats nutrient interactions and toxicity. J Nutr 133:1909–1912
Romeis J, Bartsch D, Bigler F, Candolfi MP, Gielkens MM, Hartley SE, Hellmich RL, Huesing JE, Jepson PC, Layton R, Quemada H, Raybould A, Rose RI, Schiemann J, Sears MK, Shelton AM, Sweet J, Vaituzis Z, Wolt JD (2008) Assessment of risk of insect-resistant transgenic crops to nontarget arthropods. Nat Biotechnol 26:203–208
Saxena D, Flores S, Stotzky G (1999) Insecticidal toxin in root exudates from Bt corn. Nature 402:480
Sayanova O, Smith MA, Lapinskas P, Stobart AK, Dobson G, Christie WW (1997) Expression of a borage desaturase cDNA containing an N-terminal cytochrome b5 domain results in the accumulation of high levels of delta6-desaturated fatty acids in transgenic tobacco. Proc Natl Acad Sci U S A 94:4211–4216
Schell J, van Montagu M, Holsters M, Zambryski P, Joos H, Inze D, Herrera-Estrella L, Depicker A, de Block M, Caplan A, Dhaese P, Van Haute E, Hernalsteens J-P, de Greve H, Leemans J, Deblaere R, Willmitzer L, Schroder J, Otten L (1983) Ti plasmids as experimental gene vectors for plants. Advances in gene technology: molecular genetics of plants and animals. Miami Winter Symposia 20:191–209
Schlüter K, Fütterer J, Potrykus I (1995) Horizontal gene transfer from a transgenic potato line to a bacterial pathogen (*Erwinia chrysanthemi*) occurs- if at all- at an extremely low frequency. Bio Technol 13:1094–1098
Schoenly KG, Cohen MB, Barrion AT, Zhang W, Gaolach B, Viajante VD (2003) Effects of *Bacillus thuringiensis* on non-target herbivore and natural enemy assemblages in tropical irrigated rice. Environ Biosaf Res 2:181–206
Schroder M, Poulsen M, Wilcks A et al (2007) A 90-day safety study of genetically modified rice expressing Cry1Ab protein (Bacillus thuringiensis toxin) in wistar rats. Food Chem Toxicol 45:339–349

Schubbert R, Lettman C, Doerfler W (1994) Ingested foreign (phage M13) DNA survives transiently in the gastrointestinal tract and enters the bloodstream of mice. Mol Gen Genet 242:495–504

Sears MK, Hellmich RL, Stanley-Horn DE, Oberhauser KS, Pleasants JM, Mattila HR, Siegfried BD, Dively GP (2001) Impact of Bt corn pollen on monarch butterfly populations: a risk assessment. Proc Natl Acad Sci U S A 98:11937–11942

Shelton AM, Zhao JZ, Roush RT (2002) Economic ecological food safety and social consequences of the deployment of Bt transgenic plants. Annu Rev Entomol 47:845–881

Sims SR (1995) Bacillus thuringiensis var kurstaki (Cry1Ac) protein expressed in transgenic cotton: effects on beneficial and other non-target insects. Southwest Entomol 20(4):493–500

Sims SR, Holden LR (1996) Insect bioassay for determining soil degradation of Bacillus thuringiensis subsp Kurstaki CryIA(b) protein in corn tissue. Environ Entomol 25:659–664

Singh NP, McCoy MT, Tice RR, Shneider EL (1988) A simple technique for quantification of low level of DNA damage in individual cells. Exp Cell Res 715:184–194

Singh AK, Mehta AK, Sridhara S, Gaur SN, Singh BP, Sarma PU, Arora N (2006) Allergenicity assessment of transgenic mustard (*Brassica juncea*) expressing bacterial cod A gene. Allergy 61:491–497

Stanley-Horn DE, Dively GP, Hellmich RL, Mattila HR, Sears MK, Rose R, Jesse LCH, Losey JE, Obrycki JJ, Lewis LC (2001) Assessing the impact of Cry1Ab-expressing corn pollen on monarch butterfly larvae in field studies. Proc Natl Acad Sci U S A 98:11931–11936

Stotzky G (2000) Persistence and biological activity in soil of insecticidal proteins from *Bacillus thuringiensis* and of bacterial DNA bound on clays and humic acids. J Environ Qual 29:691–705

Subedi KD, Ma BL (2007) Dry matter and nitrogen partitioning patterns in Bt and non-Bt near-isoline maize hybrids. Crop Sci 47:1186–1192

Subramanian A, Qaim M (2009) Village-wide effects of agricultural biotechnology: the case of Bt cotton in India. World Dev 37:256–267

Subramanian A, Qaim M (2010) The impact of Bt cotton on poor households in rural India. J Dev Stud 46:295–311

Sudha SN, Jayakumar R, Vaithilingam S (1999) Introduction and expression of the cry1Ac gene of *Bacillus thuringiensis* in a cereal-associated bacterium, *Bacillus polymyxa*. Curr Microbiol 38:163–167

Tabashnik BE (2010) Communal benefits of transgenic corn. Science 330:189–190

Tapp H, Stotzky G (1998) Persistence of the insecticidal toxin from *Bacillus thuringiensis* subsp kurstaki in soil. Soil Biol Biochem 30:471–476

Taylor SL, Hefle SL (2002) Genetically engineered foods: implications for food allergy. Curr Opin Allergy Clin Immunol 2:249–252

Tepfer M, Gaubert S, Leroux-Coyau M, Prince S, Houdebine LM (2004) Transient expression in mammalian cells of transgenes transcribed from the Cauliflower mosaic virus 35S promoter. Environ Biosafety Res 3:91–97

Tice RR, Agurell E, Andeson D, Burlinson B, Hartmann A, Kobayashi H, Miyamae Y, Rojas E, Ryu JC, Sasaki YF (2000) Single cell gel/comet assay: guidelines for in vitro and in vivo genetic toxicity testing. Env Mol Muta 35:206–221

Tryphonas H, Arvanitakis G, Vavasour E, Bondy G (2003) Animal models to detect allergenicity to foods and genetically modified products: workshop summary. Environ Health Persp 11:221–222

Van Lijsebettens M, Vanderhaeghen R, Van Montagu M (1991) Insertional mutagenesis in arabidopsis thaliana: isolation of a T-DNA-linked mutation that alters leaf morphology. Theor Appl Genet 81:277–284

Wang G, Wu Y, Gao W, Fok M, Liang W (2008) Impact of Bt cotton on the farmer's livelihood system in China. International cotton conference, rationales and evolutions of cotton policies in main producing countries. ISSCRI International Conference (May 13–17), Montpellier, France

Wu G, Cu HR, Shu Q, Xia YW, Xiang YB, Gao MW, Cheng X, Altosaar (2000) Stripped stem borer (*Chilo suppressalis*) resistant transgenic rice with a cry1Ab gene from Bt (*Bacillus thuringiensis*) and its rapid screening. J Zhejiang University 19:15–18

Xu-Chongren MY, Chang S (2001) Research on biosafety of transgenic Bt cotton. College of Life Sciences, Beijing University, Beijing, China
Yanni SF, Whaen JK, Ma BL (2011) Field-grown Bt and non-Bt corn: yield, chemical composition, and decomposition. Agron J 103:486–493
Yonemochi C, Fujisaki H, Harada C, Kusama T, Hanazumi M (2002) Evaluation of transgenic event CBH 351 (StarLink) corn in broiler chicks. Anim Sci J 73:221–228
Zilberman D, Ameden H, Qaim M (2007) The impact of agricultural biotechnology on yields, risks, and biodiversity in low-income countries. J Dev Stud 43:63–78

Chapter 9
Foundry Air Pollution: Hazards, Measurements and Control

R. Krishnaraj

Contents

Abstract The emission of harmful gaseous pollutants in the atmosphere is a major health issue. Here we review hazards due to foundry air pollution. We present methods to measure pollution emissions. We also list methods to limit air pollution in foundries. We found that the use of dust collectors in foundries is still rare. We also discuss the standard norms of pollution control.

Keywords Foundry air pollution · Wet scrubber · Cassette filter · Environmental management system · Emission control development

R. Krishnaraj (✉)
Associate Professor, Department of Mechanical Engineering, Institute of Technology, Ambo University, Ethiopia-19
e-mail: r.kraj009@gmail.com

© Springer International Publishing Switzerland 2015

E. Lichtfouse et al. (eds.), *CO_2 Sequestration, Biofuels and Depollution*,
Environmental Chemistry for a Sustainable World 5, DOI 10.1007/978-3-319-11906-9_9

9.1 Introduction

Many products in the world are produced by foundry practices (Biswas et al. 2001). Foundry practices are dominated by the melting of metals, pouring it into moulds, solidifying the molten metal in the moulds, removing of solidified melts from the moulds and cleaning those solidified castings (Parka et al. 2012; Ribeiro and Filho 2006). Foundry industry provides products, which are widely used in pumps, automobile windmill and manufacturing industries, as well as other industries (Wang et al. 2007; Olasupo and Omotoyinbo 2009). The contribution of foundry industry to the world's growth is vital and crucial to many economies. However, the emission by the foundry industry causes key environmental problem. These emissions are mineral dust, organic carbons emitted from melting, sand moulding, casting and cleaning the casting (Krishnaraj et al. 2012). With growing awareness on environmental problems, issues about the pollution caused by the foundry industry have to be seriously and urgently reviewed.

Till 1900s Cupola furnace was used to melt the metal in foundries. After 1900s, foundries began to install induction furnaces. In induction furnace electrical coils are used to burn metals. In the case of Cupola a large amount of scrap is burned which causes the emission of highly polluting gases. In the case of induction furnace, scrap is not burnt and hence the polluting gases are not emitted (Liang et al. 2010). Since majority of the foundries today are utilizing induction furnaces to burn the metals, the pollutant level emission during the melting is reduced considerably. However, during the other processes of foundry like mixing of sand, preparation of sand moulds and pouring metals (Zanetti and Fiove 2002; Yu et al. 2006), emission of pollutants continues to be high. High level of emission of pollutants affect the eco system (Frohling and Rentz 2010; Zanetti and Godio 2006; Taha et al. 2007; Zanetti and Fiore 2002; Subramanya 2006) on widening the reduction of emission through the use of induction furnace.

In this context, the engineers and managers have been realizing the need of absorbing cleaner production technologies in foundries (Martinez and Cabezas 2009; Nie et al. 2005; Niksa and Fujiwara 2005). This is because induction furnace serves as a cleaner production technology as the pollutants affecting the atmosphere are not emitted in the case of induction furnace. In line with this development, both engineers and managers have been attempting to evolve cleaner production technological devices for controlling the emission of pollutants during the sand preparation, melting and pouring of metals while carrying out foundry practices (Sekhar and Mahanti 2006; Strobos and Friend 2004; Huvinena et al. 1997; Andres et al. 1998; Mirasgedis et al. 2008; Rao and Khan 2009; Wu et al. 2004; Georgiadis and Kyrtopoulos 1999; Browne et al. 1999; Keshava and Ong 1999). Although good amount of time and money has been spent in this direction, the levels of pollutants emitted by the foundries have been crossing the limits stipulated by the environmental laws being promulgated in various countries, this perplexing situation is the case which needs a review of the literature arena for studying the emission control technological devices employed in Foundries, their capabilities and imperatives are involve highly improved devices. In the context of this observation the literature was reviewed. The knowledge gathered through this exercise has been briefly presented in this paper.

9.2 Classification

In order to study the pollution emitted by the foundries as well as the efforts made by the companies to control emission in foundries and future imperatives for emission control development. In foundries, the literature was searched to gather papers reporting researchers dealing with these issues. The knowledge gathered by the study of researches reported is classified under four sections of this chapter. The details of these classifications are as follows.

9.2.1 Health Hazards due to Foundry Pollution

It has been widely reported that foundry workers are exposed to polycyclic aromatic hydrocarbon (PAH), which causes damage of the blood cells (Georgiadies and Kyrtopalos 1999). Some researchers have worked in the direction of health hazards faced by foundry workers. The reports presented by these authors have been briefly reported in this section. Liu et al. (2009) investigated the various chemicals that come from foundry practices cause damages to many parts of the bodies of foundry workers. Some of those parts that are damaged due to the foundry emission are lungs, liver, kidney and brain. Besides this, the silica of the moulding sand affects the DNA of the foundry workers. In this paper a study involving foundry workers from two foundries located in Taiwan has been made the finding of the study is that foundry workers are highly subjected to higher plasma MBA and 8- hydrogen gasoline (8-OH-Dg). Lin et al. (2011) have reported that the hazards caused to the foundry workers of Taiwan. According to these authors the silica contained in the molding sand was found to cause respiratory diseases among the foundry workers.

Furthermore the silica dust exposure may create chronic obstructive pulxmory disease among the foundry workers. In this paper, the research involves in the study of the health hazards to the foundry workers in Northern Taiwan about. In this study the health condition of 22 workers was analyzed. The result of the analysis reveals that foundry sand causes DNA damage apart from an increase of urinary 8-OH-dG. Lewtas (2007) has reported that urinary 1-OHP in foundry workers has been found to be, higher level in line to this report. Wang et al. (2011) have reported a research, which involves the study of hazardous air pollutants emission from foundry practices. This research was started after gathering the information from the literature that PAH emission is highest when the thermal decomposition occurs in the core and mould material during the casting process. In this research Curie point pyrolysis emission test was employed to identify PAH that are emitted from casting process.

This result identified 19 PAH elements that emanated from casting process. This identification revealed that significant amount of emission from the casting process is harmful. Yang et al. (2002) have pointed out that according to the epiderimolgical studies the steel foundry workers are subjected to 2.5 times the chance of getting lung cancer. Similar research has been made by Liu et al. (2010). In this paper the level of PAH found in two foundries has been reported. These authors have pointed out that the foundry workers were unable to wear masks during the working hours due to the high prevalence of heat in the work environment. Hence the PAH level of foundry workers is high, In comparison to those working in other environment. These authors have also found that, even the administrative employees working in that foundry are exposed to PAHs. In total, these PAHs increase the risk of getting cancer among the workers and employee working in foundry units. Xu et al. (1996) have furnished certain data to indicate that the foundry workers are subjected to high risk of lung cancer Eder (1999) has briefly reported a study that was carried out among the Finish foundry workers. The result of this study indicates that foundry workers are highly exposed to PAHs.

Anderson et al. (2008) have presented reports found in various other papers. According to these reports foundry workers are subjected to high risk of getting several diseases like silicosis, lung cancer and heat related diseases. Subsequently, they have reported the study conducted in 11 Swedish iron industries by gathering data on repairable dust and quartz among the foundry workers. During this study, the data were gathered from the foundry workers carrying out several activities like casting, core making and moulding. The foundry workers carrying out casting and furnace ladle repair were subjected to higher degree of dust in foundries. Altogether the study reported in this paper reveals the high exposure of quartz and sand dust, which are harmful to the health of the foundry workers. However, an interesting observation of this study is that, foundries that have followed environmental norms emit less dust particles and hence the men working in these foundries are less exposed to such harmful dust than those who are working in foundries, which do not follow appropriate environmental norms.

Kuo et al. (1998) have narrated that paper reporting the high risk of foundry workers in getting the disease according to this narration, foundry workers are subjected to high risk of getting lung and stomach cancer and silicosis. There reports have also indicated the release of substance which are harmful to the humans from the foundries. Subsequent to this narration these authors have reported a survey undertaken among 718 foundry workers who had been working in 50 iron foundries situated in central Taiwan. One of the major findings is that the foundry workers are subjected to high risk of getting pneumoconiosis. The levels of risk of getting pneumoconiosis among the foundry workers varied due to various factors. For example with the number of years of work experience the chance of getting pneumoconiosis among the foundry workers was found to increase. Likewise foundry workers associated with the operation of furnace were found to have 11 times higher risk of getting pneumoconiosis than the other foundry workers working in other areas of the foundry. The different levels of risk associated with the foundry workers in pneumoconiosis among the foundry workers has been used by these author to draw the interference that the prevalence of pneumoconiosis, among the foundry workers can be reduced by following environmental norms.

The research reported in the above papers has revealed that the foundries emit highly harmful substances, which cause diseases like cancer and silicosis among the foundry workers. Though this fact has been vociferously appraised by the researchers, during the recent times some efforts like maintaining good ventilation and preventing the spreading of dust in working environment have reduced the emission of harmful substances by the foundries. This finding triggers the modern researchers to find a solution and develop the devices that would prevent the emission of the harmful substance by the foundries.

9.2.2 Pollution Emission Measurement

While surveying the literature it was found that some researchers have measured the level of pollutants emitted from the foundry. In the literature survey, reported here, few papers report such researches leading to the measurement of pollutants emitted by the foundries. The most important information derived from reading those papers has been described in this section. Andres et al. (1995) have carried out a chemical process leading to the solidation of steel foundry dust and conducted a chemical analysis. After this analysis was conducted, steel foundry dust was found to contain as many as 22 elements. Out of them (Cr, Cd, Pb and Zn) chromium, cadmium, lead and zinc are harmful to humans. Out of these harmful elements zinc constitutes 23.37 % by weight in the SFD. These authors have classified SFD as hazardous waste, which requires detoxification before it is dispersed off. Cheng et al. (2008) have reported the measurement of ultrafine particles that are immanent during melting and pouring of metals in an iron foundry. These authors have pointed out that ultrafine particles are harmful to humans.

These authors have used scanning mobility particle size (SMPs) to determine the number of concentration and sample area concentration. While these authors have found out that Ni and Si concentration varied in the iron foundry from 2.07×10^4 to 2.85×10^5 particles cm^{-3} and 67.56 to 2.13×10^3 $\mu m^{2\ cm^{-3}}$ respectively, these sizes varied due to various factors like different operations. The sizes of ultrafine particles also found to vary due to seasons like winter and summer. These authors have also reported similar studies and ultrafine particle size measurement. Lv et al. (2011), have reported research involving the measurement of polychlorinated-p-dibenzoioxins and dibenzofurans (PCDD/Fs) in iron foundry. These authors have reported that the emission of dioxine by the iron foundries is known widely. However, PCDD/Fs emission by iron foundry is rarely known. These authors have taken samples of fly ash from the iron foundries located in China and quantified the PCDD/Fs content. The contents of these PCDD/Fs and PCBs were analyzed using chromatography and mass spectrometer. The concentration of PCDD/Fs in stack gas was measured using Pg. WHO-TEQ Nm^{-3}. (WHO- world health organization- TEQ- Total Equivalent Quantity).

In the case PCDD/Fs the value ranges from 56.5 to 232 in the case PCDD/Fs concentration. These authors have compared the emission of PCDD/Fs with that released by other thermal processes. They have found out that PCDD/Fs concentration was less by the Iron foundries compared to that released by other thermal processes. These authors have also claimed that the emission of PCDD/Fs depends upon the type of foundries, raw materials and pollution control systems employed. Like these authors Yu et al. (2006) have also studied the emission of PCDD/Fs and PCBs by various metallurgy industries situated in South Korea. One of the Industries from where these emissions were studied was a ferrous foundry. According to their findings the emission of PCDD/Fs and Co PCBs are higher in pig iron foundry than that emitted in the steel foundry. In order to measure these toxic pollutants this authors used emission factors in ng TEQ/ton.

Polizzi et al. (2007) have measured the particulate matters in mg/m^3 in the iron casting and aluminum foundries situated in Taiwan district of Italy. According to the findings of these authors, the aluminum and iron particles matters from the foundry studied during the research reported in this paper are found to be significantly higher in ambient air up to 6 km.

Grochowalski et al. (2007) have reported a research involving the measurement of four pollutants, namely PCDDS, PCDFS, PCBS and HCB emission by the metallurgy industries of Poland. These authors have developed a sampling methodology to measure these pollutants simultaneously. Under this method, these authors have developed a sampling apparatus consisting of sample nozzles, probe and heating compartments. These authors have claimed that the method that was adapted ensures 95 % collection efficiency of the above mentioned four pollutants from flue gas. In order to quantify these pollutants these authors have used ng-TEQ/Nm^3 as the unit of measurement. These authors have also used a measure called emission factors to indicate the emission of the above pollutants. An interesting observation is that all the industries that exhibited highest emission factor are cast iron foundries.

Melendez et al. (2010) have reported a research in which the pollutants emitted by foundry cupola furnace were measured using the concentration of particulate matter. These authors have used mg/Nm^3 to indicate the emission of a particulate matter. The concentration of a particulate matter was measured immediately below the cupola furnace. These authors have also mentioned that the size of PM is so small that it is necessary to empower some scanning electron microscope and energy dispensive X-Ray spectroscopy. While concluding, these authors have mentioned that the maximum size of a particulate matter including Zinc needs to be specified for determining the pollution in foundries.

Altogether the information gathered from the above papers indicates that researchers have been trying to quantify the pollutants that arise from foundries. They have used several units including ng TEQ/Ton, mg/Nm^3. However, these measurement approaches are not standardized. In many industries the data are maintained in database. In spite of the lack of standardized units, the researchers and practitioners have been adopting any one of more methods to reduce the pollution that is caused by foundry units. Particularly, these researchers and practitioners have been adopting these measurement units to check the performance of foundries by employing or not employing pollution control devices.

9.2.3 *Pollution Control Devices*

During casting, lots of gases and small dust particles are generated inside the foundry. This is harmful to the workers' health and at the same time it will also damage the surrounding environment. Micron-sized dust particles create fatal effect on human bodies, especially on the heart and lungs. For avoiding these problems lots of equipment's are used in industries to avoid the dust and to collect the dust particles and also to separate the dust particles from the gas. In industries four principal types of dust collectors are used (1) Fabric filters (2) Electrostatic precipitators (3) Wet scrubbers, and (4) Inertial separators. The details of these dust collectors are as follows.

9.2.3.1 Fabric Filters

The filter cloth is mounted on a cylindrical shaped bag and suspended which is called bag house or fabric filter show as shown in the figure given below. The dust particles flow through this filter, which is then collected by the filter cloth. The collected dust particles form a thick layer around the fabric filter surfaces. Due to this reason, pressure loss occurs. The collected dust particles are removed by the mechanical shaking of filters or pulse jet. Then these particles are collected in a hopper placed at the bottom of the fabric filter (Fig. 9.1).

9.2.3.2 Electrostatic Precipitators

Electrostatic precipitators use electrostatic forces to separate dust particles from exhaust gases. The contaminated gases flow through the aerosol tube formed by the discharge and collecting electrodes. The airborne particles receive a negative charge as they pass through the ionized field between the electrodes. These charged particles are then attracted to a grounded or positively charged electrode and adhere to it. The collected material on the electrodes is then removed by rapping or vibrating the collecting electrodes either continuously or at predetermined intervals (Fig. 9.2).

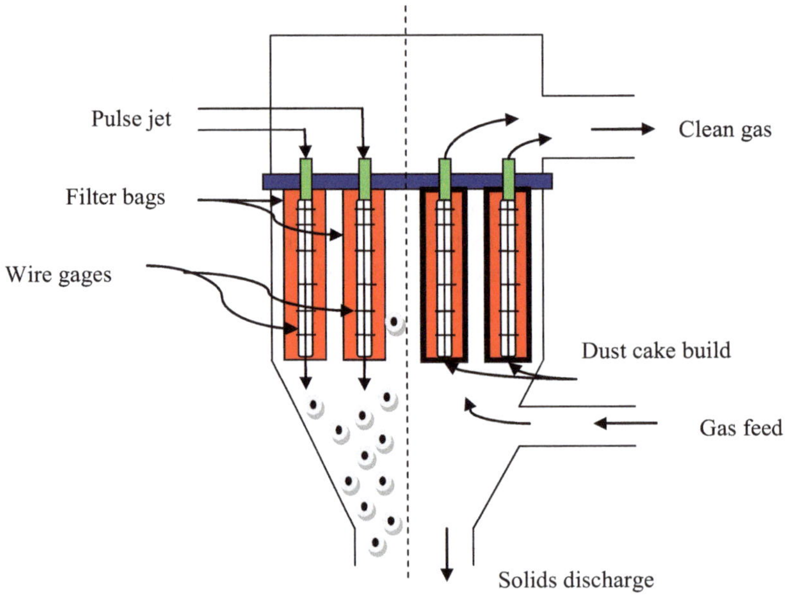

Fig. 9.1 Schematic diagram of fabric filter. (Darcovkic et al. 1997)

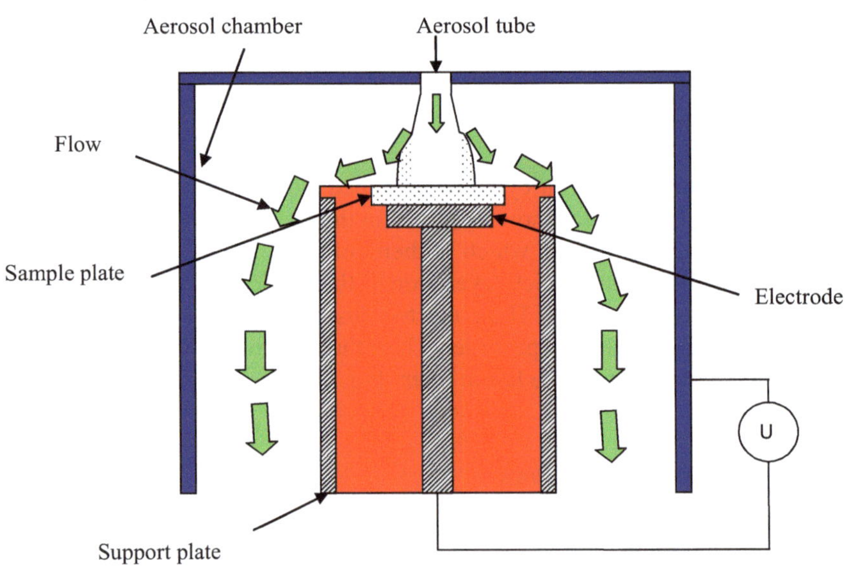

Fig. 9.2 Schematic illustration of electrostatic precipitator. (Dixkens and Fissan 1999)

9.2.3.3 Wet Scrubbers

Dust collectors that use liquid are known as wet scrubbers. In this system, the scrubbing liquid (usually water) comes in contact with a gas stream containing dust

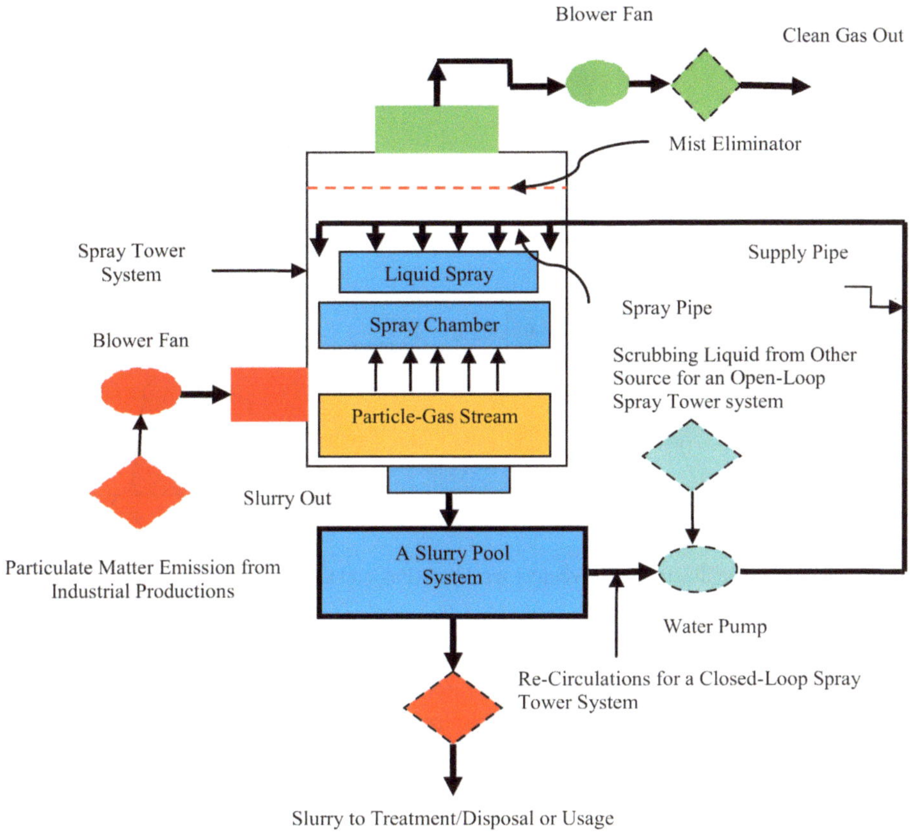

Fig. 9.3 Schematic illustration of wet scrubber. (Danzomo and Salami 2012)

particles. Greater contact of the gas and liquid streams yields higher dust removal efficiency. The simple wet scrubber arrangement is shown in the following figure. In this system the dust particles enter through the inlet port of the wet scrubber. Now the water is sprayed from the spray tower arrangement. The dust particles come in contact with the water droplets, through which the dirty water is drained. The wet scrubbers are subdivided into four types (1) Low-energy scrubbers, (2) Low- to- medium energy scrubbers, (3) Medium- to-high-energy scrubbers, and (4) High-energy scrubbers (Fig. 9.3).

9.2.3.4 Venturi Scrubbers

The high energy scrubbers are also called venturi scrubbers, which are used for dust cleaning in foundry. It has more collecting efficiency compared to cyclone scrubber, but this scrubber results in more power consumption. A venturi scrubber consists of three parts: convergence, throat and diffuser as shown in the following figure. The dust particles introduced with the air in the convergence section collide

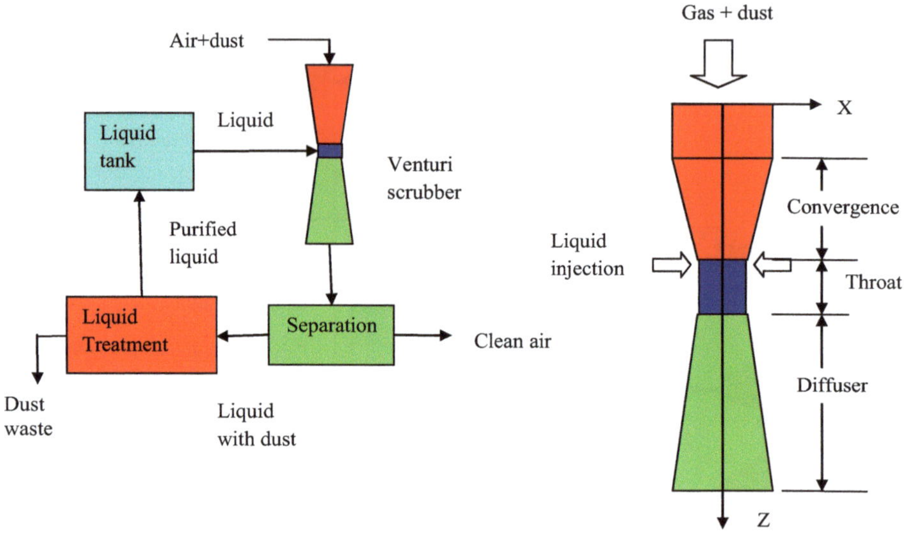

Fig. 9.4 Schematic illustration of venturi scrubber. (Pak and Chang 2006)

with the liquid droplets at the throat and then they pass to the diffuser. The droplets that captured the dust particles are separated at the separation tank connected to the Venturi scrubber, and go out through the drain hole. The clean air is discharged into the atmosphere and the drained liquid is treated for reuse (Fig. 9.4).

9.2.3.5 Inertial Separators

Inertial separators separate the dust from gas streams using a combination of forces, such as centrifugal, gravitational and inertial. These forces move the dust to an area where the forces exerted by the gas stream are minimal. The separated dust is moved by gravity into a hopper, where it is temporarily stored. Inertial separators are subdivided into three types (1) Settling chambers. (2) Baffle chambers. (3) Centrifugal collectors. The following section gives the working of a cyclone separator.

9.2.3.6 I Centrifugal Collectors

Centrifugal collectors are also known as cyclone separators. Centrifugal collectors use cyclonic action to separate the dust particles from the gas stream. In a typical cyclone, the dust gas stream enters at an inlet port rapidly. The centrifugal force created by the circular flow throws the dust particles toward the wall of the cyclone. After striking the wall, these particles fall into a hopper located underneath.

Cyclone separators had shown in Fig. 1.1 are mostly used in industries to separate the dust from gas and filtration of the dust from air. Due to its simplicity, there is less power consumption and low maintenance cost (Fig. 9.5).

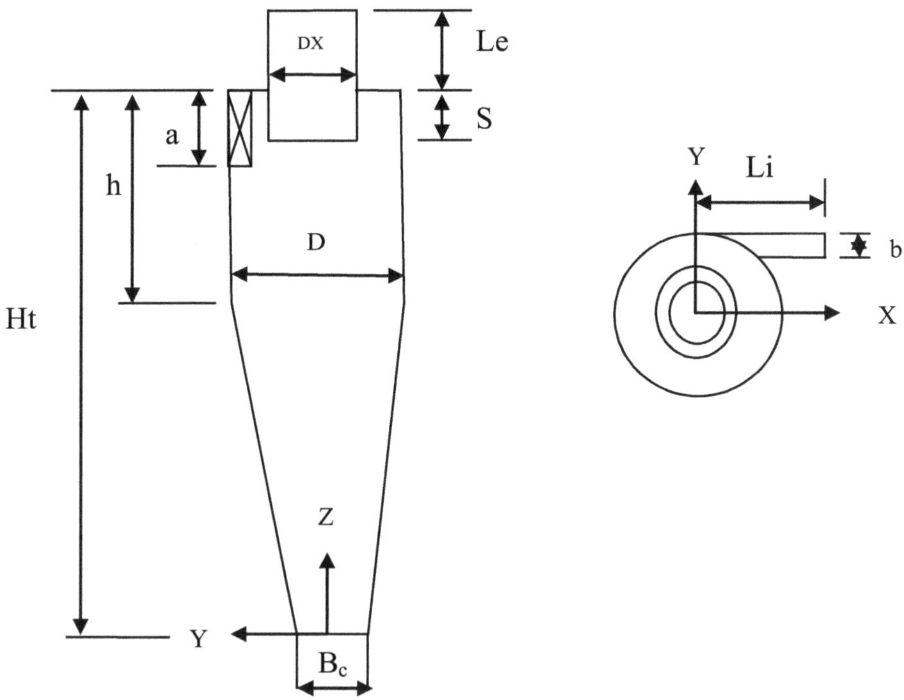

Fig. 9.5 Schematic illustration of cyclone. (Elsayed and Lacor 2011)

While the researchers, practitioners and administrators have been involved in measuring the emission of pollutants from foundries, a considerable effort has been exerted both practically and theoretically. Some of those efforts have been published in literature arena. In the literature survey reported here, 15 papers report the usage of pollution controlling devices in foundries. The knowledge gathered from these papers has been highlighted in the following parts of this section. Pal et al. (2008) have reported a research on applying pollution control devices in foundries located in Howrah district of India. These authors have mentioned that Cupola is the furnace used by major foundries in India. Because this Cupola emits harmful pollutant, Central Pollution Control Board of India (CPCB) evolved norms for controlling the suspended particulate matter (SPM). These authors have mentioned that several devices like centrifugal separators, low energy scrubbers and intensity scrubbers are available to control the pollution emitted from Cupola furnace. Out of them, these authors have considered venturi scrubber system for examining its capability in controlling the pollution from Cupola furnace and the unique feature of this research is the aspect of increasing energy through the controlling of pollution.

Since energy reduction was not effective in the case of employing venture scrubber, these authors developed a device called Divided Blast Cupola (DBC) furnace. Using this device these authors could bring down the SPM emission from Cupola furnace from 2000 to 70 mg/Nm^3 (milligram per normal cubic meter). These authors have mentioned that these DBC technologies has been used across India to reduce the SPM levels emitted by the Cupola furnace, while Neto et al. (2008, 2009a, 2009b) have described the avenue available for reducing pollutants from Aluminum pressure dicasting plants. A model for calculating the avenue for reducing the emission using different pollution controlling devices has been devised. Some of them include filters and scrubbers. Besides, some options like reducing scrap rate and modification of combustion process were suggested for reducing the emission of pollution of aluminum pressure dicasting unit. An important inference drawn frame this research is that no single device or approach will reduce the emission of pollutants from the aluminum pressure dicasting plants. Where it is required a combination of devices and opticus are applied to reduce the emission of pollution from various processes and stages of aluminum pressure dicasting plants.

Fore and Mbohwa (2010) have given suggestions to reduce the pollutants by the foundries. These authors have traced various aspects of controlling pollution from foundry units. For example, they have drawn from the literature that the particulate matter emission should be below 20 mg/m^3 from any industry. These authors also have considered PCDDS, PCBS, and CO_2as the major pollutants from the foundry industry. These authors also have mentioned about the DBC reported by Pal et al. (2008). One of the major contributions of this author is the listing of foundry processes from which air pollution occurs. These authors have listed the dust emission technologies as cyclone, scrubber, Bay houses and electrostatic precipitators as in the case of the research outcome reported by Pat et al. These authors have listed the various options like recycling sand after removing binders and reducing NOx emission as the means of controlling pollution in the foundry industry. These authors have also mentioned that care has to be taken to maintain the furnace so that emission from the foundry unit is controlled.

An important observation of these authors which corroborates with pollution controlling devices enhances the energy released by the foundry units. On the whole, these authors have inferred that several devices, technologies and approaches are required in combination to reduce pollution from foundry industry. Biswas et al. (2001) have studied the emission of pollutants from Cupola. In order to reduce this emission, these authors have mentioned the construction of four devices, namely Dry gas cleaning plant, dry gas cleaning with quencher, wet gas cleaning plant and coke less cupola. Out of these devices these authors have made a detailed analysis of gas cleaning plant and cupola. In this analysis, the levels of pollution emitted from these devices and the cost involving in installing them (converting them from some other system) was investigated. The conclusion is that the usage of coke less cupola will result in controlling the pollution requiring lesser investment.

Fatta et al. 2004, have drawn valuable information from the literature arena to prescribe the methods of reducing the emission of pollutants from the foundry units. This prescription has been given under the title guidelines for foundries. These guidelines have been listed from the perspectives of pollutants from the foundries, by using the appropriate technologies for preventing the emission of pollutants. Some of these guidelines include the suggestion to use (1) induction furnace in place of cupola furnace, (2) dust emission controlling devices like cyclones and scrubbers, and (3) natural gas fuel to reduce NOx emission of pollutants through a fabric filter to achieve emission level of10 mg/Nm^3. In line with the observations of Pal et al. (2008) and Neto et al. (2009). These authors have also mentioned that a combination of devices and approaches are necessary to reduce emission of pollutants from the foundries to the permissible levels. Rabah (1999) studied economical aspects of utilizing several devices for controlling pollution from iron foundries of Egypt. This author analyzed the emission of pollutants by making use of the term called Total Suspended Particle (TSP) and using the unit mg/m^3 to measure it.

This author listed various pollution control devices, which include cyclone fabric filter and wet scrubber. After a thorough analysis, this author came to the conclusion that wet scrubber is the most economical pollution control device that can be utilized in Egypt. On the whole, the research reported in this paper reveals the need of conducting research on economical aspects of choosing and controlling pollution controlling devices in foundries. Particularly this research is concerned with the justification in installation of some pollution controlling devices which would not get back the invested amount. Myers and Ibarreta (2009) have reported about the elimination of dust from a foundry. In order to control the dust accumulation they have suggested a dust collector system. On the whole, though pollution control in foundry has not been explicitly dealt with, this paper shows the need of controlling pollution through education and following good practices in foundries.

By applying this process, the pore structure is formed in coals. These authors have claimed that pyrolysis is an economical way of controlling pollution in green sand mould casting process commonly adopted in foundries. The review of the paper presented in this section has indicates that due to the need of controlling the emission of pollutants from the foundries, several devices and approaches have been utilized in foundries situated in certain parts of the world such as India, Crypus and Italy. These countries are developing countries, which cannot dispense with the foundries as they contribute much to wealth generation. Hence, researchers hailing from these countries have been examining the effectiveness of pollution controlling devices and approaches. On the whole these researches have examined the effectiveness of these pollution control devices and approaches from the perspective of economical effectiveness and energy efficiently. With the suggested new devices, this kind of examination is required to be continued in the years to come. Such practices are required to be guided to adopt the combination of pollution control devices and approaches that will lead to the limiting of the emission of pollutants within the admissible levels in an economical and efficient manner.

9.3 Discussion

9.3.1 Distinctive Features of Dust Collectors

This final section points out some of the most important conclusions of the research and formulates some guidelines for future research. The major role of this paper is to provide a comprehensive review of peer-reviewed journal publication on foundry air pollution. In our vision three major topics are to be emphasized in this respect.

- Secondary pollution in the wet scrubber system has to take into account a wider range of issues and, therefore, it has to look at the larger part of the pollution control devices.

This assertion emerges as one of the most distinct features of green deprivation, which also requires a remarkable approach in conducting pragmatic research. Quite a few case studies have collected data from numerous or even all stages, although rarely done in literature. This situation have been identified as an urgent need in research on dust collectors, while avoiding the risks associated with secondary pollution, foundry industries may find themselves in a situation where they must spend more attention on pollution control devices. Here, the intersection among environmental management with cleaner production starts to emerge. Up to now very few papers hold the term 'secondary pollution of wet scrubber' in their label. Yet, this issue imposes a major challenge as it would solicit a major endeavor. Leading to the argument about the assessment of standards with pollution control board to meet with European standards.

- It is a fact that foundries have to use of cassette filters in induction furnace rather than wet scrubber.

One allegation of cassette filters for companies and, therefore, for dust collector is the wider set of criteria that have to be met. Environmental and social issues are increasingly on the public agenda, which provide the need of the hour to include them in emission reduction. The adaptation of the cassette filter in induction furnace has continued for some time for single companies although their implications in foundry industries have not been fully explored. As already mentioned, several surveys report a positive correlation between environmental and cost-effective concert. Still the underlying question offers a major avenue for future research. Again, the interrelation with more suitable dust collectors proposes a full range of research issues.

- Induction furnaces deal with enhanced set of performance in emission reduction. Thereby it takes into account an assortment of emission assessment in foundries to reinstate cupola with induction.

Though emission reduction can be considered a hot topic, much of the scholastic discussion as well as findings from pragmatic research point towards the narrow integration that can be observed. Several case studies have outlined how far the teamwork has to reach and how much endeavor focal companies have to make before they can meet the green index. Involving this flipside to the performance issue, this imposes the question. How should companies follow environmental management system so as to reach a sustainable performance. These distinguishing features together place dust collectors well within the overall area of research, and offer evidence on which additional issues have to be addressed. This might also feedback into conventional cupola emission literature.

Table 9.1 Dioxin emission data for various foundry types. (Source: Brettschneider and Vennebusch 1992; Kran et al. 1995; European Commission 2005)

Product type	Furnace	Melting (t/h)	Flue-gas (m^3/h)	Abatement	O_2 (%)	PCDD/F (ng TEQ/Nm^3)
Aluminum	Hearth type	n.d	n.d		n.d	0.002
Aluminum	Hearth type	0.45	9300	None	18.8	0.002
Aluminum	Shaft	1.5	8400	None	18.4	0.01
Cast iron	CBC	3.4	15,900	Bag filter	n.d	0.04
Cast iron	CBC	3.7	14,300	Bag filter	16	0.09
Cast iron	CBC	4.5	14,300	Bag filter	n.d	0.09
Cast iron	CBC	3.4	n.d	Bag filter	n.d	0.33
Cast iron	CBC	5.5	17,400	Bag filter	15.9	0.51
Cast iron	CBC	6.5	17,500	Bag filter	n.d	0.51
Cast iron	CBC	6	27,600		n.d	3.14
Cast iron	HBC	45.5	55,000	Disintegrator	6	0.003
Cast iron	HBC	60	n.d	Disintegrator	n.d	0.003
Cast iron	HBC	40.6	75,000	Bag filter	12.5	0.05
Cast iron	HBC	50	75,000	Bag filter	n.d	0.07
Cast iron	HBC	15	36,400	Bag + PC	n.d	0.05
Cast iron	HBC	13	n.d	Bag filter	n.d	0.10
Cast iron	HBC	18.2	29,100	Bag filter	8.6	0.20
Cast iron	HBC	17.1	22,500	Bag filter	7.5	0.29
Cast iron	HBC	27	n.d	Bag + PC	n.d	1.00
Cast iron	HBC	28	37,000		n.d	2.08
Cast iron	HBC	21	32,000		n.d	3.09
Cast iron	IF	19.5	208,000	Bag filter	20.2	0.003
Cast iron	IF	n.d	n.d	Bag filter	n.d	0.01
Cast iron	RF	8	n.d		n.d	0.004
Cast iron	RF	1.4	9000	None	n.d	0.02
Cast iron	RF	2.1	18,600	Bag filter	19.9	0.45
Cast iron	RF	3.5	n.d	Bag filter	n.d	0.61
Steel	EAF	5.4	54,150	Bag filter	20.9	0.003
Steel	EAF	9	5000	Wet scrubber	n.d	0.02

CBC COld blast cupola, *HBC* hot blast cupola, *RF* rotary furnace, *IF* induction furnace, *EAF* electric arc furnace

Table 9.1 clearly focuses on abatement techniques commonly used with respect to the product type, type of furnace melting capacity, flue gas passed, oxygen percentages utilized for burning the coal and coke for melting the raw material. The Table clearly reveals that for aluminum product, Hearth and Shaft type furnace was specifically used to melt the raw material, for which the melting capacity was 0.45 and 1.5 t/h. It is observed that when the flue gas of 8400 m^3/h passes in Hearth type, 18.8 % of O_2 was utilized. On the other hand when the flue gas of 8400 m^3/h pass in Shaft type furnace, only18.4 % of $O_2$was utilized in Shaft type. The case of cast iron, 4 types of furnaces was adopted, namely cold blast furnace, hot blast furnace, induction furnace and rotary furnace. Cast iron obtained from the cold blast cupola furnace has 3.2–6.5 t/h, in which flue gas emission is 14,300–17,500 m^3/h. 16 % of O_2 supplied to the cold blast cupola furnace and PCCD/F has the range 0.09–0.51 ng TEG/Nm^3,for which bag filter used to control the emission. In the hot blast cupola type of furnace, maximum melting capacity is 19.5 T/h; flue gas emission is 20,8000 m^3/h. For any type of induction furnace the emissions controlled through bag filter ranges from 0.003 to 0.01 ng TEG/Nm^3. In the rotary furnace, maximum melting capacity is 8 t/h, and minimum is 1.4 t/h, in which flue gas emissions is 9000 m^3/h, PCDD/F ranges from 0.004 to 0.02 ng TEG/Nm^3. In the case the steel products obtained from Electric arc furnace, the maximum melting capacity is 9 t/h, minimum 5.4 t/h, flue gas emission range from 6000 to 54,150 m^3/h. It is clear that most of the cast iron product type industries use bag filter to remove the particles passing through the furnaces (Table 9.1). Though bag filter was used in Barriers towards cleaner production in one foundry that was not very effective in removing the dust particle. Since bag filter is not efficient enough for removing the SO_2, leakage of gases before entering the PCD was prominent. It was observed that lack of replacement of torn bag at the appropriate time caused the leakage and loss of filtering action. If the bags are maintained properly, they can be used in a foundry as the final collector in a series to remove the last remaining fine dust particles, followed by settling chamber or cyclone or scrubber (Mukherjee 2010).

9.4 Implementation of Environmental Management System Towards Cleaner Production

9.4.1 Environmental Management Systems for an Integrated Pollution Prevention Control (IPPC)

Environmental Management Systems (EMS) is an institutionalized system developed for the institutions and the management of people that impact the environment. In general, EMS is a useful tool to aid the prevention of pollution from industrial activities. The environmental authorities have set standards for pollutants from industries. The standards for industrial emission/effluents are urbanized based on Best Practicable Treatment technology (BPT). If the scene of the industry has high environmental compassion, it is a requisite to abide by stricter standards of pollution control and employ Best Available Technology (BAT) that can put an exorbitant high encumber on the industry. On the other hand for sustainable development, it is very important that, there are no conflicts with the environment or at least, such conflicts are minimal so that in the event of failure of pollution control systems by the industries, there still is some time for them to resolve without having any adverse impact on the environment. Hence proper information on the status of environment would help to reduce pollution threats and regulatory risks to industries and ensure their investment safety. In practice the emission of pollutants and controlling solutions have been delivering solutions and information in scattered directions. It is high time that these information and solution are realized to develop a system that would enable the sustenance of foundries through the prevention of the emissions of pollutants. In order to overcome this research and practice needs, research in the direction of developing environmental management system pertaining to the foundry practices. Sekhar and Mahanti (2006) have reported the status of foundry industry in India. These authors have mentioned that the owners of foundries in India have stated to install pollution control devices like wet scrubber, venturi scrubber, while the major scope of this paper is to report the application of six sigma concept to control the emission of pollutants from foundries, small portion is devoted for appraising the need of Implementing ISO 4001 based environmental management system in foundries. These authors have appraised the importance of implementing ISO4001 based environmental management system in foundries. Particularly these authors have pointed out that the Implementation of ISO 4001 standards based EMS would create Environmental ethics in foundries. According to this emphasis the foundries will have appropriate documentation about the emission and control of pollutants in foundries. These authors have not name any foundries situated either within India or outside India which have Implemented ISO 4001 standards based EMS. Availability only this paper in literature arena and absence of report about implementing ISO 4001 standard based EMS give rise to an impressed that enormous amount of practice oriented researchers are required to carried out in the direction of ISO 4001 standard based EMS.

9.4.2 *Techniques to Determine the Best Available Techniques*

The best environmental performance is usually achieved by the adaptation of the best technology and its operation in the most effective and efficient manner. This is recognized by the IPPC Directive definition of 'techniques' as "both the technology used and the way in which the installation is designed, built, maintained, operated and decommissioned". For IPPC installation an environmental Management system (EMS) is a tool that operators can use to address these design, construction, maintenance, operation and decommissioning issues in a systematic, demonstrable way (European Commission 2005).

An EMS includes the organizational structure, responsibilities, practices, procedures, processes and resources for developing, implementing, maintaining, reviewing and monitoring the environmental policy. Environmental Management Systems are most effective and efficient where they form an inherent part of the overall management and operation of an installation. Within the European Union, many organizations have decided on a voluntary basis to implement environmental management systems based on EN ISO 14001:1996 or Eco-Management and Audit Scheme (EMAS). EMAS includes the management system requirements of

EN ISO 14001, but places additional emphasis on legal compliance, environmental performance and employee involvement; it also requires external verification of the management system and validation of a public environmental statement (in EN ISO 14001 self-declaration is an alternative to external verification). While both standardized systems (EN ISO 14001:1996 and EMAS) and non-standardized ("customized") systems in principle take the organization as the entity, this document takes a more narrow approach, not including all activities of the organization e.g. with regard to their products and services, due to the fact that the regulated entity under the IPPC Directive is the adopted (European Commission 2005).

9.4.3 Achieved Environmental Benefits

Implementation of and adherence to an EMS focuses the attention of the operator on the environmental performance of the installation. In particular, the maintenance of and compliance with clear operating procedures for both normal and abnormal situations and the associated lines of responsibility should ensure that the installation's permit conditions and other environmental targets and objectives are met at all times (European Commission 2005).

Environmental management systems typically ensure the continuous improvement of the environmental performance of the installation. The poorer the starting point is, the more significant short-term improvements can be expected. If the installation already has a good overall environmental performance, the system helps the operator to maintain the high performance level.

9.4.4 Driving Force for Implementation

Environmental management systems can provide a number of advantages (European Commission 2005) such as:

- improved insight into the environmental aspects of the company
- improved basis for decision-making
- improved motivation of the personnel
- additional opportunities for operational cost reduction and product quality improvement
- improved environmental performance
- improved company image
- reduced liability, insurance and non-compliance costs
- increased attractiveness for employees, customers and investors
- increased trust of regulators, which could lead to reduced regulatory oversight
- Improved relationship with environmental groups.

9.4.5 Guidelines for Exhaust Air Cleaning Techniques

- **Blasting**: Blasting generates a lot of dust. Off-gas capture is unproblematic due to the total enclosure of the blasting process in a closed cabin. Customary exhaust air decontamination processes are wet washing and dry filters, usually with a cyclone as a pre-filter (European Commission 2005).
- **Slide grinding, drumming**: Both processes do not require emission reduction in normal conditions. Aerosols that can form in quickly moving drums can be neglected (European Commission 2005).
- **Cutting**: All thermal separation processes generate emissions. Exhaust capture is achieved through enclosing the work place and carrying out an intensive extraction of the arising fume. It is important to place the capture elements as close as possible to the emission source, nut without hindering the operating process. In some cases it has proven beneficial to combine a mobile arm for direct extraction and additional cabin extraction. By carefully arranging the extraction elements, the dust-filled air can be guided away from the breathing area of the worker. The usual exhaust air decontamination processes are wet washing and dry filters. It should be noted that the systems are set up for the smaller grain sizes of the emitted fumes. Inertia force separators are applied as pre-separators (European Commission 2005).
- **Abrasive cutting**: Stationary abrasive cutting installations are extracted. Customary exhaust air decontamination process is cyclones, wet washing and dry filters (European Commission (2005)).
- **Sawing, rapping, pressing**: These handling processes cause few emissions and do not require reduction measures in normal conditions (European Commission 2005).
- **Chiseling, needling**: In these processes mainly coarse particles are produced which are difficult to remove through extraction. The work is mainly carried out in cabins for safety reasons. In special cases, e.g. when removing burnt-on sand, the generated dust may be extracted by an extraction arm. The exhaust air is decontaminated in cyclones, wet washing and dry filters (European Commission (2005).
- **Grinding:** Emission capture when grinding on stationary machines takes place in a similar way as with abrasive cutting, i.e. via fixed funnels into which the abrasive blast is directed. An enclosed work place is used for manual grinding and abrasive cutting. Extraction walls can then be applied in the housing. Air decontamination processes are cyclones, wet washing and dry filters (European Commission 2005).
- **Stamping, milling**: These handling processes cause few emissions and do not require reduction measures in normal conditions (European Commission 2005).
- **Welding**: More or less emissions occur according to the type of the selected welding process, which is usually being best collected by extraction arms. For exhaust air cleaning, wet washing and dry filters and occasionally electrostatic filters are used (European Commission (2005)).

9.5 Conclusion

As mentioned by many authors in research papers, foundry industry has been playing key roles in enhancing the economy of the societies, as many products used by humans are made up of components produced in foundries. Yet many developed countries are hesitant to promote foundries, as the pollutants emitted from them affect the health of the humans. In the case of developing countries it is unaffordable to restrict the growth of foundry industry since this attempt will affect the economical growth. Hence a need is arisen to establish the way of sustaining the foundry industry particularly in developing countries. In order to meet this needs research in different direction have been carried out. In one direction, researchers on estimating the hazards caused due to the pollutant emitted by the foundries. In the other direction research involving the measurement of pollutants emitted from the foundries had been reported. In the third direction, the pollution controlling devices installed in foundries have been investigated. These research activities have not brought out many comprehensive solutions that would prevent the evaluation of pollutants from foundries. Comprehensive approach is required to ensure that this research outcome are brought out to offer standardized procedures and methodologies for ensuring the sustains of foundry Industry. This is possible if a standard for exclusive application of EMS in foundries is developed. Hence numerous trials have been taken to assure the eminence of the review reported. Future work might perk up this strategies by taking a closer look at exact publications, i.e., from a practical implementation perspective. This might allow explicit features to be recognized in enhanced aspect of Cassette filter.

Acknowledgement The work of first author is supported by DST Inspire Fellowship, Ministry of Science and Technology, Government of India-Ref. DST/Inspire Fellowship/IF 110324.

Reference

Anderson L, Bryngelsson IL, Ohlson CG, Naystron P, Lilja BG, Westberg H (2008) Quartz and dust exposure in Swedish iron foundries. J Occup Environ Hyg 6:9–18

Andres A, Ortiz I, Viguri JR, Irabien A (1995) Long-term behavior of toxic metals in stabilize steel foundry dusts. J Hazard Mater 40:31–42

Andres A, Ibfifiez R, Ortiz I, Irabien JA (1998) Experimental study of the waste binder an hydrite in the Solidification/stabilization process of heavy metal sludges. J Hazard Mater 57:155–168

Biswas DK, Asthana SR, Rau VG (2001) Pollution management with techno-economic evalution for coke based and cokeless cupolas. Applied Engineering 21: 359–379

Brettschneider D, Vennebusch B (1992) Improvement of conditions for emissions of a melting range with two rotary furnaces -final Report 1119 50:441-1/56

Browne DR, Husni A, Risk MJ (1999) Airborne lead and particulate levels in Semarang, Indonesia and potential health impacts. Sci Total Environ 227:145–154

Cheng YH, Chao YC, Wu CH, Tsai CJ, Uang SN, Shih TS (2008) Measurement of ultrafine particle concentrations and size distribution in an iron foundry. J Hazard Mater 158:24–130

Danzomo BA, Salami ME (2012) Performance evaluation of wet scrubber system for industrial air pollution control. J Eng Appl Sci 7(12):1669–1677
Darcovkic KA, Jonasson CE, Capes (1997) Developments in the control of fine particulate air emission. Adv Powder Technol 2(3):179–215
Dixkens J, Fissan H (1999) Development of an electrostatic precipitator for off-line particle analysis, aerosol. Sci Technol 30:438–453
Elsayed K, Lacor C (2011) Numerical modeling of the flow field and performance in cyclones of different cone tip diameters. Comput Fluids 51:48–59
European Commission (2005) Integrated pollution prevention and control. Reference document on best available techniques in the Smitheries and Foundry Industry
Fatta D, Marneri M, Papadopoulos A, Savvides Ch, Mentzis, Nikolaides L, Loizidou M (2004) Industrial pollution and control measures for a foundry in Cyprus. J Clean Prod 12:29–36
Fore S, Mbohwa CT (2010) Cleaner production for environmental conscious manufacturing in the foundry industry. J Eng Des Technol 8(3):314–333
Frohling M, Rentz O (2010) A case study on raw material blending for the recycling of ferrous wastes in a blast furnace. J Clean Prod 18:161–173
Georgiadis P, Kyrtopoulos SA (1999) Molecular epidemiological approaches to the study of the genotoxic effects of urban air pollution. Mutat Res 428:91–98
Grochowalski A, Lassen C, Holtzer M, Sadowski M, Hudyma T (2007) Determination of PCDDs, PCDFs, PCBs and HCB emissions from the metallurgical sector in Poland. Env Sci Pollut Res 14(5):326–332
Holtzer M (2005) Implementation of IPPC directive in foundries. Metalurgija 44:141–146
Huvinena M, Oksanen L, Kalliomaki K, Kalliomaki PL, Moilanen M (1997) Estimation of individual dust expose by magneto pneumography in stainless steel production. Sci Total Environ 199:133–139
Keshava N, Ong T (1999) Occupational exposure to genotoxicagents. Mutat Res 437:175–194
Kran HP, Rudolph A, Reinhold M (1995) Reduction of emissions of a hot blast cupola furnace by application of a fabric filter and by using of waste heat and residue—final report, Umweltbundesamt, UBA 0 441 -/ZII 1.1-50 441-5/178
Krishnaraj R, Sakthivel M, Devadassan SR (2012) Performance efficiency of wet scrubber in induction furnace towards green revolution-a case study in Indian foundry. J Environ Res Dev 6:824–833
Kuo HW, Chang CL, Lai JS, Lee FC, Chung BC, Chen CJ (1998) Prevalence of and factors related to pneumoconiosis among foundry workers in central Taiwan. Sci Total Environ 222:133–139
Lewtas J (2007) Air pollution combustion emissions: characterization of causative agents and mechanisms associated either cancer, reproductive, and cardiovascular effects. Mutat Res 636:95–133
Liang CDY, Yang KH, Lee JD, Hong GB (2010) The case study of furnace use and energy conservation in iron and steel industry. Energy 35:1665–1670
Lin MH, Liou SH, Chang CW, Huang IH, Strickland PT, Lai CH (2011) An engineering intervention resulting in improvement in lung function and change in urinary 8-hydroxydeoxyguanosine among foundry workers in Taiwan. Int Arch Occup Environ Health 84:175–183
Liu HH, Lin MH, Liu PC, Chan CI, Chen HL (2009) Health risk assessment by measuring plasma malondiadehyde (MDA) urinary hydroxydeoxyguanosine (8-OH-dG) and DNA strand breakage following metal exposure in foundry workers. J Hazard Mater 170:699–704
Liu HH, Lin MH, Chan CI, Chen HL (2010) Oxidative damage in foundry workers occupationally co-exposed to PAHs and metals. Int J Hyg Environ Health 213:93–98
Lv Pv, Zheng M, Liu G, Liu W, Xiao K (2011) Estimation and characterization of PCCD/Fs and dioxin-like PCBs from Chinese iron foundries. Chemosphere 82:759–763
Mukherjee DP (2010) Barriers towards cleaner Production for optimizing energy use and pollution control for foundry sector in Howrah India. Clean Technology Environ Policy 28:110–119
Martinez A, Cabezas J (2009) Emission control system for nitrogen oxides using enhanced oxidation, scrubbing, and biofiltration. Environ Eng Sci 26:883–890

Melendez A, Garcia E, Carnicer P, Pena E, Larrion M, Legarreta JA, Cañas CG (2010) Fine ultrafine emission dynamics from a ferrous cupola furnace. J Air Manage Assoc 60(5):556–567
Mirasgedis S, Hontou V, Georgepoulou E, Sarafidis Y, Gakis N, Lala DP, Loukatos A, Gargoulas N, Mentzis A, Economidis D, Triantafilopoulos T, Korizi K, Mavrotas G (2008) Environmental damage costs from airborne pollution of industrial activities in the greater Athens, Greece area and the resulting benefits from the introduction of BAT. Environ Impact Assess Rev 28:39–56
Myers T, Ibarreta A (2009) Investigation of the John Foundry and CTA acoustics dust explosions: similarities and differences. J Loss Prev Process Ind 22:740–745
Neto B, Kroeze C, Hordijk L, Costa C (2008) Modeling the environmental impact of an aluminum pressure die casting plant and options for control. Environ Model Softw 23:147–168
Neto B, Kroeze C, Hordijk L, Costa C (2009a) Inventory of pollution reduction options for an aluminum pressure die casting plant. Resour Conserv Recycl 53:309–320
Neto B, Kroeze C, Hordijk L, Costa C, Pulles T (2009b) Strategies to reduce the environment impact of an aluminum pressure die casting plant: a scenario analysis. J Environ Manage 90:815–830
Nie WX, Shea G, Yarnick TP (2005) Analysis estimates sulfuric acid emissions from FCCU wet gas scrubbers. Oil Gas J 103:62–64
Niksa S, Fujiwara N (2005) The impact of wet flue gas desulfurization scrubbing on mercury emissions from coal-fired power stations. J Air Waste Manage Assoc 89:970–977
Olasupo OA, Omotoyinbo JA (2009) Moulding properties of a Nigerian silica-clay mixture for foundry use. Appl Clay Sci 45:244–247
Pak SI, Chang KS (2006) Performance estimation of a Venturi scrubber using a computational model for capturing dust particles with liquid spray. J Hazard Mater 138:560–573
Pal P, Sethi G, Nath A, Swami S (2008) Towards cleaner technologies in small and micro enterprises: a process-based case study of foundry industry in India. J Cleaner Prod 16:1264–1274
Parka CL, Kim BG, Yu Y (2012) There generation of waste foundry sand residue stabilization using coal refuse. J Hazard Mater 203:176–182
Polizzi S, Ferrara M, Bugiani M, Barbero D, Baccolo T (2007) Aluminum and iron air pollution near and iron casting and aluminum foundry in Turin district Italy. J Inorg Biochem 101:1339–1343
Rabah MA (1999) Cost effectiveness of abatement options for emissions control in Egyptian iron foundries. Waste Manage 19:283–292
Rao RAK, Khan MA (2009) Biosorption of bivalent metal ions from aqueous solution by an agricultural waste: kinetics, thermodynamics and environmental effects. Coll Surf A Physicochem Eng Asp 332:121–128
Ribeiro MG, Filho WRP (2006) Risk assessment of chemicals in foundries: the international chemical toolkit pilot-project. J Hazard Mater A136:432–437
Sekhar H, Mahanti R (2006) Confluence of six sigma, simulation and environmental quality; an application in foundry industry, management of environmental quality. Int J 17:170–83
Strobos JG, Friend JFC (2004) Zinc recovery from bag house dust generated at Ferro chrome foundries. Hydrometallurgy 74:165–171
Subramanya MHB (2006) Energy intensity and economic performance small scale bricks and foundry clusters in India, does energy intensity matter. Energy Policy 34:489–497
Taha RA, Alnuaimi AS, Jabri AKS, Harthy AAS (2007) Evaluation of controlled low strength materials containing industrial by-products. Build Environ 42:3366–3372
Wang Y, Cannon FS, Salama M, Goudzwaaed J, Funrness JC (2007) Characterization of hydrocarbon emissions from green sand foundry core binders by analytical pyrolysis. Environ Sci Technol 41:7922–7927
Wang Y, Cannon FS, Li X (2011) Comparative analysis of hazardous air pollutant emissions of casting materials measured in analytical pyrolysis and conventional metal pouring emission tests. Environ Sci Technol 45:8529–8535
Wu CH, Feng CT, Lo YS, Lin TY, LO JG (2004) Determination of volatile organic compounds in workplace air by multi sorbent adsorption/thermal desorption. Chemosphere 56:71–80

Xu ZY, Brown L, Pan GW, Li G, Feng YP, Guan DX, Liu TF, Liu LM, Chao RM, Sheng JH, Gao GC (1996) Life style, environmental pollution and lung cancer in cities of Liaonung in North-eastern China. Lung Cancer 4:S149–S160
Yang HH, Lai SO, Hsieh LT, Hsueh HJ, Chi TW (2002) Profile of PAH emission from steel and iron industries. Chemosphere 48:1061–1074
Yu BW, Jin GZ, Moon YH, Kim MK, Kyoung JD, Chang YS (2006) Emission of PCDD/Fs and dioxin-like PCBs from metallurgy industries in S. Korea. Chemosphere 62:494–501
Zanetti MC, Fiore S (2002) Foundry processes: the recovery of green moulding sands for core operations. Resour Conserv Recycl 38:243–254
Zanetti M, Godio A (2006) Recovery of foundry sands and iron fractions from an industrial waste landfill. Resour Conserv Recycl 48:396–411

Chapter 10
Ligninolytic Enzymes for Water Depollution, Coal Breakdown, and Paper Industry

Meera Yadav, Sunil Kumar Singh, Sudha Yadava and Kapil Deo Singh Yadav

Contents

Abstract Lignin breakdown modifies lignocellulose structure to produce smaller carbohydrates usable for further bioconversion. White rot fungi produces ligninolytic enzymes such as lignin peroxidase, manganese peroxidase, laccases and versatile peroxidase, which efficiently mineralize lignin. We review applications of ligninolytic enzymes. Applications include delignification of lignocellulose, removal of organic pollutants, wastewater treatment, dye decolorization, soil treatment, breakdown of coal into low molecular weight fractions, biopulping and biobleaching in paper industries, and enzymatic polymerization in polymer industries.

Keywords Lignin peroxidase · Mn-peroxidase · Laccasse · Versatile peroxidase · Lignocellulose · Polyaromatic hydrocarbons · Bioremediation · Oxidoreductases

---

M. Yadav (✉) · S. K. Singh · S. Yadava · K. D. Singh Yadav
Department of Chemistry, DDU Gorakhpur University, Gorakhpur 273009, India
e-mail: drmeerayadav@rediffmail.com

© Springer International Publishing Switzerland 2015
E. Lichtfouse et al. (eds.), *CO_2 Sequestration, Biofuels and Depollution*,
Environmental Chemistry for a Sustainable World 5, DOI 10.1007/978-3-319-11906-9_10

10.1 Introduction

In recent years, environmental pollution by synthetic chemicals such as organochlorine compounds and petroleum hydrocarbons has caused great concern. These chemicals are toxic and adversely affect the health of a individual. The reaction of chemical discharge to the environment is therefore of wide concern and there is need for pollutant monitoring. Bioremediation technology is applicable to the treatment of such environmental pollution. Bioremediation involves the use of living organisms, usually, bacteria or fungi to remove pollutants from soil and water. There is therefore, intense interest in white rot fungi which has the capacity to degrade an extremely persistent or toxic environmental pollutants.

Lignocellulose biomass has a great potential as feed stock for production of more value added products such as low price chemicals, e.g. xylose, glucose, furfural, fuels, biofibres, ruminant feed, biopulp or even for enzyme production (Isroi et al. 2011). However, these carbohydrates are generally infiltrated by lignin.Breakdown of the lignin barrier will alter lignocelluloses structures and make the carbohydrates accessible for moreefficient bioconversion. White rot fungi produce ligninolytic enzymes (Lignin peroxidase, Mnperoxidase, Laccases) and efficiently mineralize lignin into CO_2 and H_2O (Hatakka 1994). Lignin peroxidase are known as strong oxidants because the iron in the porphyrin ring is more electron-defficient than in classical peroxidases (Millis et al. 1989).Recent work showed that the catalytic efficiency (Kcat/Km) for oxidation of a lignin model trimer by lignin peroxidase was only about 4% of the value found for oxidation of a monomeric model (Baciocchi et al. 2003). Mn peroxidases are more widespread than Lignin peroxidases (Hofrichter 2002; Hammel and Cullen 2008). The redox potential of the Mn peroxidase system is lower than that of lignin peroxidase and it has shown capacity for preferable oxidize in vitro phenolic substrates.Laccases are usually the first ligninolytic enzymes secreted to the surrounding media by the fungus that normally oxidizes only those lignin model compounds with a free phenolic group forming phenoxy radicals as the mediatos that are a group of low molecular weight organic compounds (Bourbonnais and Paice 1990; Call and Mucke 1997).

Ligninolytic enzymes of the basidiomycetes play a crucial role in the global carbon cycle.The demand for application of ligninolytic enzymes complexes of white rot fungi in industries and biotechnology is ever increasing due to their use in a variety of processes.Ligninolytic enzymes have potential applications in a large number of fields, including the chemical, fuel, food, agricultural, paper, textile, cosmetic industries and more. This ligninolytic system of white rot fungi is also directly involved in the degradation of various xenobiotic compounds and dyes. Their capacities to remove xenobiotic substances and produce polymeric products make them a useful tool for bioremediation purposes (Marcia et al. 2010).

Environmental biotechnology is the application of all components of biotechnology to solve environmental problems. Thus environmental use of biotechnology includes the development, biosecurity use and regulation of biological systems for remediation contaminated environments like land, water, air as well as for use environmentally sound processes leading to clean technologies and sustainable development.

10.2 Oxidative Enzymes and Their Reaction Mechanisms

Fungi are the only organisms that are able to completely mineralize lignocellulose, the most abundant recalcitrant renewable material available in nature. They produce enzymes for breaking down polysaccharides, cellulose and hemicellulose as well as lignin, a natural aromatic polymer. Ligninase or ligninolytic enzymes constitute a group of oxidoreductases that are highly specialized in polymerization as well as in the degradation of lignin. These enzymes are mostly produced by the so called white rot fungi and litter-decomposing fungi. The three main lignin modifying enzymes are lignin peroxidase, manganese peroxidase, laccase and versatile peroxidase. White rot fungi contains all three enzymes and is therefore able to breakdown and mineralize several environmental pollutants into non-toxic forms. This chapter reviews the role of ligninolytic enzymes in the growing field of bioremediation.

10.2.1 Lignin Peroxidase

The lignin decomposing basidiomycete white rot fungi secrete a hemeprotein, lignin peroxidases (LiP, EC 1.11.1.7] which in presence of H_2O_2 degrades lignin and lignin model compounds (Edwards et al. 1993). The sequence of catalysis is given below:

$$\text{LiP}(\text{Fe}^{3+})\text{P} + \text{H}_2\text{O}_2 \rightarrow \text{LiP} - \text{I}(\text{Fe}^{4+} - \text{O})\text{P}^{+\bullet} + \text{H}_2\text{O} \quad (1)$$

$$\text{LiP-I}(\text{Fe}^{4+}\text{-O})\text{P}^{+\bullet} + \text{R} \rightarrow \text{LiP-II}(\text{Fe}^{4+}\text{-O})\text{P} + \text{R}^{\bullet} \quad (2)$$

$$\text{LiP-II}(\text{Fe}^{4+}\text{-O})\text{P} + R_{\bullet} + 2\text{H}^{+} \rightarrow \text{LiP}(\text{Fe}^{3+})\text{P} + \text{R}^{\bullet} + \text{H}_2\text{O} \quad (3)$$

where R is the organic substrate and P is porphyrin. Lignin peroxidasecompound I (LiP-I) carries both oxidizing equivalents of H_2O_2, one as an oxyferryl (Fe^{4+}-O) center and one as a porphyrin π cation radical ($P^{+\bullet}$), whereas lignin peroxidase compound II (LiP-II) carries only one oxidizing equivalent. The substrate R is oxidized by compound I to an aryl cation radical which with subsequent nonenzymatic reactions yield the final products. The different enzyme intermediates could be depicted by Fig. 10.1.

Structural and functional aspects of lignin peroxidase have been studied (Harley et al. 1988; Catcheside and Ralph 1999). Lignin peroxidase is a biotechnologicaly important enzyme having wide potential applications (i) in delignification of lignocellulosic materials (Harley et al. 1988) which are seen as an alternative to the depleting oil reserves, (ii) in the conversion of coal to low molecular mass fractions (Catcheside and Ralph 1999) which could be used as a feed stock for the production of commodity chemicals, (iii) in biopulping and biobleaching (Eriksson and Kirk 1994) in paper industries, (iv) in removal of recalcitrant organic pollutants (Bumpus et al. 1985; Kwant and Chang 1998; Satwinder et al. 1998; Levin et al. 2004; Cenek et al. 2004) and (v) in the enzymatic polymerization (Hiroshi and Shiro 1999) in polymer industries. Keeping in view the biotechnological potential of lignin

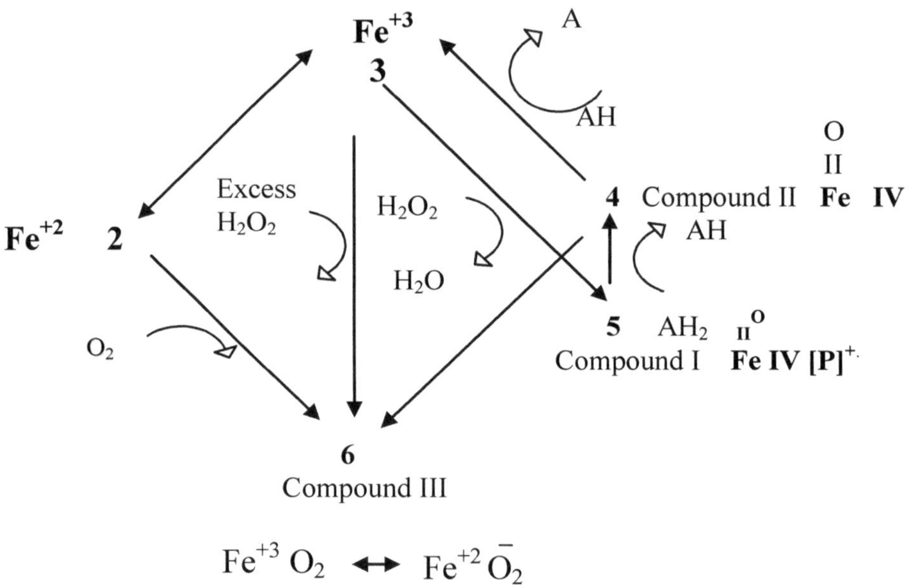

Fig. 10.1 Interrelationship between the five redox states of lignin peroxidase (Renganathan and Gold 1986). Reaction paths 3→5→4 indicate one catalytic cycle of the enzyme. AH_2 substrate

peroxidase, the authors have initiated enzymatic studies on the lignin peroxidase from indigenous fungal strains. Lignin peroxidase of *Phanerocheate chyrosoporium* has been extensively studied (Tien and Kirk 1988; Vyas et al. 1994; Kang et al. 1993) and LiP of *Trametes versicolor* (Vyas et al. 1994) [12], *Pleurotus ostreatus* (Kang et al. 1993), *P.ostroiformis* (Satyahari et al. 1994), *Ganoderma lucidum* (Perumal and Kalaichelvan 1996), *Aspergillus terreus* (Meera et al. 2002), *Fusarium oxysporum* (Meera et al. 2002), *Pencillium citrinum* (Meera et al. 2002), *Rizopus nigrican* (Shanmugan and Yadav 1997), *Pleurotus sajor-caju* (Shanmugan and Yadav 1996), *Abortiporus biennis* (Patel et al. 2007), *Pestalotia bicolor* (Patel et al. 2007), *Heterobasidium annosum* (Patel et al. 2007), *Gloephyllum striatum* (Patel et al. 2007), *Loweporus lividus* (Patel et al. 2007), *Pycnoporus sanguineus* MTCC-137 (Sharma et al. 2011), *Lenzitus seperia* MTCC-1170 (Yadav et al. 2009a), *Loweporus lividus* MTCC-1178 (Yadav et al. 2009b), *Hexagona tenuis* MTCC-1119 (Yadav et al. 2010), *Gleophyllum striatum* MTCC-1117 (Yadav et al. 2011), *Lenzitus betulina* MTCC-1183 (Yadav et al. 2012) have been reported. Our research interest in lignin peroxidases has prompted us to screen this fungal strain for the secretion of lignin peroxidase.

The lignin peroxidase isozymes are similar in structure and functions. Some of their physical properties are summarised in Table 10.1.

The lignin peroxidase H8 contains one protophorphyrin IX derived heme per enzyme molecule and is composed of 15 % by weight of carbohydrate. The electron absorption spectrum of lignin peroxidase H8 is typical of most heme proteins

Table 10.1 Physical properties of lignin peroxidase isozymes

Isozyme	Molecular weight	Carbohydrate	ε_{409} (mM⁻ cm⁻)
H1	38,000	+	169
H2	38,000	+	165
H6	43,000	+	162
H7	42,000	+	177
H8	42,000	+	168
H10	46,000	+	182

showing a Soret peak at 409 nm and visible absorption bands at 498 and 630 nm. Ferric (resting) lignin peroxidase forms complexes with cyanide and azide.

The kinetics of lignin peroxidase catalysis have been studied by both steady state and transient state techniques (Tein et al. 1986). The steady state studies of veratryl alcohol oxidation indicate that the mechanism of catalysis is Ping-Pong (Tein et al. 1986). Ping-Pong kinetics are consistent with the mechanism of other peroxidases (Kedderis and Hollenberg 1983). The initial step in catalysis is the reaction of lignin peroxidase with H_2O_2 resulting in the formation of an oxidised enzyme intermediate. This intermediate returns to the resting state by oxidising its aromatic substrates.

Transient state kinetic studies of the lignin peroxidase show that two intermediate states of the enzyme are formed during catalysis. These two states are similar to those formed by other peroxidases. These are classical intermediate compounds I and II characterised by Chance (1952). Formation of lignin peroxidase compounds I and II were detected by stopped flow rapid scan spectral analysis of the reaction between lignin peroxidase and H_2O_2 (Tein et al. 1986). The initial step in catalysis is the reaction of lignin peroxidase with H_2O_2 to form compound I, which is two electron oxidised. Compound I then reacts with a substrate molecule to form product and the compound II intermediate of lignin peroxidase which is one electron oxidised. Compound II returns to resting enzyme by reacting with another molecule of substrate.

The catalytic cycle described above where two molecules of free radical products are formed per turn over, is common for all peroxidases. Formation of free radical products during lignin peroxidase catalysis has been demonstrated by ESR spectroscopy (Kersten et al. 1985). Kersten et al. detected the cation radicals of methoxybenzene by ESR spectroscopy. These studies have led to conclusion that secondary reactions of cation radicals (Snook and Hamilton 1973) are also important in the degradation of lignins by lignin peroxidase.

10.2.2 Laccases

Laccase [E.C. 1.10.3.2] belongs to copper containing oxidases and catalyzes (Messerschmidt 1997; Yadav and Yadav 2007; Riva 2006; Baldrian 2006; Dwivedi et al. 2011) the four electron reduction of molecular oxygen to water. It is a glycoprotein

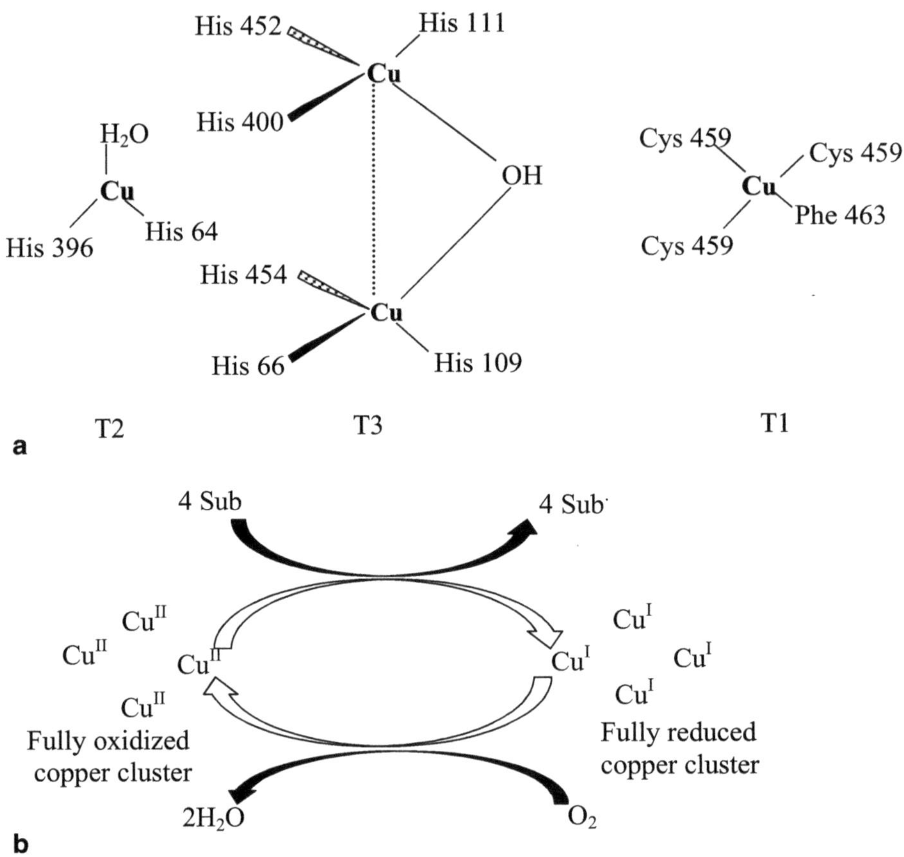

Fig. 10.2 **a** Laccase containing four copper atoms per monomer distributed in three redox sites termed as type1 (*T1*) or blue Cu, type2 (*T2*) or normal Cu and Type3 (*T3*). Type2 and Type3 coppers form a trinuclear cluster. **b** Catalytic cycle of Laccase

which contains four copper atoms per monomer distributed in three redox sites (Quintanar et al. 2005) termed as type1 (T1) or blue Cu, type2 (T2) or normal Cu and type3 (T3) or coupled binuclear Cu. Type2 and type3 coppers form a trinuclear cluster as shown in Fig. 10.2a. The catalysis can be schematically shown as in Fig. 10.2b.

Since laccase recycles on molecular oxygen as an electron acceptor and does not require any other co-substrate, it is the most promising enzyme of oxidoreductases group for industrial applications (Wandrey et al. 2000; Conto and Harrera 2006; Xu 2005).During the last two decades, laccases have turned out to be the most promising enzymes for industrial uses (Conto and Harrera 2006; Xu 2005) having applications in food, pulp and paper, textile, cosmetic industries and in synthetic organic chemistry (Fritz-Langhals and Kunath 1998; Mikolasch et al. 2002; Mikolasch et al. 2009).

The fungal strains which secrete blue laccases in submersed culture, secrete yellow laccase in solid state fermentation (Leontievsky et al. 1997). Yellow laccases lack UV/Vis absorption band near 600 nm found in blue laccases. Yellow laccase can oxidize non-phenolic substrates without mediators which are essential in case of blue laccases which are rare.

The organic substrate is oxidized by one electron at the active site of the laccase generating a reaction radical which further reacts non-enzymetically. The electron is received at type1 Cu and is shuttled to the trinuclear cluster where oxygen is reduced to water.

Ortho and para diphenols, aminophenols, polyphenols, polyamines, lignins, and arylamines and some of the inorganic ions are the substrates for laccases. Since laccase recycles on molecular oxygen as an electron acceptor and does not require any other co-substrate, it is the most promising enzyme of oxidoreductases group for industrial applications (Wandrey et al. 2000; Couto and Harrera 2006; Xu 2005). The biotechnological importance of laccases have increased after the discovery that oxidizable reaction substrate range could be further extended in the presence of small readily oxidizable molecules called mediators (Acunzo and Galli 2003; Morozova et al. 2007). During the last two decades, laccases have turned out to be most promising enzymes for industrial uses (Couto and Harrera 2006; Xu 2005) having applications in food, pulp and paper, textile, cosmetics industries and in synthetic organic chemistry (Coniglio et al. 2008; Mikolasch et al. 2002, 2007).

Laccases purified from different sources (Sahay et al. 2008, 2009) exhibit different properties and are suitable for different applications. Enguita et al. (Enguita et al. 2003) have crystallized the cotA laccase from the endospore coat of *Bacillus subtilis* in presence of the non-catalytic co-oxidant 2, 2′-azinobis-(3-ethylbenzothiazoline-6-sulfonate (ABTS) and the crystal structure has been solved.

10.2.3 Manganese Peroxidase

Manganese peroxidase, MnP [E.C.1.11.1.13] is a heme containing enzyme (Kuwahara et al. 1984). It has been shown to be present in the culture filtrates of a number of fungal strains (Hatakka 1994; Pelaez et al. 1995; Silva et al. 2008; Tanaka et al. 2009). The catalytic cycle of Mn-peroxidase resembles those of other heme peroxidases such as horseradish peroxidase (Dunford 1991) and lignin peroxidase (Wariishi et al. 1991a, b; Kirk and Cullen 1998) and includes the native ferric enzyme as well as the reactive intermediates compound I and compound II. The catalytic cycle can be shown as follows,

$$MnP + H_2O_2 \rightarrow MnP(I) + H_2O \qquad (1)$$

$$MnP(I) + Mn^{II} \rightarrow MnP(II) + Mn^{III} \qquad (2)$$

$$MnP(II) + Mn^{II} \rightarrow MnP + Mn^{III} + H_2O \qquad (3)$$

H_2O_2 oxidizes the enzyme by two electrons to form Mn-peroxidase compound (I) which is oxyferryl porphyrin cation radical $[Fe^{4+}=O\ P]^+$. Mn (II) or phenolic compounds can serve as reductants for the Mn-peroxidase compound (I) and form Mn-peroxidase compound (II) which is an oxyferryl chemical species $[Fe^{4+}=O\ P]$, one electron oxidized form of the enzyme. For the reduction of Mn-peroxidase compound (II) to the enzyme, Mn (II) is absolutely essential (Glenn et al. 1986; Wariishi et al. 1992). Kinetic and spectroscopic characterisation of oxidised intermediates, Mn-peroxidase compound (I) and (II) have indicated that the catalytic cycle of Mn-peroxidase is similar to that of lignin peroxidase. However, Mn-peroxidase is unique in its ability to oxidise Mn (II) to Mn (III) (Glenn et al. 1986). The enzyme generated Mn (III) is stabilized by chelators such as oxalate which are secreted by the fungi (Uan and Tien 1993; Frichter et al. 1999). The Mn (III) chelator complex oxidises phenolic substrates such as lignin substructures (Wariishi et al. 1989, 1991a, b; Bao et al. 1994; Tuor et al. 1992; Lackner et al. 1991; Hofrichter et al. 2001) and aromatic pollutants (Hofrichter et al. 1998; Moreira et al. 2001; Steffen et al. 2003). The crystal structure of Mn- peroxidase from *P.chryosporium* has been solved (Sundaramoorthy et al. 1994).

Recently, there has been a growing interest in studying the lignin modifying enzymes of a wider array of white-rot-fungi with the expectation of finding better lignin-degrading systems for use in various biotechnological applications such as biotransformation of raw plant materials to feed and fuels; production of enzymes, antibiotics, polysaccharides and other physiologically active compounds; biopulping, biobleaching of paper pulp and bioremediation of soils and industrial waters polluted with toxic chemicals and dyes.

10.2.4 Versatile Peroxidase

Versatile peroxidase (VP) has been recently described as a new family of ligninolytic peroxidases, together with lignin peroxidase (LiP) and manganese peroxidase (MnP) both reported for *Phenerocheate chrysosporium* for the first time (Martinez 2002). The genome of this model fungus has been sequenced revealing two families of lignin peroxidase and Mn-peroxidase genes together with a "hybrid peroxidase" gene (Martnez et al. 2004). Versatile peroxidase seems to be produced from genera *Pleurotus, Bjerkandera, Lipista* (Mester and Field 1998; Martinez et al. 1996; Ruiz-Duenas et al. 1999; Camarero et al. 1996; Sarkar et al. 1997; Martinez and Martinez 1996; Heinfling et al. 1998a, b; Zorn et al. 2003) *Panus and Trametes* (Martinez 2002; Lisov et al. 2003) species.

A catalytic cycle (Fig. 10.3) that combines the cycles of other fungal peroxidases, including lignin peroxidase and Mn-peroxidase was proposed by Ruiz-Duenas et al. The basic features are common to most peroxidases but versatile peroxidase is unique regarding the substrate that it is able to oxidizes. Versatile peroxidase oxidize high redox potential dyes such as Reactive Black 5, as well as low redox potential dyes, substituted phenols and Mn^{+2}, as reported for wild type fungal

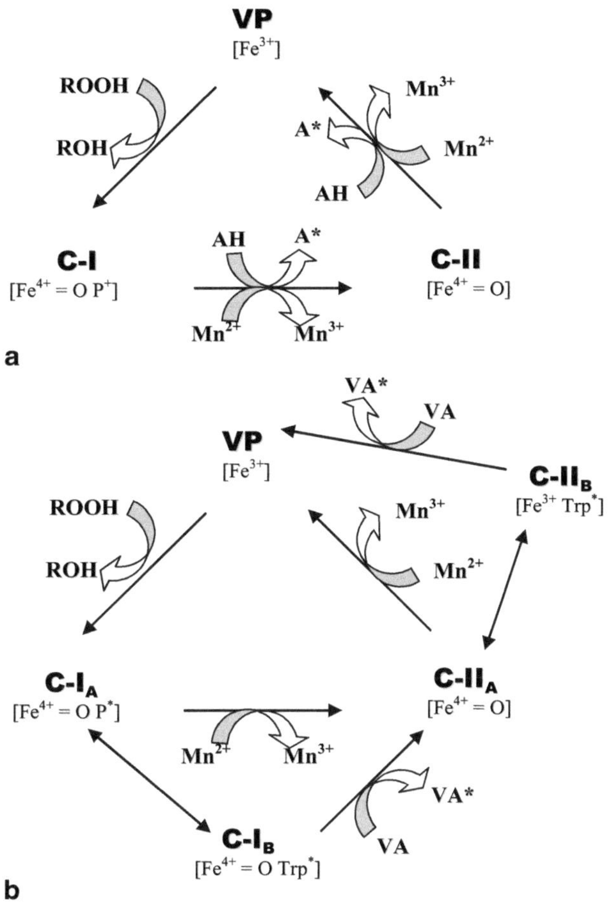

Fig. 10.3 Schemes of Versatile peroxidase (*VP*) catalytic cycle. **a** Basic cycle described by Ruiz-Duenas et al. Including two electron oxidation of the resting peroxidase (*VP*, containing Fe^{3+}) by hydroperoxide to yield compound I whose reduction in two one- electron reactions results in the intermediate compound II and then the resting form of the enzyme. As shown in the cycle, Versatile peroxidase can oxidize both: (i) Aromatic substrates (*AH*) to the corresponding radicals (*A**): and (ii) Mn^{2+} to Mn^{3+}. **b** Extended cycle including compounds IB and IIB involved in oxidation of veratryl alcohol (*VA*) and otherhigh redox potential aromatic compounds (*C-IB* and *C-IIB* are in equilibrium with *C-IA* and *C-IIA* respectively, which corresponding to *C-I* and *C-II* in (**a**)

enzyme (Heinfling et al. 1998a, b). It oxidizes variety of substrates phenolic and non-phenolic lignin dimmers, alpha-keto-y-thiomethylbutyric acid (KTBA), veratryl alcohol, dimethoxybenzene, different types of dyes, substituted phenols and hydroquinones (Heinfling et al. 1998a, b; Caramelo et al. 1999).

10.3 Application of Ligninolytic Enzymes

Biological technologies dealing with the use of oxidoreductive enzymes e.g. laccases and peroxidases may offer an efficient alternative means of addressing the clean up of phenol polluted waste water (Shutleworth and Bellog 1986; Adam et al. 1999; May 1999; Regalado et al. 2004). Lignin peroxidase from *Phenerocheate chrysosporium* (Bumpus and Aust 1987) was one of the first enzyme of basidiomycete capable of PAH degradation. PAHs are recognized as a group of chemicals that are formed during the incomplete burning of coal, oil, gas or other organic substances contain several fused benzene rings and are biotoxic compounds with potentially carcinogenic and mutagenic properties. Lignin peroxidase also catalyzes the oxidation of variety of recalcitrant aromatic substrates (Harvey et al. 1986; Valli et al. 1990; Joshi and Gold 1994). Phenol polluted water is produced by number of industrial and agricultural activities and are harmful to living organisms even at low concentrations. Catechol and substituted catechols especially chlorinated and methylated ones are by products in pulp and oil mills (Schweigert et al. 2001) When released in environment they accumulate in soil, ground water and therefore become an issue of great environmental concern.

10.3.1 Delignification of Lignocellulose

One of the research interests in the studies on lignin peroxidase is to use this enzyme for delignification of lignocellulosic materials so that delignified material can be used as a feed stock for the production of commodity and rare chemicals. Lignin is an extremely complex, three dimensional heteropolymer made up of phenyl propyl units as shown in Fig. 10.4. It is lignin which makes lignocellulose resistant to biodegradation. The removal of lignin from lignocellulosic materials will make if suitable for conversions to other chemicals. However studies so far made have not given any fruitful results regarding the delignification of lignocellulosic materials using lignin peroxidase (Breen and Singleton 1999).

Biological treatment employs wood degrading micro-organisms, including white-rot fungi, brown-rot fungi, soft-rot fungi and bacteria to modify the chemical composition and structure of the lignocellulosic biomass so that the modified biomass is more amenable to enzyme digestion. Fungi have distinct degradation characteristics on lignocellulose biomass.In general, brown and soft rot fungi mainly attack cellulose while imparting minor modifications to lignin and white rot fungi are more active in degrading the lignin components (Sun and Cheng 2002). Present research is aimed towards finding those organisms which can degrade lignin more effectively and more specifically. White- rot fungi were considered the most promising basidiomycetes for biopretreatment of biomass and were the most studied biomass degrading micro-organisms (Sanchez 2009).

The biological pretreatment appears to be a promising technique and has very evident advantages, including no chemical requirement, low energy input, mild environmental conditions and environmentally friendly working manner (Kurakale et al. 2007; Salvachua et al. 2011).

10.3.2 Removal of Recalcitrant Polyaromatic Hydrocarbons (PAHs)

Lignin peroxidase a higher redox potential than any other peroxidases (Hammel et al. 1986; Kersten et al. 1990) and has been reported to oxidize aromatic compounds

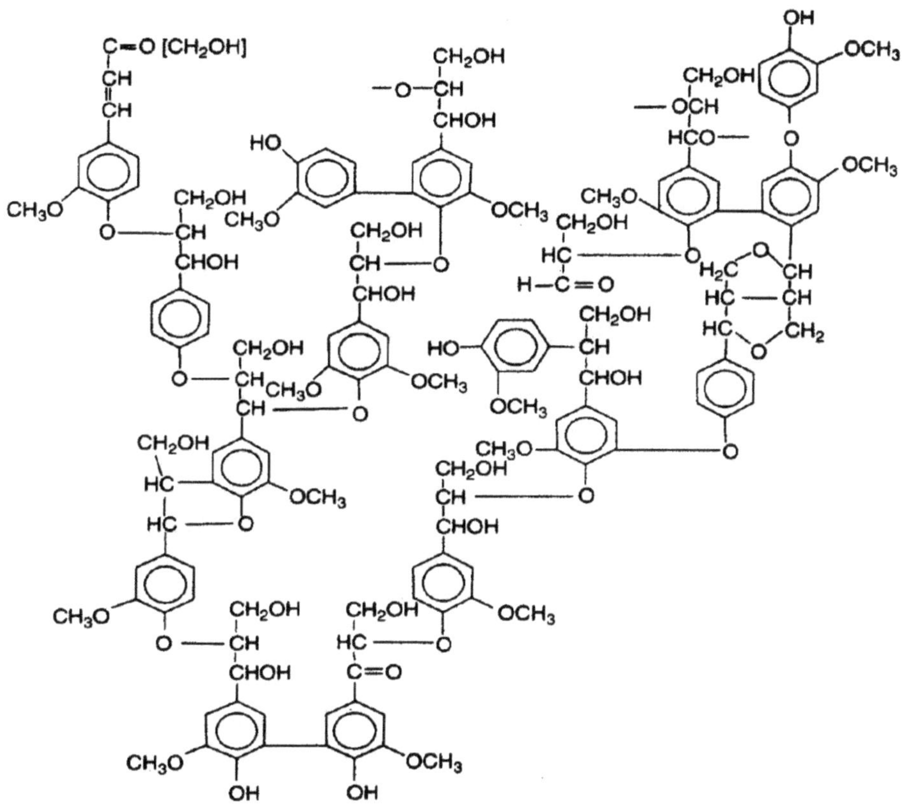

Fig. 10.4 Structure of lignin. (Adler 1977)

with calculated ionization potential values up to 9.0 ev (Have et al. 1998; Ward et al. 2003a). Lignin peroxidase is of interest in wastewater treatment and in catalyzing chemical transformations.

P. sordida was also most useful in the degradation of PAHs from soil. (Davis et al. 1993) showed that *P.sordida* was capable of degrading efficiently the three ring PAHs, but less efficiently the four-ring PAHs. *Phenerocheate chrysosporium* has been shown to degrade a number of toxic xenobiotics such as aromatic hydrocarbons (Benzoalphapyrene, Phenanthrene, Pyrene) chlorinated organics (Alkyl halide insecticide chloroanilines, DDT, Pentachlorophenols, Trichlorophenoxyacetic acid), nitrogen aromatics (2,4-Dinitrotoluene, 2,4,6-Trinitrotoluene-TNT) and several miscellaneous compounds (such as sulfonated azodyes).

Laccases oxidizes many substrates phenolic dyes (Chivukula and Renganathan 1995; Ishihara et al. 1997; Dwran et al. 2000), phenols (Dahiya et al. 1998; Gianfreda et al. 1998; Pothast et al. 1997), chlorophenols (Dec and Bollag 1994; Dec and Bollag 1995; Raper et al. 1995; Grey et al. 1998; Kadhim et al. 1999), lignin related diphenyl methanes (Xu et al. 1997; Crestini and Argyropoulos 1998), benzopyrenes (Rama et al. 1998), N-substituted p-phenylenediamines (Krikstopaitis et al. 1998), organophosphorus (Amitai et al. 1998) and non-phenolic beta-o-lignin model dimmer (Kawai et al. 1999; Majcherczyk et al. 1999). Laccase being multi-copper containing oxidase that catalysed the monoelectric oxidation of a range of inorganic and aromatic substances with relatively low redox potentials with reduction of O_2 to water (Baldrian 2006), including degradation of lignin and related compounds (Xu 1996). The redox potential of laccase is rather low and less than those of non-phenolic substrates, which would be expected to limit their role in lignin degradation. However, in the presence of an appropriate effective redox mediator or laccase-mediator system, laccases can oxidize the non-phenolic lignin structure (Bourbannais and Paice 1996; Bourbannais et al. 1997).

The combined action of intracellular cytochrome P450 and extracellular laccase and peroxidase enzyme reactions enable white-rot fungi to degrade a variety of toxic organopollutants, such as polychlorinated phenols, polychlorinated dibenzo-p-dioxins and polycyclic aromatic hydrocarbon (PAHs) and other xenobiotics (Boominathan and Reddy 1992). The *G. lucidum* laccase (Punnapayak et al. 2009) was found to degrade sixteen types of polycyclic aromatic hydrocarbons (PAHs). *G.lucidum* laccase degraded anthracene, benzo(a)pyrene, fluorine, acenapthene, acenaphthylene and benzo[a]anthracene, dibenzo[ah]anthracene, benzo(b)fluoranthene, benzo [ghi]perylene, benzo[k]fluoranthene, chrysene, fluoranthene, fluorine, indole [123-cd]pyrene, naphthalene, phenathrene and pyrene with and without the presence of 1-hydroxybenzotriazole as a redox mediator.

The oxidative activity of Manganese peroxidase is mediated through the production of manganese ions, acting as freely diffusible oxidants. In a way to reproduce the degradative action of Mn-peroxidase, manganic acetate was found to be incapable of oxidizing PAHs with IPs equal or greater than that of chrysene (~7.8 ev) which gives an idea about the threshold value for the PAHs degradation by the catalytic action of Mn-peroxidase (Cavalier and Rogan 1985). Little work has been developed about in vitro oxidation of PAHs by Mn-peroxidase and it has been focused on the determination of the limit IP mentioned above (Bogan and Lamar 1996; Bogan et al. 1996; Sack et al. 1997; Wang et al. 2003). Gunther (Gunther et al. 1998) have reported the degradation of 30% anthracene and 12% pyrene by Mn-peroxidase from *Nematoloma frowardie* after 24hrs of reactions. Degradation of three PAHs anthracene, pyrene and dibenzothiophene by Mn-peroxidase of *Bjerkandera* sp. BeS55 has been reported (Eibes et al. 2006). Summarised Table 10.2 is given for PAHs degradation.

10.3.3 Conversion of Coal to Low Molecular Mass Fraction

Cohen and Gabriele reported that fungi could grow directly on and metabolize naturally occurring coal, biological conversion of low-rank coals by bacteria, fungi or preparations of the enzymes they produce has been the subject of intensive research (Cohen and Gabriele 1982). These processes occur at ambient temperature and pressures, they represent a potential savings in the processing of certain coals and lignites.They also reported that fungi could metabolize leonardite, a naturally oxidized form of lignite coal. Wilson et al. have shown that the leonardite-biodegraded

Table 10.2 Summarized polyaromatic hydrocarbons (PAHs) degradation by different fungi

Substances	Enzyme source	Metabolites	Ref.
Polyaromatic hydrocarbons e.g. Benzo(a) pyrene, pyrene, anthacene etc.	*Phenerocheate chrysosporium*	PAH quinones	Bogan et al. (1996)
	Phenerocheate laevis	Ring fission products	Bogan et al. (1996)
	Nematoloma frowardie	PAH quinones, CO_2	Sack et al. (1997)
			Gunthar et al. (1998)
Chlorophenols (e.g. pentachlorophenols)	*Nematoloma frowardie*	Polar ring fission products, CO_2	Hofrichter et al. (1998)
Nitroaromatic compounds (e.g. 2,4,6 trinitrotoluene and its metabolites)	*Phenerocheate chrysosporium*	Quinones, NO_2- CO_2	Valli et al. (1992)
	Nematoloma frowardie	Polar ring fission products, CO_2	Van Aken et al. (2000)
	Phlebia radiate		
	Stropharia rugosoannulata	CO_2	Van Aken et al. (1999)
			Scheibner et al. (1998)
Arsenic containing warefare agents	*Nematoloma frowardie*	Unknown products	Fritsche et al. (2000)

product from C.versicolor, a white rot fungus, was water soluble and contained no detectable polycyclic aromatic hydrocarbons (Wilson et al. 1987).

Current technology for coal conversion requires both high temperatures and pressures which may result in the production of components that are more toxic than the original starting material.

Macromolecules from brown coal are decolourised and depolymerised by the white-rot fungus *Phanerochaete chrysosporium* (Ralph and Catcheside 1994; Ralph et al. 1996). The lignin peroxidase is likely to have central role in these processes. Ralph and Catcheside (Ralph and Catcheside 1994) have shown that both methylated and unmethylated coal fractions of solubilised macromolecules from brown coal were decolourised in reactions requiring H_2O_2 and veratryl alcohol. Neither coal fraction was transformed when heat inactivated enzyme was used. Gel permeation chromatography showed that methylated coal fraction and not the unmethylated coal fraction was depolymerised. Nine monomeric products have been identified by GC-MS. However, it is still not very obivious that a technology based on the above observation will be feasible for conversion of high molecular mass fraction of coal to low molecular mass fraction of coal.

10.3.4 Biopulping and Biobleaching in Paper Industry

Ligninolytic fungi have been shown to play roles in biopulping (Breen and Singleton 1999) and biobleaching (Marwaha et al. 1998). It could be inferred that lignin peroxidase might be playing an important role in the process of biopulping and biobleaching but direct evidences are not available (Bajpai 1999). Laccases are able to depolymerize lignin and delignify wood pulps, kraft pulp fibres and chlorine free in the biopolpation process (Bourbannais et al. 1997; Lund and Ragauskas 2001; Camarero et al. 2004; Rodrriguiz and Toca 2006; Vikineswary et al. 2006). One of the most studied applications in the industry is the laccasses-mediator bleaching of kraft pulp and the efficiency of which has been proven in mill-scale trials (Strebotnik and Hammel 2000).This ability could be used in the future to attach chemically versatile compounds in the fiber surfaces and let recycled pulp for new use (Rodriguez and Toca 2006; Widsten and Kandelbauer 2008). Lignin peroxidase compared with laccase are the biocatalysts of choice for bleaching (Bajpai 2004; Sigoillot et al. 2005). Lignin peroxidase and Mn-peroxidase were reported to be effective in decolourizing kraft pulp mill effluents (Ferrer et al. 1991; Moreira et al. 2003). In laboratory scale the consumption of refining energy in mechanical pulping was reduced with Mn-peroxidase pretreatment with a slight improvement in pulp properties (Kurek et al. 2001; Wasenberg et al. 2003; Maijala et al. 2007).

10.3.5 Polymerisation in Polymer Industries

One of the roles of lignin peroxidase is in polymerization (Hiroshi and Shiro 1999). It has been reported that polymers formed by the polymerisation induced by lignin peroxidase have better properties (Hiroshi and Shiro 1999). Oleg Milstein et al. reported that laccase was able to co-polymerise lignin with low molecular mass compounds of different origins, particularly with aromatic containing either carbonyl or isocyanate groups as well as acrylamide- an aliphatic monomers containing a vinyl group (Milstein et al. 1994). The reaction of the monomeric lignin model compound guaiacol and the β-o-4 type dimereral with laccase from *Trametes hirsute* was studied in the presence of the mediator ABTS (2,2′-azino-di [3 ethyl benzothiazoline-6-sulfonic acid]) to give polymeric coupling products (Rittstieg et al. 2003).

10.3.6 Biodegradation of Dyes

Different dyes and pigments are extensively used in the textile, paper, plastics, pharmaceutical and food industries. Biodegradation of dyes is being considered as an environment friendly and cost effective option which also offers environmental control. Environmental pollution due to urbanization and rapid growth of industries has an adverse effect on human health and ecology. The unadhered dye from the textile industries is left into the water bodies without any treatment which significantly affect the photosynthetic activity of the species present. Micro-organisms which are as known natures recyclers, convert toxic compounds to harmless products such as CO_2 and water (Joshi et al. 2010). The various organisms which degrade dyes are fungi, bacteria and actinomycetes. It is also known that a mixture of organisms degrade a dye better as compared to individual organisms as their complexity enables them to act on a variety of pollutants (Dafale et al. 2008). White rot fungi are better dye degraders than prokaryotes due to their extracellular non-specific system capable of degrading a wide range of dyes (Christian et al. 2005). White rot fungi such as *Phenerocheate chrysosporium*, *Trametes versicolor, Pleurotus ostreatus* and *Irpex lacteus* have gained importance in the biodegradation of dye for their ability to produce extracellular Ligninolytic enzymes such as lignin peroxidase, Mn-peroxidase, laccase and Mn-independent enzymes which are not specific and can attack a wide variety of complex aromatic dyestuffs. The degradation ability of the organisms also depends on the oxidation potential of the dye (Tauber et al. 2008). Biosorption of the dyes is also a promising technique which will enhance the color removal of the effluent steam containing a mixture of dye. Mechanisms of biodegradation and profiles of biodegradation products of different dyes have also been determined (Zhao et al. 2005, 2006; Gavril and Hodson 2007; Zhao and Hardin 2007). The decolorization ability of white rot fungi can be substantially increased by carefully optimizing the operational conditions such as initial dye concentration, nutrient, content of media, age of fungus carbon and nitrogen sources (Ozsay et al. 2005; Nilsson et al. 2006; Sanghi et al. 2006). However, other white rot fungi listed in Table 10.3 have been shown to have higher dye decolorization rates than *Phenerocheate chrysosporium* and *Trametes versicolor.*

Table 10.3 Ligninolytic enzymes of white rot fungi involved in biodegradation of different dyestuffs

White rot strains	Dyestuffs	Enzymes	Ref.
Coriolus versicolor	Malachite Green, Azure B, Poly R-478, Anthraquinone Blue, Congo red, Xylidine	Laccase	Levin et al. (2004)
Cerrena unicolor	Acid Blue 62, Acid blue 40, Reactive Blue 81, Direct Black 22, Acid Red 27	Laccase	Michniewicz et al. (2008)
Daedalea quercina	Chicago Sky Blue, Poly B-411, Remazol Brilliant Blue R	Laccase	Baldrian (2004)
Funalia trogii	Trypan Blue, Reactive Blue 2 Reactive Black 5	Laccase	Mazmanci and Unyayar (2005)
Funalia trogii	Remazol Brilliant Blue Royal (RBBR) Drimaren Blue CL-BR	Laccase	Erkurt et al. (2007)
Irpex lacteus	Reactive Blue 19 (RBBR), Reactive Black 5	MnP, Laccase	Maximo and Costa-Ferreira (2004)
Irpex lacteus	Reactive orange 16 and Remazol Brilliant Blue R	Laccase	Svobodova et al. (2008)
Irpex lacteus (immobilized)	Reactive Orange 16	Laccase, MnP	Tavèar et al. (2006)

Table 10.3 (continued)

White rot strains	Dyestuffs	Enzymes	Ref.
Lentinula edodes	Remazole Brilliant Blue R	MnP	Boer et al. (2004)
Phenerocheate chrysosporium (immobilized on $ZrOCl_2^-$ activated pumice)	Direct Blue 15	MnP	Pazarlioglu et al. (2005)
Phenerocheate chrysosporium (immobilized on Kissiris)	Methylene Blue	MnP	Karimi et al. (2006)
Phenerocheate chrysosporium BKM-F1767	Direct Blue 15, Direct Green 6, Congo Red	MnP	Urek and Pazarlioglu (2005)
Pleurotus pulmonarius	Amido Black, Congo Red, Trypan Blue, Methyl Green, Remazol Brilliant Blue R, Methyl Violet, Ethyl Violet, Brilliant Cresyl Blue, Methylene Blue, Poly R-478	Laccase	Tychanowicz et al. (2004)
Pleurotus ostreatus	Phenol Red, Orthocresol Red, Meta-cresol Purple, Bromophenol Red, Bromocresol Purple, Bromophenol Blue, Bromocresol Green	MnP	Srivastava et al. (2005)
Scyzophyllum commune IBL-06	Solar golden yellow R	MnP, Laccase	Asgher et al. (2008)
Trametes trogii	Malachite Green, Xylidine, Ponceau 2R, Anthraquinone Blue	Laccase, MnP	Levin et al. (2005)
Trametes versicolor	Remazol Brilliant Blue R	LiP	Christian et al. (2005)
Trametes versicolor CNPR8107	Remazol Brilliant Blue RR, Remazol Red RR, Remazol Yellow RR	MnP, Laccase	Toh et al. (2003)

10.3.7 Wastewater Treatment

Beyond the production of extracellular proteins, organic acids and other metabolites, fungi have been attracting a growing interest for the biotreatment of wastewater ingredients such as metals, inorganic nutrients and organic compounds (Akthar and Mohan 1995; Field et al. 1993; Palma et al. 1999). The heat treatment liquor of an activated sludge was decolourised by *Coriolus hirsutus* (Fujita et al. 2000). The fungal strain exhibited a strong ability to declourise heat treatment liquor 70 % with an accumulation of manganese independent peroxidase and manganese peroxidase.

Industries of olive oil, distillery, paper and pulp processing produce several billion litres of coloured, toxic and harmful wastewaters over the world annually. Paper and pulp industrial wastewater treatment, detoxification and decolourization rates has been observed with *Ceriporiapsis subvermispora, Phenerocheate chrysosporium, Trametes versicolor, Rhizopus oryzae* and *Rhizopus pusillus* (Manzanares et al. 1995; Van Driessel and Christov 2002; Nagarathnamma et al. 1999). The mechanisms of decolourization of agroindustrial effluents by fungi are reported to include biosorption and biodegradation (Ohmomo 1988; Christov et al. 1999; Nagarathnamma et al. 1999). Ligninolytic enzymes are also involved in the degradation of organic compounds including dyes within these effluents (Chivukula et al. 1995). White rot fungi and their ligninolytic enzyme have wide industrial potential *Phenerocheate chrysosporium* and *Coriolus versicolor* are able to efficiently decolorize and dechlorinate effluents (Archibald 1990; Fukui et al. 1992).

Colour removal from kraft effluent by immobilized lignin peroxidase (CNBr-Sepharoze 4B) was studied (Ferrer et al. 1991). The immobilized lignin peroxidase improved the decolorization by a factor of 2.9. Lignin peroxidase immobilized an activated silica produced 20% mineralization, 65% chemical oxygen demand reduction and 12% decolorization (Dezotti et al. 1995). The immobilization of lignin peroxidase on Amberlite IRA-400 resin exhibited 70% decolorization and a total organic carbon reduction of 15% in 3 h of treatment (Peralta-Zamora et al. 1998).

10.3.8 Soil Treatment

Biological treatment technologies for the remediation of soils and groundwater contaminated with organopollutants are widely used for their friendly impact combined with low cost compared to other treatment alternatives (Sasek et al. 2003). Due to low substrate specificity of their degradative enzymes e.g. laccase, lignin peroxidase and Mn-peroxidase, fungi are able to perform the breakdown of a wide range of organopollutants in contaminated soils (Duran and Esposito 2000; Field et al. 1993). The majority of remediation studies have been performed on artificially contaminated soils spiked with organic pollutants (Pointing 2001; Sasek 2003). Now it is important to investigate the use of fungal remediation under non-sterile conditions and with soils from real contaminated sites, thus making the studies potentially transferable to a field scale (Sasek 2003). Annibale et al. (Annibale et al. 2006) screened fungal strains for degradation of aromatic hydrocarbons from soil and to assess the possible use of such selected autochthomus fungi in an exsitu soil biotreatment via bioaugmentation.

Bollag (1992) suggested that it is possible to enhance the natural process of xenobiotic binding and incorporation into the humus by adding laccase to the soil (Dec and Bollag 1994, 1995). In this direction ligninolytic enzymes acting on organochlorides and its association with humic acid (Esposito et al. 1997, 1995, 1998) were studied. Polymerization of humic acid by oxidative enzymes in soil organic matters was studied by Saccomandi et al. (1998).

10.4 Conclusion

Oxidative enzymes play an important role in the removal of recalcitrant organic pollutants, wastewater treatment and decolorization of dyes and in soil treatment. It has wide potential application in delignification of lignocellolosic materials which are seen as an alternative to the depleting oil reserves, in conversion of high molecular weight coal fractions to low molecular weight coal fractions which could be used as a feed stock for the production of commodity chemicals, in biopulping and biobleaching in paper industries and in enzymatic polymerization in polymer industies.

Lignin peroxidase is an extracellular enzyme dependent of H_2O_2 with high redox potential and low optimum pH. Lignin peroxidase is capable of oxidizing a variety of reducing substrates including polymeric substrates. Due to their high redox potential and their enlarged substrate range lignin peroxidases have great potential for application in various industrial processes.Lignin peroxidase has low substrate specificity, reacting with a wide variety of lignin model compounds. It has the distinction of being able to oxidize methoxylated aromatic rings without a free phenolic group, generating cation radicals that can react further by a variety of pathways, including ring opening, demethylation and phenol dimerisation (Haglund 1999). Lignin peroxidase in contrast with laccases does not require mediators to degrade high redox potential compounds but it needs H_2O_2 to initiate the catalysis.

The high degradative potential of Mnperoxidase makes this enzyme attractive for biotechnological applications e.g. in pulping and bleaching of cellulose, in removing of hazardous wastes or in certain organic syntheses. Peroxidases and phenoloxidases can act on specific recalcitrant pollutants in wastewater or soil by precipitation or transforming to other products and permitting a better final treatment of the waste.Laccase is capable of eliminating the phenols through polymerization process.However, in presence of mediator such as ABTS and HBT degraded phenol by oxidative process (Nelson and Elisa 2000).

The ability of ligninolytic enzymes to oxidize both phenolic and non-phenolic lignin related compounds as well as highly recalcitrant environmental pollutants makes these enzymes very useful for their application to several biotechnological processes. This group of enzyme is highly versatile in nature and they find application in a wide variety of industries.Ligninolytic enzymes are promising to replace the conventional chemical processes of several industries.Thus, there will be a broad area of research open to new findings in the near future. Enzymes present environmental advantages against chemical and micro-organisms.However, the structures of enzymes, regulatory aspects and molecular biology still required thorough understanding.

Acknowledgements The financial support of UGC, New Delhi through its Project no.F:4-2/2006 (BSR)/13-149/2008 (BSR) to Dr. Meera Yadav as a UGC's Dr. D. S. Kothari Postdoctoral Fellowship is thankfully acknowledged. Dr. Sunil Kumar Singh is thankful to CSIR for awarding him RA.

References

Acunzo DF, Galli C (2003) First evidence of catalytic mediation by phenolic compounds in the laccaseinduced oxidation of lignin models. J Eur Biochem 270:3634–3640

Adam W, Lazarus M, Saha Moller CR, Weichold O, Hoch U, Haring D, Schreier P (1999) Biotransformations with peroxidases. Adv Biochem Eng Biotechnol 63:74–108

Adler E (1977) Lignin chemistry-past, present and future. J Wood Sci Technol 11:169–218

Akthar MN, Mohan PM (1995) Bioremediation of toxic metal ions from polluted lake waters and industrial effluents by fungal biosorbent. Curr Sci 69:1018–1030

Amitai G, Adami R, Moriah GS, Rabinovtz I, Vincze A, Leader H, Chefetz B, Leibovitz-Persky L, Friesem D, Hadar Y (1998) Oxidative biodgradation of phosphorothiolates by fungal laccase. FEBS Lett 438:195–200

Annibale AD, Rosetto F, Leonatdii V, Federici F, Petruccioli M (2006) Role of autochthonus filamentous fungi in bioremediation of a soil historically contaminated with aromatic hydrocarbons. Appl Environ Microbiol 72:28–36

Archibald F, Paice MG, Jurasek L (1990) Decolourization of kraft bleachery effluent chromophores by coriolus (Trametes) versicolor. Enzyme Microb Technol 12:846–853

Asgher M, Kausar S, Bhatti HN, Shah SAH, Ali M (2008) Optimization of medium for decolorization of solar golden yellow R direct textile dye by Schizophyllum commune IBL-06. Int Biodeter Biodeg 61:189–193

Baciocchi E, Fabbri C, Lanzalunga O (2003) Lignin peroxidase production by Phanerochaete chrysosporium INA-12 by temperature shifting. Appl Environ Microbiol 54:3194–3196

Bajpai P (1999) Application of enzymes in the pulp and paper industry. Biotechnol Prog 15:147–157

Bajpai P (2004) Biological bleaching of chemical pulps. Critical Rev in Biotechnol 24:1–58

Baldrian P (2004) Purification and characterization of laccase from the white rot fungus Daedalea quercina and decolorization of synthetic dyes by the enzyme. Appl Microbiol Biotechnol 63:560–563

Baldrian P (2006) Fungal laccases-occurance and properties. FEMS Microbiol Rev 30:215–242

Bao W, Eukushima Y, Jensen KA, Moen MA, Hammel KE (1994) Oxidative degradation of nonphenolic lignin during lipid peroxidation by fungal manganese peroxidase. FEBS Lett 354:297–300

Boer CG, Obici L, Giatti C, Souza CG, Rosane M, Peralta M (2004) Decolorization of synthetic dyes by solid-state cultures of Lentinula (Lentinus) edodes producing manganese peroxidase as the main ligninolytic enzymes. Bioresour Technol 94:107–112

Bogan BW, Lamar RT (1995) One electron oxidation in the degradation of creosote polycyclic aromatic hydrocarbons by *Phenerochaete chrysosporium*. Appl Environ Microbiol 61:2631–2835

Bogan BW, Lamar RT (1996) Polycyclic aromatic hydrocarbons degrading capabilities of Phenerochaete laevis HHB-1625 and its extracellular ligninolytic enzymes. Appl Environ Microbiol 62:1597–1603

Bogan BW, Lamar RT, Hammel KE (1996) Fluorene oxidation in vivo by *Phanerochaete chrysosporium* and in vitro during manganese peroxidase-dependent lipid peroxidation. Appl Environ Microbiol 62:1788–1792

Bollag JM (1992) Decontaminating soil with enzymes. Environ Sci Technol 26:1876

Boominathan K, Reddy CA (1992) Fungal degradation of lignin: biotechnological applications. In: Arora DK, Elander RP, Mukerji KG (ed) Handbook of applied mycology, vol 4. Marcel Dekker Inc., New York, pp 763–822

Bourbonnais R, Paice MG (1990) Oxidation of non-phenolic substrates: an expanded role of laccase in lignin biodegradation. FEBS Lett 267:99–102

Bourbannais R, Paice MG (1996) Enzymatic delignification of kraft pulp using laccase and a mediator. TAPPI J 79:199–204

Bourbannais R, Paice MG, Freiermuth B, Brodie E, Borneman S (1997) Reactivities of various mediators and laccases with kraft pulp and lignin model compounds. Appl Environ Microbiol 63:4627–4623

Breen A, Singleton FL (1999) Fungi in lignocellulose breakdown and biopulping. Current Opin Biotechnol 10:252–258
Bumpus JA, Aust SD (1987) Biodegradation of DDT (1,1,1-trichloro-2,2-bis (4-chlorophenyl) ethane) by the white-rot fungus *Phenerochaete chrysosporium*. Appl Environ Microbiol 53:2001–2008
Bumpus JA, Tien M, Wright D, Aust SD (1985) Oxidation of persistent environmental pollutants by a white rot fungus. Science 228:1434–1436
Call HP, Mucke I (1997) History, overview and applications of mediated ligninolytic systems, especially laccase-mediator systems (Lignozym(R)-Process). J Biotechnol 53:163–202
Camarero S, Bockle B, Martinez MJ, Martinez AT (1996) Manganese-mediated lignin degradation by *Pleurotus pulmonarius*. Appl Environ Microbiol 62:1070–1072
Camarero S, Gareia, Vidal T, Colom J, Delrio JC, Gutiertez A, Gras JM, Monje R, Martinez MJ, Martinez AT (2004) Efficient bleaching of non wood high quality paper pulp using laccase-mediator system. Enzyme Microbiol Technol 35:113–120
Caramelo L, Martinez MJ, Martinez AT (1999) A search for ligninolytic peroxidases in the fungus *Pleurotus eryngii* involving alpha-keto-y-thiomethylbutyric acid and lignin model dimers. Appl Environ Microbiol 65:916–922
Catcheside DEA, Ralph JP (1999) Biological processing of coal. Appl Microbiol Biotechnol 53:16–24
Cavalieri EL, Rogan EG (1985) Role of radical cations in aromatic hydrocarbon carcinogenesis. Envion Health Perspect 64:69–84
Cenek N, Katerina S, Pavla E, Tomas C, Aparna K, Elke L, Vaclav S (2004) Ligninolytic fungi in bioremediation: extracellular enzyme production and degradation rate. Soil Biol Biochem 36:1545–1551
Chance B (1952) The kinetics and stoichiometry of the transition from primary to the secondary peroxidase peroxide complex Arch Biochem Biophys 41:416–424
Chivukula M, Renganathan V (1995) Phenolic azo-dye oxidation by laccase from *Pyricularia orizai*. Appl Environ Microbiol 61:4374–4377
Chivukula M, Spadaro JT, Renganathan V (1995) Lignin peroxidase catalysed oxidation of sulfonated azo dyes generates novel sulfophenyl hydroperoxidases. Biochemistry 34:7765–7772
Christian V, Shrivastava R, Shukla D, Modi H, Rajiv B, Vyas M (2005) Mediator role of veratryl alcohol in the lignin peroxidase catalysed oxidative decolorization of remazol brilliant blue R. Enzyme Microb Technol 36:426–431
Christov LP, Van Driessel B, Du Plessis CA (1999) Fungal biomass from *Rhizomucor pulsillus* as adsorbent of chromophores from a bleach plant effluent. Process Biochem 35:91–95
Cohen MS, Gabriele PB (1982) Degradation of coal by the fungi *Polyporus versicplor* and *Poria monticola*. Appl Environ Microbiol. 44: 23–27
Coniglio A, Galli C, Gentili P (2008) Oxidation of amides by laccase- generated aminoxyl radicals. J Mol Catal B Enzymatic 50:40–49
Conto SR, Harrera JLT (2006) Industrial and biotechnological appliations of laccase: a review. Biotecnol Adv 24:500–513
Crestini C, Argyropoulos DS (1998) The early oxidative biodegradation steps of residual kraft lignin models with laccase. Biorg Med Chem 6:2161–2169
Dafale N, Nageswara RN, Sudhir U (2008) Decolorization of azodyes and simulated dye bath wastewater using acclimatized micro-bial consortium biostimulation and halo tolerance. Bioresour Technol 99:2552–2558
Dahiya JS, Singh D, Nigam P (1998) Characterization of laccase produced by *Coniothyrium minitans*. J Basic Microbiol 38:349–359
Davis JM, Lohmann RC, Phillips FM, Wilson JL, Love DW (1993) Architecture of the sierra ladrones formation, central new mexico: depositional controls on the permeability correlation structure. Geol Soc Amer Bull 105(8):998–1007
Dec J, Bollag JM (1994) Use of plant material for the decontamination of water polluted with phenols. Biotechnol Bioeng 44:1132–1139
Dec J, Bollag JM (1995) Effect of various factors on dehalogenation of chlorinated phenols and anilines during oxidative coupling. Environ Sci Technol 29:657–663

Dezotti M, Innocentini-Mei LH, Duran N (1995) Silica immobilized enzyme catalysed removal of chlorolignin from Eucalyptus kraft effluent. J Biotechnol 43:161–167
Dunford HB (1991) Horseradish peroxidase: Structure and kinetic properties. In: Everse J, Everse KE, Grisham MB (eds). CRC Press, Boca Raton, pp 1–24
Duran N, Esposito E (2000) Potential applications of oxidative enzymes and phenoloxidase like compounds in waste water and soil treatment. Appl Catal B 28:83–89
Dwivedi UN, Singh P, Pandey VP, Kumar A (2011) Structure-function relationship among bacterial, fungal and plant laccases. J Mol Catal B Enzymatic 68:117–128
Dwran N, Minussi RC, Pastore GM, Alves OL, Gimenes IF, Peralta-zamora P, Moraes SG (2000) Laccase production and its environmental applications in the presence of mediators In: Soares CH (ed) Proceedings of the second national meeting of Environmental Applied Microbiology, vol 2. Florianopolis, p 12
Edwards SL, Raag R, Wariishi H, Michael HG, Thomas LP (1993) Crystal structure of lignin peroxidase. Proc Natl Acad Sci U S A 90:750–754
Eibes G, Cajthaml T, Moreira MT, Feijoo G, Lema JM (2006) Enzymatic degradation of anthracene, dibenzothiophene and pyrene by manganese peroxidase in media containing acetone. Chemosphere 64:408–414
Enguita FJ, Martins LO, Henriques AO, Carrondo MA (2003) Crystal structure of a bacterial endospore coat component: a laccase with enhanced thermostability properties. J Biol Chem 278:19416–19425
Eriksson KE, Kirk TK (1994) Biopulping: an overview of developments in an environmentally safe paper making technology. FEMS Microbiol Rev 13:351–364
Erkurt EA, Ünyayar A, Kumbur H (2007) Decolorization of synthetic dyes by white rot fungi, involving laccase enzyme in the process. Process Biochem 42(10):1429–1435
Esposito E, Resende MOO, Freer J, Baeza J, Carvalho ME (1995) Ligninnolytic enzymes application on humic acid organochloride compounds association. In: Melo EHM, Pimentel MCB (eds) Proceedings of the fourth Brazilian Symposium on Chemical Lignins other Wood Comp., Brazil, vol 5, p. 251. Chem Abstr. 126:108249
Esposito E, Manfio G, Villas-Boas S, Antunes R, Paulillo S, Souza JA (1997) Microbiological strategies on remediation of contaminated soil with organochlorides. In: Esposito E (ed) Proceedings of the First National Meeting of Environmental Applied Microbiology Campinas, S.P., Brazil, vol 1, p 80. Chem. Abstr. 127: 358427
Esposito E, Manfio G, Villas-Boas S, Manfio G (1998) Fungal potential for soil bioremediation. In: Gaylarde CC, Barbosa TCP, Gabilan NH (eds) Proceedings of the Third Latin American Biodegradation & Biodeterioration Symposium-LABS-3, Florianopolis, Brazil, CD-Rom Paper 25
Ferrer M, Dezotli ND (1991) Decolorization of kraft effluent by free and immobilized lignin peroxidase and horseradish peroxidase. Biotechnol Lett 13:577–582
Ferrer I, Dezotti M, Duran N (1991) Decolorization of kraft effluent by free and immobilized lignin peroxidase and horseradish peroxidase. Biotechnol Lett 13:577–582
Field JA, De Jong E, Costa GF, De Bont JAM (1993) Screening of ligninolytic fungi applicable to biodegradation of xenobiotics. Trends Biotechnol 11:44–49
Frichter M, Vares T, Kalsi M, Galkin S, Scheibner K, Fritsche W, Hatakka A (1999) Production of MnP and organic acids and mineralization of 14 C-labelled lignin (14 C-DHP) during solid-state fermentation of wheat-straw with the white rot fungus *Nematoloma frowardii*. Appl Environ Microbiol 65:1864–1870
Fritsche W, Scheibner K, Herre A, Hofrichter M (2000) Fungal degradation of explosives: TNT and related nitroaromatic compounds. In: Spain JC, Hughes JB, Knackmuss HJ (eds) Nitroaromatic compounds and explosives. Lewis, Boca Raton, pp 213–237
Fritz-Langhals E, Kunath B (1998) Synthesis of aromatic aldehydes by laccase-mediater assisted oxidation. Tetrahedron Lett 39:5955–5956
Fujita M, Era A, Lke M, Soda S, Miyata N, Hirao T (2000) Decolorization of heat treatment liquor of waste sludge by a bioreactor using polyurethane foam immobilized white rot fungi equipped with an ultramembrane filtration unit. J Biosc Bioeng 90:387–394

Fukui HTL, Presnell TW, Joyce, Chang H (1992) Dechlorination and detoxification of bleach plant effluent by Phanerochaete chrysosporium. J Biotechnol 24:267–275.

Gavril M, Hodson PV (2007) Chemical evidence for the mechanism of the biodecoloration of Amaranth by *Trametes versicolor*. World J Microbiol Biotechnol 23:103–124

Gianfreda L, Sannina F, Filazzola MT, Leonowicz A (1998) Catalytic behaviour and detoxifying ability of a laccase from the fungal strain *Cerrena unicolor*. J Mol Catal B Enzymatic 4:13–23

Glenn JK, Akileswaran L, Gold MH (1986) Mn(II) oxidation is the principal function of the extracellular Mn-Peroxidase from *Phanerochaete chrysosporium*. Arch Biochem Biophy 251:688–696

Grey R, Hafer C, Schlosser D (1998) Degradation of 2-chlorophenol and formation of 2-chloro-14-benzoquinone by mycelia and cell free crude culture liquids of *Trametes versicolor* in relation to extracellular laccase activity. J Basic Microbiol 38:371–382

Gunther T, Sack U, Hofrichter M, Latz M (1998) Oxidation of PAH and PAH-derivatives by fungal and plant oxidoreductases. J Basic Microbiol 38:113–122

Haglund C (1999) Biodegradation of xenobiotic compounds by the white rot fungus *Trametes trogii*. Molecular Biotechnology Programme. Uppsela University School of Engineering, p 30

Hammel KE, Cullen D (2008) Role of fungal peroxidases in biological ligninolysis. Curr opin plant Biol 11:1–7

Hammel KS, Kalyanaraman B, Kirk TK (1986) Oxidation of polycyclic aromatic hydrocarbons and dibenzo[p]-dioxins by *Phenerochaete chrysosporium* ligninase. J Biol Chem 26:948–995

Hatakka A (1994) Lignin-modifying enzymes from selected white-rot fungi: production and role in lignin degradation. FEMS Microbiol Rev 13:125–135

Harley BS, Brodo PMA, Senior PJ (1988) Proceeding of Royal Society discussion meeting on utilisation of lignocellulosic wastes. Cambridge Univ Press, Cambridge

Have RT, Rietjens IMCM, Hartmans S, Swarts HJ, Field JA (1998) Calculated ionization potentials determine the oxidation of vanillin precursors by lignin peroxidase. Federation Eur Microbiol Soc Rev 13:125–135

Harvey PJ, Schoemaker HE, Palmer JM (1986) Veratryl alcohol as a mediator and the role of radical cations in lignin biodegradation by *Phenerochaete chrysosporium*. FEBS Lett 195:242–246

Heinfling A, Martinez MJ, Martinez AT, Bergbauer M, Szewzyk U (1998a) Purification and characterization of peroxidases from the dye-decolrizing fungus *Bjerkandera adusta*. FEMS Microbiol Lett 165:43–50

Heinfling A, Ruiz-Duenas FJ, Martinez MJ, Bergbauer M, Szewzyk U, Martinez AT (1998b) A study on reducing substrates of manganese oxidizing peroxidases from *Pleurotus eryngii* and *Bjerkandera adusta*. FEBS Lett 428:141–146

Hiroshi U, Shiro K (1999) Enzymatic polymerization yields useful polyphenols. Chemtech 29:22–28

Hofrichter M (2002) Review: lignin conversion by manganese peroxidase (MnP). Enzyme Microb Technol 30:454–466

Hofrichter M, Scheibner K, Schneegab V, Fritsche W (1998) Enzymatic combustion of aromatic and aliphatic compounds by MnP from *Nematoloma frowardii*. Appl Environ Microbiol 64:399–404

Hofrichter M, Lundell T, Hatakka A (2001) Conversion of milled pine wood by manganese peroxidase from *Phlebia radiata*. Appl Environ Microbiol 67:4588–4593

Ishihara A, Maeshima Y, Veta H, Oishi K (1997) Decolorization of dyes by laccase, Shizuoka-ken Hammatsu Kogyo Gijuku, Senta kenkyu Hokoku 721. Chem Abstr 128:4654.

Isroi, Rai M, Siti S, Claes N, Muhammad NC, Knut L, Mohammad JT (2011) Biological pretreatment of lignocelluloses with white rot fungi and its applications: a review. BioResources 6:5224–5259

Joshi DK, Gold MH (1994) Oxidation of dibenzo-p-dioxin by lignin peroxidase from the basidiomycete *Phenerochaete chrysosporium*. Biochemistry 33:1096–10976

Joshi SM, Inamalor SA, Telke AA (2010) Exploring the potential of natural bacterial consortium to degrade mixture of dyes and textile effluent. Int J Biodeter Biodegrad 64:622–628

Kadhim H, Graham C, Barratt P, Evans CS, Rastall RA (1999) Removal of phenolic compounds in water using *Cariolus versicolor* grown on wheat bran. Enzyme Microbiol Technol 24:303–307
Kang S, Shin KS, Han YH, Youn HD, Hah YC (1993) Purification and characterisation of an extracellular peroxidase from white-rot fungus *Pleurotus ostreatus*. Biochem Biophys Acta 1163:158–164
Karimi A, Vahabzadeh F, Bonakdarpour B (2006) Use of *Phanerochaete chrysosporium* immobilized on Kissiris for synthetic dye decolourization: involvement of manganese peroxidase. World J Microbiol Biotechnol 22:1251–1257
Kawai S, Nakagawa M, Ohashi H (1999) Aromatic ring cleavage a non-phenolic bête-O-4 lignin model dimmer by laccase of Trametes versicolor in the presence of 1-hydroxybenzotriazole. FEBS Lett 446:355–358
Kedderis GL, Hollenberg PF (1983) Steady-state kinetics of chloroperoxidase-catalyzed N-demethylation reactions. J Biol chem 258:12413–12419
Kersten P, Tein M, Kalyanaraman B, Kirk TK (1985) The ligninase of *Phanerochaete chrysosporium* generates cation radical from methoxy benzenes. J Biol chem 260:2609–2612
Kersten PJ, Kalyanaraman B, Hammel KE, Reinhammar B, Kirk TK (1990) Comparison of lignin peroxidase, horseradish peroxidase and laccase in the oxidation of methoxybenzenes. Biochem J 268:475–480
Kirk TK, Cullen D (1998) Enzymology and molecular genetics of wood degradation by white-rot fungi. In: Young RA, Akthar M (eds) Environmentally friendly technologies for the pulp and paper industry. Wiley, New York, pp 273–307
Krikstopaitis K, Kulys J, Pederson AH, Schneider P (1998) N-substituted p-phenylenediamines as peroxidase and laccase substrates. Acta Chem Scan 52:469–474
Kurakale M, Ide N, Komaki T (2007) Biological pretreatment with two bacterial strains for enzymatic hydrolysis of office paper. Curr Microbiol 54:424–428
Kurek B, Petit-conil M, Sigoillot JC, Herpoel I, Ruel K, Moukha S, Joseleav JP, Pennincks M, Asther M, Gazza G, Dechoudens C (2001) Treatment of high yield pulp with fungal peoxidases from laboratory to pilot scale study. In: Argyropoulos D (ed) ACS symposium series 785, oxidative delignification chemistry, fundamental and catalysis, vol 785. American Chemical Society, Washigton DC, 474–486 (Chap. 30)
Kuwahara M, Glenn JK, Morgan MA, Gold MH (1984) Separation and characterization of two extracellular H2O2-dependent oxidases from ligninolytic cultures of *Phanerochaete chrysosporium*. FEBS Lett 169:247–250
Kwant SS, Chang JK (1998) Decolorisation of artifi cial dyes by peroxidase from the white-rot fungus *Pleurotus ostreatus*. Biotechnol Lett 20:569–572
Lackner R, Srebotnik E, Messner K (1991) Oxidative degradation of high molecular weight chlorolignin by MnP of *Phanerochaete chrysosporium*. Biochem Biophys Res Comm 178:1092–1098
Leontievsky A, Vares T, Lankinen P, Shergill JK, Pozdnyakova N, Myasoedova NM (1997) Blue and yellow laccases of ligninolytic fungi. FEMS Microbiol Lett 156:9–14
Levin L, Papinutti L, Forchiassin F (2004) Evaluation of four Argentinean white rot fungi; their ability to produce lignin modifying enzymes and decolorize industrial dyes. Bioresour Technol 2:169–176
Levin L, Forchiassin FF, Viale A (2005) Ligninolytic enzyme production and dye decolorization by *Trametes trogii*: application of the Plackett-Burman experimental design to evaluate nutritional requirements. Process Biochem 40:1381–1387
Lisov AV, Leontievsky AA, Golovleva LA (2003) Hybrid Mn-peroxidase from the ligninolytic fungus Panus tigrinus 8/18. Isolation, substrate specificity and catalytic cycle. Biochem (Moscow) 68:1027–1035
Lund M, Ragauskas AJ (2001) Enzymatic modifications of kraft lignin through oxidation coupling with water-soluble phenols. Appl Microbiol Biotechnol 55:699–703
Maijala P, Mettalo A, Kleen M, Westin C, Poppius-Levin K, Herranen K, Lehto JH, Reponen P, Maentausta O, Hatakka A (2007) Treatment of softwood chips with enzymes may reduce refining energy consumption and increase surface charge of fibers. In: 10th international congress on biotechnology in the pulp and paper Industry. Madison Wisconsin, Book of Abstracts. p 65

Majcherczyk A, Johannes C, Huttermann A (1999) Oxidation of aromatic alcohols by laccase from Trametes versicolor mediated by 22′-azino-bis-(3-ethylbenzothiazoline-6-sulphonic acid) cation radical and dication. Appl Microbiol Biotechnol 51:267–276

Manzanares P, Fajardo S, Martin C (1995) Production of ligninolytic activities wheen treating paper pulp effluents by Trametes versicolor. J Biotechnol 4B:125–132

Marcia JMM, Ademir CES, Helena CTR (2010) Industrial and biotechnological applications of ligninolytic enzymes of the basidiomycota: A review. E J Biotechnol. doi:10.2225/vol13-issue6-fulltext-2

Martinez AT (2002) Molecular biology and structure and function of lignin-degrading heme peroxidases. Enzym Microb Technol 30:425–444

Martinez MJ, Martinez AT (1996) Characterization of MnP isoenzymes of *Pleurotus eryngii* exhibiting Mn-independent activities on 2,6-dimethoxyphenol and veratryl alcohol. In: Messner K, Srebotnik E (eds) Biotechnology in the pulp and paper Industries: Recent advances in applied and fundamental research. Facultas-Universitats, Vienna, pp 417–420

Martinez MJ, Ruiz-Duenas FJ, Guillen F, Martinez AT (1996) Purification and catalytic properties of two manganese-peroxidase isoenzymes from *Pleurotus eryngii*. Eur J Biochem 237:424–432

Martnez D, Larrondo LF, Putnam N, Gelpke MD, Huang K, Chapman J (2004) Genome sequence of the lignocellulose degrading fungus *Phenerocheate chrysosporium* strain RP78. Nature Biotechnol 22:695–700

Marwaha SS, Grover R, Prakash C., Kennedy JF (1998) Continuous biobleaching of black liquor from the pulp and paper industry using an immobilised cell system. J Chem Technol Biotechnol 73:292–296

Maximo C, Costa-Ferreira MC (2004) Decolourization of reactive textile dyes by Irpex lacteus and lignin modifying enzymes. Process Biochem 39:1475–1479

May SW (1999) Applications of oxidoreductase. Curr Opin Biotechnol 10:370–375

Mazmanci MA, Unyayar A (2005) Decolourization of reactive black 5 by *Funalia trogii* immobilized on Luffa cylindrical sponge. Proc Biochem 40:337–342

Meera K, Yadav RSS, Yadav KDS (2002) Secretion of ligninperoxidase by *Penicillium citrinum*, *Fusarium oxysporum* and *Aspergillus terreus*. Indian J Exp Biol 40:802–806

Messerschmidt A (1997) Multi-copper oxidases. World Scientific, Singapore

Mester T, Field JA (1998) Characterization of a novel manganese peroxidase-lignin peroxidase hybrid isozyme produced by Bjerkandera species strain BOS55 in the absence of manganese. J Biol Chem 273:15412–15417

Michniewicz A, Ledakowicz S, Ullrich R, Hofrichter M (2008) Kinetics of the enzymatic decolorization of textile dyes by laccase from *Cerrena unicolor*. Dyes Pigm 77:295–302

Mikolasch A, Hammer E, Jones U, Popowski K, Stielow A, Schauer F (2002) Synthesis of 3-(3,4-dihydroxy phenyl)-propionic acid derivatives by N-coupling of amines using laccase. Tetrahedron 58:7589–7593

Mikolasch A, Niedermeyer THJ, Lalk M, Witt S, Seefeld S, Hammer E, Schauer F, Salazar MG, Hessel S, Julich WD, Lindoquist U (2007) Novel cephalosporins synthesized by amination of 2,5-dihydroxybenzoic acid derivatives using fungal laccases II. Chem and Pharm Bull 55:412–416

Mikolasch A, Schauer F (2009) Fungal laccases as tools for the synthesis of new hybrid molecules and biomaterials. Appl Microbiol Biotechnol 82:605–624.

Millis CD, Cai D, Stankovich MT, Tien M (1989) Oxidation-reduction potentials and ionization states of extracellular peroxidases from the lignin degrading fungus *Phanerochaete chrysosporium*. Biochem 28:8484–8489

Milstein O, Huttermann A, Frund R, Ludemann HD (1994) Enzymatic copolymerization of lignin with low molecular mass compounds. Appl Miicrobiol Biotechnol 40:760–767

Moreira MT, Palma C, Mielgo I, Feijoo G, Lema JM (2001) In vitro degradation of a polymeric dye (Poly R-478) by MnP. Biotechnol Bioeng 75:362–368

Moreira MT, Feijoo G, Canoval J, Lema JM (2003) Semipilot-scale bleaching of kraft pulp with MnP. Wood Sci Technol 37:117–123

Morozova OV, Shumakovich GP, Shleev SV, Yaropolov YaI (2007) Laccase-mediator systems and their applications: a review. Appl Biochem Microbiol 43:523–535
Nagarathnamma R, Bajpai P (1999) Decolorization and detoxification of extraction stage effluent from chlorine bleaching of kraft pulp by *Rhizopus oryzae*. Appl Environ Microbiol 65:1078–1082
Nagarathnamma R, Bajpai P, Bajpai PK (1999) Studies on decolorization and detoxification of chlorinated lignin compounds in kraft bleaching effluents by *Ceriporiopsis subvermispora*. Process Biochem 34:939–948
Nelson D, Elisa E (2000) Potential applications of oxidative enzymes and phenoloxidase like compounds in wastewater and soil treatment: a review. Appl Catal B: Environ 28:83–99
Nilsson I, Moller A, Mattiasson B, Rubindamayugi MST, Welander U (2006) Decolorization of synthetic and real textile wastewater by the use of white-rot fungi. Enzyme Microb Technol 38:94–100
Ohmomo S, Kainuma M, Sirianuntapiboon S, Aoshima I, Atthasampunna P (1988) Adsorption of melanoidin to the mycelia of *Aspergillus oryzae* Y-32. Agric Biol Chem 52:381–386
Ozsoy HD, Unyayar A, Mazmanci MA (2005) Decolorization of reactive textile dyes drimarene blue X3LR and Remazol brilliant blue R by *Funalia trogii* ATCC 200800. Biodegradation 16:195–204
Palma C, Moreira MT, Mielgo I, Feijoo G, Lema IM (1999) Use of a fungal bioreactor as a post treatment step for continuous decolorisation of dyes. Wat Sci Technol 40:131–136
Patel VK, Yadav RSS, Yadav KDS (2007) Enzymatic characteristics of lignin peroxidases of indigenous ligninolytic fungal strains—Part I. Indian J Biotechnol 6:553–556
Pazarlioglu NK, Sariisik M, Telefoncu A (2005) Laccase production by *Trametes versicolor* and application to denim washing. Process Biochem 40:1673–1678
Pelaez F, Martinez MJ, Martinez AT (1995) Screening of 68 species of basidiomycetes for enzymes involved in lignin degradation. Mycological Research 99:37–42
Peralta-Zamora P, Moares SG, Esposito E, Antunes R, Refus J, Duran N (1998) Decolorization of pulp mill effluents with immobilized lignin and manganese peroxidase from *Phenerocheate chrysosporium*. Environ Technol 19:521–528
Perumal K, Kalaichelvan PT (1996) Production of extracellular lignin peroxidase & laccase by *Ganoderma lucidum* PTK3 on sugarcane bagasse lignin. Indian J Expt Biol 34:1121–1125
Pointing SB (2001) Feasibility of bioremediation by white-rot fungi. Appl Microbiol Biotechnol 57:20–33
Pothast A, Rosanau T, Koch H, Fisher K (1999) The reaction of phenolic model compounds in the laccase mediator system (LMS) investigations by matrix assisted laser desorption ionization time of flight mass spectrometry [MALDI-TOF-MS]. Holzfrschung 53:175–180
Punna Payak H, Prasongsuk S, Messner K, Danmek K, Lotrakul P (2009) Polycyclic aromatic hydrocarbons (PAHs) degradation by laccase from a tropical white rot fungus *Ganaderma lucidum*. Afr J Biotechnol 8:5897–5900
Quintanar L, Yoon J, Aznar CP, Palmer AE, Andersson KK, Britt RD, Solomon EI (2005) Spectroscopic and electronic structure studies of the trinuclear as cluster active site of the multicopper oxidase laccase: nature of its co-ordination unsaturation. J Am Chem Soc 127:13832–13845
Ralph JP, Catcheside DEA (1994) Decolourisation and depolymerisation of solubilised low-rank coal by the white-rot basidiomycete *Phanerochaete chrysosporium*. Appl Microbiol Biotechnol 42:536–542
Ralph JP, Graham LA, Catcheside DEA (1996) Extracellular oxidase and the transformation of solubilised fractions of low rank coal by wood rot fungi. Appl Microbiol Biotechnol 46:226–232
Rama R, Mougin C, Boyer FD, Kollmann A, Molosse C, Sigoillot JC (1998) Biotransformation of benzo[a]pyrene in bench scale reactor using laccase of *Pycnoporus cinnabarinus*. Biotechnol Lett 20:1101–1104
Raper JC, Sarkar JM, Bollag JM (1995) Enhanced enzymatic removal of chlorophenols in the presence of co-substrates. Water Res 29:2720–2724
Regalado C, Garcia-Almendarez BE, Duarte-Vazquez MA (2004) Biotechnological applications of peroxidases. Phytochem Rev 3:243–256

Renganathan V, Gold MH (1986) Spectral characterisation of the oxidized states of lignin peroxidase, an extracellular heme enzyme from white rot basidiomycete *Phanerochaete chrysosporium*. Biochemistry 25:1626–1631

Rittstieg K, Suurnakki A, Suortti T, Kruus K, Guabitz AM, Buchart J (2003) Polymerisation of guaiacol and a phenolic β-o-4 substructure by *Trametes hirsuta* laccase in the presence of ABTS. Biotechnol Progess 19:1505–1509

Riva S (2006) Laccases: blue enzymes for green chemistry. Trends Biotechnol 24:219–226

Rodriguez S, Toca JL (2006) Industrial and biotechnological applications of laccases: a review. Biotechnol Adv 24:500–513

Ruiz-Duenas FJ, Martinez MJ, Martinez AT (1999) Molecular characterization of a novel peroxidase isolated from the ligninolytic fungus *Pleurotus eryngii*. Mol Microbiol 31:223–236

Saccomandi F, Conte P, Piccolo A (1998) Use of oxidase enzyme to increase polymerization of soil organic matter. Fresenius Environ Bull 7:537–543

Sack U, Hofrichter M, Fritsche W (1997) Degradation of polycyclic aromatic hydrocarbons by manganese peroxidase of *Nematoloma frowardii*. FEMS Microbiol Lett 152:227–234

Sahay R, Yadav RSS, Yadav KDS (2008) Purification and characterization of extracellular laccase secreted by *Pleurotus sajor-caju* MTCC-141. Chin J of Biotechnol 24:2068–2073

Sahay R, Yadav RSS, Yadav KDS (2009) Purification and Characterization of Laccase Secreted by *L. lividus*. Appl Biochem Biotechnol 157:311–320

Salvachua D, Prieto A, Lopez-Abelairas M, Lu-Chau T, Martinez AT, Martinez MJ (2011) Fungal pretreatment: an alternative in second-generation ethanol from wheat straw. Bioresour Technol 102:7500–7506

Sanchez C (2009) Lignocellulosic residues: biodegradation and bioconversion by fungi. Biotech Adv 27:185–194

Sanghi R, Dixit A, Guha S (2006) Sequential batch culture studies for the decolorization of reactive dye by *Coriolus versicolor*. Bioresour Technol 97:396–400

Sarkar S, Martinez AT, Martinez MJ (1997) Biochemical and molecular characterization of a manganese peroxidase isoenzyme from *Pleurotus ostreatus*. Biochem Biophys Acta 1339:23–30

Sasek V (2003) Why mycoremediations have not yet come into practice.In: Sasek V, Glaser JA, Baveye P (ed) The utilization of bioremediation to reduce soil contamination: problems and solutions. Kluwer Academic Publishers, Amsterdam, pp 247–260

Satwinder SM, Rajesh G, Chand P, John FK (1998) Continuous biobleaching of black liquor from the pulp and paper industry using an immobilised cell system. J Chem Technol Biotechnol 73:292–296

Satyahari D, Maiti TK, Bhattacharyya BC (1994) Production of some extracellular enzymes by a lignin peroxidase-producing brown rot fungus, *Polyporus ostreiformis*, and its comparative abilities for lignin degradation and dye decolorization. Appl Environ Microbiol 60:4216–4218

Scheibner K, Hofrichter M (1998) Conversion of aminonitrotoluenes by fungal manganese peroxidases. J Basic Microbiol 38:51–59

Schweigert N, Zehnder AJB, Eggen RIL (2001) Chemical properties of catechols and their molecular modes of toxic action in cells from micro-organisms to mammals. Environ Microb 3:81–91

Shanmugan V, Yadav KDS (1996) Production of extracellular lignin peroxidase by *Pleurotus sajor-caju*. Indian J Exp Biol 34:1164–1165

Shanmugan V, Yadav KDS (1997) Production of lignin peroxidase by *Rhizopus nigricans*. Indian J Microbiol 37:105–106

Sharma JK, Yadav M, Singh NP, Yadav KDS (2011) Purification and characterisation of lignin peroxidase from *Pycnoporus sanguineus* MTCC-137. Appl Biochem Microbiol 47:532–537

Shuttleworth KL, Bollag JM (1986) Soluble and immobilized laccase as catalysts for the transformation of substituted phenols. Enzyme Microbial Technol 8:171–177

Sigoillot C, Camarero S, Vidal T, Record E, Asther M, Colom JF and Martinez AT (2005) Comparison of different fungal enzymes for bleaching high-quality paper pulps. J Biotechnol 115:333–343

Silva EM, Martins SF, Milagres AMF (2008) Extraction of manganese peroxidase produced by *Lentinula edodes*. Bioresour Technol 99:2471–2475

Snook ME, Hamilton GA (1973) Oxidation and fragmentation of some phenyl- substituted alcohols and ethers by peroxydisulfate and Fenton's reagent. J Am Chem Soc 96:860–869

Srivastava R, Christian V, Vyas BRM (2005) Enzymatic decolorization of sulfonphthalein dyes. Enzyme Microb Technol 36:333–337

Steffen KT, Hatakka A, Hofrichter M (2003) Degradation of Benzo[a] pyrene by the litter decomposing basidiomycete *Stropharia coronilla*: role of MnP. Appl Environ Microbiol 69:3957–3964

Strebotnik E, Hammel KE (2000) Degradation of nonphenolic lignin by the laccase/1-hidroxy benzotriazole system. J Biotechnol 83:1–11.

Sundaramoorthy M, Kishni K, Gold MH, Poulos TL (1994) The crystal structure of MnP from P.chrysosporium at 2.06-A resolution. J Biol Chem 269:32759–32767

Sun Y, Cheng J (2002) Hydrolysis of lignocellulose materials for ethanol production: a review. Bioresour Technol 83:1–11

Svobodova K, Majcherczyk A, Novotn C, Kues U (2008) Implication of mycelium-associated laccase from Irpex lacteus in the decolorization of synthetic dyes. Bioresour Technol 99:463–471

Tanaka H, Koike K, Itakura S, Enoki A (2009) Degradation of wood and enzyme production by *Ceriporiopsis subvermispora*. Enzyme Microb Technol 45:384–390

Tauber MM, Gubitz GM, Rehorek A (2008) Degradation of azo dyes by oxidative processes: laccase and ultrasound treatment. Biores Technol 99:4213–4220

Tavèar M, Svobodová K, Kuplenk J, Novotný C, Pavko A (2006) Biodegradation of azo dye RO16 in different reactors by immobilized Irpex lacteus. Acta Chim Slovenica 53:338–343.

Tien M, Kirk TK (1988) Lignin Peroxidase of *Phanerochaete chrysosporium*: Methods in Enzymology 161:238–249

Tein M, Kirk TK, Bull C, Fee JA (1986) Steady-state and transient-state kinetic studies on the oxidation of 3,4-dimethoxybenzyl alcohol catalyzed by the ligninase of *Phanerocheate chrysosporium* Burds. J Biol chem 261:1687–1693

Toh Y, Jia J, Yen L, Obbard JP, Ting Y (2003) Decolorization of azo dyes by white-rot fungi (WRF) isolated in Singapore. Enzyme Microb Technol 33:569–575

Tuor U, Wariishi H, Schoemaker HE, Gold MH (1992) Oxidation of phenolic arylglycerol β-aryl ether lignin model compounds by manganese peroxidase from Phanerochaete chrysosporium: oxidative cleavage of an α-carbonyl model compound. Biochemistry 31:4986–4995

Tychanowicz GK, Zilly A, Giatti C, Marques de Souza M, Peralta RM (2004) Decolorization of industrial dyes by solid-state cultures of *Pleurotus pulmonarius*. Proc Biochem 39:855–859

Uan IC, Tien M (1993) Stimulation of Mn peroxidase activity: A possible role for oxalate in lignin biodegradation. Biochemistry 90:1242–1246

Urek RO, Pazarlioglu NK (2005) Production and stimulation of manganese peroxidase by immobilized *Phanerochaete chrysosporium*. Proc Biochem 40:83–87

Valli K, Wariishi H, Gold MH (1990) Oxidation of monomethoxylated aromatic compounds by lignin peroxidase: role of veratryl alcohol in lignin biodegradation. Biochemistry. 29:8535–8539.

Valli K, Brock BJ, Joshi DK, Gold MH (1992). Degradation of 2,4-dinitrotoluene by the lignin-degrading fungus *Phanerochaete chrysosporium*. Appl Environ Microbiol 58:221–228

Van Aken B, Hofrichter M, Scheibner K, Hatakka A, Naveau H, Agathos SN (1999) Transformation and mineralization of 2,4,6-trinitrotoluene (TNT) by manganese peroxidase from the white rot basidiomycete *Phlebia radiate*. Biodegradation 10:83–91

Van Aken B, Cameron MD, Stahl JD, Plumat A, Naveau H, Aust SD, Agathos SN (2000) Glutathione-mediated mineralization of 14 C-labeled 2-amino-4,6-dinitrotoluene by manganese dependent peroxidase H5 from the white-rot fungus *Phanerochaete chrysosporium*. Appl Microbiol Biotecnol 54:659–664

Van Driessel B, Christov L (2002) Adsorption of colour from a bleach plant effluent using biomass and cell wall fractions from *Rhizomucor pusillus*. J Chem Technol Biotechnol 77:155–158

Vikineswary S, Abbullah N, Renuvathani M, Sekaran M, Pandey A and Jones GEB (2006) Productivity of laccase in solid substrate fermutation of selected agro-residues by *Pycnoporus sanguineus*. Bioresour Technol 97:171–177

Vyas BRM, Vole J, Saaek V (1994) Ligninolytic enzymes of selected white-rot fungi cultivated on wheat-straw. Folia Microbiol 39:235–240

Wandrey C, Lies A, Kihumbu, D (2000) Industrial biocatalysis: past, present and future. Org Proc Res Dev 4:285–290

Wang Y, Vazquez-Duhalt R, Pickard MA (2003) Manganese-lignin peroxidase hybridf from *Bjerkandera adust* oxidizes polycyclic aromatic hydrocarbons more activity in the absence of manganese. Can J Microbiol 49:675–682

Ward G, Hadar Y, Bilkis I, Dosoretz CG, (2003a) Mechanistic features of lignin peroxidase-catalysed oxidation of substituted phenols and 1,2-dimethoxyarenes. J Biol Chem 278:39726–39734

Wariishi H, Valli K, Renganathan V, Gold MH (1989) Thiol mediated oxidation of nonphenolic lignin model compounds by MnP of *Phanerochaete chrysosporium*. J Biol Chem 264:14185–14191

Wariishi H, Huang J, Dunford HB, Gold MH (1991a) Reactions of lignin peroxidase Compounds I and II with veratryl alcohol: Transient-state kinetics characterization. J Biol Chem 266:20694–20699

Wariishi H, Valli K, Gold MH (1991b) In vitro depolymerization of lignin by MnP of *Phanerochaete chrysosporium*. Biochem Biophys Res Comm 176:269–275

Wariishi H, Valli K, Gold MH (1992) Manganese(II) oxidation by manganese peroxidase from the basidiomycete *Phanerochaete chrysosporium*. Kinetic mechanism and role of chelators. J Biol Chem 267:23688–23695

Wasenberg O, Kryriakides I and Agathos SN (2003) White rot fungi and their enzymes for the treatment of industrial dye effluents. Biotechnol Adv 22:161–187

Widsten P, Kandelbauer A (2008) Laccase applications in the forest products industry; a review. Enzyme and Microbiol Technol 42:293–307

Wilson BW, Bean RM, Franz JA, Thomas BL, Cohen MS, Aronson H, Gray ET Jr. (1987) Microbial conversion of low rank coal: Characterisation of biodegraded product. Energy Fuel 1:80–84

Xu F (1996) Oxidation of phenols, anilines and benzenethiols by fungal laccases: correlation between activity and redox potentials as well as halide inhibition. Biochemistry 35:7607–7614

Xu F (2005) Application of oxidoreductase: recent progress. Ind Biotechnol 1:38–50

Xu H, Lai YZ, Slomezynski D, Nakas JP, Tanenbaum SW (1997) Mediator-assisted selective oxidation of lignin model compounds by laccase from *Botrytis anerea*. Biotechnol Lett 19:957–960

Yadav M, Yadav KDS (2007) In: Kuhad RC, Singh A (eds) Lignocellulose biotechnology: future prospects. International publishing House Pvt. Ltd, New Delhi, pp 63–88

Yadav M, Yadav P, Yadav KDS (2009a) Characterisation and coal depolymerizing activity of lignin peroxidase from *Lenzitus seperia* MTCC-1170. Biochemistry (Moscow) 74:1125–1131

Yadav M, Yadav P, Yadav KDS (2009b) Production, purification and characterisation of lignin peroxidase from *Loweporus lividus* MTCC-1178. Eng life Sci 9:124–129

Yadav M, Singh SK, Yadav KS, Yadav KDS (2010) Purification of lignin peroxidase from Hexagona tenuis MTCC-1119 and its kinetic properties in aqueous medium containing miscible organic solvents. Indian J Chem (Sec-B) 49B:489–494

Yadav M, Singh SK, Sharma JK, Yadav KDS (2011) Oxidation of polyaromatic hydrocarbons in system containing organic solvent by lignin peroxidase from *Gleophyllum striatum* MTCC-1117. Environ Technol 32:1287–1294

Yadav M, Singh SK, Yadava S (In press 2012) Purification, characterisation and coal depolymerising activity of lignin peroxidase from *Lenzitus betulina* MTCC-1183. Appl Biochem Microbiol 48: 583–589.

Zhao X, Hardin IR (2007) HPLC and spectrophotometric analysis of biodegradation of azo dyes by *Pleurotus ostreatus*. Dyes Pigm 73:322–325

Zhao X, Yiping LY, Hardin I (2005) Determination of biodegradation products from sulfonated dyes by *Pleurotus ostreatus* using capillary electrophoresis coupled with mass spectrophoresis coupled with mass spectrometry. Biotechnol Lett 27:69–72

Zhao X, Hardin IR, Hwang HM (2006) Biodegradation of a model azo disperse dye by the white rot fungus *Pleurotus ostreatus*. Int Biodet Biodeg 57:1–6

Zorn H, Langhoff S, Scheibner M, Nimtz M, Berger RG (2003) A peroxidase from *Lepista irina* cleaves ß, ß- carotene to flavor compounds. J Biol Chem 384:1049–1056

Index

© Springer International Publishing Switzerland 2015
E. Lichtfouse et al. (eds.), *CO_2 Sequestration, Biofuels and Depollution*,
Environmental Chemistry for a Sustainable World 5, DOI 10.1007/978-3-319-11906-9

MIX
Papier aus verantwortungsvollen Quellen
Paper from responsible sources
FSC® C105338

If you have any concerns about our products,
you can contact us on
ProductSafety@springernature.com

In case Publisher is established outside the EU,
the EU authorized representative is:
Springer Nature Customer Service Center GmbH
Europaplatz 3, 69115 Heidelberg, Germany

Printed by Libri Plureos GmbH
in Hamburg, Germany